TWO-PHASE COOLING AND CORROSION IN NUCLEAR POWER PLANTS

TWO–PHASE COOLING AND CORROSION IN NUCLEAR POWER PLANTS

M. A. Styrikovich
V. S. Polonsky
G. V. Tsiklauri

High Temperature Institute
of the USSR Academy of Sciences, Moscow

English-Edition Editor

G. F. Hewitt
United Kingdom Atomic Energy
Research Establishment, Harwell

HEMISPHERE PUBLISHING CORPORATION
A subsidiary of Harper & Row, Publishers, Inc.

Washington   New York   London

DISTRIBUTION OUTSIDE NORTH AMERICA

SPRINGER–VERLAG

Berlin   Heidelberg   New York   London   Paris   Tokyo

Two-Phase Cooling and Corrosion in Nuclear Power Plants

Originally published by Nauka Press, as Teplomassoobmen i Gidrodinamika v Dvukhfaznykh Potokakh Atomnykh Elektricheskikh Stantsiy.

1 2 3 4 5 6 7 8 9 0   E B E B   8 9 8 7

This book was set in English Times by Hemisphere Publishing Corporation.
The cover designer was Don Wittig; the production supervisor
was Peggy M. Rote; and the typesetter was Sandra F. Watts.
Edwards Brothers, Inc. was printer and binder.

Library of Congress Cataloging-in-Publication Data

Styrikovich, M. A. (Mikhail Adol'fovich)
  Two-Phase cooling and corrosion in nuclear
power plants

  Translation of: Teplomassoobmen i gidrodinamika v
dvukhfaznykh potokakh atomnykh elektricheskikh stantsii.
  "High Temperature Institute of the USSR Academy of
Sciences, Moscow."
  Bibliography: p.
  Includes index.
  1. Nuclear power plants. 2. Two-phase flow.
3. Heat—Transmission. 4. Mass transfer. I. Polonsky,
V. S. (Vladimir Sergeevich) II. Tsiklauri, G. V.
(Georgii Viktorovich) III. Title.
TK1078.S7913 1987   621.48'3   87–8709
ISBN 0-0-89116-424-3   Hemisphere Publishing Corporation

DISTRIBUTION OUTSIDE NORTH AMERICA:
ISBN 3-540-18106-7 Springer-Verlag Berlin

CONTENTS

PREFACE

The rapid development of nuclear power production and the extremely large amount of attention given to development of scientific (pure and applied) problems in this field has resulted in the appearance of an extensive scientific and engineering literature on this topic.

In addition to the huge number of individual articles and reports at conferences and symposia, there have appeared a large number of surveys and monographs concerned with different aspects of the problem. These surveys and monographs pertain to hydrodynamic, thermophysical, water-chemistry, corrosion and other processes occurring in nuclear power plant equipment.

However, these publications do not illuminate the problem as a whole. There are almost no publications concerned with the interaction of heat- and mass-transfer with the behavior of impurities contained in the primary coolant and working fluid of the majority of nuclear power plants.

At the same time, it became obvious during the past several yeas that it is the behavior of corrosive impurities in two-phase systems (where under certain conditions the concentration of such impurities in the liquid layer in direct contact with the metal surface may differ by several orders of magnitude from their concentration in the bulk flow) is a factor controlling to a large degree the reliability and continuity of the operation of a number of major components of nuclear power plants.

Damage due to high local concentrations of impurities such as NaOH and NaCl are now responsible for a very significant part of the outages of large nuclear power units, which involves huge losses.

Until recently attention was paid primarily to failure of steam generators of two-loop pressurized water reactor (PWR). It has become clear that other components of thermomechanical equipment of nuclear power plants, particularly turbines, frequently go out of service due to high local concentrations of corrosive impurities in the liquid phase and correspondingly rapid corrosion of even the most resistant materials, particularly those subjected to high mechanical stresses.

This book is an attempt at examining the processes occurring in nuclear power plant equipment from the point of view of heat and mass transfer and of the behavior of impurities contained in water and in steam in the center of the actual water conditions in nuclear power plants.

Since even the thermodynamic parameters, and the more so the kinetics of such processes, have been poorly explored, the authors are aware of the fact that the book will inevitably be in a number of cases the statement of a given problem, rather than its solution. The authors hope, however, that even in this form the book will be useful to some extent, in attaining a greater degree of understanding between specialists in the fields of mass and heat transfer and hydrodynamics of two-phase fluids, and in the field of the water regime, corrosion and design of thermomechanical equipment of nuclear power plants.

With reference to the above, it is very important to correlate data on heat and mass transfer and hydrodynamics in two-phase flows, arising in power generating equipment, particularly for nuclear power plants; this data is available to different specific groups of investigators. It is natural that in a problem so little explored and additionally so complicated investigators pay the greatest attention to studies performed by themselves and to comparison of their results with those published by others.

However, before considering in detail the actual data obtained under laboratory and full-size conditions, and comparing these data with published information on the operation of nuclear power plants equipment, it appears advisable to present a general survey of the working processes occurring in equipment of nuclear power plants of various types, from the point of view of the possible effect of heat and mass transfer and hydrodynamics of two-phase flows on the efficiency and, most importantly, continuity of operation of nuclear power plant equipment.

Chapter 1 describes the principal problems of heat transfer and corrosion and their effect on the availability of nuclear power plant equipment, including the core, steam generator, separating reheaters, regenerative preheaters, turbine and condenser. Chapter 1, and also Section 5.5 (together with V. S. Polonskiy) and Section 7.6 (together with O. A. Povarov) were written by M. A. Styrikovich. Chapter 2 and Sections 3.2, 3.4, 4.2, and 4.3 were written by G. V. Tsiklauri and he, together with V. S. Polonskiy wrote Section 2.1, Chapters 5 and 6. Sections 3.1, 3.3, 3.5–3.7, 4.1, 4.4, and 4.5 were written by V. S. Polonskiy. Chapters 7 and 8, concerned with wet-steam turbines and separators of nuclear power plants, were written by O. A. Povarov.

The book describes the results of experiments performed in which the authors were assisted by Yu, V. Baryshev, M. Ye. Grigor'yeva, A. S. Zuykov, P. V. Besfamil'nyy, I. I. Malashkin and others, to whom the authors wish to express their thanks.

M. A. Styrikovich
V. S. Polonskiy
G. V. Tsiklauri

TWO–PHASE COOLING AND CORROSION IN NUCLEAR POWER PLANTS

ROLE OF TWO–PHASE FLOWS
AT NUCLEAR POWER PLANTS

As a result of the rapid development of nuclear power technology a significant portion of the power generated in a number of countries or regions is now produced at nuclear power plants. In the near future, the entire growth in the base load will be economically taken up by nuclear power plants in most parts of the world with the exception of individual regions which have available very large deposits of cheap, not very easily transported fossil fuel, or extremely cheap hydroelectric power.

The capital investment for constructing a nuclear power plant is high, whereas the fuel cost is relatively low. For this reason any shutdown of a nuclear power plant unit is very expensive. This situation has become particularly acute during the past several years when the cost of construction of nuclear power plants has risen significantly (basically due to increasing safety and environmental protection requirements) and when, simultaneously a several fold rise in the cost of liquid fuel has occurred. In the current fuel situation, when the coal industry cannot supply the ever growing demand in the majority of countries, including the European USSR, any undergeneration at nuclear power plants is covered, when sufficient reserve capacity is available, by additional generation by conventional units which burn expensive liquid fuel. This results in very significant losses due to nonutilization of the nuclear power generating capacity; for a 1000 MW unit, for example, the downtime loss according to current estimates are from $300,000 to 500,000 per day (if substituted by coal) up to 1 million dollars per day (if substituted by oil).

In addition, due to the nature of nuclear power plant operation, the repair of many equipment components is not only more expensive than at conventional thermal power plants, but generally requires more time. In the case of plants which have been in operation for a long period of time, the situation is made more difficult due to radiation fields affecting equipment, resulting in increasing repair costs and repair time. For this reason a great deal of attention is paid to questions of availability of nuclear plants.

It is important to note that many nuclear power plant outages are due to malfunctions of thermomechanical plant components, e.g., electrochemical corrosion resulting from various water impurities. Prevention of corrosion damage is a problem of particular complexity in two-phase systems, in which the concentration of corrosive impurities in the layer of the liquid in direct contact with the metal may differ by several orders of magnitude from the mean concentration in the two-phase flow. Moreover, the determination of the true concentration of any impurities in the liquid near the heating surface is extremely complicated.

The equilibrium distribution (between the water and the steam) of the majority of impurities characteristic of the water-steam passages of nuclear power plants at high pressures has been researched. However, the validity of extending these data to the low-pressure region requires an experimental check. In addition, most of the experimental data pertains to solutions of single electrolytes. It is unclear whether they are applicable to real cases, e.g., when the system simultaneously contains a number of electrolytes which are dissociated to varying degrees.

Another poorly explored problem is that of the behavior of mixtures in nonequilibrium conditions which are usually present during heat and mass transfer. Only elementary cases—particularly heat and mass transfer in steady-state boiling water flowing in straight circular tubes—have been completely investigated. However, even in these cases, when friable permeable deposits, e.g., products of corrosion, are present on the heating surfaces, there exist very significant differences between the data of various investigators.

Quantitative studies of the concentration of water impurities in the wall layer of the liquid in intricately shaped passages (e.g., narrow, variable-cross section crevices), which are frequently encountered in many parts of the steam generating equipment of nuclear power plants, are very sparsely covered in the literature.

Finally, studies of the kinetics of impurity distribution between the steam and water in rapidly-occurring processes, which are of particular importance in assessing the operating conditions of different elements of turbine flow passages, are apparently only just starting.

1.1 PRINCIPAL THERMAL AND HYDRODYNAMIC PARAMETERS OF NUCLEAR REACTORS AND NUCLEAR POWER PLANT STEAM GENERATORS

Experience in nuclear power generation shows that the following reactor types appear to be most promising for use in the near future: (1) pressurized-water pressure vessel type reactors; (2) boiling-water pressure tube type reactors; (3) boiling-water pressure vessel type reactors; (4) liquid-metal fast breeder reactor.

All these reactor types can be regarded as a basis for implementation of the nuclear-power generating program as well as for developing short-term forecasts in the development of nuclear power. Each of the reactors has a number of specific features which make it particularly suitable for specific locations and scale. thus, the availability of two reactors of different design (vessel- and pressure tube-type) provides a freedom of choice in selecting the nuclear power plant construction site, since the construction of pressure tube reactors involves less transportation and materials handling work. As more nuclear power plants are constructed, location problems will definitely increase.

In this chapter we present a brief description of the above types of reactors and their further improvement. Naturally, this assessment reflects the level of knowledge of the authors concerning the engineering and economical progress at the present stage in the development of nuclear power.

1.1.1 Nuclear Power Plants with Water-Moderated Water-Cooled Power Reactors

Type PWR-440 and PWR-1000 (water-moderated water-cooled power reactors) are the types most used in the Soviet Union. Water-moderated water-cooled power plants in the West are primarily constructed by Westinghouse Electric, Combustion Engineering, and Babcock and Wilcox. They exhibit high core power densities, are compact and have a relatively low overall construction cost.

The operation of such a reactor is illustrated by Fig. 1.1, which is a schematic of the PWR-1000. Pressurized water enters the annulus between the shell and the tube-bundle case, descends, as shown in Fig. 1.1, and then rises through the core. The principal thermal and hydrodynamic specifications of this reactor are shown in Table 1.1. All the reactors listed in the table employ UO_2 (clad in zirconium-alloy sheaths to which 1% niobium has been added) as the fuel. The reactors operate with intermittent partial refuelling; the fuel in the center is unloaded, the outer fuel shifted to the center and fresh fuel is loaded at the periphery. The mean time between fuel loadings is 6,500 to 7,000 effective hours, which makes it possible to perform a single reloading per year at a time most convenient from the point of view of load on the given power system.

Nuclear system generators produced in the Soviet Union are primarily of the horizontal type. The first-generation nuclear power plants employing PWR-440 nuclear reactors used six horizontal steam generators with an electrical rating of 73.5 MW and steam capacity of 450 tons/hr. The PWR-1000 reactors use four 250 MW steam generators with a steam capacity of 1470 tons/hr. The U-shaped tubes in horizontal steam generators are fastened to two vertical tube sheets, located in the center part of the drum. A perforated plate, submerged beneath the evaporation surface serves to equalize the steam load; steam separation occurs in a single-stage louver-type separating device at the upper generatrix of the drum.

The PWR-1000 also employs a horizontal steam generator of design identical to those used in the PWR-440, but with a greater shell diameter (4000 instead of 3200

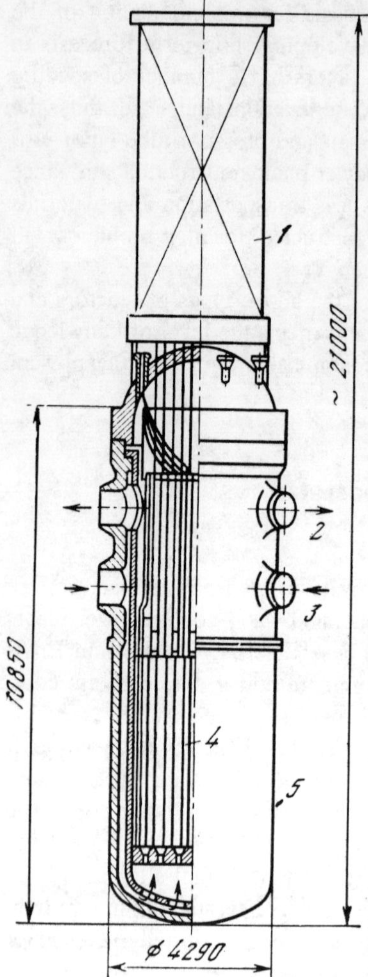

Figure 1.1 The PWR (water-moderated water cooled power reactor)—1000. 1, Upper unit with the drive for control and safety rods; 2 and 3, coolant exit and inlet, respectively; 4, core; 5, vessel.

mm for the PWR-440) and with heat exchanger tubes of smaller diameter. The principal parameters of the PWR-440 and 1000 steam generators are listed in Table 1.2.

Many Western companies employ vertical steam generators with U-shaped tube bundles, horizontal lower tube sheet and two-stage steam separation. The 1100 to 1300 MW units constructed by Westinghouse and Kraftwerk Union (KWU), for example, employ four steam generators with an electric rating between 300 and 325 MW per one reactor (Fig. 1.2); the 900–1150 MW units, produced by Combustion Engineering and Babcock and Wilcox employ two 450–575 MW steam generators per reactor. These units exhibit simplicity of design and manufacture. However, in operation damaged tubes were encountered in a number of cases above the hot half of the tube sheet and in the locations of tube contact with support plates, resulting from corrosion in the points of slag deposition in crevices. Another group of reactors are Canadian CANDU of the pressure-tube type and employ heavy water as the modera-

Table 1.1

Quantity	Troyan (Westinghouse)	Fork River (Combustion Engineering)	Oconee (Babcock and Wilcox)	PWR-1000
Thermal power	3411	3390	2568	3000
Number of loops	4	2	2	4
Pressure in first loop, Pa	15.8	15.8	15.4	16.0
Overall coolant flowrate, 10^6 kg/hr	60.25	67.10	59.6	76.0
Mean mass velocity in the core, kg/m^2-sec	3350	3634	3420	—
Coolant temperature, °C:				
At the inlet	289	289.5	290	289
At the exit from the reactor shell	324	321.7	318	324
Mean heat flux, W/m^2	685,050	646,885	540,816	615,600
Maximum heat flux, W/m^2	1,828,058	1,732,492	1,685,624	—
Number of fuel elements	39,372	38,192	36,816	49,981
Outer diameter of fuel element, mm	10.72	11.18	10.92	9.1
Length of fuel element, mm	3657.6	3657.6	3657.6	3500
Cladding thickness, mm	0.62	0.63	0.67	0.62
Gas gap	0.17	0.22	0.18	—
Mean specific power density, kW/m^2	108	114	110	111
Equivalent core diameter, m	3.0	2.95	2.95	3.2

The data in Tables 1.1 through 1.3 are given according to data of the 4th Geneva Conference [1.1].

Table 1.2

Quantity	VVER-440	VVER-1000
Electrical rating, MW	73.3	250
Steaming rate, tons/hr	452	1470
Discharge steam pressure, MPa	4.7	6.4
Feedwater temperature, °C	226	220
Inlet coolant temperature, °C	268	289
Coolant velocity, m/sec	2.7	4.89
Temperature difference, °C	21.2	24.7
Mean heat flux, W/m$^2 \cdot 10^{-3}$	80	158
Number of tubes	5146	15,648
Wall diameter and thickness, mm	16 × 1.4	12 × 1.5
Steam percentage moisture, %	0.005	0.2
Steam yield of heating surface, kg/m^2-hr	180	282
Temperature difference (minimum/maximum)	10/41	11/44
Mean log difference, °C	21.2	24.7
Steam-generator heating surface, m^2	2500	5200

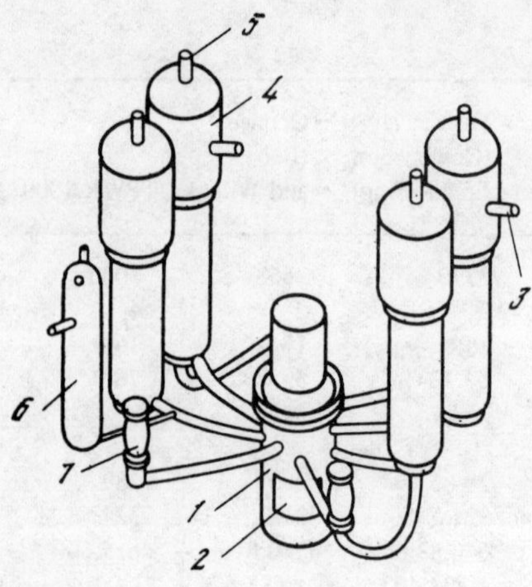

Figure 1.2 The Westinghouse pressurized water reactor. 1, Shell; 2, core; 3, feedwater; 4, steam generator; 5, steam to turbine; 6 pressurizer; 7, pump.

tor and coolant. This makes it possible to use non-enriched uranium as the fuel. The steam generators of these nuclear power plants are also vertical with U-shaped tubes.

1.1.2 Channel-Type, Boiling-Water Reactor Nuclear Power Plants

In the USSR, graphite-moderated nuclear power plants of the pressure-tube type, have been in operation for a number of years. Batch-produced 1000 MW reactors of this type (termed Type RBM-K—i.e., High Power Reactor Channel Type—in the USSR) started going on line in 1973 and are in operation in a number of nuclear power plants.

The main advantages of the RBM-K, compared to other channel-type reactors, are:

1. High reliability of the system as a whole due to channel-by-channel monitoring and possibility of partial repair of an individual channel without reactor stoppage.
2. Feasibility of attaining a high safety level by subdividing the coolant circulation loop into small units.
3. Feasibility of increasing the power rating by using design components of the same type.
4. Feasibility of fuel reloading with the reactor in operation.

The channel-type reactor has shortcomings in its large overall dimensions, including the highly branched and relatively cumbersome circulation system.

The Leningrad Nuclear Power Plant employs four RBM-K-1000 reactors. This plant is the first of a series of nuclear power plants currently operating and under construction (Chernobyl', Kursk, Smolensk, and others) of the same type, which are expected to supply a significant portion of the power demand of European USSR at present and in the near future.

A schematic of a nuclear power plant with an RBM-K-1000 reactor is shown in Fig. 1.3. Each power-generating unit includes reactor (*1*) with circulation system and (*2*) auxiliary systems, steam and condensate-feed lines and two 500 MW (or one 1000 MW) turbogenerators (*5*). The circulation system consists of two parallel loops. Each of the latter includes two drum separators (*3*), four circulation pumps (*4*) with their associated valves and piping (*7*) of medium (D_y = 300 mm) and large (D_y = 500 mm) diameters and 22 distributing cluster headers (*6*), which supply the reactor channels. The ion-exchange purification system for the primary water has a flowrate of 4% of the steam rate of the reactor.

Saturated steam at 7 MPa is supplied from the separators along eight steam lines to two K-500-65 (or one 1000 MW) turbines. The condensate is returned through a system of mixed-bed ionite filters and low-pressure preheaters to a deaerator and then by feedwater pumps through level-control valves to separator drums.

The remaining parameters of the RBM-K reactors already in service and under design are listed in Table 1.3.

1.1.3 Pressure Vessel Type Boiling-Water Reactors

Boiling-water pressure vessel type reactors (BWRs) employ light water in the core as the coolant and moderator. The wet steam generated in the core is directed to the turbine through a separator (Fig. 1.4). Internal circulation is ensured by a jet pump, which diverts water from the downcurve into the core as shown in Fig. 1.4. The driving fluid in the jet pump is reactor water, pumped by a high-pressure pump external to the vessel as shown. A number of companies, e.g., KWU (West Germany) and ASEA-Atom (Sweden), use axial pumps, built into the reactor vessel as circulation pumps, which makes it possible to avoid the use of large-diameter external piping.

The principal manufacturer of vessel-type boiling-water reactors is the General Electric Company (USA). Table 1.4 lists the principal parameters of the BWRs in two nuclear power plants: Duane Arnold and La Salle County. It is seen that, on the one hand, these reactors have a twofold smaller specific energy release in the core than

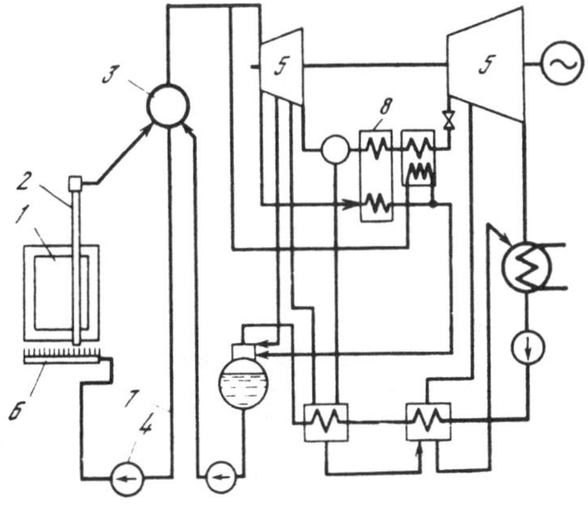

Figure 1.3 Thermal loop of nuclear power plant with channel-type reactor. 1, Reactor; 2, evaporation channel; 3, drum; 4, circulation pump; 5, turbine; 6, header; 7, circulation line; 8, reheater.

Table 1.3

Reactor parameter	RBM-K-1000	RBM-K-1500
Rating, MW:		
Thermal	3200	4800
Electrical	1000	1500
Parameters of steam entering the turbine:		
Pressure, MPa	6.5	6.5
Temperature, °C	280	280
Number of fuel channels	1693	1661
Reheater pressure, MPa	0.33	0.552
Reheat temperature upstream of low-pressure		
section, °C	264	263
Steaming rate of reactor, ton/hr	5800	8800
Maximum heat flux, 10^{-6} MW/m^2	0.7	0.9
Steam flowrate per turbine	5400	8200
Flowrate of water circulating in the reactor loop, ton/hr	37,500	29,000
Fuel-element cladding dimensions (thickness/diameter),		
mm	0.9/13.5	0.9/13.5
Fuel-element cladding material	Zirconium alloy	Zirconium alloy
Core height, m	7	7

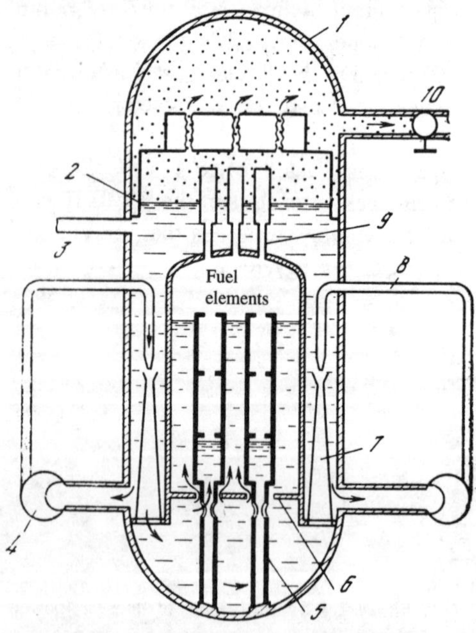

Figure 1.4 Vessel-type boiling-water reactor. 1, Reactor shell; 2, water level; 3, feedwater; 4, pump; 5, fuel-element assembly supports; 6, grid; 7, jet pump; 8, recirculation loop; 9, cyclone separator; 10, steam delivery.

Table 1.4

Parameters of boiling-water reactors	Duane Arnold, started up in 1969	La Salle-County, started up in 1972
Thermal rating, MW	1,593	3,293
Electrical rating, MW	545	1,052
Reactor shell:		
Inside diameter, m	4.68	6.38
Length, m	20.2	23.2
Design pressure, MPa	8.79	8.79
Design temperature, °C	301.7	301.7
Steam line diameter, mm	508.0	660.4
Steam flowrate, kg/sec	9,280	19,210
Number of feedwater loops	4	6
Piping diameter, mm	254	304.8
Core coolant flowrate, kg/sec	65,770	167,060
Steam pressure, MPa	7.17	7.17
Steam temperature, °C	286.1	286.1
Feedwater pressure, MPa	7.733	7.733
Feedwater temperature, °C	215.6	215.6
Inlet subcooling	13.56	13.56
Mean heat flux, W/m^2	517,045	517,040
Maximum heat flux, W/m^2	1,351,173	1,328,570
Number of fuel assemblies	386	764
Number of rods	18,023	36,672
Core length, m	3.66	3.66
Cladding jacket diameter, mm	14.30	14.30
Cladding thickness, mm	0.81	0.81
Fuel cylinder diameter, mm	12.37	12.37
Core power density, kW/liter	50.9	1

PWRs, i.e., in order to generate the same power as an equivalent PWR unit they require a larger diameter vessel. Thus, reducing the pressure in the vessel gives practically no reduction in vessel weight. However, nuclear power plants with boiling water shell-type reactors do not need steam generators. On the other hand, the entire turbine unit uses radioactive steam, i.e., it requires protection.

1.1.4 Sodium-Cooled Fast-Neutron Reactors

The prospect for very extensive development of nuclear power depends on the use of fast breeder reactors. Sodium-cooled fast breeder reactors have been investigated worldwide over a period of 25 years and a great deal of experience has been accumulated, which now allows the construction of the first commercial fast-reactor nuclear power plants. In the USSR these are nuclear power plants employing the 150 MW BN-350 reactor and the BN-600 reactor (which also produces up to 120,000 m^3 per day of desalinated water in Shevchenko City). The temperature and hydraulic characteristics of liquid-metal loops are such that the steam-water cycle of nuclear power plants with BN-600 reactor can be operated under conditions close to those in conventional thermal power plants (pressure of 13 MPa and temperature of 500 °C with reheating to a temperature of 500 °C). The temperature of fuel element cladding is then 680 to 700 °C, the temperature of Na leaving the reactor is 550 °C and the mean temperature rise is 170 °C.

The outstanding thermophysical properties of the sodium coolant ensure a high power density in fast reactors: it is 470 kW/liter in BN-350, 550 kW/liter in BN-600 and 560 kW/liter in SuperPHENIX. The fast neutron breeding reactor is usually of the swimming-pool type with several submerged circulation pumps (the SuperPHENIX reactor is equipped with four pumps) in the primary loop. To ensure high reliability and operating safety heat is removed from the reactor by intermediate sodium loops using heat exchangers in which, as in the primary loop, the boiling of sodium is suppressed. The sodium temperature in the primary loop does not exceed 580 °C (Table 1.5).

The power-generating steam-water loop consists of a steam generator, turbine, condenser, reheater with sodium superheat and other thermomechanical equipment. Reheating in the SuperPHENIX reactor is achieved by steam bled from the exit of the high-pressure cylinder. The steam generator is a counterflow unit with water in the tubes flowing upward; the economizer is at the bottom, the evaporator is in the middle part, whereas the supereheater and reheater are located in the upper part. Reheating in BN-600 steam generators occurs at 1.08 MPa to the live-steam temperature; the intermediate pressure in the SuperPHENIX is 1.2 MPa, and the reheat temperature gain is low.

The steam generator at the Creys-Melville Nuclear Power Plant with a SuperPHENIX reactor, currently under construction, is also a counterflow unit with spiral arrangement of steam-generating tubes. Each tube passes through sealing rings welded to the steam-generator housing. In the opinion of the designers, this makes it possible to reduce the danger of leakage in the tube sheets. A 45 MW (electrical) steam-generator prototype, in operation since 1973, has demonstrated the reliability of the above solution.

In addition to the USSR and France, fast-breeder nuclear power plants are under development in other countries. The PFR, 250 MW (electrical) is operating in England, a 300 MW (electrical) SNR reactor is under construction in West Germany, and the 250 MW (electrical) Monju unit is being constructed in Japan. The main parameters of these reactors are listed in Table 1.5.

1.2 EFFECT OF HEAT AND MASS TRANSFER ON THE AVAILABILITY OF NUCLEAR POWER PLANT EQUIPMENT. CORROSION IN TWO–PHASE SYSTEMS

Virtually all kinds of boiling-water reactor (BWR) nuclear power plant equipment—including the core, steam separator, wet-steam turbine, intermediate separator and

Table 1.5

Parameter	BN-350 (USSR)	BN-600 (USSR)	PFR (England)	PHENIX (France)	SNR (West Germany)	Monju (Japan)	Super-PHENIX
Rating:							
Thermal, MW	1000	1470	600	560	730	714	3000
Electrical, MW	350	600	250	250	300	300	1200
Power density, kW/liter	470	550	500	420	400	450	560
Core volume, liter	1900	2300	1320	1150	1600	—	5350
Temperature:							
At the core inlet, °C	300	380	425	400	380	390	395
At the core exit, °C	530	550	585	560	580	540	545
Number of steam generators	3	3	3	3	3	2	4
Steam generator rating, MW	120	200	200	186.3	243.3	225	750
Sodium temperature:							
At the steam-generator inlet, °C	500	550	565	550	565	520	525
At the steam-generator exit, °C	300	380	350	350	350	345	345
Of superheated steam, °C	500	500	505	512	500	495	490
Steam pressure upstream of the turbine, MPa	5	13.0	—	17.0	—	13.0	17.7
Reheater pressure, MPa	—	1.08	—	1.2	—	—	1.2
Temperature past the reheater, °C	—	550	—	535	—	—	535

steam reheater, the steam side of regenerative preheaters and of the turbine condenser—operate under two-phase flow conditions. In pressurized water cooled nuclear power plants, this situation applies fully to the entire second loop, including the low-pressure side of the steam generator. Even the primary loops of the newest PWR plants are partially operating under surface boiling conditions, and there is a tendency toward the use of developed low-quality boiling in subsequent designs. For example, the new designs of the CANDU-600 reactor are calculated on the assumption to an outlet quality of 3%. In this case the high-pressure side of the steam generator will also operate under two-phase conditions.

The sodium loops in liquid-metal cooled reactors are single-phase, but almost the entire steam-water loop, including the subcritical pressure steam generators used in all the operating and projected nuclear power plants of this type, are subjected to two-phase conditions. The same applies to gas-cooled reactor nuclear power plants, both operating and projected.

Unlike conventional fossil fueled thermal power plants, water-cooled nuclear

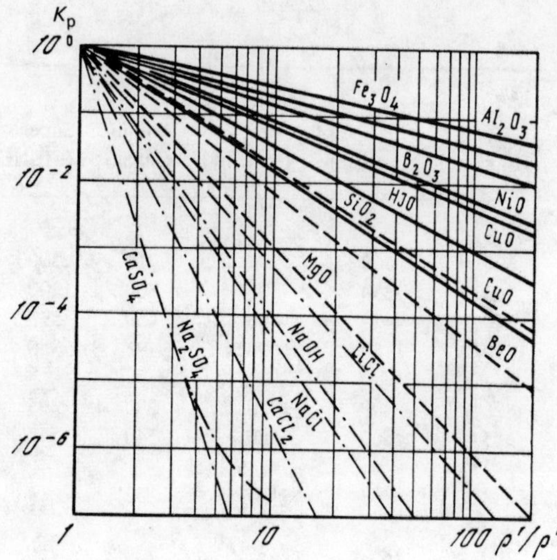

Figure 1.5 Distribution factor vs. density ratio.

power plants employ low-pressure, low-temperature steam which means that the difference between the thermophysical and physico-chemical properties of the steam and water is particularly great. This requires careful attention in considering processes occurring in the two-phase region and, to a large extent, controls the availability of the equipment.

The two-phase conditions have a particularly significant effect on electrochemical corrosion. In single-phase systems the concentration of any corrosive impurity in water is virtually constant throughout the entire coolant loop, both in the core of the flow, and hear the heated or cooled surfaces. Hence if the concentration of a given impurity is maintained in the flow core within limits corresponding to a very low (permissible) rate of metal corrosion, one can be sure that no damage occurs in equipment due to electrochemical corrosion. This conclusion is virtually independent of the conditions of flow over heat transfer surfaces and of heat and mass transfer processes as such.

The equilibrium distribution of any impurity between the liquid and vapor phases in two-phase systems is highly nonuniform and is characterized by the "distribution factor" which is the ratio of the concentrations in the liquid and vapor phases respectively. Figure 1.5, for example, shows the dependence of the distribution factor for a number of impurities typical of the steam-water flowpath of nuclear power plants between water and steam. This result, established more than 25 years ago [1.2, 1.3], shows that the distribution factor may differ from unity by several orders of magnitude, even at the high pressures characteristic of the primary loop (15 to 17 MPa). At pressures typical of the secondary loop or of boiling reactors (7.5 MPa), and the moreso at pressures typical of reheaters (0.5 to 1.2 MPa) factor rises to extremely high value.

It is important to note that the distribution factors are extremely high for a number of highly corrosive compounds (NaOH, NaCl, etc.). Hence, in situations with a very low moisture content, the residual water droplets may contain corrosion-active impurities at very high concentrations. Examples of situations where this concentration effect may occur are once-through steam generators (where the entire liquid phase is ultimately evaporated), the inlet zones of superheated and reheated, and turbine stages in which slightly superheated steam is expanded. The thermodynamic equilibrium concentrations of these impurities in water which is in contact with metal may be high even when, in the entire coolant flow, it is very low. Furthermore, even at low vapor quality in the bulk flow, stagnant regions may exist on the heating surface (narrow slots, crevices, thick layers of porous deposits) where virtually complete evaporation occurs with the corresponding appearance of high impurity concentrations in the liquid which is in contact with the metal.

Obviously, under conditions when heating occurs by means of a working fluid whose temperature slightly exceeds the boiling temperature (as, for example, in the steam-generators of PWR nuclear power plants), the maximum concentration of the vaporizing liquid cannot exceed that at which the boiling temperature of the solution would rise to the level of the heating-fluid temperature. However, this limitation is not too effective for the majority of typical corrosive impurities [1.4]. Even at low temperature differences (Δt), typical of steam generators of the newest nuclear power plants ($\Delta t_{max} = 35-40\,°C$, $\Delta t_{min} = 12-15\,°C$) concentrations may occur which are not permissible even for highly corrosion resistant alloys.

Conversely, the equilibrium concentrations for certain gases, which are significant to corrosion processes (ammonia, oxygen) are several orders of magnitude higher in the vapor phase than in the liquid phase. Hence their equilibrium concentration in liquid in parts of the system with low moisture content can be exceptionally low (at low pressures by 5–6 orders of magnitude lower than in the flow core), which in a number of cases may have a significant effect on the corrosion rate.

At the same time, if the bulk of the condensate with low impurity content (for example, ammonia) separates from the remainders of noncondensed steam with a high NH_3 concentration, then at the end of condensation the condensate of this residual steam will have a high ammonia concentration. This concentration manifests itself even more strongly in the case of the condensation of oxygen-containing steam.

It should be noted that on the whole, there is virtually nothing to be gained by knowing and maintaining a certain concentration of a given impurity in the bulk of the two-phase flow in terms of estimating the corrosion effect due to this concentration. It is necessary to know the real concentrations in the liquid in direct contact with the metal, and these may differ by many orders of magnitude from the concentrations measured in the flow core. At the same time, the measurement of the real concentrations in the liquid film adjoining the metal and their distribution over individual elements of the heating surfaces of equipment in actual operation is a very difficult task, at least at the present time.

The effects noted above impart particular importance to the study of the heat and mass transfer processes which determine the extent to which the distribution between both phases approaches the equilibrium value. It is necessary to be able to calculate the actual degree of impurity concentration in the liquid film in contact with the metal surface, which is basically controlled by the behavior of mass transfer in the boundary layer. Only then is it possible to draw specific conclusions on the reliability of the performance of materials under the given water chemistry conditions in equipment components subjected to given hydrodynamic and heat and mass transfer conditions.

In a number of cases, when the heating surface is in contact with a superheated steam flow, it is important to know whether actual contact will occur between the surface between any liquid droplets entrained in the flow. Only if such contact occurs, is it possible for electrochemical corrosion, associated with the concentration of impurities in droplets settling on the hot wall, to occur.

1.3 HEAT AND MASS TRANSFER IN THE REACTOR CORE

Heat transfer in the nuclear core virtually controls the power which the reactor can reliably generate. In spite of the large number and variety of experimental studies and computational methods, allowing the determination to a certain extent of the heat transfer processes on the intensely heated fuel element surfaces, this problem cannot be regarded as fully exhausted. Overheating of even a small part of the heating surface can lead to the possibility of failure. Thus, the engineering design calculations should take account of the very complex geometric configurations, and allow for all the local conditions, both with respect to nonuniformity of the heat flux (including heat generation surges), and with respect to the nature of flow over surfaces. The flow is perturbed by the presence of uneven clearances between the fuel elements and also by the presences of various spacing and anti-vibration devices.

For BWR units, artificial turbilization devices are often used to suppress dryout of the fuel. The relevance of these devices to the formation of scale on fuel elements and corrosion of their surface, has not been significantly explored.

It was rather reliably established, by studies of hydrodynamics and heat and mass transfer on the inside of straight circular tubes, that mass transfer occurs at a very high rate between the boiling wall layer and the flow core for flow conditions up to the dryout zone (for more details see Chapter 5). However, detailed studies for more complex cases of flow are virtually nonexistent. It is particularly important to investigate mass transfer under dryout conditions, when the dryout has been artificially delayed by means of various turbulence promoters. In this case there is a danger that satisfactory heat transfer rates may not be accompanied by satisfactory mass transfer conditions, which may result in concentration of impurities in boundary layer, corrosion or even the appearance of scale on the fuel-element heating surfaces.

In modern PWR units, where surface boiling occurs in the core, steam appears in the boundary layer of the fuel elements and subsequently in the water flow core. Since, under these conditions, the rate of mass transfer between the boiling boundary layer and the flow core is quite high, one needs not fear a significant concentration of salts or strong bases near the heating surfaces and the consequent acceleration of corrosion of fuel element cladding. The migration of hydrogen into the steam, accompanied by a steep drop in its concentration in the water, and with attendant amplification of radiolysis, is a more significant danger. The question of the extent of this

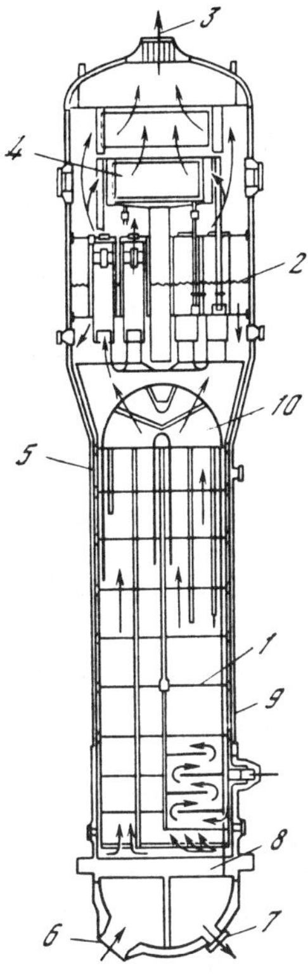

Figure 1.6 The Westinghouse vertical-tube steam generator. 1, Antivibration bars; 2, water level; 3, steam outlet; 4, separator; 5, recirculation of water; 6 and 7, exit and inlet, respectively; 8, tube sheet; 9, shell; 10, tube bundle.

phenomenon of its possible effect on the corrosion of the primary loop, and of its significance on the selection of the water chemistry conditions, all require further study [1.5].

1.4 PROBLEMS OF HEAT AND MASS TRANSFER AND CORROSION IN NUCLEAR POWER PLANT STEAM GENERATORS

1.4.1 Water-Cooled Steam Generators

The steam generator is one of the most vulnerable components of nuclear power plants with PWR reactors. In such steam generators, high-pressure heating water flows inside of small-diameter tube bundles, whereas medium-pressure (5–75 MPa) steam is generated in the space between the tubes (the ''secondary side''). Steam generator failure is the most common cause of breakdowns in PWRs.

Although many steam-generator designs have been proposed, only three of these are in actual use. The most frequently used design is the vertical type with natural circulation on the secondary side (Fig. 1.6). Such steam generators are produced by

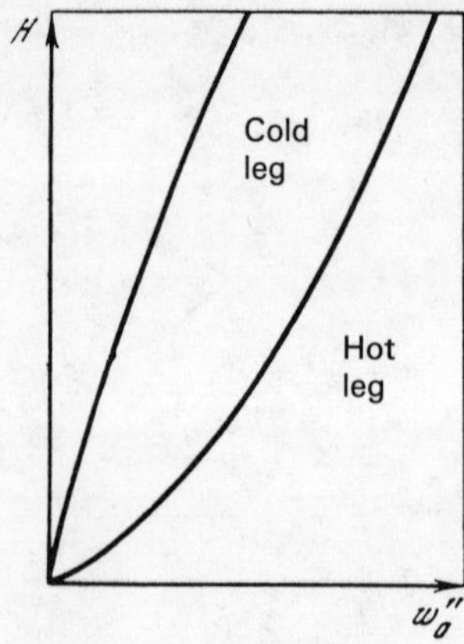

Figure 1.7 Distribution of reduced steam velocity over the height of a vertical type generator.

the Westinghouse company (or by its licensees) with the rating per unit being 300–500 MW (1,500–1,700 tons/hr steam) and by Combustion Engineering whose units have ratings up to 650 MW (3,900 tons/hr steam). Such units require very large heating surfaces (10.4×10^3 m^2 for the 3,900 tons/hr generator) and a very large number of small diameter (d = 16–22 mm) tubes in the bundle. To reduce the overall dimensions and span of the horizontal tube sheet (about 600 mm thick) the tubes are staggered and tightly packed (s/d = 1.35–1.5). The circulating water is separated from steam in separators and then mixed with feedwater whose temperature is usually 50–70 °C below saturation temperature. The water then enters the lower part of the tube bundle directly above the tube sheet and over the entire outer perimeter.

The tube bundle is heated very nonuniformly—the temperature difference at the inlet to the heating tubes is approximately three times higher than at the exit from them. For this reason the rate of vaporization along the hot leg of the U-shaped tubes is significantly higher than at the second half of the bundle cross section, the so-called cold leg. It is important to note that the steam generation rate in the hot leg is highest at the bottom of the bundle, whereas for the cold leg the highest steam generation occurs at the top, which amplifies the difference in temperature driving force (Fig. 1.7). The overall circulation head in the hot part is much higher, which results in a horizontal shift of the inlet which is difficult to calculate [1.5]. In addition, at 5–10 vertical locations along the bundle the spacing between tubes is maintained using perforated plates or by a system intersecting plates passed between the tubes; this significantly blocks the flow area. The need for spaced devices complicates even further the geometry of the flow-channel, introducing additional complexity in any calculation of the distribution of steam-water flows over the large volume (to 20–25 m^3) of the evaporation zone.

Subcooling of the feedwater, results in a significant lag in steam generation (15% and more of the entire heating power is spent for heating the water to saturation). At the same time, it is extremely difficult to ensure that the subcooled water will reach all the spaces within a tight multirow bundle and that it will be uniformly distributed over the (very large) cross section. The overall effect (particularly in many nuclear power plant steam generators ordered up to 1972) is a pronounced nonuniformity of circulation over the bundle cross section. This may go as far as to cause circulation reversal zones within the bundle, and stagnant regions between these zones and the uptake flow (which is the probable cause of the "banana-shaped" damage zone on the lower tube sheet [1.6]).

The latest designs of steam generators of this type improve the circulation, particularly when heating feedwater, by supplying the water at the top of the steam generator along a perforated annular tube. This ensures satisfactory mixing with circulating water and its partial heating due to condensation of steam which enters the water through cyclone separators. In other designs, a part of the lower bundle (cold leg) is used for heating of feedwater (the so-called integral economizer), which means that the water fed to the main stream is heated to saturation. In addition, the use of such an economizer made it possible to raise the pressure of the secondary steam at the given heating-water temperature, or to reduce the temperature of water returning to the reactor at the given secondary-loop pressure.

The main difficulties in operating steam generators of this type were corrosion damage in locations where the heating surfaces were insufficiently washed by the coolant. In such locations, almost all the water has evaporated and the concentration of corrosive impurities, which tended to remain in the water was very high (several orders of magnitude higher than the bulk mean concentration). Initially, such damage took the form of corrosion cracking of the stressed tube material, starting with the low-pressure side. It arose near the tube sheet and even inside the latter. It must be remembered that the first-generation steam generators provided for sealing of the tubes only in the lower part of the sheet, whereas at the top there was a deep (up to 400–500 mm) thin slot between the tube and the opening in the tube sheet [1.6]. Loose corrosion products gradually accumulated on the horizontal surface of the tube sheet, pierced by several thousand tubes. Such deposits were particularly significant in the so-called "banana-shaped" zone, where the boundary between uptake and downtake flow apparently passed, and where the circulation velocities were close to zero. The damage was typical of corrosion cracking of Inconel-600 under stress in a concentrated alkaline solution; such a solution formed by almost complete evaporation of the feedwater concentrating the impurities resulting from cooling-water leakage in the condenser.

Some investigators [1.7, 1.8] attribute the presence of high impurity concentration near the heating surface to evaporation of almost all the water in thin porous ("ventilated") magnetite deposits; this phenomenon was first discovered by Macbeth [1.9]. However, experiments by the present authors [1.10, 1.11] showed that in thin deposit layers the concentration ratio due to evaporation under normal flow conditions over heating surface, does not exceed 10 over the entire range of nucleate boiling. Very high concentration ratios are possible only in very thick layers of deposits or upon disturbance of normal water circulation. These conclusions are in agreement with experience of operation of nuclear power plant steam generators, where corrosion damage occurred either in narrow crevices (gaps between the tubes and the tube sheet, between the tubes and the spacing devices), or beneath a very thick layer of loose deposits (often 10–15 cm and more in thickness) on horizontal tube sheets [1.12, 1.13].

When several nuclear power plant steam generators suffered from tube damage by stress corrosion cracking, which is characteristic of tubes of Inconel-600 in contact with concentrated alkali solutions, it was suggested by the Westinghouse Company that the accumulation of free alkalies could be prevented by using phosphated water with high phosphate concentration (70 ppm). However, this resulted in increasing the damage of tubes at certain nuclear power plants where the tubes were already damaged, and at the same time tube thinning occurred at many nuclear power plants, associated with concentration of free acid. Following the identification of these problems, Westinghouse suggested returning to the hydrazine-ammonia system and, starting in April 1974, almost all the nuclear power plants with PWR reactors in the USA and the majority of such reactors in other countries converted to this system (the so-called "all-volatile treatment"—AVT). Theoretically this should have eliminated high local concentrations of corrosive phosphates. However, damage persisted, particularly in steam generators which previously operated in the phosphate regime, but also in those where the steam-generator tubes were carefully checked by eddy-current techniques for absence of cracks before conversion to the hydrazine-ammonia regime, and where the steam generators were washed by powerful water jets.

Moreover, there appeared a new type of damage, termed "denting." This kind of damage occurs in places where the tubes pass through the spacing plate. This results in a rapid growth of corrosion in the gap between the tube and the passage in the baffle, causing jamming and damage of the tube. During the past few years a great deal of attention has been paid in the USA to this new type of damage. The problem of denting has been discussed in detail at several conferences [1.5, 1.14].

It is currently assumed that the corrosion products filling the gap between the tube and the spacing grid (the thickness of the latter is close to the tube diameter), are not transported from the outside by the boiling water, but are rather the result of corrosion of the grid material—low-carbon steel. The corrosion is caused by the fact that when the tube is pressed to one side of the hole, whose diameter is only slightly larger than the outer diameter of the tube, there forms an eccentric annulus, which very slowly widens from the point of contact over the perimeter of the hole (see Fig. 1.8). In this small gap there is virtually no circulation of water and evaporation is close to total which results in the formation of a solution sufficiently concentrated to induce intensive corrosion of the carbon steel of which the plate is manufactured. The products of corrosion grow in thickness, squeezing the Inconel-600 tube and thus damaging it. Denting was not observed in steam generators manufactured by the KWU (West Germany) and Breda (Italy) companies, which use a system of intersecting flat stainless steel bars ("egg crate") as the spacing device; this may be due to the more favorable shape of the space between the tube and the spacing device.

The denting type of damage occurs not only on the hot leg of the U-shaped tubes, but also on the cold leg, even at the lower spacing grid of the latter, i.e., when the difference between the heating-water and secondary-steam-side boiling temperature is about 10 °C. Solomon in his report at the First Bournemouth Conference [1.14] presented results of checks of the concentration of solutions in gaps, obtained with single-tube models of a steam generator with different types of gaps. It was noted that in some types of gap, high concentrations were measured based on the corrosion

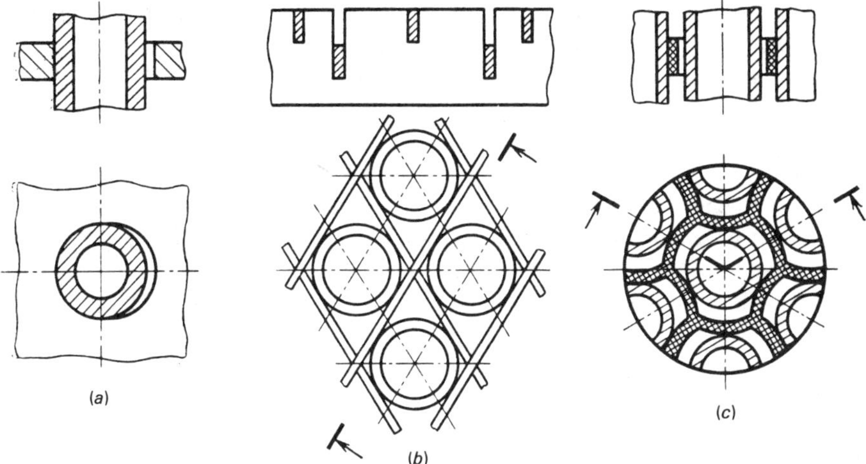

Figure 1.8 Devices for spacing of steam-generator tubes, employed by different manufacturers. (*a*) Westinghouse. (*b*) Kraftwerk Union. (*c*) Babcock and Wilcox.

rate*, even at very low salt concentrations in the bulk of the water. The evaporation proceeded to the thermodynamically possible level, when the boiling temperature of the concentrated solution rose to the temperature of the heating water. In other cases the corrosion rate increased linearly with the rise of concentration in the bulk of the liquid, i.e., there occurred a certain mass transfer from the gap which limited the rise of concentration there (apparently the concentration was maintained at a certain level).

Investigations performed at five operating nuclear power plants by Nuclear Water and Waste Technology (USA) (one of the five EPRI** programs) showed that a phenomenon known as "salt hideout" occurred at all the nuclear power plants with 530–810 MW PWR reactors which were put into service between May 1973 and May 1975. This indicates that all steam generators have parts of heating surfaces which are insufficiently washed. These serve as sites for salt deposits which are then reabsorbed by the water when the steam generation is discontinued. Significant concentrations of calcium, magnesium, sodium and chlorides were observed upon wet shutdown of steam generators even at Prairie Island No. 1, where the condensers are very leak-tight and are cooled by low-salt-content water. This was notwithstanding the fact that in the course of normal operation, the content of these minerals in the steam-generator water were below the level which could be sensed by instrumentation. The locations of poorly-wetted regions could not easily be ascertained, but, judging by the extent of "salt hideout", it is assumed that their total surface was rather significant.

It is important to note that although a number of nuclear power plant steam generators operating in a number of countries, produced by Westinghouse and Com-

---

*This concentration measuring technique is regarded as approximate.
**Electric Power Research Institute (USA).

bustion Engineering or by their licensees, suffer from corrosion cracking and denting, no such damage was observed in all the second and third generation nuclear power plant steam generators manufactured by KWU (West Germany). This company employs phosphate addition in its nuclear power plant units, starting with 650 MW Stade unit, put on line in 1972 and ending with the 1200 MW Biblis-A and the 1300 MW Biblis-B units. This runs contrary not only to the USA practice, but also to earlier recommendations of the highly-influential VGB scientific and engineering society. In their units, the phosphate concentration is very low (2–6 ppm of PO_4), and the mole ratio of Na/P is held within the narrow limits of 2.0 to 2.6, i.e., the phosphate is in the form of a dilute acid. It is claimed by Solomon [1.14] that no steam generator corrosion problems are encountered at these nuclear power plants, for which it is now recommended by the VGB that the phosphate regime be used with KWU steam generators.

The absence of denting in the steam generators of these KWU nuclear power plants may also stem from the fact that they do not use Inconel-600 tubes, but rather tubes of the Incoloy-800 alloy, which is significantly more resistant to stress corrosion cracking. In addition, the KWU steam generators employ a different tube spacing method—not perforated horizontal plates, but flat stainless-steel bars ("egg-crate") (see Fig. 1.8*b*). The latter circumstance may be significant.

As seen from Fig. 1.8, the contact between the tube and a straight flat plate produces a gap which, unlike the gap between the hole and the tube in the Westinghouse design, widens much more rapidly. The probability of flow stagnation in this case is significantly lower. It is interesting to note that the gap in the spacing grids used by the Babcock and Wilcox Company in its once-through steam generators (see Fig. 1.8*c*) widens even more rapidly, since the broached trifoil hole in the plate (more precisely the three of its projections that center the tube) have a curvature in the points of contact which is inverse to that of the tube (similar type of hole used in new Westinghouse and Framatom steam generators—quadzofoil broached).

No corrosion problems have been found with the steam generators of Canadian CANDU nuclear power plants. In particular, all four units of the Pickering Plant [1.5] have performed satisfactorily; they employ Incoloy-800 tubes and have better designed spacing plates similar to BKW-type. Recently, all firms switched from low-carbon steel to more corrosion-resistant alloys as material for the plates.

Soviet-manufactured nuclear power plants, used within the USSR and abroad, employ horizontal steam generators with natural circulations (Fig. 1.9). For VVER-440 reactors these units are rated only at 75 MW (e); however, the VVER-1000 employs units rated up to 250 MW (e) with a steaming rate of 1300 ton/hr. These steam generators typically employ vertical cylindrical manifolds and horizontal steam-generating tube bundles. This design eliminates the deposition of thick porous deposit layers on the hot surfaces of the tube sheet, which occur in vertical steam generators with U-shaped tubes. However, the site where the relatively thick-walled hot-water tube sheet passes through the water-steam interface, where the metal temperature may be alternately heated or cooled with fluctuations in water level by as much as 20–30 °C, requires special protection. This is particularly important when the manifolds are manufactured of austenitic stainless steel (VVER-440 steam generators) which is more sensitive to temperature fluctuations than the low-alloy perlitic steel, used in the VVER-1000 steam generators. This protection can be provided by welding a thin-walled jacket on the low-pressure side. The jacket, however, must be very leak-proof, since otherwise water will leak in and be continuously evaporated within the jacket, which poses the danger of accelerated corrosion cracking of the manifold.

It is much better to provide protection from the inner side, since then the entire upper part of the tube sheet becomes a "dead" (stagnant) region, and the wall temperature in the vicinity of the [water] level approaches the saturation temperature. It is important that this makes the appearance of the steam phase or of high [corrosives] concentration impossible.

The possibility of almost complete evaporation of water from the horizontal gaps between the spacing bars and tubes and of local accumulation of loose deposits in these spaces still remains, as in all other horizontal steam generators, but up to now no corrosion was observed in these zones possibly because the bars are stainless steel. A specific shortcoming of the horizontal design is difficulty of separation, due to the small available height of the vapor space above the submerged tube bundle and the significant nonuniformity of the steam load on the vaporization surface. These difficulties become amplified with increasing unit rating. In addition, the assembly work involved in the manufacture of a highly rated steam generator (which is of necessity very long) inside the containment cylinder or sphere is much more difficult than for vertical steam generators. As a result there have appeared a large number of natural-circulation, vertical steam generator designs which do not employ a horizontal tube sheet.

This design trend is represented by a large number of suggested designs of once-through PWR steam generators, of which only the straight-tube Babcock and Wilcox units are as of now in practical use (Fig. 1.10). These steam generators use a bundle of straight tubes, inside of which high-pressure water flows down, whereas the secondary-side boiling water flows up in the space between the tubes. Here the diameter of the tube sheets, each of which holds only one half of tube ends, is smaller than in a steam generator with U-shaped tubes; the tube sheet diameter is accordingly also smaller. The upper part of the steam generator operates under film boiling conditions. This is followed by region where heat is transmitted to slightly superheated steam over a very small temperature difference. The mass velocities in the secondary loop are moderate—of the order of 200–300 kg/m$^2 \cdot$ sec, which means that the coefficients of heat transfer to the superheated steam are low.

The heat transfer is improved somewhat by the presence of moisture droplets in the steam. However, the region with significant droplet precipitation on the surface is relatively small. Heat transfer should have been significantly enhanced by spacing plates (see Fig. 1.8), because the mass velocity in the space between them and the tubes increases approximately threefold. However, these plates are spaced approximately 1 m apart, i.e., more than 50 hydraulic diameters of the channels between the tubes. For this reason the heat transfer enhancement produced by them (Fig. 1.10) is apparently not too significant.

The central part of the steam generator operates under annular flow evaporation

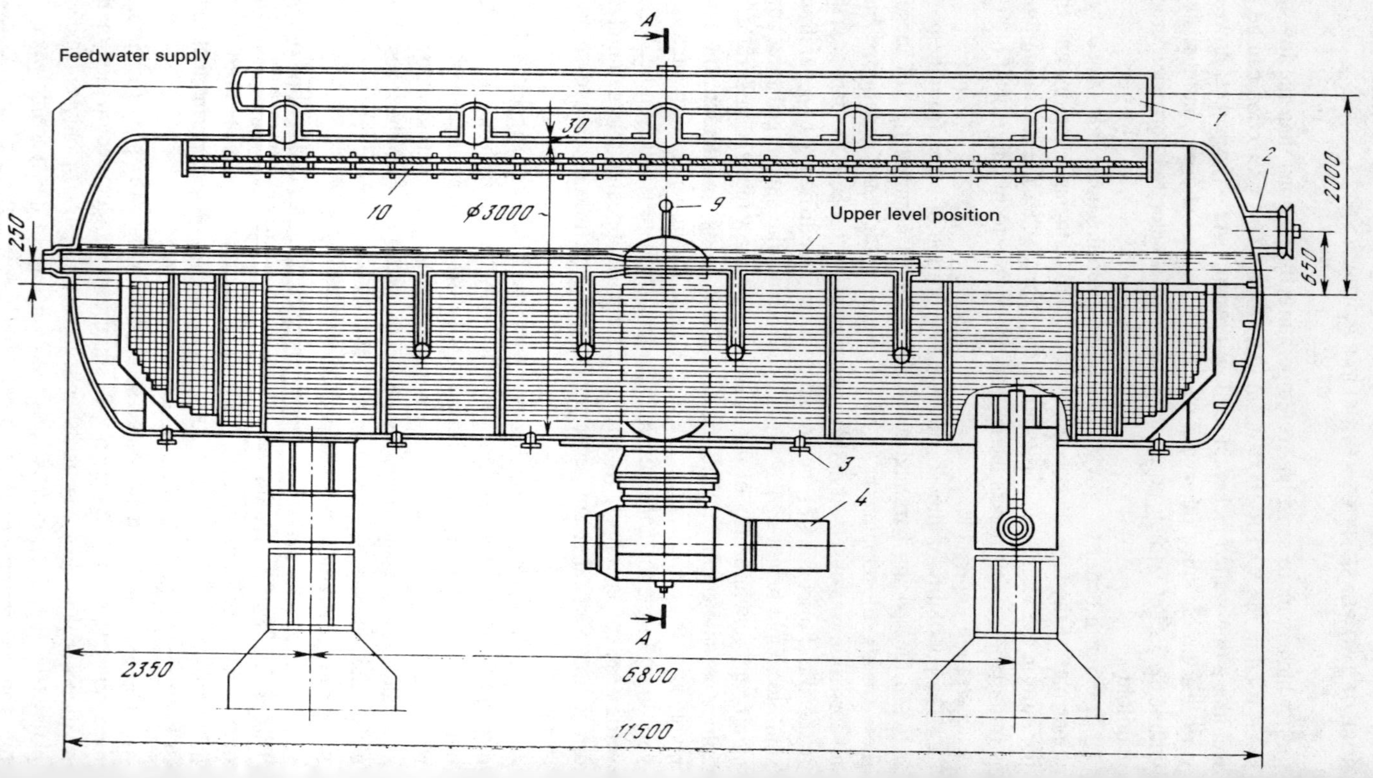

Feedwater supply
A
30
Upper level position
2000
250
650
10
Ø3000
9
1
2
3
4
A
2350
6800
11500

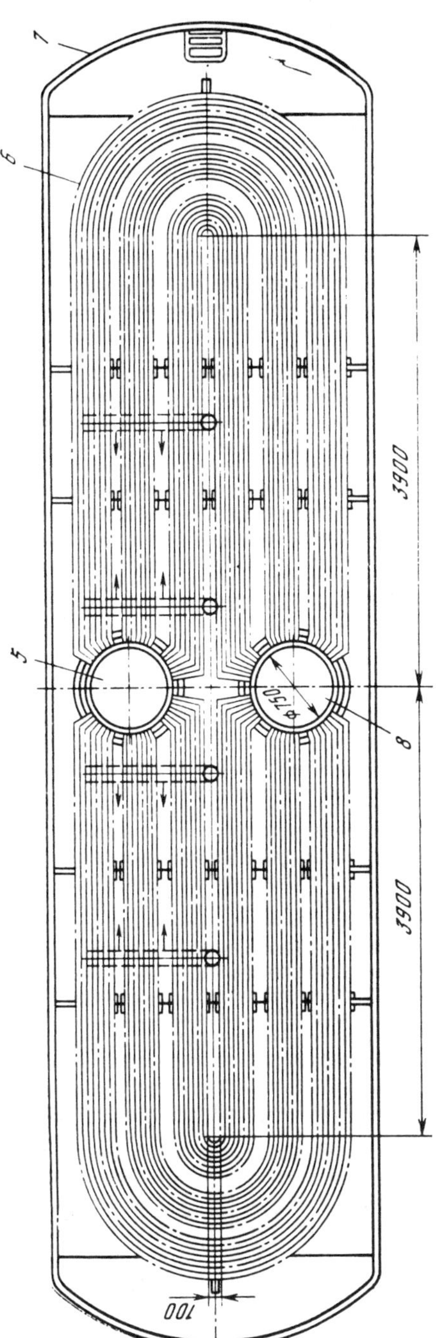

Figure 1.9 Horizontal PWR steam generator for the VVER-210. 1, Steam delivery manifold; 2, manway; 3, blowdown and drain pipes; 4, coolant pipe connection; 5, coolant inlet manifold; 6, heating surface; 7, steam-generator shell; 8, coolant discharge manifold; 9, primary loop air vent; 10, louvered steam separator.

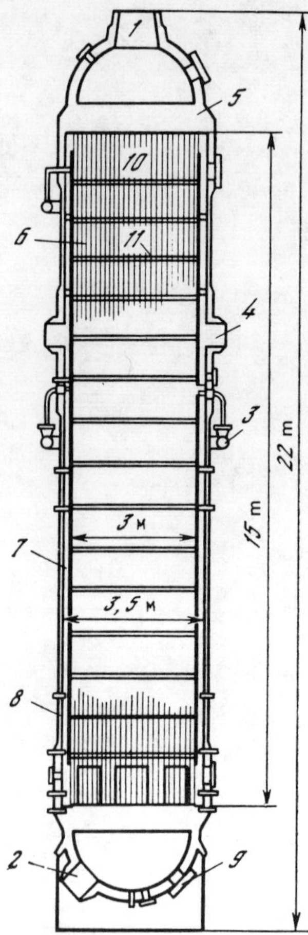

Figure 1.10 A Babcock and Wilcox nuclear steam generator. 1 and 2, Reactor coolant inlet and outlet; 3, feedwater inlet; 4, steam outlet; 5, upper tubesheet; 6, tube bundle; 7, downtake plenum; 8, shell; 9, manway; 10, superheated steam region; 11, tube spacing grid.

conditions, whereas the lower part (approximately one-third of the unit) operates under nucleate boiling conditions.

The throttling produced by the spacing grids improves the stability of the upward directed steam-water mixture flow which might have otherwise been unstable, given the low mass velocity and pressure. However, it should be noted that in this design, unlike the case of U-tube steam generators, the heating of all the tubes in a bundle in any horizontal cross section is virtually the same, and this reduces the danger of circulation reversal and of formation of downtake water flows be it even between adjoining spacing plates.

It is important to note that in the Babcock and Wilcox design the feedwater of the secondary loop is preheated before entering the steam generator tube bundle to a temperature close to that at saturation, by mixing it with a part of the steam. As a result, the temperature of the primary-loop water drops only slightly under partial load, whereas the feedwater temperature does drop. At the same time, supply of water at close to saturation temperature directly to the steam generator improves the stability of the uptake flow of the steam-water mixture at partial load, when the danger of water flow direction reversal increases.

Obviously, when feedwater is heated by steam injection, one must forego some additional cooling of the water in the primary loop, which is possible even at full load had the lower part of the steam generator performed as an economizer. Dispensing with the economizer results in some reduction in nominal power output in the core. However, this reduction is insignificant, whereas constancy of temperature at the steam-generator inlet is very desirable in order to reduce thermal stresses.

In view of the high stiffness of the system, consisting of straight tubes and a massive cylindrical shell, it is necessary that these two components be held at the same mean temperature. This is achieved in the following manner. Slightly super-heated steam ($\sim 30\,°C$), discharged from the upper part of the bundle, descends along the space between the tube-bundle jacket and the shell over a part of the length of the latter, before leaving the steam generator only then. The lower part of the shell length is filled with feedwater, preheated to close to saturation by injection of steam.

The wall temperature of the lower part of the tube bundle where nucleate boiling occurs is only slightly above the saturation temperature; however, in the upper part, after the onset of film boiling, the tube wall temperature is close to the temperature of primary-loop water.

Proper selection of the lengths of the upper and lower parts of the jacket with allowance for differences in the thermal expansion coefficients of the tube and shell materials, make it possible to obtain equality between the overall elongation of the housing and tubes, and to thus eliminate thermal stresses when starting up the steam generator. However, for this one must know the heat transfer coefficients for all the parts of the steam-generating tube: nucleate boiling, dispersed-annular flow, transition region, and post-crisis boiling.

The cold steam generator, on the other hand, is not subject to significant stresses, since after all the welding operations are completed the entire steam generator is subjected to stress relief at $\sim 620\,°C$.

The above equalization of the thermal expansions of the tubes and the shell may be achieved only for a certain set of conditions (for example, as shown above, at full load). When the load is reduced, the fraction of the steam generator which is performing as a superheater will increase, and accordingly, the mean temperature of the tubes and their overall length will also increase. At the same time the temperature of the superheated steam and the mean temperature of the shell will increase only insignificantly; all this will produce thermal stresses. For this reason any deviation of steam generator load from the nominal may result in thermal stresses and as time goes on in the appearance of fatigue cracks. The temperature of tube metal in the droplet precipitation region may also undergo moderate, but very frequent oscillations.

From the electrochemical corrosion point of view, the once-through steam generator presents stiff requirements on feedwater quality, since the ratio of the distribution of a number of corrosive impurities, in particular NaOH and NaCl in water and steam is very high—of the order of 10^5. For this reason, the equilibrium concentration of these impurities in the last water droplets to evaporate may be higher than is permitted even for corrosion-resistant materials. At moderate wall temperatures, it should be expected that the droplets will deposit in the heated surface and evaporate to a concentration at which the salts will be captured by the steam.

When this concentration is attained, the droplets evaporate until both the water and the salt vanish entirely; this means that when the feedwater contains any impurity in concentrations lower than the solubility of the impurity in steam, there will be no accumulation of the solid phase or concentrated solution.

In the case of very pure feedwater the fraction of droplets with high impurity concentration will amount to 10^{-2}–$10^{-3}\%$, and the mean statistical time of their contact with any part of the surface will be very small. It can then be expected that the corrosion rates will be low and the tube life-time will be high. Naturally, it is difficult to quantitatively calculate the corrosion rate even if the composition of the water is known, and it can be determined only be practical testing.

It is important to note that corrosion damage of tubes in the nucleate boiling zone is highly improbable, since the water is very pure, and there are no stagnant zones, as in the case of the U-tube steam generators. Complete vaporization is also highly improbable in places where the tubes pass through spacing plates, since only shallow crevices can form in the points of contact between the tube and the plate inside the specially configured holes, and these crevices open very rapidly in both directions. Mass transfer in the crevices is also facilitated by the high velocities prevailing in the spaces between the tubes and the spacing plate.

Only a limited body of published data is currently available on the operation of once-through straight-tube nuclear steam generators. Only a few units have been in operation over an extended time period: i.e., at the Oconee nuclear power plant, the No. 1 unit at the Three-Mile Island nuclear power plant, the Arkansas power No. 1 unit, and the Rancho Seco nuclear power plant. For about 3 years all these steam generators operated without much trouble. The Oconee 1 unit sprang its first leaks in 1976, and toward the end of 1977 all of the three Oconee Plant steam generators leaked, apparently due to fatigue failure of tubes. So far no detailed description of the situation is available, and it is difficult to know whether the damage is due to pure metal fatigue, or what appears more probable, to corrosion fatigue. Note that the quality of the feedwater, all of which was desalinated by precoat filters, was very high. The sodium concentration was as low as 4–5 ppb. Under these conditions no local precipitation should be expected (the solubility of NaCl and NaOH in steam in terms of Na^{+} is 50–60 ppb). However, the chloride and alkali concentrations in the droplets evaporating last may be sufficiently high to significantly affect the development of fatigue cracks in Inconel-600 tubes.

The fact that droplets of concentrated solutions are a factor in tube damage is also confirmed by the fact that the bulk of tube damage occurred along the edges of the tube-free radial clearance, where more water was present than in the tightly-packed bundle itself. It should, however, be noted that the accelerated cooling of tubes arranged along the clearance may have resulted in additional tensile stresses.

The high stiffness of the entire system of once-through straight-tube steam generators and the low heat fluxes at the upper part of the tube bundle resulted in the appearance of a large number of designs of once-through steam generators with specially-shaped (undulated, helical, etc.) tubes. An example is the steam generator manufactured by the Trepand Company, which has not yet come into practical use. These steam generators have a number of advantages over straight-tube steam generators; (1) the use of undulated tubes eliminates the problem of thermal stresses; (2) the presence of a central tube decreases the pressure on the tubesheets, allowing the use of thinner material; (3) the use of transverse flow in regions with nonboiling water and particularly of slightly superheated steam makes it possible to enhance heat transfer in these zones and thus reduce the heating surface needed; (4) the presence of a built-in economizer allows better utilization of the temperature difference.

At the same time, it is unclear how it will be possible, in the presence of horizontal baffles, to eliminate poorly ventilated gaps between them and the tubes, particularly if it is remembered that it is more difficult to pass bent tubes through openings in the baffles. In any case, the advantages of a moderate-pressure once-through steam generator will be offset by the increased requirements for feedwater quality. However, the extent to which these conditions will differ from those needed for reliable operation of the conventional multi-circulation steam generator, depends entirely on the local conditions of heat and mass transfer in the respective designs. The requirements with respect to steam purity for reliable operation of the turbines as such and the entire steam flow passage to and through the turbines are easier met in the case of multiple circulation, where the steam-generator blowdown system ensures removal of salts from the cycle at a lower cost than in nuclear power plants with once-through steam generators.

One of the features of once-through steam generators that requires special attention is their low water volume, which requires particular care in providing an uninterrupted supply of feedwater. It is important to note that most of the water is situated in the small lower part of the steam generator. Calculations show that in the case of feedwater pump failure, as was the case in the Three-Mile Island plant, dryout of the heating surface occurs very rapidly. Although water may still remain in the lower part of the steam generator, the heating surface operating under evaporation conditions becomes so small, that even if the reactor is shut down immediately, the residual heat release in the core (resulting primarily from heat accumulated in the fuel elements), will be greater than the heat absorption capacity of the steam generator. This results in a temperature and pressure rise in the primary loop.

1.4.2 Steam Generators at Liquid Metal Cooled Fast-Breeder Reactor [LMFBR] Nuclear Power Plants

The experience in operating such steam generators is very limited and pertains to relatively prolonged (10–15 thousand hours) tests of experimental models, and also to operation of three pilot-plant units: the BN-350 (Shevchenko City, USSR), PHENIX (France), and Dounreay (England). Research and development work is being carried out in these countries in order to design larger LMFBRs. These are both of the once-through type (BN-600, SuperPHENIX) and also recirculation units (Dounreay).

In addition to problems associated with interaction between sodium and water (basically contamination of the reactor coolant by metal oxides), the main difficulty is

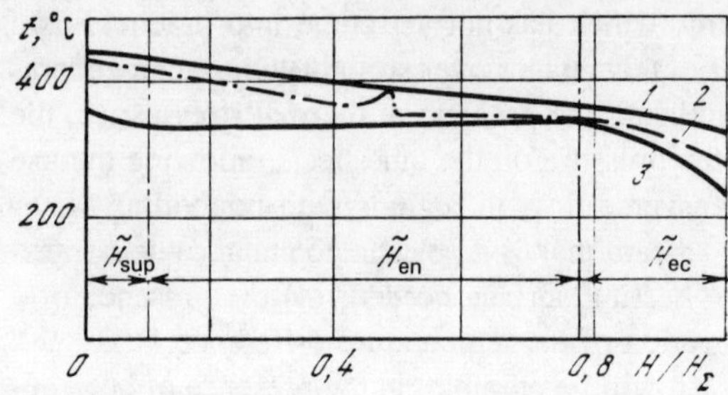

Figure 1.11 Temperature variation along a steam generator. 1, Reactor coolant temperature; 2, temperature of steam generating surface; 3, steam temperature.

the large drop in the liquid-metal temperature along the steam generator. This results in a significant increase in heat flux in the region near the liquid metal inlet to the steam generator. When metal-water counterflow, which is desirable for better utilization of the economizer zone and greater ease of assembly, is used, the rise in the temperature difference in the direction of the water flow involves a reduction in q_{cr} (Fig. 1.11). As a result, dryout* occurs at low qualities ($x = 0.3–0.4$), and the subsequent portions of the surface operate under film boiling conditions. This significantly increases heating surface required and poses the danger of development of fatigue phenomena due to fluctuations of 30 °C in the metal temperature in the region near the dryout point. The danger of the appearance of concentrated salt solutions at the end of the evaporation region in this case is less than for once-through PWR steam generators, since the pressure in fast reactor evaporation is much higher.

Thus, for the BN-600 at turbine pressure of 13 MPa the pressure at the end of the vaporator is about 15–16 MPa, whereas in PHENIX the pressure in the critical region is 17–18 MPa and higher while the turbine pressure is 16 MPa. In addition, due to the high temperature difference, particularly at the end of the evaporator, the fraction of droplets that reach the wall will be lower, although droplets may wet the wall and their residues may evaporate from it even at high temperatures. From the point of view of boiling heat transfer, parallel metal and water flow may be preferable for once-through, sodium cooled steam generators. In this case the zone of maximum temperature differences corresponds to the zone of minimum enthalpies, and it is possible to maintain the bulk of the heating surface under nucleate boiling (pre-dryout) conditions, particularly if the heat flux in the zone of highest sodium temperatures is reduced. The latter can be attained either by reducing the velocity of the metal in the hot zone or by recirculating a part of the liquid metal from the steam generator and mixing it with the hot metal upstream of the steam generator.

It is much simpler to provide satisfactory mass transfer and prevent corrosion of the steam-water flow passage in liquid-metal heated steam generators than in generators heated by high-pressure water, since when water boils and wet-steam moves inside of smooth cylindrical tubes it is easier to ensure reliable flushing (by the evaporating crate) of the entire tube perimeter. When slightly inclined or horizontal

---

*In this book the term "dryout" has been used to describe the critical conditions in boiling. It is synonymous with the terms such as "burnout," "critical heat flux," "DNB," and "boiling crises."

tubes are used it is necessary to prevent stratification of flow, which creates conditions for close to complete evaporation at the junction between the wet and dryout zones. This may, if the system is not flexible enough, result in the appearance of fatigue cracks due to nonuniform temperature distribution of the metal over the tube perimeter.

The problem of equalization of expansion of the tube bundle and the shell (which is a difficult problem in PWR steam generators), is easily solved in once-through, straight-tube liquid-metal steam generators by providing the low pressure thin-walled shell with a corrugated expansion joint. However, the danger still remains of temperature stresses in the tubes when their expansion is not uniform. For this reason it is very important to ensure complete stability of the hydraulic performance of steam-generating tubes, particularly in view of the high feedwater subcooling. In fact, in the presence of uneven distribution of water, dryout in individual tubes may begin significantly later than the bundle average, and such tubes, whose metal is at a temperature significantly lower than the bundle mean, will be subjected to greater tensile stresses, since the difference in metal temperature between the nucleate boiling zone and the post-critical region is significantly greater than in PWR steam generators.

1.5 HEAT AND MASS TRANSFER IN MOISTURE SEPARATOR REHEATERS OF NUCLEAR POWER PLANTS (THE LOW–PRESSURE SIDE)

Nuclear power plants with water cooled reactors are, as a rule, equipped with a moisture separator, situated between the turbine cylinders; in the majority of cases the separator is followed by a reheater, which forms a single unit with the former. Reheating is obtained in a single or, more usually, two stages; in the first stage reheating by steam bled from the high-pressure cylinder and, in the second, by fresh steam. Due to the fact that the pressure of the reheated steam is low (usually 1 to 1.2 MPa) the coefficients of heat transfer to it are low and the temperature of the reheater wall is close to that of condensing steam ($\sim 220\,^{\circ}$C in the first and $\sim 240\,^{\circ}$C in the second stage). The low pressure and the resulting huge volumes of steam passing through the separator make the attaining of low residual percentage moisture difficult.

At an initial percentage moisture of 12–14% and separator efficiency $\eta_{\text{sep}} = 0.95$ the residual percentage moisture is 0.6–0.7% and at $\eta_{\text{sep}} = 0.98$ it is 0.25%. The residual percentage moisture as such has relatively little effect on operations and has an insignificant effect on the required reheater surface). However, η_{sep} may have a significant effect on the operating conditions of the remaining part of the turbine and its piping. In low-pressure steam, the ratio of the content of most impurities (with the exception of iron oxides) in the steam and water is many-fold lower than any attainable moisture content. For this reason the concentration of any corrosive impurity in the wet steam leaving the separator is virtually controlled by the residual moisture content. True, it is not entirely clear whether close-to-equilibrium distribution of salts between the steam and water droplets is attained at the time the droplets separate from the steam flow, since the time of contact in the turbine is exceedingly small, whereas in the piping and in the separator proper, the moisture is for most part in the form of relatively large droplets. This question must be considered separately. In any case, the effect of separator efficiency on the purity of steam entering the reheater must definitely be considered in assessing the operating conditions of the latter.

Any impurities entering the reheater dissolved in the moisture drops will precipitate on the reheater surfaces and, at sufficiently high wall temperatures, the majority of impurities, including NaCl are fully dehydrated and remain on the heating surface. When these impurities have undesirable properties, electrochemical corrosion may occur in regions of their precipitation. However, the steam supplied to the turbine is of high purity, and the bulk of impurities that enter the separator leave it with the separated moisture. The steam flowing to the reheater contains only extremely small amounts of impurities, which are insufficient for any significant general corrosion of the metal. It is hence desirable to use reheater materials which are insensitive to local forms of corrosion (corrosion cracking, etc.). From this point of view it is best to use ordinary steel, which, in addition has a rather low rate of overall (uniform) corrosion at temperatures in excess of 200 °C and, what is most important, corrosion products from such steel are transferred to the steam at a rather low rate.

In addition to the operating conditions of the reheater proper, consideration must also be given the effect of the reheater on the quality of steam flowing to the low-pressure part of the turbine. The moderate temperature differences, low superheated-steam side heat transfer coefficients and the resulting low heat flux density are responsible for the fact that, apparently, virtually all the vaporization occurs on the heating surfaces, rather than in the flow core. Under such conditions almost all the impurities precipitate on the heating surface, forming solid deposits. Compounds with a high solubility in water, in particular NaOH, leave a highly concentrated residue, which remains in the form of a viscous liquid at superheats as high as 250–260 °C, these supereheaters being characteristic of high-power used in water-cooled reactors. Here, due to the very low solubility of NaOH in steam at a pressure of ~ 1 MPa, even when the concentration of NaOH in the droplets is as high as 60%, which corresponds to thermodynamic equilibrium at 1 MPa and 250 °C, not all the NaOH can be absorbed by the departing steam particularly at the elevated (> 10 ppb) NaOH concentrations in the feedwater, which are permitted for short times by many operation rules.

In spite of the high viscosity of the film of concentrated NaOH solution which forms in these cases on the heating surface of the first stage of the reheater, this film will be gradually carried into the second stage, where it is entrained from the heating surface into the steam from which carries it out of the reheater. This may result in corrosion damage of flow connecting piping and the low-pressure side of the turbine. At lower reheating pressures (for example, 0.35 mPa in the first units of the RBM-K reactors) liquid droplets containing high concentrations of NaOH may be present in the steam leaving the reheater even at inlet NaOH concentrations below 10 ppb. All the other compounds present in original water droplets carried from the separator to the reheater will be almost completely precipitated in the latter and only a very small fraction may be carried with the steam in the form of solid dust particles.

Thus, compounds having low solubility in water (silicates, calcium salts, etc.) will accumulate during the entire time of operation, being only insignificantly removed during changes in load or shutdown of the unit, when the steam in the whole reheater may be wet. Deposits of easily soluble compounds (for example NaCl) may, upon change in load, be washed out by the wet steam, giving a transient "surge" of salt concentration which evaporates again on the heating surfaces of the second stage of the reheater. This effect may be responsible for the appearance of significant discharges of dust-like particles at the reheater exit; however, these discharges are insignificant, their duration being short and hence not dangerous from the point of view of turbine corrosion.

1.6 HEAT AND MASS TRANSFER IN THE REHEATER, REGENERATIVE PREHEATERS AND CONDENSER OF THE TURBINE (HEATING SIDE)

The conditions of heat and mass transfer on the heating side of regenerative preheaters are characterized, as in the reheater and main condenser of the turbine, by almost complete condensation of the steam entering the unit. Accumulation of noncondensing gases in the residual parts of the steam is detrimental to heat transfer, and in BWR plants, where the steam contains a rather large quantity of radiolysis products, it may result in formation of an explosive mixture of hydrogen and oxygen. For this reason, in regenerative preheaters and reheaters, some steam is usually bled together with the condensate to the lower pressure parts of the system in the main condenser and sucked out by an ejector and discharged to the atmosphere through a cooler (in a BWR system, the discharge is through the exhaust purification system).

The conditions of heat and mass transfer along the steam condensation path are highly nonuniform. The first portions of the condensate absorb the bulk of the salts contained in the steam, whereas the gases dissolved in the steam are present in the condensate only in small fractions, with their bulk held in the remaining portions of the steam. For this reason the concentration of impurities in the condensate film is very different from their concentration in the steam, which may have a significant effect on electrochemical corrosion of the metal.

Oxygen reacts mainly with heating surfaces of carbon or low-alloy steel, whereas ammonia is most damaging to copper-containing alloy surfaces. In the initial stages of condensation, there is virtually no oxygen in the liquid film. Even in BWR systems, where the oxygen content of the steam is as high as 20–40 ppm, its concentration in the initial condensate, particularly at low temperatures, may be less than 10 ppb. Under these conditions, particularly at low temperature, one may have to deal with the problem of "oxygen starvation." Here, the absence of oxygen in the high purity water makes difficult the formation of a stable protective film, leading to an acceleration of corrosion of carbon or low-alloy steel.

As more condensate accumulates, the oxygen concentration in the steam and in the newly produced condensate rises. However, initially this occurs at a very low rate and, only after 90% of the steam condensed, do the oxygen concentrations rise almost 10-fold and, after 99% of the steam has condensed—almost 100-fold. If the condensate film falls freely down vertical tubes, diffusion of oxygen through the thickness of the liquid film to the metal surface does not occur at a sufficient pace. As a result, the oxygen concentration in the wall layer remains significantly lower than in the last portions of the condensate entering the film. Hence, as a rule, there is no danger of excessive oxygen concentration near the metal surface.

Evidently, the real situation may be significantly more complicated, particularly in the main condenser of the turbine, where the steam velocities are very high, and the configuration of the horizontal tube bundle at which condensation occurs is very complicated. Under real condensation conditions, the condensate film separates, comes into contact with the steam flow, and receives additional amounts of residual gases from the steam in which the content of these is high, thus increasing the gas concentration in the condensate film.

The last steam to condense, and the associated condensate, contain significant quantities of residual gases, including (when the unit operates in the hydrazine-ammonia regime) a large quantity of NH_3. This may result in accelerated corrosion of copper-alloy condenser tubes in the so-called air-cooling section, which is the last section of condensation surface along the steam path. It is hence necessary to provide steam and condensate flow patterns ensuring that water with high NH_3 content will not flow over heating surfaces made from copper alloys sensitive to NH_3. Considerations need be given only to mass transfer and one must be able to calculate the concentration of the given impurities in the condensate precipitating at different parts of heating surfaces. This makes it possible to forego the use of expensive material such as titanium, which is capable of resisting attack by various impurities present in the liquid over a very wide range of concentrations.

Obviously, the use of titanium even for the fresh-water-cooled main condensers can be justified if this is required by the cooling-water-side corrosion situation, but this problem is outside the scope of this book.

1.7 HEAT AND MASS TRANSFER AND HYDRODYNAMICS IN NUCLEAR POWER PLANT TURBINES

The turbines employed in water-cooled nuclear power plants typically operate on low pressure steam and, what is most important, the steam entering the turbine is, as a rule, at saturation temperature. Only units with once-through steam generators have a moderate initial superheat ($\sim 30\,°C$). This means that with the exception of the first stages of the low-pressure cylinder following the reheater (and, in units with once-through steam generators, the first stage of the high-pressure part) the remaining part of the turbine operates on wet steam. This has a significant effect on the turbine efficiency, creates conditions for erosion, particularly in the last stages of the high- and intermediate-pressure parts. This imparts great significance to the study of hydrodynamics of individual turbine stages and to measures for mitigating the detrimental effect of moisture in the steam.

On the other hand, the low operation temperatures virtually eliminate the problem of creep of the blade and rotor metal, and for wet-steam stages, also the problem of deposits, since these stages operate at all times under well washed conditions.

At the same time several recent investigators [1.16, 1.17] have identified various types of corrosion damage at low pressure, particularly in the region near the Wilson line (i.e., when wetness starts to be formed in the steam). Very little experience has accumulated as of now in the operation of the high-pressure part of units associated with once-through steam generators. In analyzing the problems of corrosion in the initial condensation zone, it is very important to investigate the behavior of impurities contained in the steam. Due to the presence of the separator, the concentration of these contaminants entering the reheater is very low; in addition, the concentration of most impurities decreases significantly upon passing the reheater. However, due to the very low pressures, the distribution of a given impurity between the steam and liquid phase may be highly unbalanced. In principle this may lead to concentrations in the liquid phase, upon attainment of thermodynamic equilibrium, which exceed the mean flow concentrations by a factor of 10^8–10^9. The complexity of the processes occurring in the turbine make it very difficult to investigate the problem.

On the other hand, under conditions prevailing in steam generators or reheaters (i.e., where the processes are relatively slow, it is possible to use, with some confidence, the thermodynamic equilibrium values for the ratio of contaminants in the steam and water. In spite of certain difficulties arising from lack of information on thermodynamic quantities at low pressures (and particularly, lack of data needed for calculating the kinetics of heat and mass transfer) the concept of treating corrosion from the point of view of local equilibrium concentration of impurities in the liquid phase is currently more or less the accepted practice.

However, under the conditions found in turbine operation, where the processes occur at a very high rate, and spontaneous condensation of even pure steam starts only at high supersaturations, it is extremely difficult to calculate the behavior of impurities, the concentration of which in the steam amounts to several ppb. As a result of the presence of dissolved impurities, the formation of droplets of saturated solution starts in principle at temperatures higher than the saturation temperature for perfectly pure steam! This excess, which is insignificant for compounds with low solubility in water (metal oxides, silica, calcium and magnesium salts, etc.), is rather significant for NaCl (in the zone of intersection of saturation curves in the low-pressure part of the turbine the excess is approximately 20–25 °C). It is extremely high for NaOH which forms very concentrated aqueous solutions over the entire range of turbine operation not only of water-cooled nuclear power plants, but also in turbines of liquid-metal-cooled nuclear power plants with steam superheated up to 500 °C (and even higher in future plants).

However, even at elevated pressures, the contents of NaOH and NaCl in the water phase are usually higher than their contents in the steam; thus, these impurities are present at the inlet to saturated-steam turbines (carried by the water droplets). In the high-pressure turbines of fast-neutron nuclear power plants, however, the impurities are present in a truly dissolved state in the steam. On the other hand, spontaneous condensation of impurities which are present in the steam in exceedingly small quantities is very unlikely to occur.

It can be assumed that the condensation of truly dissolved impurities from steam in the course of rapid expansion of the latter will require a significantly greater supersaturation (relative to the boiling temperature of the saturated solution of these impurities) than the well-explored supersaturation of perfectly pure steam. At present it is difficult to determine this supersaturation quantitatively, particularly for true steam solutions with exceedingly low impurity concentrations (of the order of 0.1 to 1 ppb).

The situation changes if the steam contains particles of solid salt ("dust carryover" from the reheater) or droplets of concentrated NaOH solution, surrounded by H_2O molecules; these particles and droplets serve as readily-available condensation sites and may start absorbing moisture long before the steam attains the saturation temperature. Hence corrosion phenomena may occur prior to the attainment of supersaturation conditions corresponding to the Wilson curve, but only under certain conditions. Here the corrosion induced by NaOH can occur along the entire flowpath from the inlet to the low-pressure part to Wilson line position, below which the impurity concentration in the droplets will drop rapidly due to their dilution by the condensate. For NaCl and other compounds, where the boiling temperature of the saturated solution excess only slightly the boiling temperature of pure water, the zone of possible corrosion will be very narrow.

In those rare cases when there is no reheater (new nuclear power plant units almost invariably have a reheater), the purity of the steam entering the turbine depends significantly on the separator efficiency. However, in such cases the purity of the steam does not have a significant effect on the danger of corrosion damage to the turbine, since the latter operates completely under wet steam conditions and, as expansion proceeds the wetness of the steam continuously rises, and consequently, the impurity concentration in moisture droplets, which is initially very low, falls rapidly. In the final analysis, even when the steam contains corrosive impurities (for example, free NaOH) their concentrations along the entire turbine flowpath remains much lower than those required to give corrosion conditions. The situation is different in the majority of cases, where the separator is followed by a reheater and the steam in the first stages of the low-pressure zone of the turbine is superheated. Thus, individual impurities in the steam entering the turbine can be present in the form of highly vaporized droplets (for example, NaOH solution), solid particles of salt (for example, NaCl particles), and in the form of a true solution in the steam. Naturally, under these conditions the purity of steam significantly affects the reliability of the operation of the low-pressure part of the turbine. This applies to high-temperature and high-pressure turbines, which do have a reheater (as used in fossil-fuel units or liquid-metal or gas-cooled nuclear power plants).

During the past several years corrosion damage was observed extensively not only of blades, but also of turbine rotor wheels [1.17]. The damage was of different kinds (stress corrosion cracking, corrosion fatigue, corrosion-erosion), but all were, as a rule, concentrated at the initial condensation zone, close to the Wilson line. This was attributed to the fact that in this zone, even a moderate contamination of the steam, when concentrated in a very small amount of moisture could produce high, in a number of cases corrosive, concentrations. For this reason a number of investigators [1.17] have recently suggested that separators should be designed with very high efficiency (as close as possible to 100%).

However, this approach does not pay sufficient attention to the behavior of impurities as they pass through the reheater. It is important to note that reheat at nuclear power plants with water cooled reactors is, as a rule, produced by condensing steam, and hence the reheated steam temperature is not very high (not above 260 °C). For this reason the reheat pressure at such nuclear power plants is low (usually 10–12 MPa, and for a number of units, even only 0.3–0.5 MPa), i.e., significantly lower than in large fossil-fuel fired units. The behavior of NaOH in turbines employing high-temperature, high-pressure steam, where reheat occurs at pressures of the order of 3 MPa, is characterized by the presence in the reheater of only truly dissolved NaOH (even when its concentration in the steam is relatively high, of the order of 100 ppb). The appearance of concentrated NaOH solutions in the turbine flowpath is possible only under conditions outside normal practice (inadequate regeneration of condensate-purification filters, injection of moisture from the steam-generator drum when the boiler water contains free alkalies, etc.), when the NaOH concentration in the steam undergoes a short but steep surge.

Droplets in the reheaters of nuclear power plants with water-cooled reactors may evaporate to very high concentration (40–50%) of NaOH without complete varporization of the latter even when the NaOH content of the steam is relatively low (10 ppb).

Naturally, this conclusion is to some extent purely theoretical, because it is virtually impossible to expect that the steam contains NaOH as the only impurity, whereas in the presence of other substances, the majority of which give small depression of the boiling point. Thus, there may form more complex compounds, capable of complete evaporation at moderate concentrations in the water. In these cases the reheater serves as an additional "trap" for the impurities, and the content of any impurity past the reheater will not exceed the solubility of the given impurity in steam at the given temperature and pressure. Even when the overall content of corrosive impurities is relatively high, for example, 20 ppb, and the separator efficiency low (80%), the steam flowing to the reheater will have an impurities content of about 4 ppb. Even if it is assumed that the true solubility of these impurities in 260 °C steam is only 0.1 ppb, then about 3.9 kg of impurities will accumulate in the reheater per 10^6 tons of steam. In large units this is equivalent to deposits of about 100 kg/year, which is a very significant amount for a turbine, but insignificant for a reheater which has a heating surface of 10,000 m^2 (although it is possible that the impurities may precipitate on only part of the heating surface).

Note that all the above considerations are purely qualitative, since they neglect the removal of salts from the reheater in the form of dust; in addition, the solubility of NaOH in steam at 0.5–1.2 MPa has been estimated approximately (by extrapolation into the low-pressure region), and also no consideration was given to the possibility of formation of complex compounds and the manner of their behavior in the steam.

It should be noted however, that it was shown in a large number of laboratory studies, and also by experience in operating low-pressure once-through boilers in the USSR, that when salts are deposited on heating surfaces due to complete evaporation of their solvent fluid (particularly at moderate heat flux densities) virtually all the impurities precipitate on the heating surface. On the other hand, data on the solubility in steam of various classes of compounds usually indicate that the solubilities of complex compounds is low.

TWO–PHASE FLOW HYDRODYNAMICS

2.1 TWO–PHASE FLOW STRUCTURES AND FLOW MAPS

The structures of two-phase flow encountered in the core of boiling-water reactors (BWRs) and in their steam generators range from bubbly to dispersed-annular. The structure of two-phase flow is one of its most important features, for which reason it is little wonder that a large number of investigators have concerned themselves with this problem. Such studies were performed by Kosterin [2.1] on air-water flows in horizontal, inclined and vertical tubes, Styrikovich and Surkov [2.2] on flows of steam-water mixtures in vertical, horizontal, inclined heated and adiabatic tubes, by Baker [2.3] of air-oil flows in horizontal tubes by Hewitt (1965) of steam-water flows in vertical tubes, and many other investigations.

These authors point to the existence of various two-phase flow regimes: bubbly, foam, plug, slug, rod, intermittent, annular, dispersed-annular, stratified (only in horizontal and slightly inclined tubes) and drop. A large difference in two-phase flow pattern is observed between isothermal two-component flows and single-component flow in heated channels (in particular, in steam-generating tubes). First we consider two-component isothermal flows.

2.1.1 Two-Component Isothermal Flows

Two-phase flow patterns were initially investigated for conditions prevailing in the petroleum and chemical industries with typically low pressures and under isothermal conditions. It was found that four basic flow modes prevail in vertical tubes (Fig. 2.1): *bubbly flow*, in which the gas phase is dispersed in the form of discrete bubbles

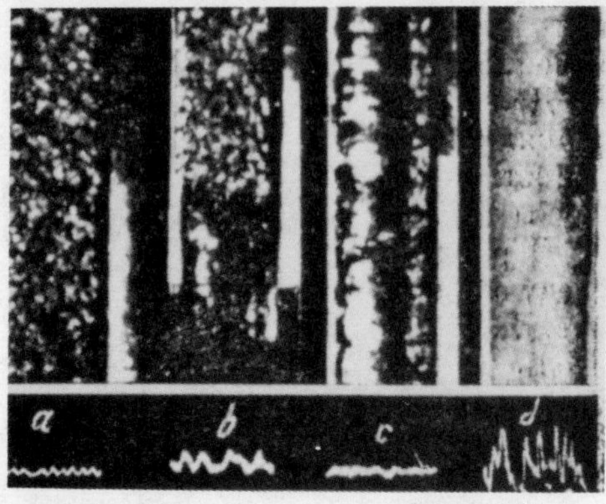

Figure 2.1 Principal forms of two-phase flow. (*a*) Bubbly; (*b*) slug; (*c*) annular; (*d*) droplet.

in a liquid continuum (Fig. 2.1*a*); *slug flow*, where large portions of gas (slugs) alternate periodically with liquid plug, within which there exist small bubbles (Fig. 2.1*b*); *annular flow*, in which the liquid moves along the channel walls as an annular film whose surface may be undulated by a complex wave system (Fig. 2.1*c*) whereas gas flows in the core, and *droplet** flow, in which the bulk of the liquid moves in the form of discrete droplets in the gas continuum, whereas a thin liquid film, the flowrate of which is several percents of the overall flowrate, moves along the wall (Fig. 2.1*d*). The remaining flow modes can be regarded as transient between the four basic regimes.

In horizontal and slightly inclined tubes there occurs at low velocities an additional flow mode due to phase stratification, meaning that gas flows in the upper part of the tube, while liquid moves along its bottom part.

Visual observations and high-speed photography do not provide unequivocal information on the type of two-phase flow structure, because the optical glasses [serving as viewing ports in these methods] are coated by a wave-covered liquid film. The flow modes are more reliably determined by analyzing the spectral density of fluctuations of the light intensity in transparent objects, and also by spectral analysis of the fluctuations of the static pressure or temperature in nontransparent channels. Figure 2.1 [sic] shows examples of the principal structures of two-phase flows and oscillograms of static-pressure fluctuations, whereas Fig. 2.2 presents oscillograms of light intensity for the same flow modes.

In bubbly flow the gas phase is uniformly distributed in the liquid in the form of individual rather small bubbles, moving at a velocity different from that of the liquid, the velocity difference being smaller, the smaller the bubble diameter (see Fig. 2.1*a*). The static-pressure fluctuations for this mode, shown in the lower portion of Fig. 2.1, have high frequency and low amplitudes.

---

*Editor's Note: In the Western literature the ''annular flow'' and ''droplet'' flow regimes referred to here are often classified together and given the name ''annular flow'' or ''annular dispersed flow.'' This has some logical basis in that annular flow without entrainment is a rarity and the extent of entrainment varies considerably. However, the USSR terms are retained in the present text.

When using transparent channels, the oscillograph screen (Fig. 2.2*a*) displays peaks and valleys in the intensity of light; it can be assumed that each light beam intensity peak fixes in the first approximation the location of the bubble, with the peak width corresponding to its size. Using the above statistical analysis of random fluctuating quantities, it is possible to construct the distribution function of the spectral density of fluctuations, and to determine from its minimum the flow mode, the probable mean bubble dimensions and velocity. Below each moving picture frame in Fig. 2.2 are given measured fluctuations of the intensity of light passing through the two-phase flow.

When the gas content increases, a large number of spherical bubbles merge which results in the formation of vapor bubbles of irregular shape, which are typical or intermittent or frothy flow (Fig. 2.2*b*). The light-intensity oscillograms show a rather large number of wide peaks, whereas the pressure fluctuations occur with increasing amplitude and accordingly decreasing frequency.

A further increase in the gas content results in formation of slug flow, in which elongated gas cavities in the flow core alternate with bubbly flow along the tube (Fig. 2.1*b*, 2.2*c*).

Liquid plugs, occupying, as seen from Fig. 2.1*b*, the entire cross section of the tube, appear periodically in a sequence at a frequency of approximately 1 Hz. The slugs move rather rapidly along the tube, which is responsible for the pulsatile nature of the entire flow. This circumstance is recorded by oscillograms of wall pressure fluctuations, which consist of a sequence of constant-frequency high-amplitude pressure surges.

On transition from slug to dispersed-annular flow (Fig. 2.1*c*) the amplitude decreases, the frequency increases and in the limiting case of developed dispersed-annular flow the oscillogram is similar to that produced by dispersed bubbly flow. As the quality [in the case of steam] increases, the flow mode becomes dropwise-dispersed with a thin film on the wall, a typical oscillogram of pressure fluctuations for this flow is given in Fig. 2.1*d*.

Regime map for isothermal flows. Various investigators suggested a large number of flow regime maps, based on their experimental data. However, the generality of

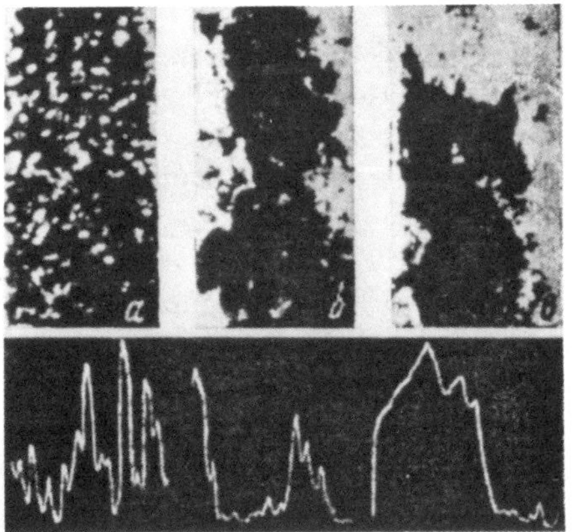

Figure 2.2 Oscillograms of light intensity for different two-phase flow structures at $p = 0.2$ MPa. (*a*) Bubbly; (*b*) intermittent; (*c*) slug.

these charts is doubtful due to the limited range of fluid physical properties and pressures used in their experiments (the majority of investigators used air-water mixtures at near atmospheric pressure.

The most extensively used flow regime map for isothermal conditions is that due to Baker [2.13], obtained initially for mixtures of air and petroleum products in horizontal tubes. Baker used the mass velocities of the liquid and gas as the coordinates. Subsequently the Baker diagram was modified and is plotted as j_1/λ vs. $j_2\lambda\psi/j$. Here

$$\lambda = \left[\left(\frac{\rho_1}{0.075}\right)\left(\frac{\rho_2}{62.3}\right)\right]^{0.5}; \quad \psi = \left(\frac{[0.005}{\sigma}\right)\left[\frac{\mu_2}{2.42}\left(\frac{62.3}{\rho_2}\right)^2\right]^{3/2}$$

are parameters of the physical properties of the flow. The boundaries between these structures, as seen in Fig. 2.3, are not clearly delineated and are comprised of regions of transition from one two-phase flow mode to another (shaded regions in Fig. 2.3). The reason for the transition is hydrodynamic instability of the interface which, in addition exhibits hysteresis.

For two-phase two-component flow under isothermal conditions in vertical tubes the transitions between regimes are best described by equations suggested by Kozlov [2.14]. He found that transition from bubbly to slug flow occurs along the curve

$$\beta = 0.05 \ \mathrm{Fr}^{0.2}$$

here $\beta = Q_1/Q$, $\mathrm{Fr} = ((Q_1 + Q_2)/A)^2 \ 1/gD$

According to Kozlov, transition from slug to bubbly flow occurs at $\beta = 0.5\mathrm{Fr}^{0.1}$, and from bubbly to dispersed-annular—at $\beta = 0.85\mathrm{Fr}^{0.02}$.

Subsequently these relationships were used for constructing the two-phase flow pattern maps for vertical tube flow, shown in Fig. 2.4.

2.1.2 Single-Component Two-Phase Flow in a Heated Tube

Studies in this area were performed in connection with developments in boiler and reactor design, for which reason the experiments were performed with high-pressure wet steam. A large number of investigators [2.5–2.29] concerned themselves with steam-water flows. We shall now consider the motion of a steam-water mixture in a vertical steam-generating tube.

The structure of a liquid flow moving in a sufficiently long heated vertical tube changes both along the channel, and over its cross section. The region of existence of each flow pattern is controlled by the relevant combination of the parameters x (quality), p (pressure), ρw (mass flux), q_w (heat flux), channel geometry and boundary conditions. Figure 2.5 shows the behavior of flow structure in a steam generating channel. The same figure shows curves of certain principal parameters along the tube.

The first region I (or 1) of purely single-phase flow is followed by a vaporization region II, which includes sections with subcooled boiling 2 and pool boiling of saturated liquid 3–5. The region of the channel with pool boiling is comprised of segments with bubbly (3), slug (4) and dispersed-annular (5) flow patterns. In the bubbly flow regime the liquid is saturated by fine vapor bubbles. As the steam quality is increased, some of them merge, forming large bubbles (slugs), separated from the tube walls by a thin liquid layer. A rise in the pressure with its attendant reduction in surface tension due to the fact that the phase densities become closer results in a steep reduction in slug lengths at pressures as low as $p \geq 3.0$ MPa [2.1]. According to Kozlov [2.4], slug flow no longer exists at $p = 13$ MPa. A vapor quality increases and the order of magnitude of the volumetric flows of the respective phases is close, the large bubbles merge and the dispersed-annular flow pattern occur, in which wet steam moves in the flow core and a thin annular layer of liquid moves at the wall. The thickness of this layer gradually decreases with rise in vapor quality. Following complete vaporization of the liquid annulus, the post-dryout region (III) is entered consisting of dispersed flow (6) and wet steam/fog (7) regions. Finally, the region of single-phase steam flow (IV-8) is reached.

Naturally, each flow regime has corresponding to it specific heat transfer patterns, which depend on the flow variables, and primarily on the heat flux and vapor quality. All the heat transfer modes in forced-convection coolant flow are depicted schematically in Fig. 2.6, which also shows by a dashed line [sic] the region of the dryout of the second kind, which exists over a narrow range of parameters. As seen from the chart, heat transfer due to forced-convection evaporation of the liquid is limited to relatively low heat fluxes. Other heat transfer modes exist over wider limits.

The principal steam/water flow patterns encountered in the heated channels of modern power plants, which operate basically at pressures between 6 and 20 MPa, are bubbly and annular dispersed. There are very little data on the conditions controlling the transition from one flow mode to another. It is seen from the paper by Bennet et al.[2.16] that the difference Δx in steam quality between the inception of surface boiling and the inset of annular flow at moderate heat fluxes ($q \leq 0.75$ MW/m^2) does not exceed 3–5%. Milashenko et al. [2.12] point out that at $p = 2.8$–6.8 MPa and mass flux of $\rho w = 500$–2000 kg/m$^2 \cdot$sec transition to dispersed-annular flow occurred at flow quality $x = 0.05$–0.09.

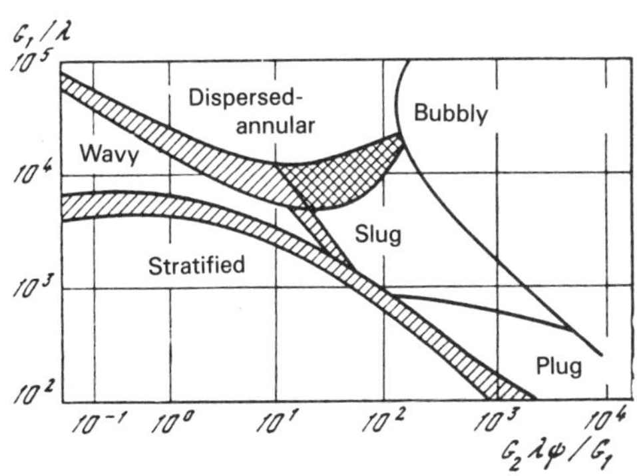

Figure 2.3 Baker's flow map.

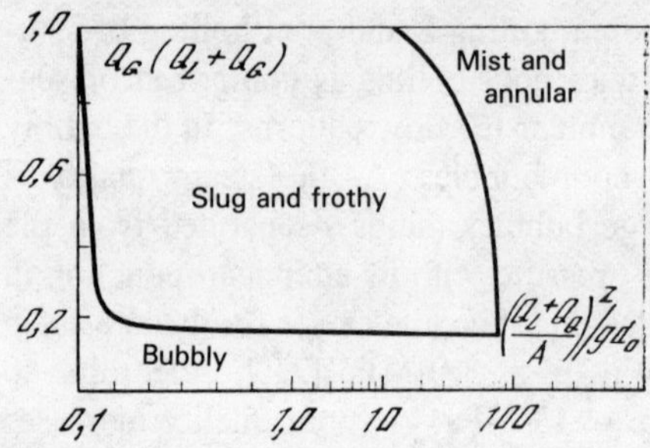

Figure 2.4 Flow map for air-water flows in vertical tubes at $p = 0.1$ MPa.

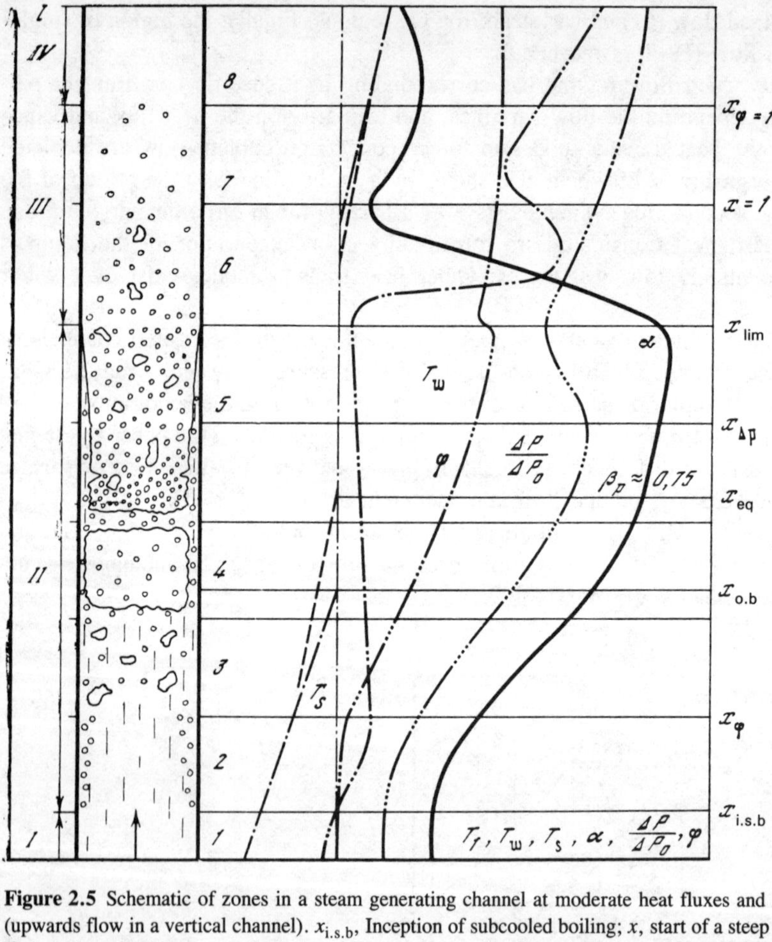

Figure 2.5 Schematic of zones in a steam generating channel at moderate heat fluxes and high pressures (upwards flow in a vertical channel). $x_{\text{i.s.b}}$, Inception of subcooled boiling; x, start of a steep rise in φ; $x_{\text{o.b}}$, region where the balance quality is equal to zero; x_{eq}, zone where the subcooled liquid vanishes; $x_{\Delta p}$, zone of microfilm start; x, liquid-film dryout zone.

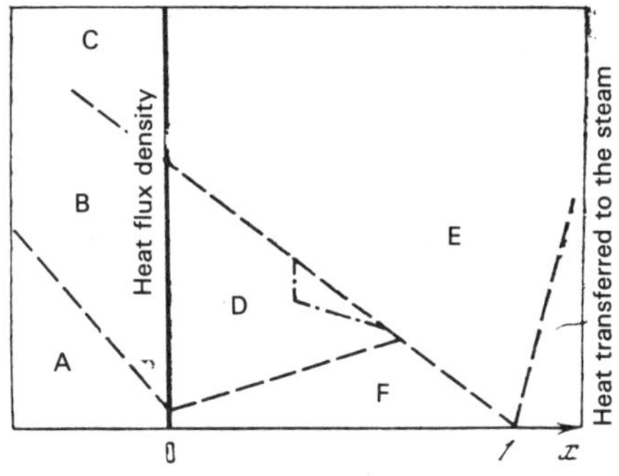

Figure 2.6 Chart of heat transfer in forced convection channel flow of coolant [2.14]. A, Forced-convection heat transfer from the liquid; B, subcooled surface boiling; C, film boiling; D, saturated nucleate boiling; E, forced-convection vaporization; F, film boiling.

In dispersed-annular flow the liquid film contains vapor inclusions, whereas the vapor core contains liquid droplets. The phase interface is more or less clearly defined. The mean interface parallels approximately the channel surface. The flow consists of three zones: (1) a relatively slowly flowing liquid film (laminar or turbulent film flow is possible); (2) liquid droplets in the core, moving at a velocity which exceeds by a large factor the velocity of the liquid in the film; and (3) an even more rapidly moving steam core [2.13], the interaction with the liquid film leads to the droplet entrainment.

Near the onset of dispersed-annular flow the liquid film is still relatively thick [2.14, 2.15]. A rise in vapor quality increases its linear velocity, which results in the appearance of a complex system of waves on the surface of the liquid film. The ripple pattern, which is an ensemble of waves of relatively low amplitude, is virtually always present on the liquid surface. Under certain conditions there arise disturbance waves with amplitude exceeding the minimum film thickness by a large factor ($\delta_{max}/\delta_{min} = 10$–$50$). These disturbance waves move at a velocity higher than the liquid film. The axial length of the disturbance is of the order 1–5 times the mean film thickness.

The disturbance waves make a major contribution to the increase in pressure drop of the steam flow and, in addition, frequently induce the generation of bubbles on the heating surface. These waves are one of the major factors in the entrainment of liquid droplets. It is concluded by Milashenko et al. [2.12] that at $p = 6.8$ MPa, $\rho w = 1000$ kg/m^2·sec and $x = 0.4$ the minimum film thickness (δ_{min}) is 0.1 mm, i.e., a perceptible part of the liquid in the wall film is transported in the wave crests. Droplets are entrained at significant steam velocities, when the disturbance waves are broken as a result of "undercutting" of the wave crest, its curling up and peripheral breakup. As a result the flow entrains droplets of different diameter, depending on the flow conditions at the point of their formation.

The second reason for the appearance of droplets in the core flow is the arrival of a vapor bubble at the phase interface and its subsequent collapse. The resultant break in the bubble shell is accompanied by separation of a part of the film with subsequent conversion into droplets. At the time of bursting of the vapor bubble, the thickness of the surrounding liquid above and below its upper part and at the base is different, which results in the formation of droplets of different size.

A third reason for the formation of droplets in the core of dispersed-annular flow is the breakup of liquid plugs upon transition from slug to dispersed-annular flow. And, finally, droplets appear due to sputtering of the liquid film when it flows onto a hot surface at a temperature higher than the Leidenfrost temperature.

The liquid film thins out as a result of droplet entrainment and vaporization and, at a certain quality $x_{\Delta p}$ it becomes so thin that its motion is now controlled by viscosity and surface tension forces. For example, at $p = 6.8$ MPa and $x = 0.8$ the film thickness is several microns [2.15]. The film surface becomes smooth, and droplet entrainment from it ceases. A number of investigators assume that at this quality $x_{\Delta p}$ there occurs a minimum in hydraulic drag (pressure gradient).

At high heat fluxes there may occur conditions where the outwards flow of steam from the film surface will prevent deposition of droplets at the surface. However, at moderate heat fluxes the liquid droplets diffuse to the wall through the steam curtain and irrigate the microfilm.

The flow pattern in the post-critical region, naturally, is controlled to a large degree by the physical situation preceding the onset of dryout, i.e., the flowrate of liquid in the film, its rate of evaporation, entrainment of liquid from the film by vapor bubbles, mechanical entrainment of droplets, diffusion of liquid droplets from the flow core to the film.

Following the evaporation of the liquid rivulets and drying of the channel wall, the flow becomes dispersed; this flow regime is not encountered under adiabatic conditions, when a liquid film is always present on the channel surface. This change in flow pattern is accompanied by a significant deterioration in heat transfer. The flow pattern in the vicinity of dryout becomes restructured. Dryout is preceded by the so-called hydraulic crisis when $\Delta p_{\mathrm{fr}}/\Delta p_0 = p(x)$ passes through a maximum and then passes through a minimum. The maximum of the temperature curve $T_{\mathrm{w}} = T(x)$ is shifted somewhat relative to the minimum of curve $\Delta p_{\mathrm{fr}}/\Delta p_0 = p(x)$ [2.15]. The void fraction decreases somewhat, depending on the variation in the quality along the tube. Thus, the various parameters undergo significant changes in the vicinity of dryout. The rate and extent of these changes are controlled by flow and structural parameters.

Steam-generating channels heated by high-temperature fluids exhibit the same heat transfer zones as electrically or nuclear-heated channels [2.19]. However, the transition from pre-dryout to post-dryout heat transfer occurs differently. The rise in wall temperature is limited by the temperature of the heating liquid, and this produces temperature fluctuations in the wall in the dryout region.

Thermal non-equilibrium is typical of two-phase flows in steam-generating channels. A distinction is made between thermal non-equilibrium of the first and second kind. Non-equilibrium of the first kind arises in surface and nucleate boiling. Boiling at the wall starts much before the flow core attains its saturation temperature. Steam already exists in the boundary layer, whereas the liquid in the flow core is significantly subcooled. By virtue of this, rapid condensation occurs in the quality range between $x_{\mathrm{i.b}}$ where boiling is initiated and a quality x_φ; in this range of quality, the mean void fraction φ increases slowly. At the point where the quality reaches x_φ, the value of φ increases more rapidly. At the point where the thermodynamic quality $x_\delta = 0$, in channels with high heat flux, the value of φ may be as high as 0.5 and even higher. This means that under conditions when $x_\delta = 0$, the liquid is very significantly subcooled. The liquid phase reaches saturation temperature at the point where $x = x_{eq}$; then, both the liquid and steam are at saturation equilibrium conditions. This means that non-equilibrium of the first kind exists over the quality range from $x_{i.b}$ to x_{eq}.

Thermal non-equilibrium of the second kind manifests itself in steam superheated at vapor qualities $x = h_{mix}/h''$ is less than unit, and, in the presence of liquid droplets in the flow, at $x > 1$. Liquid droplets have been detected in flows of superheated steam at $x = 2.2, 3$, and even 7.5, and also when the steam was superheated by 50 and 100 °C [2.20–2.22]. Thermal non-equilibrium has been found to exist in experiments over a wide range of pressures and mass velocities with water, Freon-12, nitrogen, hydrogen, and oxygen, both in tubes and in annuli.

2.1.3 Pattern Maps for Two-Phase Flow in Heated Channels

Investigations of the flow structure of steam-water mixtures in forced convection evaporation in tubes, annuli and rectangular ducts show that significant differences exist between two-component isothermal flows and single component evaporating flows. As a result of this, attempts to use the Baker, Griffith and similar flow maps for steam/water flows in heated tubes did not succeed. As far as we know, no general flow pattern map has yet been published for evaporating flows which covers a wide range of pressure and fluids properties.

Bennet et al. [2.16] obtained a flow pattern chart for wet steam at 7 MPa, shown in Fig. 2.7, which also describes satisfactorily the experimental points of Bergles and Suo [2.8], Milashenko [2.9], and Dzarasov et al. [2.10], obtained under similar conditions. An attempt at correlating experimental data obtained at variable pressure and in different fluids was made by Mayinger [2.29] who introduced the critical pressure ratio into the Bennet diagram. On this modified Bennet diagram the abscissa represents the parameter $Z = x/(p/p_{cr})^n$, where n is an experimentally determined exponent, ranging from 0.98 to 1.4. The data obtained by various investigators for steam-water mixtures and freons are correlated in Fig. 2.8.

2.2 PRINCIPAL TWO–PHASE FLOW EQUATIONS

Two-phase flows are a particular case of heterogeneous medium flows, which may be described on the basis of many-velocity continua. Each phase fills a part of the total volume of the mixture and is characterized in the given point by the true phase density, ρ_i (mass of the ith constituent in unit volume of the ith phase), velocity w_i, temperature T_i and enthalpy i_i. When the surface tension force is low, the phase pressures are equal to one another $p_i = p_k = p$.

In a heterogeneous mixture of total volume V the ith phase is characterized by a volume V_i and a volume fraction φ_i ($i = 1, 2, \ldots, m$), occupied by it, so that

$$V_1 + V_2 + V_3 + \cdots + V_i + \cdots + V_m = V$$

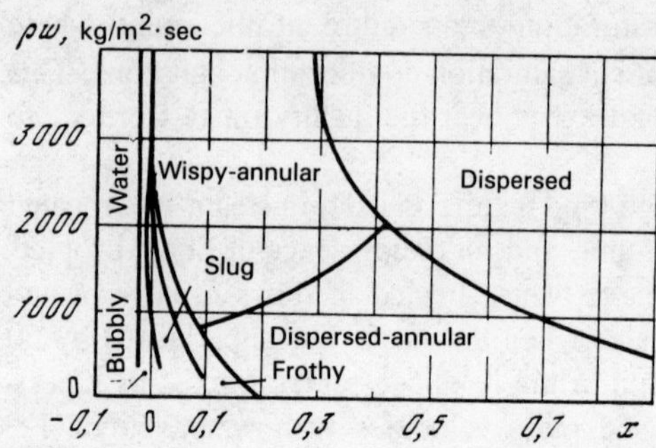

Figure 2.7 Flow patterns in boiling in a vertical tube in upwards flow [2.16].

or

$$\varphi_1 + \varphi_2 + \cdots + \varphi_i + \cdots + \varphi_m = 1 \tag{2.1}$$

The density of the ith phase is then written as

$$\rho_i = \delta m_i / dV_i \tag{2.2}$$

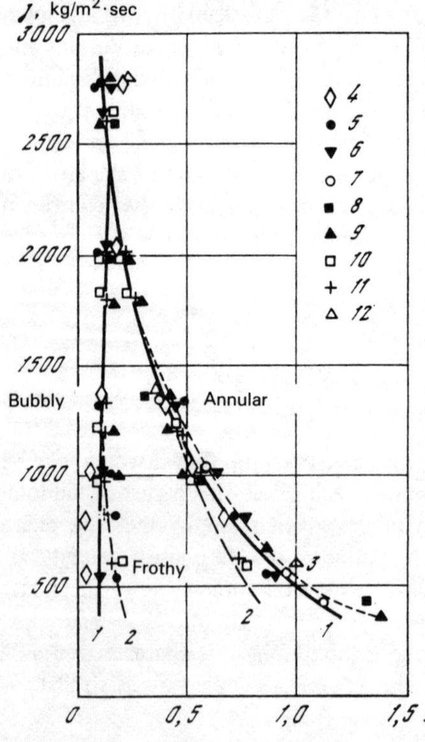

Figure 2.8 Sideman's flow pattern chart [2.19]. 1, Water; 2, Freon-12; 3, Freon-14; 4 and 5, for p = 6.8 MPa [2.8]; 6, for 3.5 MPa [2.8]; 7, for 6.8 MPa [2.16]; 8, for 3.5 MPa [2.16]; 9, for 1.1 MPa [2.55]; 10, for 0.8 MPa [2.29]; 11, for 1.36 MPa [2.29]; 12, for 1.8 MPa [2.29].

The total mass of the medium in volume dV is equal to the sum of masses taken over all the phases, i.e.,

$$\delta m = \rho dV = \sum_i \rho_i \, dV_i$$

or

$$\rho = \sum_1^m \rho_i \varphi_i \tag{2.3}$$

The mass concentration of a phase is defined as the ratio of the mass of the ith phase m_i to the entire mass of the volume, i.e.,

$$x_i = \frac{\delta m_i}{\delta m} = \rho_i \varphi_i \frac{1}{\rho} \tag{2.4}$$

The specific weight of the mixture is defined by the sum

$$v = \Sigma v_i x_i$$

It is self-evident $x_i + x_2 + \ldots + x_i + \ldots + x_m = 1$

The phase interfaces in heterogeneous mixtures are acted upon by surface forces (pressure, drag, surface-tension force, etc.) and each phase within a given elementary volume is acted upon by body forces (gravity, electromagnetic force, etc.); also the individual phases may exist in the form of macroscopic inclusions (droplets, bubbles, plugs, etc.), which in general change in time and space. Thus, the relationships describing the interphase interaction become extremely complicated, and the derivation of the principal equations of conservation of the many-phase medium reduces basically to the correct specification of forces and energy and mass fluxes at the phase interface.

The equations of multi-phase media have been formulated by a large number of scientists. Landau and Lifshits [2.30] wrote them for analyzing the flow of liquid helium, Frankl [2.31] for soils, and Slezkin [2.33] for the flow of pulp. Equations of two-phase media were obtained by averaging by Teletov in 1953 [2.32], in 1955 Rakhmatulin [2.34] suggested a closed set of equations of interpenetrating motion of multi-phase mixture of compressible fluids without heat transfer; in 1958 Kutateladze and Styrikovich [2.35] employed the two-velocity model for hydraulics of gas-liquid flows. The theory of multi-phase media was further developed by Krayko and Sternin [2.36] (nonreacting mixture without heat transfer), by Nigmatullin [2.37] for a dispersed two-phase medium with phase transitions, and also by Vernier and Delhaye [2.38], Boure and Reocreux [2.39] and others.

The interphase interaction force $\mathbf{F}_{13}$ can be represented as a sum of four components: the Stokes force $\mathbf{F}_{13}^{\mu}$, which incorporates the interaction of viscosity forces at the phase interface; the virtual mass force $\mathbf{F}_{13}^{m}$, associated with the induced mass due to slip of particles relative to the continuous fluid; the Archimedes force $\varphi_3 \nabla \mathbf{p}$, senting the pressure fields in the conveying phase, and the Magnus force $\mathbf{F}_{13}^r$ induced by the transverse velocity gradient:

$$\mathbf{F}_{13} = \varphi_3 \nabla p + \mathbf{F}_{13}^\mu + \mathbf{F}_{13}^m + \mathbf{F}_{13}^r \tag{2.5}$$

where

$$\mathbf{F}_{13}^\mu = C_D\,(\mathrm{Re})\,\varphi_3 \rho_1\,\frac{|\,w_1 - w_3\,|\,(\mathbf{w}_1 - \mathbf{w}_3)}{2r}$$

$$\mathbf{F}_{13}^m = \varphi_3 \rho_2 x^m \left(\frac{d\mathbf{w}_1}{dt} - \frac{d\mathbf{w}_2}{dt} \right)$$

$$\mathbf{F}_{13}^r = \varphi_3 \rho_1 x^r\,(\mathbf{w}_3 - \mathbf{w}_1) \times \mathrm{rot}\,\mathbf{w}_1$$

x^m and x^r are shape parameters. The work done by phase interaction forces is

$$N = \mathbf{F}_{13}\mathbf{w}_{13} \tag{2.6}$$

In general the work done by the interphase forces associated with the virtual mass $\mathbf{F}_{13}^r$ and Magnus force $\mathbf{F}_{13}^m$ is not entirely used in dissipation (i.e., irreversibly transformed into heat), but is partially converted into kinetic energy of the conveying phase. This conversion is accounted for by introducing coefficients κ_i, which are the fractions of the dispersed kinetic energy of the mixture transformed due to interphase interaction into internal energy of the ith phase.

Nigmatulin [2.37] used balance equations written for each phase within which there exists local thermodynamic equilibrium, derived a set of equations of two-velocity motion of a dispersed medium in the form:

$$\frac{\partial \rho_1}{\partial t} + \nabla \varphi_1 \rho_1 \mathbf{w}_1 = -J_{13}; \qquad \frac{\partial \rho_3}{\partial t} + \nabla \varphi_3 \rho_3 \mathbf{w}_3 = J_{13} \tag{2.7}$$

$$\varphi_1 \rho_1 \frac{D\mathbf{w}_1}{Dt} = -\varphi_1 \nabla p + \nabla^k \tau_{1*}^k - \mathbf{F}_{12} - J_{12}\,(\mathbf{w}_{13} - \mathbf{w}_1) + \varphi_1 \rho_1 g$$

$$\varphi_3 \rho_3 \frac{D\mathbf{w}_3}{Dt} = -\varphi_3 \nabla p + \mathbf{F}_{13} + J_{12}\,(\mathbf{w}_{13} - \mathbf{w}_3) + \varphi_3 \rho_1 g$$

$$\rho_1 \rho_1 \frac{Du_1}{Dt} = \frac{\varphi_1 p}{\rho_1} \frac{d\rho_1}{dt} + \varkappa_1 \mathbf{F}_{13}\,(\mathbf{w}_1 - \mathbf{w}_3) + Q_{31} + \tau_{1*}^{kl} \cdot \mathbf{e}_1^{kl} -$$
$$- J_{13} \frac{(\mathbf{w}_{12} - \mathbf{w}_1)^2}{2} - \nabla q_v - J_{13}\,(i_{13} - i_1)$$

$$\varphi_3 \rho_3 \frac{Du_3}{Dt} = \frac{\varphi_3 p}{\rho_2} \frac{d\rho_3}{dt} + \varkappa_3 \mathbf{F}_{13}\,(\mathbf{w}_1 - \mathbf{w}_3) + Q_{13} +$$
$$+ J_{13} \frac{(\mathbf{w}_{13} - \mathbf{w}_3)^2}{2} - J_{13}\,(i_i - i_{13})$$

where $\varkappa_1 + \varkappa_3 = 1$ and τ_* is the strain rate tensor.

The set of equations (2.7) is closed by expressions for heat and mass transfer between the phases and dynamic interaction between the phases, together with the thermodynamic equations of state for each phase.

We now consider in more detail problems of interphase interaction.

2.2.1 Mechanical Interaction

The value of C_D for a single particle in a steady flow of a viscous fluid can be determined only in two limiting cases, which were investigated by Stokes and Newton, respectively. Stokes obtained a solution corresponding to very low relative velocities, by dropping terms in the Navier-Stokes equations, which are associated with inertia forces (Re $\to$ 0). A flow which has corresponding to it Reynolds numbers from 0 to 0.1 is termed a "Stokes flow" and is characterized by a symmetric pattern of flow over the sphere both up and downstream of the body. The approximation obtained by Stokes yields the following expression for the resultant drag force

$$\mathbf{F_{13}} = 6\pi\mu_1 r_3 \Delta\mathbf{w} \tag{2.8}$$

or, in accordance with the definition of the drag coefficient,

$$C_D = F_{13} \Big/ \left(\frac{1}{2}\, \pi r_3^2 \rho_1 \Delta w^2 \right) \tag{2.9}$$

$$C_D = 24/\text{Re} \tag{2.10}$$

The value of the drag force in this case is composed of viscous (friction) forces (two-thirds of the total force) and pressure forces (form drag) (one-third).

Newton investigated the case of high relatively velocities, when inertia forces predominate, and obtained

$$\mathbf{F_{13}} = 0.22\, \pi r_3^2 \rho_1\, (\Delta\mathbf{w})^2 \tag{2.11}$$

which corresponds to a constant value of the drag coefficient

$$C_D = 0.44 \tag{2.12}$$

The analytic determination of the drag coefficient outside of the Stokes flow mode is based primarily on the Oseen approximation, which incorporates inertia terms in the flow only far from the sphere and yields the expression

$$C_D = \frac{24}{\text{Re}}\left(1 + \frac{3}{16}\,\text{Re}\right) \tag{2.13}$$

Proudman and Pearson [2.41] modified the Oseen formula to the form

$$C_D = \frac{24}{\text{Re}}\left[1 + \frac{3}{16}\,\text{Re} + \frac{9}{160}\,\text{Re}^2 \ln\left(\frac{1}{2}\,\text{Re}\right) + 0\,(\text{Re}^2)\right] \tag{2.14}$$

Correlation of a large volume of experimental data over a wide range of Reynolds numbers (from 0.1 to $\sim 10^7$) yields the so-called standard drag curve for a

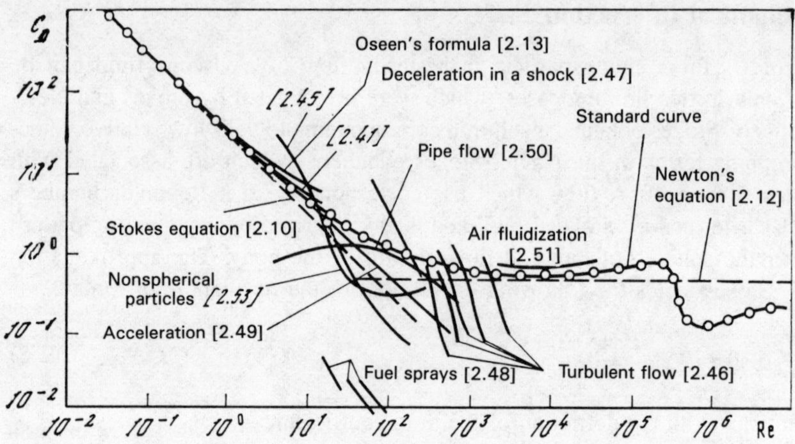

Figure 2.9 Drag coefficient of particles as a function of the Reynolds number.

single nondeforming spherical particle moving at constant velocity in a nonmoving isothermal infinite incompressible fluid.

Figure 2.9 shows this standard curve together with curves calculated from Eqs. (2.10), (2.12), and (2.13) and also lines representing the data of numerous experimental studies for different specific flow conditions. It is seen from Fig. 2.9 that the analytic results and the entire empirical standard curve are in agreement within the following regions: (a) for the Stokes formula Re $\leq$ 0.1; (b) for the Oseen formula Re $\leq$ 1; (c) for Eq. (2.14) Re $\leq$ 4; (d) for the Newton formula (2.12) over the range $7 \cdot 10^2 \leq$ Re $< 10^5$.

The most exact approximation of the standard curve at $0.1 \leq$ Re $\leq 2 \cdot 10^4$ is given by Stonecypher [2.42]:

$$\ln C_D = 3.271 - 0.8893 \ln \mathrm{Re} + 0.03471\,(\ln \mathrm{Re})^2 + 0.001443\,(\ln \mathrm{Re})^3$$

$$(2.15)$$

For individual segments of this curve one also frequently employs a number of approximate equations, for example, the Klyachko formula [2.43] for $1 < \mathrm{Re} < 500$.

$$C_D = \frac{24}{\mathrm{Re}}\,(1 + \mathrm{Re}^{2/3}/6)$$

$$(2.16)$$

or the expression given by Levich [2.44] for $10 < \mathrm{Re} < 10^3$:

$$C_D = 13/\sqrt{\mathrm{Re}}$$

$$(2.17)$$

and also other expressions. It is seen from Fig. 2.9 that at conditions not similar to those for which the standard curve was obtained, many experimental data deviate from it. These deviations are discussed by Toborin and Canvin [2.45], Hoglund [2.46], and Rudinger [2.47], and also by the last of the present authors with his co-workers [2.40] and by Prem with Parkins [2.48]. The principal factors apparently are the flow turbulence level [2.46], the effect of pressure and velocity gradients [2.47, 2.49], the volumetric concentration of particles [2.48], and also the possible asphericity of the particles, particularly of deformable bodies (liquid droplets, bubbles).

As noted by Hoglund [2.46], experimental data on the drag coefficient of spherical particles in turbulent flows range from values which exceed by threefold those obtained from the standard curve, to values which are 100-fold smaller. The physical reasons for the effect of the turbulence level on the particle drag are due to changes in the flow pattern over the particles. At high turbulence level a decrease may occur in the upper critical Reynolds number corresponding to a steep drop in drag and a transition from laminar to turbulent boundary-layer flow (Re $\cong 10^5$–10^6); the drag coefficient then becomes smaller. At low turbulence level the drag coefficient may be somewhat higher than the values obtained from the standard curve as a result of dissipation of energy in the region of the wake. Upon reduction in Re the effect of free-stream turbulence becomes less.

The effect of steep pressure and velocity gradients in gas flow on the particle drag coefficient can be illustrated by the experimental studies of Rudinger [2.47] and Saltanov [2.49].

The former determined the drag coefficients of glass beads in the flow of gas upon the passage of weak shock waves. The results of studies plotted in Fig. 2.10 point to a "steeper" curve of C_D(Re); Rudinger suggested the following expression for the range of Re (50–300) under study:

$$C_D = 600 \,/ \mathrm{Re}^{1,7} \tag{2.18}$$

The data in Fig. 2.10 are compared with the standard curve, the Stokes formula and the experimental formula, obtained by Ingebo [2.50] in investigating the motion of a particle cloud injected into a wind tunnel (particle loading ratio about 1%),

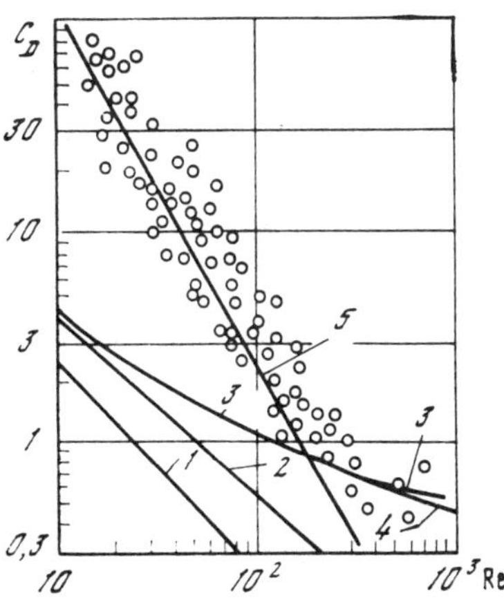

Figure 2.10 Comparison of predictions for C_D (Re) with experimental data of Rudinger [2.47]. 1, Stokes' formula; 2, Ingebo's formula; 3, standard curve; 4, Klyachko's formula; 5, Eq. (2.18).

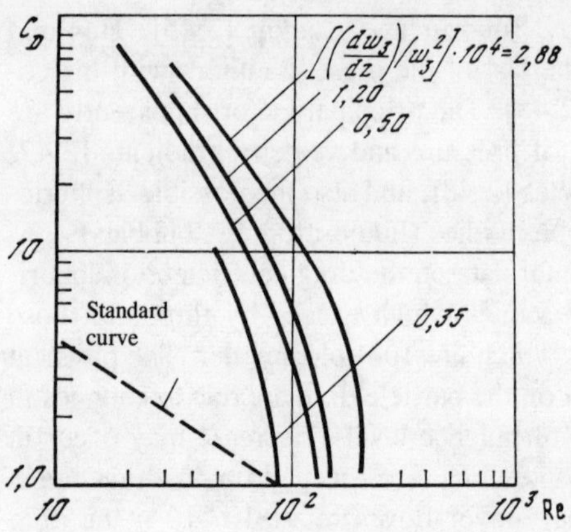

Figure 2.11 Plot of C_D (Re) for particles injected into a nozzle [2.46].

$$C_D = 27/\mathrm{Re}^{0,84}, \quad 6 < \mathrm{Re} < 400 \tag{2.19}$$

Rudinger explained his results by a special flow model, which incorporates longitudinal and transverse perturbations of the local motion of particles.

Hoglund [2.46] investigated the acceleration of spherical sand particles, injected into sub- and supersonic air-blowing nozzles. His results, plotted in Fig. 2.11, show that the drag coefficient of particles is a function of their acceleration, i.e., it depends, in addition to Re, also on the distance from the point of injection.

At high particle concentrations, when the distance between particles becomes of the same order as their size, a significant positive deviation of C_D from the standard curve occurs. This effect was observed in a large number of studies of precipitation and air fluidization. Thus, Kaye showed [2.51] that at bulk concentration of particles in excess of 5%, constrained sedimentation of particles occurs and their final velocity cannot be calculated from Stokes law.

Typical data on the effect of bulk concentration of particles on the behavior of $C_D(\mathrm{Re})$ (obtained in investigating air fluidized and in dense beds) are plotted in Fig. 2.12. Analysis of experimental data yields the following expression:

$$C_D = 200 \left(\frac{1-\varphi}{\varphi} \right) \frac{1}{\mathrm{Re}} + \frac{7}{3\varphi} \tag{2.20}$$

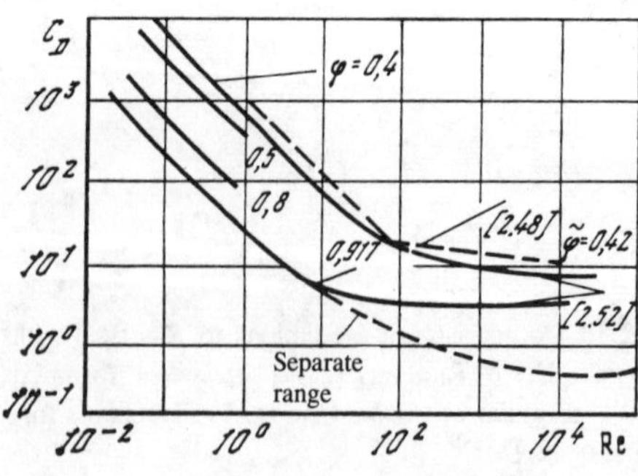

Figure 2.12 Effect of bulk concentration of particles on C_D [2.48].

It should be noted, however, that this equation is limited in some ranges of Re and for Re < 10 as $\varphi \rightarrow 1$ does not conform to the standard curve for a single sphere.

The shape of particles is an important factor in their interaction with flow over them. Shape is particularly important in the case of deformable particles. Thus, a liquid droplet accelerated by a gas flow is, due to the characteristic distribution of pressure over its circumference, deformed in such a manner, that it takes the shape of an ellipse whose major axis is oriented normal to the flow [2.53]. As the elliptical droplet continues moving it may break up. It has been established that the principal criterion affecting the breakup of low-viscosity liquids is the Weber number

$$We = \frac{r_3 \rho_1 (w_1 - w_3)^2}{\sigma_2}$$

which has a critical value, We_{cr}, which determines the maximum stable droplet radius. We_{cr} is most frequently assumed to be equal to 6.

The effect of change in the shape of droplet on its drag can be incorporated by using the correction function

$$\psi (We) = f_D^l C_D^l / f C_D \tag{2.21}$$

where f_D^l and C_D^l are the mid-plane cross sectional area and drag coefficient of the deformed droplet, whereas f and C_D are the same for a spherical droplet.

According to Raushenbakh [2.53], the experimental results are satisfactorily approximated by the expression

$$\psi = \exp [0.03 (2We)^{1,5}] \tag{2.22}$$

Rotation of the droplet or bubble may have a significant effect on the viscous drag. In the case of Stokes flow (Re $\ll$ 1) and $Re_w = \rho_1 w_3 r_3^2 / \mu_1 \ll 1$, the particle drag force F_{13} can be found from the expression

$$F_{13} = F_{13}^\mu + F_w \tag{2.23}$$

where $F_w = \rho_1 \pi r^3 [\Delta w \times w_3]$ and $F_{12}^\mu = 6\pi r_3 \mu_1 w_1$. Gol'dshtik and Sorokin [2.54] obtained the following expressions for the Magnus force at high Re (Re $\gg$ 1 and $Re_w \gg$ 1):

$$F_w = 8/3 \cdot \pi r_3^3 \rho_1 [\Delta w \times w_3] \tag{2.24}$$

2.2.2 Heat Transfer from A Particle to A Gas

Heat transfer from a particle to a gas is controlled by the physical properties of the gas and by the flow field surrounding the particle. Outside the range of applicability of the Stokes law (for which the solution is Nu = 2) the data for heat transfer, as for

$C_D(\mathrm{Re})$, are entirely empirical, with analytic studies serving primarily for obtaining a suitable form in data reduction. Usually the coefficient of heat transfer from a single solid sphere is calculated from nondimensional equations in the form

$$\mathrm{Nu} = 2 + C\mathrm{Re}^m\mathrm{Pr}^n \tag{2.25}$$

where C, m, and n are constants, the values of which, from a number of investigations, are more or less the same. The frequently used expression is

$$\mathrm{Nu} = 2 + 0{,}6\ \mathrm{Re}^{0,5}\mathrm{Pr}^{0,3} \tag{2.26}$$

or

$$\alpha = [2 + 0{,}6\ \mathrm{Re}^{0,5}\mathrm{Pr}^{0,2}] \cdot k_1/2r_3 \tag{2.27}$$

The total quantity of heat exchanged between the phases per unit time per unit volume is defined as

$$Q = 3\frac{\tilde{\Phi}_3}{r_3}\ \alpha\,(T_3 - T_1) \tag{2.28}$$

2.2.3 Phase Transitions

The problem of phase transitions in single-component two-phase flows can be subdivided into two parts: (1) formation of a new phase within an existing phase (spontaneous condensation); (2) phase transitions at the phase interfaces.

Spontaneous condensation in flowing steam. Spontaneous condensation has been given a great deal of attention by many investigators (see, among others, the book by Saltanov [2.49]). Henceforth, when appropriate, we shall employ the Frenkel' nucleation theory [2.56], according to which the formation of the liquid phase from vapor occurs as a result of local fluctuations, which take the system outside the limits of the initial state. In a thermodynamically stable system ($\Phi_2 > \Phi_1$) randomly appearing nuclei of the new phase vanish, i.e., the fluctuations are "dissipated." It is known that the metastable systems ($\Phi_2 < \Phi_1$), when it is the new phase which is stable, local fluctuations with radius smaller than a critical size r_*, are unstable and break up. Conversely, fluctuations with dimensions greater than r_* are stable and are potentially capable of growth.

The size of the critical nucleus can be determined from the conditions of equilibrium of a system consisting of vapor and liquid droplets $r_* = 2\sigma/(\rho_3 R_1 T)\ 1/(\ln p_{\mathrm{sat}}/p_{\mathrm{sat}\infty})$, where p_{sat} is the saturation pressure at temperature T and droplet radius r_*, whereas $p_{\mathrm{sat}\infty}$ is the saturation pressure at T and droplet radius $r_* = \infty$. The rate of formation of critical nuclei capable of further growth can be obtained by solving the principal kinetic equation, the particular solution of which, according to the Frenkel'-Zel'dovich theory, is:

$$J = \left(\frac{p_1}{kT_1}\right)^2 v_2 \sqrt{\frac{2\sigma}{\mu_1 \pi}} \exp\left[-\frac{\Delta W}{kT}\right] \qquad (2.29)$$

where ΔW is the work of formation of a critical droplet, equal to $^4/_3\, \pi r_*^3 \sigma$.

Condensation on the nuclei occurs in accordance with the theory of free-molecular flow, the condensation rate J_{13} is then found as

$$J_{13} = \frac{1}{A(z)} \int_{z_0}^{z} J(\xi) \frac{\lceil dm\, (\xi_1 m)}{dz} A(\xi)\, d\xi + J(Z)\, m_* \qquad (2.30)$$

where $z_0 \le \zeta \le z$ and m_* is the mass of the critical droplet in section z.

Heat transfer between the condensate and the steam is defined by the equation

$$Q = \int_{z_0}^{z} \bar{q}\,(\xi, z) \frac{J(\xi)}{C(\xi)}\, d\xi.$$

where $\bar{q}(\xi, z)$ is the quantity of heat absorbed or given up by the single particle, which originate in section ξ, in section z. Droplets of the condensate spectrum flow in the free-molecular mode and $\bar{q}(\xi z)$ is found as

$$\bar{q}\,(\xi, z) = 2{,}34 \sqrt{\frac{2R_1 T_1}{\pi}}\, R_1 \alpha_T \frac{\gamma}{\gamma - 1} r^2(\xi, z)(T_3 - T_1) \qquad (2.31)$$

Here α_T is the thermal accommodation coefficient.

Calculations show that the effect of heat transfer on spontaneous condensation is insignificant and it can be neglected in the first approximation.

Phase transitions at the interface. Mass transfer between droplets and the steam surrounding them in the continuum region ($M/\sqrt{Re} < 0.10$) has been investigated very extensively (see, among others, the paper by Boure with Reocreux [2.39] and the book by the last of the present authors with his coworkers [2.40]).

Calculations in the one-dimensional formulation assume constancy of temperature inside the droplet. Estimates of conductive cooling of the droplet show that for water at $t_3 = 100\,°C$, $\Delta t = 30\,°C$, $r_3 = 100\ \mu m$ and residence time in the nozzle of $\sim 10^{-3}$ sec the difference in temperatures in the center of the droplet and on its surface is less than 10%. Actually the temperature equalization is even more rapid due to circulation flows inside the droplet.

The evaporation of hot droplets into a gaseous substance other than the vapor of the droplet substance is usually calculated on the basis of the analogy between heat and mass transfer; here mass transfer is treated as a purely diffusive process and the nondimensional equation for the mass transfer coefficient α_D is

$$\alpha_D = [2 + C_1 Re^m Sc^n] \cdot D/2r_3 \qquad (2.32)$$

The advantage in using such equations in practical calculations is their agreement with a rather large body of experimental data, although it is very doubtful whether they reflect with sufficient accuracy the mechanism of the actual process.

According to Ranz and Marshall [2.57], it is usually assumed that

$$\text{Nu}_D = 2 + 0,6 \ \text{Re}^{0,5} \text{Sh}^{0,33} \tag{2.33}$$

or

$$a_D = [2 + 0,6 \text{Re}^{0,5} \text{Sc}^{0,33}] \frac{D}{2r_3} \tag{2.34}$$

where $\text{Sc} = \mu_1/\rho_1 D$ is the Schmidt number, and D is the diffusion coefficient. The use of Eq. (2.34) is made difficult by the need to determine the value of D, which cannot always be done with the required accuracy. It can be assumed in the first, rough, approximation that the diffusion coefficient is equal to the kinematic viscosity, i.e., $D = \nu_1 = \mu_1/\rho_1$, and the error of this assumption depends in the first place on the kind of molecules of the substance and on the parameters of the fluid.

The rate of phase transitions in this case is calculated as the product of the coefficients of mass transfer α_D and the potential difference (or driving force) for mass transfer. However, as correctly noted by Eckert [2.58], there is at present no concensus as to "which parameter should be used for expressing the transfer potential." Three equations are encountered in the literature for the rate of mass transfer

$$dm_2/dt = \alpha_D \rho_1 (w_2 - w_1) \tag{2.35}$$

for a fluid with almost constant properties

$$dm_3/dt = \alpha_D [(\rho_1)_{T_3} - (\rho_1)_{T_1}] \tag{2.36}$$

and for a small temperature difference

$$dm_3/dt = \frac{\alpha_D}{R_1 T_1} [p_s (T_3) - p_1] \tag{2.37}$$

Here w_3 and w_1 are the mass content at temperatures T_3 and T_1, respectively, $(\rho_1)_{T_3}$ are the vapor phase densities at pressure p and temperature T_1 and T_3.

Calculation of mass transfer for single-component fluids is based on the general principles of statistical physics. This method is quite general, since it is suitable both for small-diameter droplets ($r_3 \gtrsim \lambda$), and for continuum flow ($r_3 \gtrsim \lambda$).

For large droplets ($r_3 \gtrsim \lambda$), the mass flux from the droplet surface is defined as the difference of the vaporization flux from the droplet surface and the molecular flux captures by the droplet (condensation flux). The temperatures of the liquid and vapor are usually different, for which reason the molecules that evaporated from the droplet surface will have a higher energy (temperature), than the molecule of the surrounding vapor. However, in the majority of cases it can be assumed that molecules undergoing

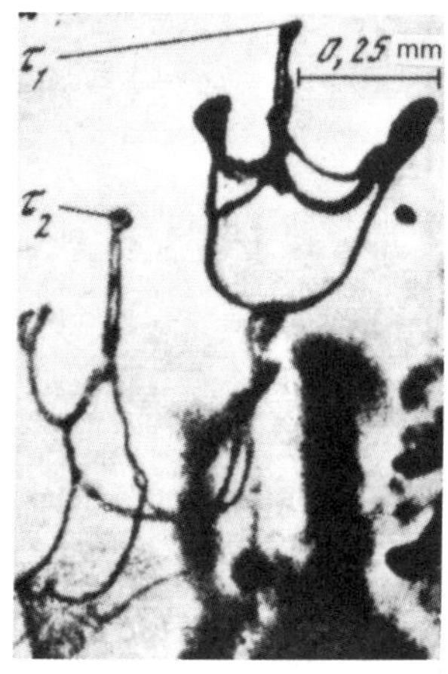

Figure 2.13 Deformation of droplets in the throat of a Venturi tube.

transition from one phase to another rather rapidly acquire the temperature of their acceptor phase.

In this case the rate of evaporation from the droplet surface is given by the Maxwell formula [2.59]

$$dm_3/dt = \frac{4\pi r_3^2 \xi}{V\,2\pi\mu_1 R T_1}\,[p_s\,(T_3) - p]$$ (2.38)

The Maxwell formula was derived on the assumption that the thermal accommodation coefficient (α_T) is equal to unity, and the evaporation and condensation coefficients are equal to one another $\alpha_{ev} = \beta = \xi$.

When the temperature difference $(T_3 - T_1)$ is small, Eq. (2.38) becomes

$$\frac{dm_3}{dt} = 4\pi r_3^2 \xi \sqrt{\frac{RT_1}{2\pi\mu_1}}\,[\rho_1\,(T_3) - \rho_1\,(T_1)]$$ (2.39)

or

$$\frac{1}{4\pi r_3^2}\frac{dm_3}{dt} = \alpha_{ev}\,[(\rho_1)_{T_3} - (\rho_1)_{T_1}]$$ (2.40)

where $\alpha_{ev} = \xi\,[(RT_1)/(2\pi\mu_1)]^{1/2}$

It can be seen by comparing Eqs. (2.36) and (2.40) that the expressions for the evaporation rate both in the case of two- and single-component fluids have the same structure: one can use as the transfer potential the difference in the vapor-phase density at droplet temperature T_3 and vapor temperature T_1 and the vapor pressure p. The mass transfer coefficient in the first case is determined from the diffusion model [Eq. (2.34)], and the coefficient vaporization of a droplet into a vapor medium of its own substance is determined from the kinetic theory of gases in accordance with the Maxwell formula.

Droplets flowing in nozzle blading or in Venturi tubes at significant gas-flow velocities do not remain spherical. Figure 2.13 shows a moving-picture frame, ob-

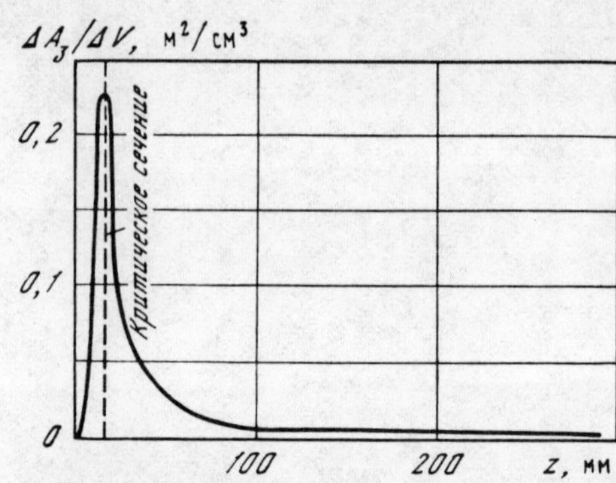

Figure 2.14 Variation in the interphase interaction surface $\Delta A_3/\Delta V$ along a Venturi tube [2.60] at $w_1 = 80$ m/sec.

tained by Numann [2.60] at a filming rate of 100,000 frame/sec, in a water-air flow in a Venturi tube at air velocity of 80 m/sec. It is seen that the droplet moving in the throat cross section take the shape of parachute-like membranes, with the interphase interaction surface per unit volume $\Delta A_3/\Delta V$ increasing at an extremely high pace. After leaving the throat the droplets revert to the spherical shape. Figure 2.14 shows the variation in $\Delta A_3/\Delta V$ per unit volume along a Venturi tube according to Numann. It is seen that the interaction cross section attains a maximum near the throat and then decreases to zero. It is precisely this region near the nozzle throat that determines the effectiveness of operation of purification equipment employing Venturi tubes. Numann suggested that mass transfer coefficient β and the interaction cross section A be determined from the empirical equation

$$\frac{\beta A_3}{V_1} = 94{,}9 \, \mathrm{Re}^{0,1} \mathrm{Eu}^{-0,2} \mathrm{We}^{0,24} \frac{1-x}{x} \left(\frac{w_s}{w_1}\right)^{0,2} \tag{2.41}$$

where

$$\mathrm{We} = \frac{d_3 \rho_1 w_1^2}{\sigma_3}, \qquad w_s = \frac{d_3^2 \rho_1 g}{18 \nu_1 \rho_1}, \qquad \mathrm{Re} = \frac{w_1 d_3}{\nu_1}, \qquad \mathrm{Eu} = \frac{\Delta p}{\rho_1 w_j^2}$$

Heat and mass transfer in a two-phase dropwise fluid with dense packing deteriorates due to the mutual effect of the particles. In the case when the distance between the particles is small, Miura et al. [2.61] suggested the formula

$$\frac{\mathrm{Nu}}{\mathrm{Nu}_0} = \frac{\mathrm{Sh}}{\mathrm{Sh}_0} = 0{,}71 \left(\frac{b}{d_1}\right)^{0,25} \left(\frac{d_3}{d_1}\right)^{0,17} + 0{,}07 \quad \text{for} \quad b/d_1 < 2$$

$$\frac{\mathrm{Nu}}{\mathrm{Nu}_0} = \frac{\mathrm{Sh}}{\mathrm{Sh}_0} = 0{,}42 \left(\frac{b}{d_1}\right)^{0,125} + 0{,}41 \quad \text{for} \quad b/d_1 > 2 \tag{2.42}$$

Here Nu_0 and Sh_0 are the Nusselt and Sherwood numbers for a solitary particle.

According to Schmidt-Traub [2.62], mass transfer in bubbly flow near the spherical boundary can be written as

$$\mathrm{Sh} = 2 + \frac{0{,}651 \, (\mathrm{Re} \, \mathrm{Sc})^{1,72}}{1 + (\mathrm{Re} \, \mathrm{Sc})^{1,22}} \tag{2.43}$$

2.3 PRESSURE GRADIENT IN ISOTHERMAL FLOWS OF TWO–COMPONENT MIXTURES

A large number of investigations have been carried out on two-phase flows; however, up to now, reliable results are available only for isothermal flows. In boiling of water in tubes, given the fact that the boiling modes, structure, and consequently also the local variables change along the tube, the relationships governing the pressure gradient have still not been sufficiently explored.

In the general case of forced convection motion of wet steam in channels the pressure gradient consists of a large number of components, among them losses due to skin friction, acceleration of droplets by the flow, wave losses on the film surface, etc. In experiments it is actually possible to single out from the overall pressure loss only the loss for acceleration of the two-phase flow $(\Delta p/\Delta z)_{ac}$ and, in the case of vertical (inclined) tubes the pressure drop needed for overcoming the static head $(\Delta p/\Delta z)_{lev}$. Consequently, it must be remembered in analyzing the experimental data of most investigators that the measured friction losses, determined as the total pressure loss minus losses for acceleration and overcoming the static head, actually consist of losses associated with skin friction, acceleration and formation of liquid droplets, wave generation and the flow of the vapor phase over the wavy surface.

To find each component of the pressure gradient in the two-phase system, let us consider the energy equation in integral form for two-phase pipe flow without phase transitions. The internal energy U changes over time $d\tau$ along a tube segment from location I to location II by the amount

$$[(\bar{U}_{II_1} - \bar{U}_{I_1})G_1 + (\bar{U}_{II_2} - \bar{U}_{I_2})G_2]\, d\tau$$

where the integral mean values $\bar{U}$ of the internal energy are

$$\bar{U}_1 = \frac{\displaystyle\int_0^{R_0} \varphi w_1 U_1 R\, dR}{\displaystyle\int_0^{R} R w_1\, dR}, \qquad \bar{U}_2 = \frac{\displaystyle\int_0^{R_0} R(1-\varphi) w_2 U_2\, dR}{\displaystyle\int_0^{R_1} R(1-\varphi) w_2\, dR}$$

The change in kinetic energy over time $d\tau$ is

$$\left(G_1\, \frac{k_{II_1}\bar{w}_{II_1}^2 - k_{I_1}\bar{w}_{I_1}^2}{2} + G_2\, \frac{k_{II_2}\bar{w}_{II_2}^2 + k_{I_2}\bar{w}_{I_2}^2}{2} \right) d\tau$$

where

$$k_1 = (2\pi/w_1)^3 \left[\int_0^{R_0} R\varphi\, dR \right]^2 \int_0^{R} R\varphi u_1^3\, dR$$

$$k_2 = \left(\frac{2\pi}{w_2}\right)^3 \left[\int_0^{R_0} R\,(1-\varphi)\,dR\right]^2 \int_0^R R\,(1-\varphi)\,w_2^3\,dR$$

The work done by the pressure forces and the lift force is expressed as:

$$p_1\,[G_1/\rho_{I1} + G_2/\rho_{I2}]\,d\tau - p_{II}\,[G_1/\rho_{II1} + G_2/\rho_{II2}]\,d\tau - (G_1 + G_2)\,\Delta z\,d\tau$$

In the final analysis we obtain

$$G_1\,[(\bar{U}_{II1} - \bar{U}_{I1}) + (k_{II1}\bar{w}_{II1}^2 - k_{I1}\bar{w}_{I1}^2)/2] +$$
$$+ G_2\,[(U_{II2} - U_{I2}) + (k_{II2}\bar{w}_{II2}^2 - k_{I2}\bar{w}_{II2}^2)/2] =$$
$$= G_1\left(\frac{p_I}{\rho_{I1}} - \frac{p_{II}}{\rho_{II1}}\right) + G_2\left(\frac{p_I}{\rho_{I2}} - \frac{p_{II}}{\rho_{II2}}\right) - \Delta z\,(G_1 + G_2) \qquad (2.44)$$

According to the first law of thermodynamics, the elementary work performed over a liquid with mass dm_2 is

$$dL = dm_2\left[(\bar{w}_{II2}^2 - \bar{w}_{I2}^2)/2 - \int_I^{II} p\,dv_2 + Q_2\right] \qquad (2.45)$$

where Q_2 is the irreversible heat release due to interphase friction and skin friction. Analogously, we have for the gas phase

$$dL = dm_1\left[(\bar{w}_{II_1}^2 - \bar{w}_{I1}^2)/2 - \int_I^{II} p\,dv_1 + Q_1\right] \qquad (2.46)$$

Equation (2.46) can be written as

$$U_{II1} - U_{I1} = Q_1 - \int_I^{II} p\,dv_1$$

Integrating the last equation over the cross section, we find

$$\int_0^{R_0} U_{II1}R\varphi_{II}w_{II1}\rho_{II1}\,dR - \int_0^{R_0} U_{I1}R\varphi w_{I1}\rho_{I1}\,dR =$$

$$= \int_0^R Q_1 R\varphi w_{I1}\rho_{I1}\,dR - \left[\int_0^{R_0} R\varphi_I w_{I1}\rho_{I1}\int_I^{II} \rho\,dv_1\right]dR$$

or

$$\bar{U}_{\mathrm{II1}} - \bar{U}_{\mathrm{I1}} = \bar{Q}_1 - \int_{\mathrm{I}}^{\mathrm{II}} p\, dv_1.$$

(2.47)

where

$$\bar{Q}_1 = \int_0^{R_0} Q_1 R \varphi w_1\, dR \Big/ \int_0^{R_0} R \varphi w_1\, dR$$

Analogously for each phase

$$\bar{U}_{\mathrm{II2}} - \bar{U}_{\mathrm{I2}} = \bar{Q}_2 - \int_{\mathrm{I}}^{\mathrm{II}} p\, dv_2.$$

(2.48)

Substitution of Eqs. (2.46) and (2.47) into (2.48), yields

$$G_1 \bar{Q}_1 + G_2 \bar{Q}_2 = - \Big\{ G_1 \int_{\mathrm{I}}^{\mathrm{II}} v_1\, dp + G_2 \int_{\mathrm{I}}^{\mathrm{II}} v_2\, dp + \Delta z (G_1 + G_2) + $$
$$+ G_1 (k_{\mathrm{II1}} \bar{w}_{\mathrm{II1}}^2 - k_{\mathrm{I1}} \bar{w}_{\mathrm{I1}}^2)/2 + G_2 (k_{\mathrm{II2}} \bar{w}_{\mathrm{II2}}^2 - k_{\mathrm{I2}} \bar{w}_{\mathrm{I2}}^2)/2 \Big\}$$

For relatively moderate velocities both phases can be assumed to be incompressible. The last equation then becomes

$$\Delta p = - \frac{G_1 \bar{Q}_1 + G_2 \bar{Q}_2}{V_1 + V_2} + \rho \Delta z + \rho_1 \varphi \frac{k_{\mathrm{II1}} \bar{w}_{\mathrm{II1}}^2 - k_{\mathrm{I1}} \bar{w}_{\mathrm{I1}}^2}{2} + $$
$$+ \rho_2 (1 - \varphi) \frac{k_{\mathrm{II2}} \bar{w}_{\mathrm{II2}}^2 - k_{\mathrm{I2}} \bar{w}_{\mathrm{I2}}^2}{2}$$

(2.49)

The first term in the right-hand side of the equation characterizes the irreversible loss of pressure due to skin and interphase friction

$$\Delta p_{\mathrm{fr}} = (G_1 \bar{Q}_1 + G_2 \bar{Q}_2)/(V_1 + V_2)$$

(2.49a)

The second term is equal to change in pressure due to the static head (leveling component)

$$\Delta p_{\mathrm{lev}} = \rho \Delta z$$

(2.49b)

where $\rho = \rho_1 \varphi + (1 - \varphi) \rho_2$

The third term, obviously, reflects the irreversible change in kinetic energy

$$\Delta p_{\text{kin}} = \left[\frac{\rho_1 \varphi}{2} \left(k_{\text{II}1} \overline{w}_{\text{II}1}^2 - k_{\text{I}1} \overline{w}_{\text{I}1}^2 \right) + \frac{\rho_2 (1 - \varphi)}{2} \left(k_{\text{II}2} \overline{w}_{\text{II}2}^2 - k_{\text{I}2} \overline{w}_{\text{I}2}^2 \right) \right]$$

$$(2.49c)$$

The friction losses for one-dimensional steady two-phase flow [Eq. (2.49)] can be written as

$$\left(\frac{dp}{dz} \right)_{\text{t.ph}} = \tau_w \frac{S}{A} + \frac{d}{dz} \left[\frac{j_2 (1 - \varphi) w_2}{2} + \frac{j_1 \varphi w_1}{2} \right] + $$
$$ + g \sin \theta \left[\varphi \rho_1 + (1 - \varphi) \rho_2 \right]$$

$$(2.50)$$

or, making the substitutions

$$j_2 = \frac{j (1 - x)}{(1 - \varphi)} \quad \text{and} \quad j_1 = \frac{jx}{\varphi}$$

$$w_2 = \frac{j (1 - x)}{\rho_2 (1 - \varphi)} \quad \text{and} \quad w_1 = \frac{jx}{\rho_1 \varphi}$$

we find

$$\left(\frac{dp}{dz} \right)_{\text{t.ph}} = \left(\frac{dp}{dz} \right)_{\text{fr}} + j^2 \frac{d}{dz} \left[\frac{x^2}{\rho_1} + \frac{(1 - x)^2}{\rho_2 (1 - \varphi)} \right] + g \sin \theta \left[\varphi \rho_1 + (1 - \varphi) \rho_2 \right]$$

$$(2.51)$$

In the general case of forced convective channel flow of two-phase fluids the identification of the separate components of the overall loss is very difficult due to the structural nonhomogeneity of the flow, slip and the presence of thermodynamic non-equilibrium. Practical calculations are performed by numerous semiempirical models which, evidently, should differ depending on the two-phase flow structure. The most important models used for calculating the hydraulic drag are examined below.

2.3.1 Homogeneous Two-Phase Flow

This case is approximately implemented in practice in bubbly and dispersed-droplet flow structures. It will be assumed that the phases are uniformly mixed and that they are in thermodynamic equilibrium with one another. Such a fluid can be treated as quasihomogeneous with variable physical properties such as density, viscosity, etc. Bankoff [2.63] suggested the following power-law expressions for the fluid's velocity profiles and the voidage (density)

$$w/w_{\text{max}} = (y/R)^{1/m} \quad \text{and} \quad \varphi/\varphi_{\text{max}} = (y/R)^{1/n} \tag{2.52}$$

The relationship between the vapor quality x and the void fraction in the Bankoff model is written as

$$x = \left[\left(1 - \frac{\rho_1}{\rho_2 S}\right) + \left(\frac{\rho_2}{\rho_1 S}\frac{K}{\varphi}\right)\right]^{-1} \tag{2.53}$$

is the phase slip, where parameter K, defined as

$$K = \frac{2(m+n+mn)(m+n+2mn)}{(n+1)(2n+1)(m+1)(2m+1)_1} \tag{2.54}$$

was measured for the first time by Armand [2.64]. He suggested that K be expressed as

$$K = 0,833 + 0,05 \ \ln p \tag{2.55}$$

According to Bankoff $K = 0.71 - 0.0142\,p$.

On assumption that liquid is always present on the wall, the shear $(\mu_{wt.ph} \sim \mu_{w2})$ stress at the tube wall is found from the Blasius law

$$\tau_{t.ph} = 0,04\ \bar{\rho}\bar{w}^2\ (\mathrm{Re})^{-0,25}$$

The ratio of the shear stress in the two-phase flow to that in the flow of liquid is

$$\frac{\tau_{t.ph}}{\tau_2} = (\bar{\rho}/\rho_2)^{3/4}\,(\bar{w}/w_2)^{7/4} = [1 + \bar{\varphi}\,(1 - \rho_1/\rho_2)]^{3/4}\,[1 - x\,(1 - \rho_2/\rho_1)]^{7/4} \tag{2.56}$$

The Bankoff model is valid for turbulent flow when the Blasius equation is used for shear stresses.

Levy in his study of two-phase flow [2.65] used the Deissler approximation for the shear stress in the turbulent wall region

$$\tau = -\,(\mu + \varepsilon\rho)\cdot dw/dy$$

where ϵ is the eddy viscosity.

In the developed turbulence region the equation for the shear stress at the wall is usually expressed as

$$\tau_w = -\,\mu\,\frac{d\bar{w}}{dy} - n^2\bar{w}y\rho\,\left(1 - e^{-\rho n^2 \bar{u} y/\mu}\right)\frac{dw}{dy} \qquad (n = 0,10) \tag{2.57}$$

The velocity profile is specified by the density profile thus

$$\frac{\rho}{\rho_2} = \frac{\rho_{max}\,w_{max}}{\rho_2\bar{u} + \rho_{max}\,(\bar{w}_{max} - \bar{w})} \tag{2.58}$$

Table 2.1

Re	$\eta=0{,}04$		$\eta=0{,}1$		$\eta=0{,}3$	
	$\bar{\rho}/\rho_2$	$\Phi_{20}^2 f_{20}$	$\bar{\rho}/\rho_2$	$\Phi_{20}^2 f_{20}$	$\bar{\rho}/\rho_2$	$\Phi_{20}^2 f_{20}$
2 000	0.845	0.0844	0.681	0.126	0.387	0.307
5 000	0.815	0.0600	0.626	0.0917	0.311	0.242
7 000	0.805	0.0538	0.607	0.0830	0.286	0.226
10 000	0.794	0.0483	0.588	0.0754	0.261	0.209
20 000	0.773	0.0399	0.522	0.0639	0.219	0.189
50 000	0.746	0.0323	0.508	0.0529	0.197	0.167
70 000	0.739	0.0299	0.492	0.0497	0.156	0.162
100 000	0.730	0.0280	0.477	0.0468	0.141	0.156
300 000	0.701	0.0229	0.430	0.0405	0.104	0.143
500 000	0.688	0.0210	0.411	0.0369	0.090	0.139

Re	$\eta=0{,}5$		$\eta=0{,}6$		$\eta=0{,}7$	
	$\bar{\rho}/\rho_2$	$\Phi_{20}^2 f_{20}$	$\bar{\rho}/\rho_2$	$\Phi_{20}^2 f_{20}$	$\bar{\rho}/\rho_2$	$\Phi_{20}^2 f_{20}$
2 000	0.250	0.568	0.206	0.731	0.173	0.919
5 000	0.180	0.475	0.142	0.626	0.114	0.799
7 000	0.157	0.451	0.122	0.598	0.0969	0.769
10 000	0.137	0.430	0.104	0.574	0.0807	0.742
20 000	0.104	0.399	0.0755	0.539	0.0563	0.704
50 000	0.0704	0.370	0.0617	0.508	0.0452	0.672
70 000	0.0607	0.363	0.0398	0.500	0.0280	0.664
100 000	0.0516	0.356	0.0335	0.493	0.0227	0.651
300 000	0.0312	0.341	0.0185	0.478	0.0117	0.642
500 000	0.0245	0.336	0.0139	0.474	0.0085	0.064

Eliminating the velocity from Eq. (2.57) and substituting Eq. (2.58) into (2.57), Levy performed numerical integrations for different Re, density ratios $\bar{\rho}/\rho_2$, Martinelli parameter Φ_2^2 and friction velocity η; here $\eta = -(\rho_{max} - \rho_2/\rho_{max} w_{max} x) \sqrt{\tau_{st}/\rho_2}$ and κ is the Prandtl constant, equal to 0.4. The results of Levy's calculations are shown in Table 2.1. It can be concluded by comparing these results with experimental data of various investigators that at high pressures ($p \geqslant 5$ MPa) and low qualities, corresponding to bubble and stratified structures of two-phase flow, which have typically low slip ratios, the agreement between analytic and experimental results is rather satisfactory.

2.3.2 Stratified Flow

Many semiempirical relationships were suggested for calculating the pressure drop in annular flow. The most extensively used model is that due to Lockhart and Martinelli, since it gives acceptable agreement with experimental data. In addition, further refinements of computational techniques is, as a rule, based on various modifications of the Martinelli method [2.66, 2.67]. For this reason we shall consider this model and the studies based on the Martinelli-Lockhart model.

Let us consider fully developed steady-state two-phase, two-component adiabatic flow in horizontal tubes, irrespective of the flow mode. The pressure gradient in such a flow is controlled by the shear stress at the wall τ_w. We assume that the gas and the liquid move separately and that the physical parameters of the gas and liquid phases are independent of the longitudinal coordinate and time. Let us consider one-dimensional flow, where the static pressure is constant in each cross section, and where the gas and the liquid have velocities w_1 and w_2, which are independent of the radius. In addition, we assume that the pressure drop in each phase can be expressed in a form analogous to the equation of single-phase flow, namely

$$\left(\frac{dp}{dz}\right)_1 = \xi_1 \frac{\rho_1 w_1^2}{2} \frac{1}{2R}, \qquad \left(\frac{\partial p}{\partial z}\right)_2 = \xi_2 \frac{\rho_2 w_2^2}{2} \frac{1}{2R_2} \tag{2.59}$$

where ξ_1 and ξ_2 are the friction coefficients for the gas and liquid.

Using the above assumptions, Martinelli and Lockart obtained the expression

$$\Delta p_{дв} = \Delta p_1 \Phi_1^2, \qquad \Delta p_{t.ph} = \Delta p_2 \Phi_2^2 \tag{2.60}$$

If the void fraction $\varphi \to 0$, then for the liquid flow

$$1/\Phi_1^2 = 0, \qquad 1/\Phi_2^2 = 1$$

For single-phase flow $\varphi = 1$, we have

$$1/\Phi_1^2 = 1, \qquad 1/\Phi_2^2 = 0$$

At the critical point, where the phase densities are equal ($\rho_1 = \rho_2$), there the following equalities should be satisfied for laminar flow

$$1/\Phi_1^2 + 1/\Phi_2^2 = 1 \mid$$

and for turbulent flow in accordance with the Blasius friction law for smooth tubes

$$1/(\Phi_1)^{7/8} + 1/(\Phi_2)^{7/8} = 1$$

The pressure gradient $(dp/dz)_1$ in the turbulent gas flow is, in accordance with Eq. (2.59), written as

$$\left(\frac{dp}{dz}\right)_1 = \xi_1 \frac{1}{R_1 f_1} \left(\frac{j_1}{\varphi}\right)^2 \tag{2.61}$$

The volumetric gas and liquid contents are, respectively

$$\varphi_1 = \left(\frac{R_1}{R}\right)^2 \quad \text{and} \quad \varphi_2 = \left(\frac{R_2}{R}\right)^2$$

The gas-phase pressure gradient is

$$\left(\frac{dp}{dz}\right)_1 = \xi_1 \frac{i_1^2}{4R\rho_1\varphi_1^{5/2}} \tag{2.62}$$

Analogously for the liquid phase

$$\left(\frac{dp}{dz}\right)_2 = \xi_2 \frac{i_2^2}{4R\rho_2(1-\varphi)^{5/2}} \tag{2.63}$$

Equating Eqs. (2.62) and (2.62) with (2.60), we find

$$\Phi_1^2 = 1/\varphi^{5/2}, \qquad \Phi_2^2 = 1/(1-\varphi)^{5/2}$$

or

$$1/\Phi_1^{2/5} + 1/\Phi_2^{2/5} = 1 \tag{2.64}$$

It can be shown that for laminar flow one can write

$$1/\Phi_1^{1/2} + 1/\Phi_2^{1/2} = 1$$

In the general case of a two-phase flow we write

$$1/\Phi_1^{1/n} + 1/\Phi_2^{1/n} = 1 \tag{2.65}$$

Equation (2.65) is a single-parameter family of curves in which the exponent n is selected on the basis of experimental data. It follows from Martinelli's experimental data that for laminar flow $n = 2\text{–}3$ and for the turbulent $n = 3.5\text{–}4$.

Martinelli postulates further that the nondimensional quantities Φ_1 and Φ_2 are a function only of the ratio X. This assumption cannot be verified analytically; however, numerous experimental data obtained over a wide range of variables confirm the validity of this assumption.

The friction coefficients in laminar and turbulent flows for each phase can be expressed as

$$\xi_1 = K_1/\mathrm{Re}_1^{m_1}, \qquad \xi = k_2/\mathrm{Re}_2^{m_2} \tag{2.66}$$

where Re is based on the hydraulic diameter, and K_1 and K_2 are constants.

One considers four characteristic modes of two-phase flow: (a) turbulent flow of liquid and gas (turbulent-turbulent); (b) the flow of liquid is laminar, and of gas—turbulent (laminar-turbulent); (c) the flow of liquid is turbulent and of gas—laminar (turbulent-laminar); (d) laminar flow of liquid gas (laminar-laminar). The exponents for the above four flow modes are listed in Table 2.2.

Figure 2.15 shows a plot of the Martinelli parameter Φ_1 vs. ratio X together with

Table 2.2

Flow mode	Wall surface	m_2	m_1	Re_2	Re_1
Turbulent-turbulent	Smooth	0.2	0.2	More than 2000	More than 2000
Turbulent-turbulent	Rough	0	0	More than 2000	More than 2000
Laminar-turbulent	Smooth	1.0	0.2	Less than 1000	More than 2000
Turbulent-laminar	Smooth	0.2	1.0	More than 2000	Less than 1000
Laminar-laminar	Smooth	1.0	1.0	Less than 2000	Less than 1000

correlation curves of family (2.65). The correlations are in better agreement with experiments at $n = 2.75$ than in turbulent-turbulent flow at $n = 4$. At the same time, it should be pointed out that the experimental data for the curve of $\Phi_1 = f(X)$ are described rather satisfactorily by a single curve of the family at $n = 3.5$ for all the flow modes. The maximum deviation of experimental points from the correlating curve is $\pm 20\%$.

Figure 2.16 shows curves of Φ_1 and φ_1 vs. X_1; the corresponding empirical relationship for the void fraction is

$$\varphi_1 = (1 - x)^{-0,378} \tag{2.67}$$

The frictional pressure drop in two-phase flows is calculated as follows. Using data in Table 2.2, one finds values of Re_1 and Re_2 from the specified gas and liquid flowrates, and then the friction coefficients ξ_1 and ξ_2 and the pressure drop in the gas and liquid phases Δp_1 and Δp_2. The square root of ratio $\Delta p_2/\Delta p_1$ yields the value of parameter X.

2.4 HYDRAULIC DRAG IN TWO–PHASE MIXTURES FLOWING IN HEATED TUBES

The vapor quality and void fraction of boiling liquids flowing in pipe change along the channel axis. In general the local values of void fraction may differ significantly from the equilibrium values, for which reason it becomes impossible to use the isothermal Martinelli model utilizing the equilibrium values of the void fraction for steam-generating channels.

2.4.1 Flow of Boiling Liquid in a Horizontal Tube

The pressure drop for this case can be written as

$$\Delta p = \frac{2}{Re} \int_0^L \tau_w\, dz + \int_0^L j\, [xw_1 + (1 - x)\, w_2]\, dz \tag{2.68}$$

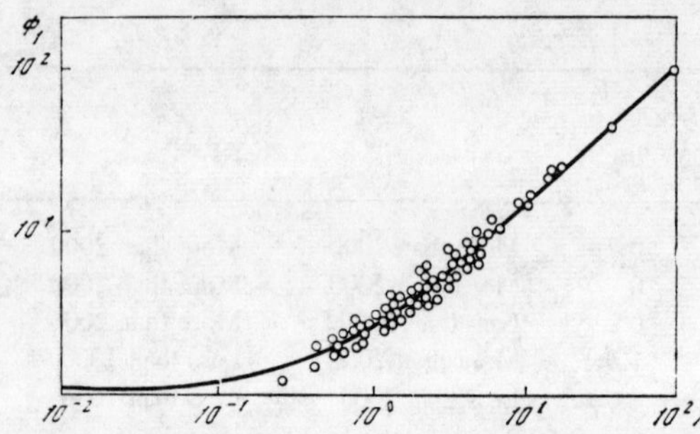

Figure 2.15 Martinelli parameter vs. X.

Let us consider for simplicity the case of flow of boiling liquid in a horizontal, constant cross section tube with constant wall heat flux. We write the energy balance for this case

$$G \frac{d}{dz}\left[xh_{01} + (1-x)h_{02}\right] = Q_{w}$$

Here h_{01} and h_{02} are the stagnation enthalpies for the gas and liquid, respectively. The last equation can be represented approximately, by assuming the physical properties and phase velocities constant, in the form:

$$dx/dz = 2q_{w}/R_0 jr \tag{2.69}$$

It follows from Eq. (2.69) that at constant wall heat flux the quality x changes linearly with distance. At the tube inlet ($z = 0$) we have saturated liquid ($x = 0$), at $z = L$, $x = x_L$. We express the shear stress at the wall in terms of the pressure gradient

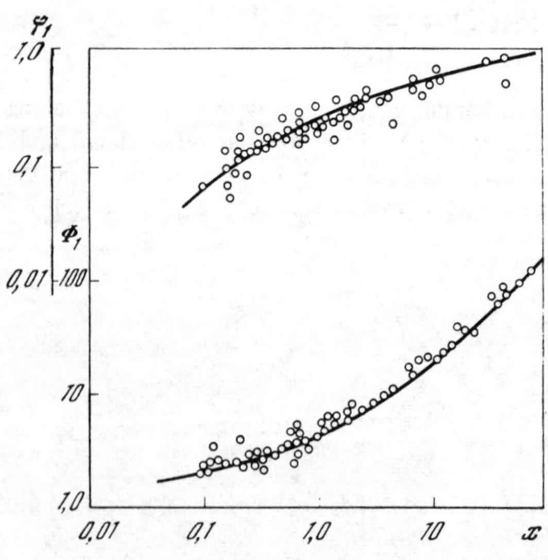

Figure 2.16 Martinelli parameter vs. X for all the flow modes and φ_1 vs. X according to Martinelli-Nelson data.

in the gas phase and the Martinelli parameter

$$\tau_w = -\frac{R_0}{2}\left(\frac{\partial p}{\partial z}\right)_{\text{t.ph}} = -\frac{R_0}{2}\left(\frac{dp}{dz}\right)_2 \Phi_2^2.$$

Writing the pressure gradient in the liquid phase in accordance with Eq. (2.59) and replacing the variables of integration, we obtain

$$\Delta p = -\xi_2(\text{Re})\,\frac{j_2^2 x}{\rho_2 R_0}\left(\frac{1}{x}\int_0^x \Phi_2^2\,dx\right) + j^2\left\{\frac{x^2}{\varphi\rho_1} + \frac{1}{\rho_2}\left[\frac{(1-x)^2}{1-\varphi} - 1\right]\right\}$$

$$(2.70)$$

It is seen from Eq. (2.70) that the pressure drop in a horizontal tube is comprised of two components: frictional pressure drop Δp_{fr} (first term) and the change in pressure Δp_x, induced by the change in vapor quality due to boiling. The first term of the equation is found by integrating the Martinelli-Nelson data

$$\Delta p_{\text{fr}} = -\xi_2\,\frac{j_2^2}{\rho_2}\,\frac{R_0}{4}\,\frac{rx}{q_w}\,\overline{\Phi}_2^2. \qquad (2.71)$$

where

$$\overline{\Phi}_2^2 = \frac{1}{x}\int_0^x \Phi_2^2\,dx$$

The second term of the equation is calculated from the specified pressure, at which both φ and x are related uniquely

$$\Delta p = j^2\overline{v}_2. \qquad (2.72)$$

where

$$\overline{v}_2 = \frac{1}{\rho_2}\left[\frac{x^2}{\varphi}\frac{\rho_1}{\rho_2} + \frac{(1-x)^2}{1-\varphi} - 1\right]$$

The results of this integration for water are plotted in Figs. 2.17 and 2.18.

Baroczy [2.66] and Chien with Ibele [2.67] suggested that Martinelli parameter Φ_2 be considered as a function of a nondimensional product, comprised of physical properties $\rho_1/\rho_2(\mu_1/\mu_2)^{0.2}$. It should be noted here that the above nondimensional product does not contain the surface tension, although the latter definitely plays an important role. Figure 2.19 shows a plot of φ_2 vs. $\rho_1/\rho_2(\mu_1/\mu_2)^{0.2}$ for different liquids. It should, however, be noted that the graph was constructed for a specified specific flowrate $j = 1.35 \times 10^3$ kg/m$^2 \cdot$sec. [sic]

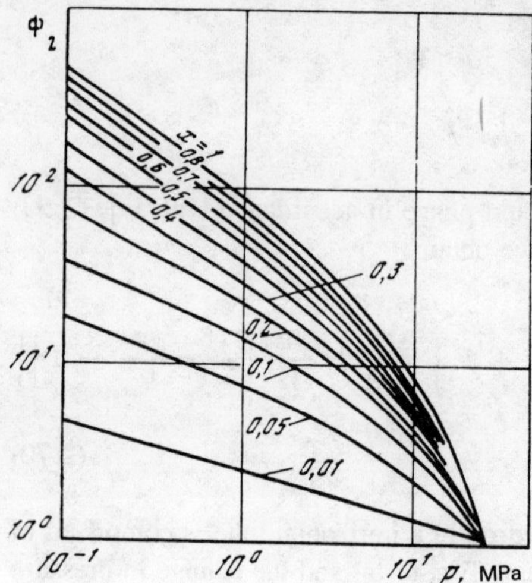

Figure 2.17 Plot of Φ_2 vs. p for water.

The Baroczy method is not suitable for practical use, since in order to find Δp_{fr} at specific flowrates differing from $j = 1.35 \times 10^3$ kg/m²·sec it is necessary to use graphical data on the effect of j on the pressure losses. Chisholm [2.68] reduced Baroczy's graphical data and suggested an expression for the pressure drop of two-phase turbulent flow in smooth tubes

$$\frac{(\Delta p/\Delta z)_{\text{t.ph}}}{(\Delta p/\Delta z)_2} = 1 + (\Gamma^2 - 1)\left[Bx^{0,875}(1-x)^{0,875} + x^{1,75}\right] \tag{2.73}$$

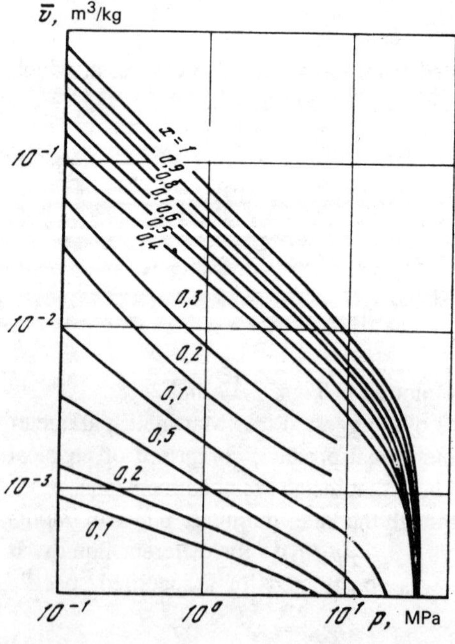

Figure 2.18 Mean specific volume $\bar{v}$ of water according to Martinelli and Nelson.

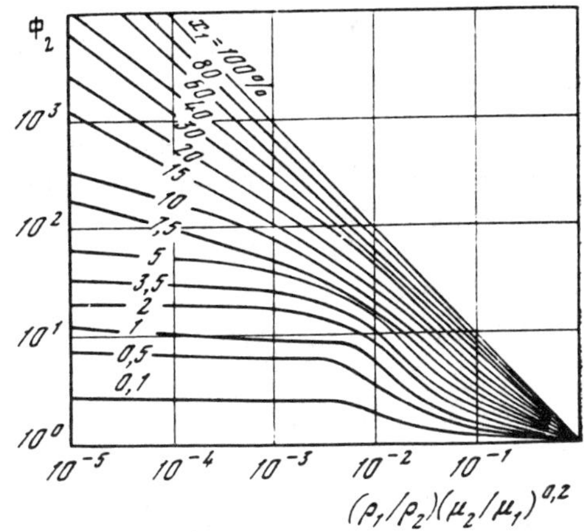

Figure 2.19 Plot of Φ_2 vs. the product $(\rho_1/\rho_2)(\mu_2/\mu_1)^{0.2}$ for different substances; specific flow $j = 1.35 \cdot 10^3$ kg/cm^2.

where $\Gamma = (\rho_2/\rho_1)^{0.5}(\mu_1/\mu_2)^{0.12}$ is a parameter incorporating the liquid properties. According to Chisholm's suggestion, coefficient B is a function of the specific flowrate j and properties of the vapor-liquid mixture and are found from data in Table 2.3.

2.4.2 Hydraulic Drag of Steam-Generating Channels

The coolant flowing in steam-generating channels undergoes a change in state from subcooled water to superheated steam, i.e., the quality changes from negative values to values greater than unity. Let us consider in more detail the friction in a steam generating channel for all the flow structures which develop in sequence along the channel.

Flow of subcooled water in surface boiling. Tarasova [2.114], Miropol'skiy [2.113] and others showed that the vapor bubbles forming on the wall in the course of surface boiling result in a rise in the effective roughness of the tube, with the result that the momentum transfer between the wall layer and the flow core increases. The drag coefficient ξ is then a function not only of Re, but also of the heat flux. The hydraulic drag is higher in surface boiling than in flows of water under the same conditions, but without boiling. The following formula was suggested for boiling of subcooled water in tubes and annulus at pressures between 5 and 19.6 MPa:

Table 2.3

Value of Γ	Specific flowrate j, kg/cm^2	Value of B	Value of Γ	Specific flowrate j, kg/cm^2	Value of B
$1 \div 9,5$	$0 \leqslant j \leqslant 500$	$4,8$	$9,5 \div 28$	$0 < j \leqslant 600$	$520/\Gamma \cdot j^{0,5}$
	$500 \leqslant j \leqslant 1900$	$2400/j$		$j > 600$	$21/\Gamma$
	$j \geqslant 1900$	$55/j^{0,5}$	$\Gamma \geqslant 28$	$j > 600$	$15\,000/\Gamma^2 \cdot j^{0,5}$

$$\left(\frac{\xi_B}{\xi_0} - 1\right) = 3{,}09 \left(\frac{q_w}{r\rho''w}\right)^{0,7} \left[7 - \left(1 + \frac{\Delta h}{\Delta h_1}\right)^{0,5}\right] \frac{U_B}{U_{\text{wet}}} \qquad (2.74)$$

where ξ_0 is the hydraulic drag coefficient for liquid; U_B/U_{wet} is the ratio of the heated to the wetted parameter; $U_B/U_{\text{wet}} = 1$ for tubes and annuli with both surfaces heated

$$\Delta h = 34 \frac{q_w\, D_{\text{eq}}}{(\rho w)^{0,9}} \left(\frac{\rho'}{\rho''}\right)^{-0,3}$$

Δh_1 is the subcooling at the channel exit.

Flow with developed nucleate boiling (foamy and slug flow). For steam-generating channels operating at high pressures this zone is relatively small. The hydraulic drag in it is also calculated with Eq. (2.74). Some investigators [2.119] suggest that Δp_{fr} for flow with developed nucleate boiling be approximated by a straight line, passing through the point of boiling inception and the start of dispersed annular flow.

Dispersed-annular flow. This region occupies the greatest length of the steam-generating tube from $x_{\text{d.a}}$ to x_{cr}. Investigations by Przhiyalkovskiy with Petrova [2.123], Miropol'skiy et al. [2.113], and also Tarasova [2.114], performed with steam-water mixtures, have established that the hydraulic drag in this zone to the start of film dryout, i.e., of the second crisis, exhibits an anomalous behavior, namely, the hydraulic drag drops rather steeply with rise in voidage, passes through a minimum, and then starts rising again. This fact is illustrated by experimental data due to Tarasova in Fig. 2.20, which shows curves of $\Delta p_{\text{fr}}/\Delta p_0$ vs. the mean quality $\bar{x}$. It is seen that heating has a significant effect on the hydraulic drag of the wet steam. Up to the region of anomal behavior of $\Delta p_{\text{fr}}/\Delta p_0$ heating increases the relative pressure drop. This is apparently because the steam-water mixture flows in this region in the dispersed annular mode with a thick film, the heated wall being covered with bubbles, which increases the skin friction in the wall region. As the quality increases, however, the friction drag decreases significantly; eventually, at high values of $\bar{x}$, the effect of the heat flux becomes less perceptible and at $\bar{x} \cong 1$, $\Delta p_{\text{fr}}/\Delta p_0$ becomes virtually identical for the heated and nonheated walls.

The experimental data of Doroshchuk [2.111] show that the wall temperature rises steeply at the minimum of the curve of $\Delta p_{\text{fr}}/\Delta p_0$ vs. x, which is due to deterioration of heat transfer in conjunction with breakdown of the film, i.e., the point of minimum virtually coincides with $\bar{x}_{\text{cr}}$. The two-phase flow downstream of the anomalous region has a dispersed droplet structure.

The reduction in hydraulic drag in the region with developed dispersed-annular flow in the steam-generating channel can be attributed to decay of waves on the surface of the liquid film when the thickness of the latter attains some minimum thickness ($\text{Re}_{\text{fm}} < 30$). However, this problem has not been sufficiently explored. Kirillov with his coworkers [2.119] suggested the following computational formulas for the hydraulic drag in dispersed-annular flow:

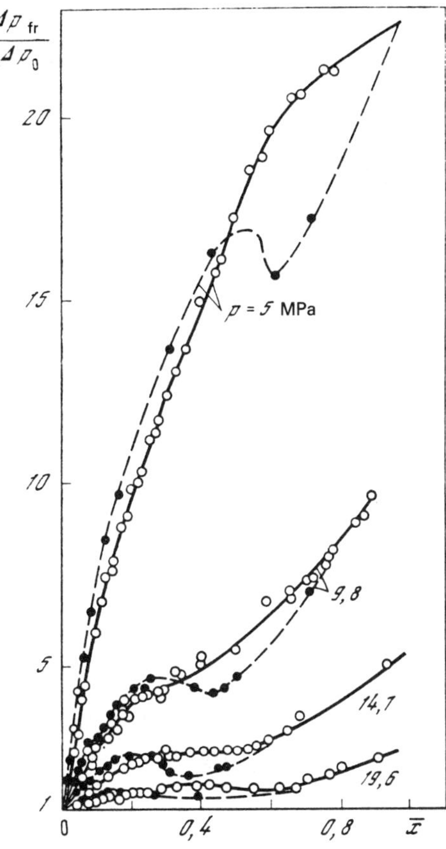

Figure 2.20 Plot of $\Delta p_{\text{fr}}/\Delta p_0$ vs. $\bar{x}$ for flow of steam-water mixtures in heated and nonheated tubes at $\rho w = 2000$ kg/m$^2 \cdot$sec [2.113]; the solid curve represents data without and the dashed curve with heating.

$$\Delta p_{\text{fr}} = \frac{1}{a_6} \frac{\Delta L \rho''}{2D} \left(\frac{\rho w x}{\varphi \rho''}\right)^2 a_{11} \tag{2.75}$$

where

$$a_1 = 10^{-3} \left[0,7 + 2,1 \left(10^{-2}p - 1,05\right)^2\right] \left(10^{-2}\rho w\right)^{1,79 \cdot 10^{-2}p} \times$$
$$\times \left\{1 + 200q \left[\left(10^{-2}p - 0,4\right)^2 - \left(10^{-2}p - 0,4\right)^3\right]\right\}^{0,33};$$

$$a_2 = \left[0,47 - \left(1 - 10^{-2}p\right)^2 + \frac{1}{1 + \left(10^{-2}p\right)^4}\right] \times$$
$$\times \left(\frac{5}{10^{-2}\varphi w}\right)^{0,711 \cdot 10^{-2}p} - \left[\left(1,5 \cdot 10^{-3}q\right)\left(1,4 - 10^{-2}p\right)\right]^{0,33};$$

$$a_3 = \left(a_5 + 8 \cdot 10^{-2}p\right) + 10^{-2}\rho w \left[0,35 + 15\left(10^{-2}p - 0,7\right)^6\right];$$

$$a_4 = 0,13 + 7,4\left(10^{-2}p - 0,7\right)^6;$$

$$a_5 = 7 - \left[170\left(10^{-2}p - 1\right)^4 + x\left(10^{-2}\rho w - 5\right)\right]\left(4,35q\right)^{0,33};$$

$$a_6 = \left(\frac{R - 3,75}{\sqrt{8}} + \frac{1}{x\sqrt{8}} \ln \frac{17 \cdot 10^{-3}}{2\delta_{\text{lim}}}\right)^2;$$

$$a_7 = \frac{10^{-3}}{x}\left(44,2 - 1,96p + 7 \cdot 10^{-3}p^2\right);$$

$$a_8 = \frac{10^{-4}}{x^2}(20{,}2 + 0{,}16p); \quad a_9 = 4{,}3\,(1 - 0{,}44\cdot 10^{-2}p);$$

$$a_{10} = (0{,}85 - 10{,}88D\cdot 10^3) + (0{,}0875 - 0{,}035\cdot 10^{-3}\rho w)$$
$$\text{at}\quad D \geqslant 17\,\text{mm}; \quad a_{10} = 0;$$
$$a_{11} = 0{,}78\,(\rho w\cdot 10^{-3} - 1) + (3{,}1 - 124D) \quad \text{at}\quad D \geqslant 17\,\text{mm}; \quad a_{11} = 1,$$
$$a_{12} = 1 - \sin a_{13}/\cos a_{13}\,(\rho w\cdot 10^{-3} - 1{,}5); \quad a_{13} = 1 - 0{,}71p\cdot 10^{-2};$$
$$R = a_3 + a_4/x\,(1 - 1/x), \quad \delta_{\lim} = a_1\,(a_2 - x)^{0{,}9};$$
$$\varphi = 1{,}06 - 8{,}3\cdot 10^{-4}p + a_7 + a_8$$

Equation (2.75) is valid at: $7 \leq p \leq 14$ MPa, $q \leq 0.6$ MW/m^2 and $\rho w < 1500$ kg/m$^2\cdot$sec, and is in satisfactory agreement with the tabulated data of [2.120].

The boundary of the dispersed-annular flow mode is found from the empirical formula

$$x_{\text{d.a}} = a_2 - (8/a_1)^{1/a_9} + a_{10} \tag{2.76}$$

and it is suggested that the critical vapor quality be calculated from the expression

$$x_{\text{cr}} = a_2 - (0.1/a_1\cdot 10^{-3})^{1/a_9} \tag{2.77}$$

The total pressure loss in the steam-generating channel is comprised of three components: friction losses Δp_{fr}, gravitational component Δp_{lev} and pressure drop for acceleration Δp_{ac}; Δp_{lev} for the entire range of qualities is found in the form

$$\Delta p_{\text{lev}} = g\bar{\rho}_{\text{mix}}\,\Delta H \tag{2.78}$$

where ΔH is the distance between the pressure taps; $\bar{\rho}_{\text{mix}} = (1 - \bar{\varphi})\,\rho' + \bar{\varphi}\rho''$ is the mean density of the mixture; Δp_{ac} is determined from the familiar expression $\Delta p_{\text{ac}} = \rho w(w_{\text{mix.out}} - w_{\text{mix.in}})$.

The void fraction φ can be found from the expression [2.64] $\varphi = (0.833 + 0.051\text{g}\,p)\,\beta$ or [2.116]

$$\varphi = \beta \left\{ 1 - \frac{m\left(1 - \dfrac{\rho''}{\rho'}\right)(1 - \beta)}{[1 + x\sqrt{\text{Fr}/(1 - p/p_{\text{cr}})^2}]^2\,(n - \beta)} \right\} \tag{2.79}$$

Here $\text{Fr} = w^2/gD$ is the Froude number; $m = 0.342$, $n = 1.04$ and $k = 0.035$ are nondimensional coefficients, whereas β is the vapor volumetric flow ratio

$$\beta = 1 \bigg/ \left(1 + \frac{1 - x}{x}\,\rho''/\rho'\right) \tag{2.80}$$

Equations (2.75)–(2.80) comprise a technique for calculating the hydraulic drag of upwards dispersed-annular flow. Kirillov and Smogalev [2.119] used the suggested technique for compiling a program of numerical calculation of hydraulic drag and for

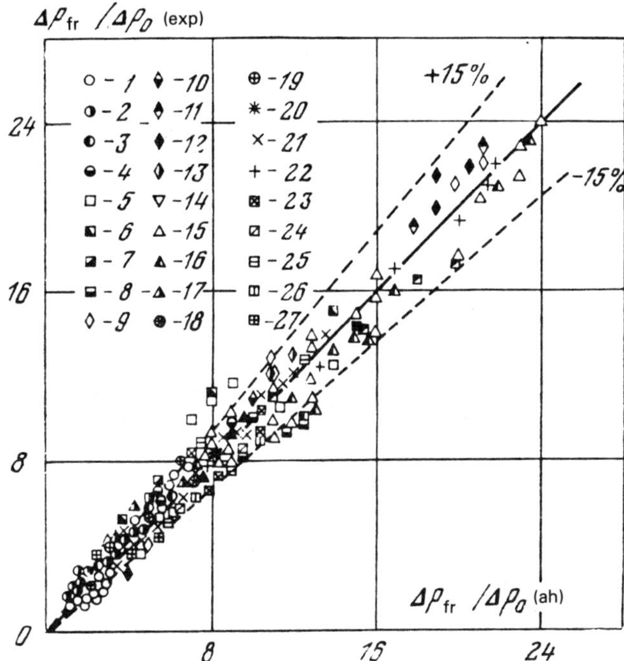

Figure 2.21 Comparison of experimental data and analytic results; after Kirillov and Smogalev [2.119]. 1–4, experimental data of Baldina with Lokshin [2.121] ($\rho w = 500$–1500 kg/m$^2 \cdot$sec, $p = 10$–18 MPa, $D = 20$ mm); 5 and 6, data of Krasyakova [2.122] ($\rho w = 100$ kg/m$^2 \cdot$sec, $p = 10$–14 MPa, $D = 20$ mm); 7, data of Tong [2.124] ($\rho w = 540$ kg/m$^2 \cdot$sec, $p = 7$ MPa, $D = 20$ mm); 8, data of Borishanskiy et al. [2.125] ($\rho w = 1200$ kg/m$^2 \cdot$sec, $p = 8$ MPa, $D = 30$ mm); 9–17, data of Kirillov and Smogalev [2.119] ($\rho w = 500$–1200 kg/m$^2 \cdot$sec, $p = 7$–14 MPa, $D = 17$ mm); 18–22, data of Tarasova [2.114] ($\rho w = 500$–2000 kg/m$^2 \cdot$sec, $p = 10$ MPa, $D = 8$ mm); 23, data of Przhiyalkovskiy and Petrova [2.123] ($\rho w = 1000$–3000 kg/m$^2 \cdot$sec, $p = 7$–8 MPa, $D = 5.2$ mm).

comparison with experimental data of various investigators. The results of comparison of the computational and experimental data on the hydraulic drag are shown in Fig. 2.21, from which it is seen that the majority of points is in agreement with the computational curve to within $\pm 15\%$. Comparison of the calculated curves for $\Delta p_{\mathrm{fr}}/\Delta p_0$ vs. x, with experimental data of the Dzherzhinskiy Heat Engineering Institute [2.111, 2.114] also attest to the acceptable accuracy for the calculations.

Simplified estimates of hydraulic drag can also be obtained from the formula [2.119]

$$\left(\frac{\Delta p_{\mathrm{fr}}^{\mathrm{h}}}{\Delta p_{\mathrm{nh}}} - 1 \right) = 4{,}4 \cdot 10^{-3} \left(\frac{q_{\mathrm{w}}}{\rho w} \right)^{0,7}. \tag{2.81}$$

The friction losses in a nonheated tube in Eq. (2.81) are calculated from the empirical correlation

$$\frac{\Delta p_{nh}}{\Delta p_0} = \frac{1,05 \cdot 10^{0,C149\rho''/\rho'}}{(1-\varphi)} \frac{1}{Fr^{7,5/p}} \qquad (2.82)$$

which agrees within $\pm 10\%$ with experimental drag data in the flow of wet steam in nonheated tubes.

2.5 DISPERSED–ANNULAR FLOW

With increasing quality, the flow structure in the tube changes from bubbly to slug and then to dispersed-annular flow due to collisions between small bubbles and their coalescence into large ones.

Experiments performed by Haberstroh and Griffith [2.69] with air-water, steam-water, air-water and alcohol, air-water-glycerine, and CO_2-water showed that the most important transition parameter is the void fraction φ_{cr}, which in all experiments was in the range 0.81–0.89 for the onset of dispersed-annular flow.

In addition to the void fraction, another major parameter is also the velocity of the gas and liquid phases, which, for critical transition from slug to dispersed-annular flow, are interrelated by the expression

$$u_1^+ = 0.4 + 06u_2^+$$

where

$$u_1^+ = \frac{w_1}{[(\rho_2/\rho_1)\, g\mathit{f}_{12}]^{1/2}} \quad \text{and} \quad u_2^+ = \frac{w_2}{[(\rho_2/\rho_1)\, gF_{12}]^{1/2}}$$

In dispersed-annular flow the liquid moves in the form of fine droplets in the vapor core and of a film at the wall. The velocities and temperatures of the droplet, film and the vapor core are in general significantly different from one another. Obviously, in order to describe the dispersed-annular flow mode one must use conservation equations written separately for each of the three flow components: film, droplets and gas [see Eqs. (2.7)]. For the one-dimensional steady-state case (see the schematic in Fig. 2.22) the conservation equations for the masses are:

$$\frac{1}{\varphi_1}\frac{d\varphi_1}{dz} + \frac{1}{\rho_1}\frac{d\rho_1}{dz} + \frac{1}{w_1}\frac{dw_1}{dz} = -\frac{1}{A}\frac{dA}{dz} + \frac{1}{m_1}J_{21} + \frac{1}{m_1}J_{31}$$

$$\frac{1}{\varphi_2}\frac{d\varphi_2}{dz} + \frac{1}{w_2}\frac{dw_2}{dz} = -\frac{1}{A}\frac{dA}{dz} - \frac{1}{m_2}J_{21} - \frac{1}{m_2}E + \frac{1}{m_2}D \qquad (2.83)$$

$$\frac{1}{\varphi_3}\frac{d\varphi_3}{dz} + \frac{1}{w_3}\frac{dw_3}{dz} = -\frac{1}{A}\frac{dA}{dz} - \frac{1}{m_3}J_{31} + \frac{1}{m_3}E - \frac{1}{m_3}D$$

Here J_{21} is the flux of vapor evaporating from the film surface per unit time per unit length; J_{31} is the flux of vapor evaporating from the droplet surfaces; E is the entrainment flux per unit channel length; D is the precipitation flux per unit channel length.

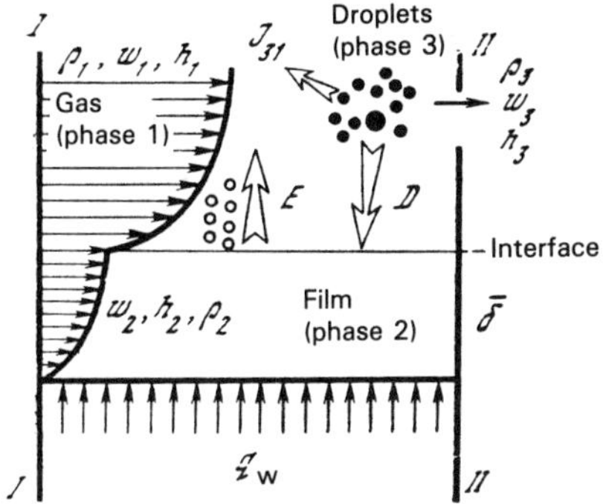

Figure 2.22 Schematic representation of heat and mass transfer between the film and the vapor-droplet core flow.

φ_1, φ_2 and φ_3 are the fractions of the cross section occupied by the vapor, film, and droplets, respectively, so that $\varphi_1 + \varphi_2 + \varphi_3 = 1$; $m_1 + m_2 + m_3 = m$ is the two-phase mixture flowrate.

As follows from Eqs. (2.7), the conservation equations for momentum can be expressed as

$$\rho_1\varphi_1 w_1 \frac{dw_1}{dz} + J_{21}(w_1 - w_{12}) + J_{31}(w_1 - w_{13}) = -\varphi_1 \frac{dp}{dz} - F_{12} - F_{13}$$

$$\rho\,\varphi_2 w_2 \frac{dw_2}{dz} - J_{21}(w_1 - w_{12}) + D(w_3 - w_{12}) - Ew_{12} =$$

$$= \varphi_2 \frac{dp}{dz} + F_{12} + \tau_w \frac{\pi}{A} \tag{2.84}$$

$$\rho_3\varphi_3 v_3 \frac{dw_3}{dz} - J_{31}(w_1 - w_{13}) + Ew_{12} - D(w_3 - w_{12}) = \varphi_3 \frac{dp}{dz} + F_{13}$$

Here w_{12} is the velocity at the film surface, equal to aw_2; a is a coefficient which is a function of the velocity profile in the film and the phase velocity of the wave (ranges from 1.5 to 2.5); F_{12} is a force produced by the shear stress at the vapor-film interface, equal to τ_{12}/d, and F_{13} is the drag force of droplets in the vapor.

The conservation equations for energy, written for the total phase enthalpies ($h_i + w_i^2/2$) become (see the energy transfer schematic in Fig. 2.22)

$$\rho_1\varphi_1 w_1 \frac{d}{dz}\left(h_1'' + \frac{w_1^2}{2}\right) = q_{13} - F_{13}w_3 - F_{12}w_2 \tag{2.85}$$

$$\rho_2\varphi_2 w_2 \frac{dh_2}{dz} = q_w - Ei_2 + D\left(i_3 + \frac{w_3^2}{2}\right) + F_{12}w_2 - \tau_w \frac{S}{A} w_2$$

$$\rho_3\varphi_3 w_3 \frac{d}{dz}\left(h_3 + \frac{w_3^2}{2}\right) = -q_{13} + E(h_3 - h_2) -$$

$$- D\left[\left(h_3 + \frac{w_3^2}{2}\right) - h_2\right] + F_{13}w_3$$

In addition to uniformity and steadiness, the following assumptions were made in

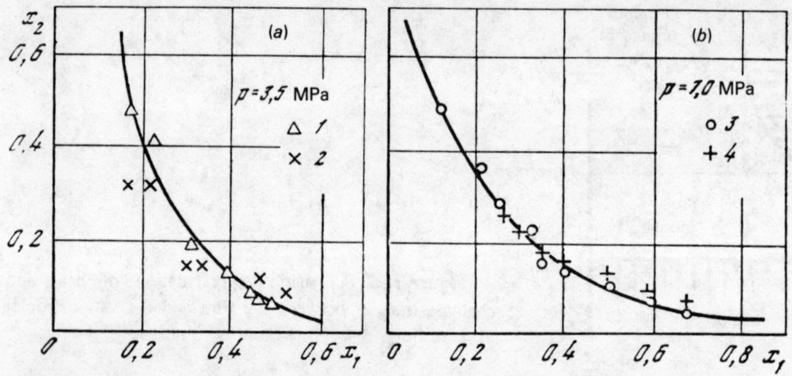

Figure 2.23 Fraction x_2 of the liquid flowrate in the film of the quality x according to data of Subbotin et al. [2.71]. (*a*) For p = 3.5 MPa: 1, data of Subbotin et al. [2.71]; 2, data of Keeys et al. [2.70]. (*b*) For p = 7 MPa; 3, data of Subbotin et al. [2.71]; 4, of Keeys et al. [2.70].

deriving Eqs. (2.83)–(2.85): (1) the temperature of the vapor-film interface is equal to the saturation temperature; (2) the droplet size is uniform (monodisperse fog); (3) coalescence and collisions between particles are neglected; (4) the pressure in the film, droplets and vapor is the same; (5) the liquid is incompressible, i.e., ρ_2 = const.

The set of differential equations (2.83)–(2.85) together with the equations of state for each of the phase can form a closed set of equations only if relationships are available for the interphase force interaction F_{12}, F_{13}, τ_w and mass transfer (J_{21}, J_{31}, E and D), and also heat transfer (q_w, q_{13}). We now consider the vapor-droplet core and the film.

2.5.1 Vapor-Droplet Core

Under steady conditions of dispersed-annular flow mass transfer between the film on the wall and the vapor-droplet core is fully developed when the deposition flux D is equal to the entrainment flux E. In this case the flowrates of the liquid in the film and core are sufficiently conservative characteristics of the two-phase liquid. Naturally, these equilibrium conditions are attained at the flow hydrodynamic stabilization length which, according to measurements by Keeys et al. [2.70] and Subbotin et al. [2.71], amount to ~ 150 tube diameters.

Figure 2.23 shows the variation in the relative flowrate of liquid in the film x_2, equal to the ratio of the flowrate of liquid in the film m_2 to the total flowrate of the mixture in the channel. The graph shows the flow quality x_2 on the abscissa. It is seen that the experimental points of the Harwell group [2.70] are in satisfactory agreement with those of Subbotin et al. [2.71]. Nigmutalin et al. [2.72] performed a comprehensive cycle of experimental studies of dispersed-annular flow at pressures from 4 to 7 MPa and specific mass flowrates 500–4000 $kg/m^2 \cdot sec$ and flow qualities x from 0.1 to 0.9 in 13.3 mm diameter tubes. Data on the flowrate of liquid in the film are correlated by the following expressions:

$$x_2 = 2{,}11 \cdot 10^{-5} \left(\frac{w_1 \mu_1}{\sigma} \right)^{-2,1} \quad \text{at} \quad \left(\frac{w_1 \mu_1}{\sigma} \right) \geqslant 10^{-2}$$

$$x_2 = 1{,}075 \cdot 10^{-2} \left(\frac{\rho_2}{\rho_1} \right)^{0,12} \left(\frac{w_1 \mu_1}{\sigma} \right)^{-0,67} \quad \text{at} \quad \left(\frac{w_1 \mu_1}{\sigma} \right) < 10^{-2}$$

$$(2.86)$$

Here σ is the surface tension, and w_1 is the vapor velocity.

In analyzing dispersed-annular flow it is important to learn to determine the fluxes of droplets entrained from the film and of deposition on it.

Theoretical analysis of the deposition of droplets on the wall is usually based on a turbulent diffusion model in which it is assumed that

$$D = (D^T + D^\mu) \frac{d\,(\varphi_3 \rho_3)}{dy} \tag{2.87}$$

where D^T and D^μ are coefficients of molecular and turbulent diffusion of particles in the gas phase, respectively. Assuming further that the droplet concentration changes significantly only in the diffusion boundary layer Δ^D near the wall, whereas the droplet concentration at the film surface is zero ($\varphi_3 = 0$) (complete capture of the droplets by the liquid film), we find

$$D \approx (D^T + D^\mu) \frac{\rho_2 \varphi_{30}}{\delta_D}$$

or

$$D = k_D \rho_2 \varphi_{30} \tag{2.88}$$

Note that in order for a droplet to be deposited on the wall it must overcome the resistance in the viscous laminar sublayer, i.e., have a sufficiently high velocity fluctuation in the y direction, which is transverse to the mainstream flow. In considering the behavior of droplets in turbulent flow one can conditionally single out three regions.

Diffusion region for submicron diameter particles. In this region the coefficient k_D of turbulent diffusion of the particles is equal to the coefficient of eddy diffusion of the gas ϵ_D. Such a droplet, upon entering the laminar viscous sublayer, deposits on the walls as a result of Brownian motion or gravitation. However, consideration must be given to transverse velocity and temperature gradients, due to which the droplet may be removed from the laminar sublayer.

Inertia region for droplets up to 10 μm in diameter. The particles in the flow core, as in the diffusion region, follow the turbulent fluctuations in the gas, i.e., $k_D \cong \epsilon_D$. However, the momentum acquired in the turbulent boundary layer may be found sufficient for the particle to overcome the resistance of the viscous sublayer and reach the wall by inertia. The Magnus force and the pushing of the droplet away from the

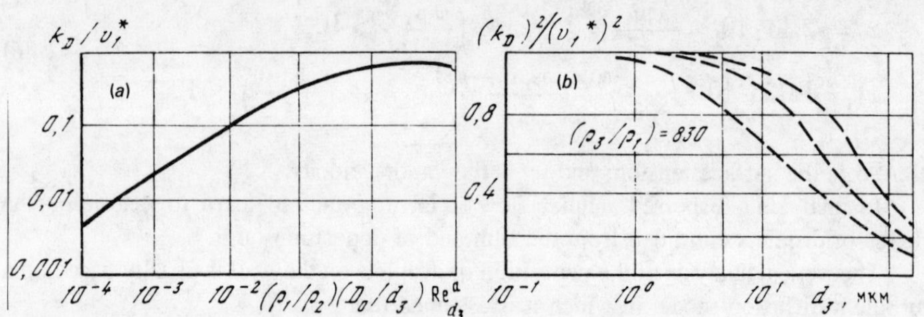

Figure 2.24 Rate of particle deposition in the inertial/region according to data of Gardner [2.73] (a) and as a function of the droplet diameter according to Hutchinson et al. [2.74] (b).

wall due to the flux of steam caused by evaporation from the interface may also be of significance for these droplets.

Collision region for particles with $d_3 > 10\ \mu$m. In this region the scale of turbulent fluctuations in the gas is small as compared with the particle size. The most probable cause for velocity fluctuations of droplets in this region are eddies forming in the trailing region of particles moving ahead of the given droplet. All the particles entering the laminar sublayer deposit on the wall.

The boundaries between the regions are determined by means of the nondimensional particle relaxation time τ^+, defined by the expression

$$t^+ = [t\,(v_1^*)^2]/\nu_1 \tag{2.89}$$

where v_1^* is the friction velocity of the gas phase, equal to $(\tau_1/\mu_1)^{1/2}$, τ is the shear stress; t is the time needed for decelerating particles that move at velocity v_1^* in the nonmoving gas.

If Stokes law is assumed to hold, time t^+ is

$$t^+ = \frac{d_3^2 v_1^{*2} \rho_1^2}{18\mu_1^2}\,\frac{\rho_3}{\rho_1} \tag{2.90}$$

In the diffusion region for submicron particles ($t^+ < 0.1$) they are deposited on the wall by virtue of Brownian motion. The mass transfer coefficient K_D for high Schmidt numbers is expressed as

$$k_D/v_1^* = 0{,}089\,(\mathrm{Sc})^{-0,7} \tag{2.91}$$

For the inertia region $0.1 < t^+ < 10$ the deposition rate is controlled by eddy diffusion. Gardner suggested the following equation for this region (Fig. 2.24a)

$$\frac{k_D}{v_1^*} = f\left[\frac{\rho_1}{\rho_2}\frac{D_0}{d_3}\mathrm{Re}_{d_3}^a\right] \tag{2.92}$$

where $\mathrm{Re}_{d_3} = (w_1 d_3)/v_1$; $a = 0.023\ln(k_D/v_1^*) - 0.68$ and D_0 is the tube diameter.

As the size of droplets in the collision region $t^+ > 10$ is increased, the radial particle velocity differs increasingly from the radial velocity of turbulent eddies and, as is seen from Fig. 2.24b, constructed from analytical results of Hutchinson et al. [2.74], the velocity ratio differs significantly from unity. In this region the coefficient of deposition of large particles k_D/v_1^* depends little on t^+. The experimental points lie within the range of $0.07 \leq k_D/v^* \leq 0.7$ at $20 \leq t_1^* \leq 10^5$.

As a first approximation for this region we can use the expression

$$k_D/v_1^* = 0{,}017 \tag{2.93}$$

Eddy diffusion is not the sole mechanism of particle deposition on the wall. Kirillov and Smogalev present the most complete listing of forces [2.75] acting on a droplet in the boundary layer and laminar sublayer. Let us consider the principal forces.

Gravity Force F_g.

Drag force in the longitudinal $\mathbf{F}_{13}^a$ ***and transverse*** $\mathbf{F}_{13}^y$ ***directions.*** The drag coefficient C_D was analyzed in detail in §2.2; it should only be noted here that, due to evaporation in the superheated boundary layer of the vapor, the droplet's drag coefficient decreases $C_D^{ev} = C_D/1 + B$, where B is the Spalding correction.

Magnus lift force. It is known that a rotating spherical particle in a gas flow with transverse velocity gradient is acted upon by a transverse force, equal to

$$\mathbf{F}_w = \frac{\pi d_3^3}{8}\rho_1[\boldsymbol{\omega}\times(\mathbf{w_1} - \mathbf{w_3})][1 + O(\mathrm{Re}_d)] \tag{2.94}$$

where ω is the vector of the angular velocity of the particle, equal to

$$\boldsymbol{\omega} = -\frac{1}{2}\frac{dw_3}{dy}$$

At low relatively liquid velocities ($w_3 < w_1$) in the boundary layer (Fig. 2.25) the Magnus force pushes the particle into the flow core.

For the case when the droplet entering the boundary layer by inertia is at a higher velocity than the local gas velocity (Fig. 2.25), Gardner obtained the following expression for the lift force. He assumed a freely-rotating sphere and a linear velocity profile in the gas with [2.76]

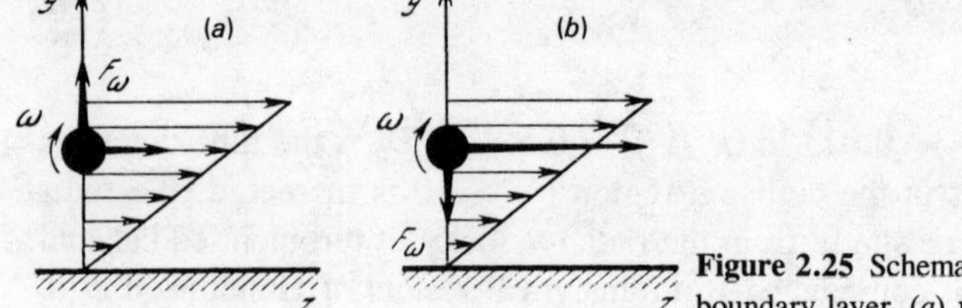

Figure 2.25 Schematic of particle motion in the boundary layer. (a) $w_3 < w_5$; (b) $w_3 > w_1$.

$$\mathbf{F}_w = C \, \frac{d_3^2}{4} \, (\rho_1 \mu_1)^{0,5} \left(\frac{dw_1}{dy} \right)^{0,5} (\mathbf{w}_1 - \mathbf{w}_3) \tag{2.95}$$

C is a constant, approximately equal to 81.2.

Force of repulsion due to the longitudinal pressure gradient (Archimedes force), equal to $F_z = \varphi_1(dp/dz)$. The transverse force does not arise under steady conditions, since the pressure across the boundary layer is identical.

Transverse fluctuating force. A particle in turbulent flow with fluctuating shear is acted upon by a transverse fluctuating force. The fluctuations of the shear stress in the wall layer are as high as 30% of the mean shear stresses, and the particle will move somewhat out of phase relative to the gas flow. Such a motion is evidently associated with the appearance of additional force F', whose direction will be the same as that of the lift force [2.77].

There exist a large number of other forces. These include the radiometric forces (thermophoresis), the Basset force and forces induced by molecular diffusion (molecular phoresis). Kirillov and Smogalev [2.75] showed that in the majority of cases of practical importance for sufficiently large ($d_3 > 5 \ \mu$m) droplets these forces may be neglected. They solved numerically the equations of motion of particles in wet-steam flow in tubes at pressures between 7 and 14 MPa. They assumed that the droplet entering the wall layer has a lower longitudinal velocity than the vapor surrounding it, i.e., the lift force prevents the motion of the droplet to the wall. They estimated the threshold transverse droplet velocity upon attaining which the droplet is able to overcome the repulsive effect of the Magnus and drag forces and deposit on the film. They showed that there exists a range of conditions where there is no deposition of droplets on the wall. It should, however, be noted that in practice the motion of large droplets in the boundary layer above the liquid film occurs under conditions when the local vapor velocity is lower than the particle velocity, and the Magnus force presses the latter to the film. This fact was confirmed by the experimental data of Farmer et al. [2.79] for downward mixture flow, as well as by data of Cousin and Hewitt [2.78] for upward dispersed-annular flow in a circular tube. High-speed cinephotography of droplets, performed in Harwell [2.78] (see Fig. 2.26), showed that the droplets reach the film surface without being perceptibly decelerated in the boundary layer, their "landing" on the film in a number of cases is similar to the landing of an aircraft on a landing strip.

A large number of experimental studies have been carried out on the deposition of droplets in dispersed-annular. Cousin and Hewitt [2.78] used the method of deposition on a dry film, i.e., they directly measured the flowrate of liquid from the flow core. Their experiments were performed in long vertical tubes with diameters of 9.52 and 31.8 mm at pressures from 2 to 5 MPa, the air velocity changed from 20 to 40 m/sec. The hydraulic stabilization length ranged from 0.6 to 2.2 m, whereas the length of the deposition segment was from 0.15 to 2 m. At the end of the deposition segment the film was completely sucked away and its flowrate was measured.

The deposition rate was determined by integrating the equations of conservation of the liquid mass in a flow core, which, in the absence of entrainment $E = 0$ (dry wall) is

$$- dm_3/dz = J_{32} \tag{2.96}$$

or

$$m_3 = \frac{\pi}{4} (d_0 - 2\bar{\delta})^2 \rho_3 \varphi_3 w_3 \tag{2.96a}$$

where w_3 is velocity of droplets in the flow core.

The slip of droplets in long tubes is small, i.e., $w_3 \simeq w_2$. Since film thickness $\bar{\delta}$ is significantly smaller than the tube diameter $\bar{\delta}/d_0 \ll$, the void fraction can be written in the form

$$\varphi_3 = | (m_3\rho_1)/(m_1\rho_3). \tag{2.97}$$

Substituting of $\tilde{\alpha}_3$ into Eq. (2.89a) yields

$$dm_3/dz = - \pi d_0 k_D \frac{m_3}{m_1} \rho_1. \tag{2.98}$$

Integration of Eq. (2.98) for $k_D = $ const yields the following expression for the deposition coefficient:

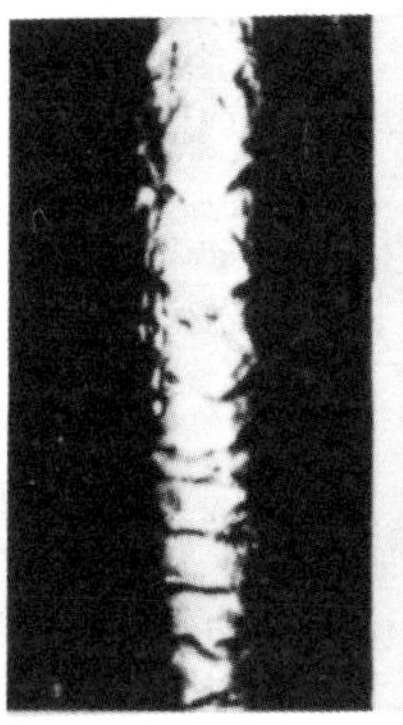

Figure 2.26 Deposition of droplets on a liquid film.

$$k_D = \frac{m_1}{\pi d_0 f_1 L} \ln \frac{m_{30}}{m_2 L},$$

(2.99)

where m_{30} and m_{3L} are the droplet flowrates entering and leaving the deposition segment, respectively; L is the length of the deposition segment.

Calculations of the deposition coefficient k_D from Eq. (2.99), from the experimental data of Cousin and Hewitt [2.78] show that k_D is independent of the particle concentration in the flow core, i.e., of φ_{30}. It is important that it is precisely the independence of k_D on the droplet concentration in the flow and makes it possible to use Eq. (2.88). Experiments show [2.78] that k_D is independent also of the tube diameter, gas velocity and pressure, but depends significantly on the channel length; with increasing L the value of k_D drops, tending to an asymptotic value.

Later publications [2.80–2.83] show that mass transfer coefficient k_D is a function both of the particle concentration and the longitudinal velocity w_3, slip ratio and other factors. Analysis of experimental data points to the contradictory nature of the suggested empirical relationships for k_D. Under the current conditions, it is apparently more reliable to describe experimental data, written directly for the flux of droplets from the core to the film J_{32}, rather than for the mass transfer coefficient k_D.

Rachkov [2.81] performed a series of experiments with wet steam at pressures between 1 and 10 MPa, mass fluxes between 500 and 2000 kg/m^2·sec and flow qualities between 0.1 and 0.3 in a 13.3 mm diameter tube. The experimental results were described by the nondimensional rate I^* defined as the ratio of the amount of liquid deposited by the flow onto the film, per unit length, to the total flowrate of droplets devoling upon unit channel perimeter:

$$I^*_{32} = (I_{32}/m_3)\pi D$$

The experimental points are correlated within $\pm 25\%$ by the expression

$$I^*_{32} = 0{,}012 \varphi_{30}^{-0,17} f(\Pi)$$

(2.100)

where

$$\Pi = 0{,}2 \frac{\sigma_2}{\sqrt{\mu_1 \mu_2 w_3}} \left(\frac{\rho_1}{\rho_2}\right)^{0,12}, \quad \Pi^* = \left(\frac{\rho_2}{\rho_1}\right)^{0,22}, \quad f(\Pi) = \Pi, \quad \Pi \geqslant \Pi^*,$$

$$f(\Pi) = \Pi^{0,5}, \quad \Pi < \Pi^*$$

Note that Eq. (2.100) quite satisfactorily describes Rachkov's data, however, other experimental points, among them those obtained by Cousin and Hewitt, fit the curve with a significantly greater scatter.

Ivandayev reduced experimental data obtained by himself [2.84] and others [2.85–2.87], and suggested the expression

$$J^*_{32} = \frac{J_{32}}{f_3 \varphi_{30} w_s d_0} = 2{,}05 \cdot 10^{-4} \left(\frac{w_3 \mu_3}{\sigma}\right)^{-12} \left(\frac{\rho_3}{\rho_1}\right)^{045}$$

(2.101)

which is compared in Fig. 2.27 with experimental results of other investigators. It is seen from the figure that the scatter of experimental points about the straight line is $\pm 35\%$.

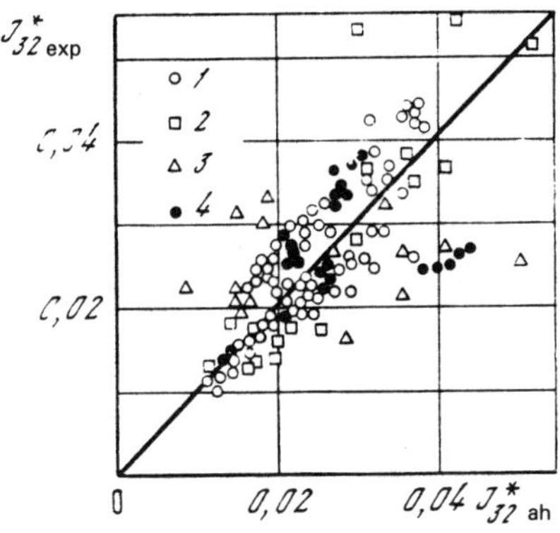

Figure 2.27 Comparison of experimental data of different investigators with correlation [2.101]. 1, Data of Cousin and Hewitt [2.78]; 2, of Jagota [2.86]; 3, of Alexander and Coldien [2.85]; 4, of Rachkov [2.81].

2.5.2 Wave Motion in Liquid Films

It is known that the film in dispersed-annular flow has waves on its surface. These waves, depending on the flow modes in the liquid and vapor phase (or gas) may have a different structure, which varies along the channel. Basically wave motion is a highly disordered three-dimensional phenomenon. However, at relatively low liquid flowrates one observes in the film two-dimensional rolling waves, with amplitude several times greater than the mean film thickness. Note that it is precisely these waves that control a number of important processes such as droplet entrainment, pressure drop in the channel and, in certain cases, for example, in the inlet region of the tube, affect the critical heat flux and mass transfer in post-dryout region.

The film flow regime and the wave structure on its surface are governed, in the case of falling films with a free surface, by the film Reynolds number

$$\mathrm{Re}_2 = \frac{\rho_2 \overline{w}_2 \overline{\delta}}{\mu_2} = \frac{G_2}{\mu_2}$$

where G_2 is the mass flowrate of liquid per unit width of wetted surface, or in terms of the volumetric flowrate Q_2, $\mathrm{Re}_2 = Q_2/\nu_2$.

Figure 2.28 shows photographs of various flow modes of liquid films falling freely down a vertical plate, according to data of Nakoryakov and Pokusayev. According to published data [2.88–2.100], laminar flow with a smooth film surface is observed to Reynolds numbers below 30–50, laminar-wavy flow occurs at $50 < \mathrm{Re}_2 < 400$ whereas a turbulent film with waves appears at $\mathrm{Re}_2 \geqq 400$.

The flow modes and waves on films with cocurrent and countercurrent gas flows are significantly more complicated, the threshold for transition from laminar-wavy flow in this case are controlled not only by Re_2, but also by the flow of the gas (vapor) phase, i.e., Re_1.

In the Institute of High Temperatures of the USSR Academy of Sciences [2.101], we performed a study of the wave parameters of a liquid film (mean, minimum, maximum film thickness, phase velocity and wave frequency) using the electrical-conduction method, based on measuring the resistance of a liquid film between two sensor electrodes mounted flush with the wall of the channel.

The use of a reference sensor in the measuring system allows compensation for changes in the chemical composition and temperature of the liquid film under study. The wave parameters of the film were measured by sensors with rod-type electrodes. The diameter of the latter and the distance between them were selected in the course of preliminary experiments in such a manner that as to ensure, over the greatest possible range of expected film thicknesses, a close to linear output signal of the instrument as a function of the film thickness. We used 0.9 mm diameter stainless steel electrodes placed with centers 4 mm apart. The sensors were placed at 350, 650, 925, and 950 mm from the inlet slot. It was found that the wave parameters of the film stabilize at $L \cong 800$ mm for $Re_2 = 800$ and the velocity profile of the air flow stabilizes at $L = 700$ mm for $Re = 12,000$. This means that at $L > 800$ mm steady unidirectional horizontal air-water stratified flow occurs in the channel at any flow condition. The wave parameters were measured by sensors placed at 925 and 950 mm. As shown in Fig. 2.29a, the signal from the electrical conductivity sensor is fed to an ITP-1 [magnetic densitometer], which measured the film thickness. A loop oscillograph, which records local instantaneous liquid-film thicknesses is connected to the output of this instrument. Various types of wire sensors, shown in Fig. 2.29b, were used.

The values of the mean, minimum and maximum film thicknesses were determined by statistical analysis of the oscillograms. The frequency characteristics of the waves were determined from analysis of the oscillograms. The phase velocity of the waves was determined by means of sensors $C\ III$ and $C\ IV$, placed at 25 mm from one another. In this case the output signals from both sensors were simultaneously fed to the loop oscillograph, and the phase velocity was calculated from oscillograms of local film thicknesses, obtained for each sensor. Figure 2.30 shows one of the oscillograms used for finding the wave velocity c_{ph}. If the distance Δz between the sensors and the time Δt of wave travel through the distance is known, c_{ph} is defined as $c_{ph} = L < (\Delta z/v)$, where L is the distance between identical peaks, measured from the oscillogram, whereas v is the rate of travel of the oscillograph tape.

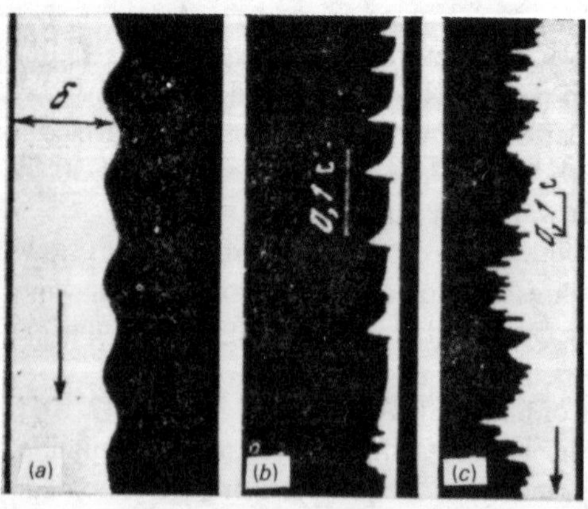

Figure 2.28 Fluctuations in the liquid film thickness. (a) Re = 6.1, δ_0 = 0.53 mm; (b) 40 and 0.2; (c) 870 and 0.7.

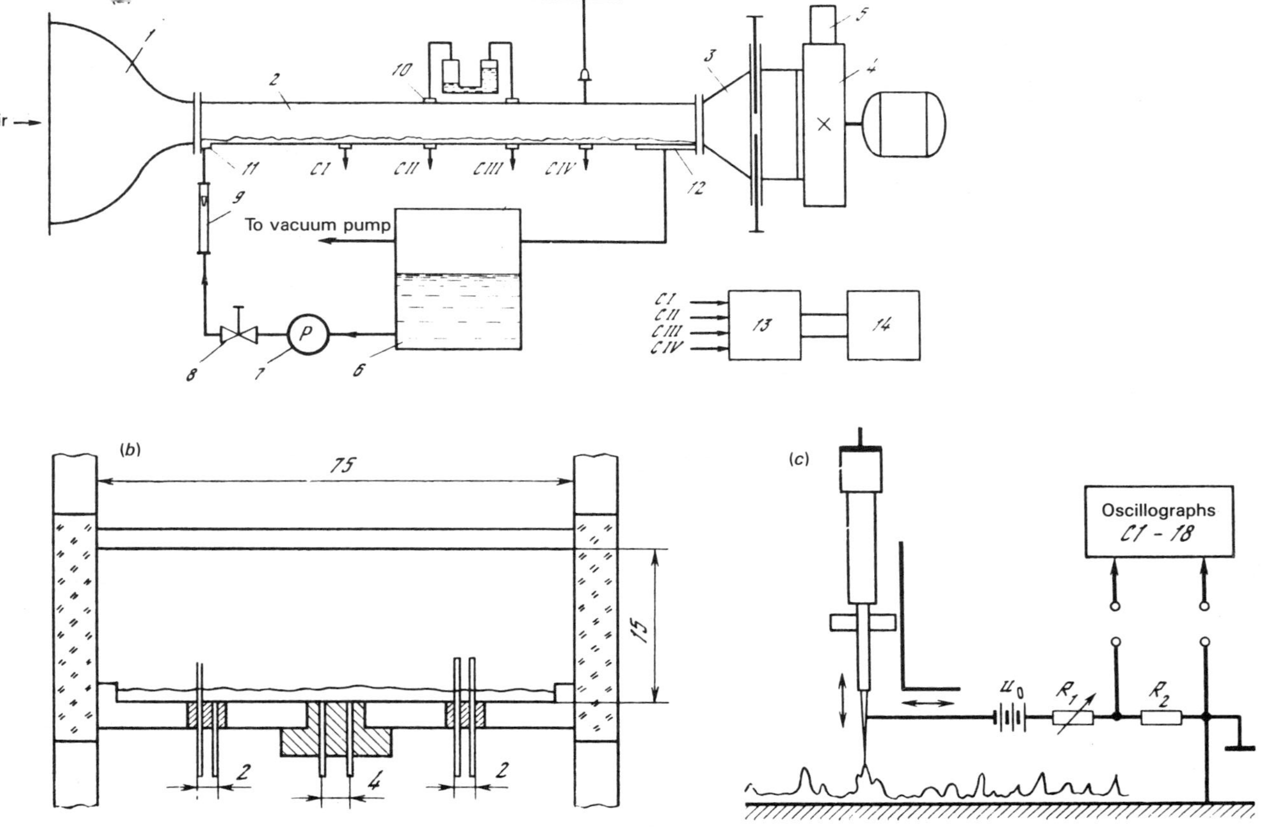

Figure 2.29 Schematic of experimental arrangement. (*a*) 1, Inlet nozzle; 2, test section; 3, exit diffuser; 4, compressor; 5, air discharge stub; 6, vacuum tank; 7, pump; 8, control valve; 9, rotameter; 10, static-pressure tap; 11, liquid film producing slot; 12, porous segment of pump; 13, film-thickness measuring instrument; 14, loop oscillograph; *CI–CIV* are film thickness sensors for measuring the wave parameters of the film. (*b*) Section through test channel. (*c*) Circuit for measuring the film thickness by the contact needle technique.

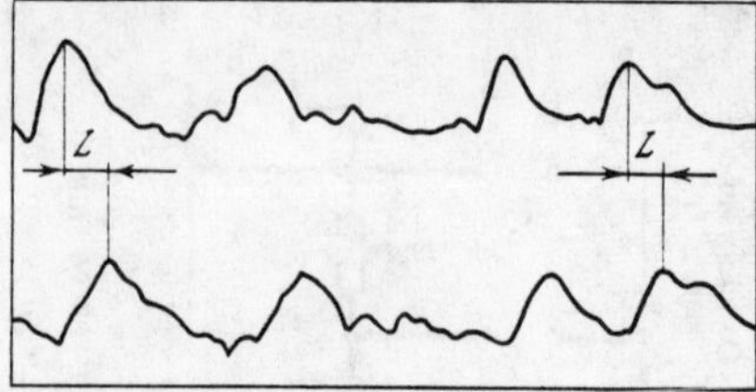

Figure 2.30 Determination of the phase velocity of the wave.

Simultaneously with the above conductometer method, the minimum and maximum wave thickness was measured by the familiar "contact needle" technique (Fig. 2.29c). The air flow velocity was measured by a DISA hot-element anemometer, with 5 μm sensor wire diameter.

Figure 2.31 shows the wave pattern chart obtained experimentally by Besfamil'nyy. This figure also shows the limits of transition from a smooth surface to one covered with two-dimensional waves, according to Hanratty's data [2.93]. Small-amplitude two-dimensional waves were observed in the present experiments only at $w_1 = 3$–5 m/sec over the entire range of Re_2 under study. The velocity of these waves is 0.2–0.3 m/sec, i.e., is approximately equal to the velocity of the film at the interface. It is seen from the wave regime chart that, for the range of variables studied, boundary between the smooth film and a surface with two-dimensional waves depends little on Re_2, being primarily controlled by the gas velocity. Hanratty and Engen [2.93] assume that the appearance of perturbation waves at the phase interface coincides with transition from laminar to turbulent gas flow ($Re_1 = 2000$–5000). Two-dimensional waves exist over a relatively narrow range of Re_1 (in experiments at the High Temperature Institute of the USSR Academy of Sciences they were found to exist up to $w_1 = 6$ m/sec).

A further rise in the gas velocity results in the breakup of the two-dimensional wave structure and appearance of three-dimensional waves on the film surface; this transition is sudden, without a transition region.

It is seen from Fig. 2.31 that the region of existence of three-dimensional waves is significantly wider than that of two-dimensional waves. It was established in the High Temperature Institute experiments that three-dimensional waves exist stably

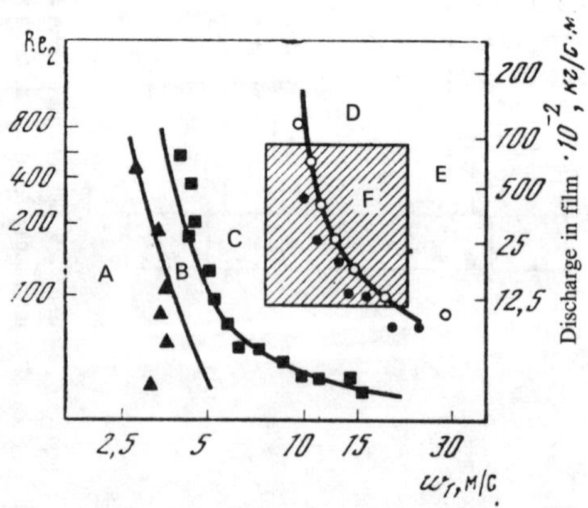

Figure 2.31 Film flow pattern chart. A, Smooth surface; B, two-dimensional waves; C, three-dimensional waves; D, rolling-over waves; E, droplet entrainment; F, region of modes investigated by the last of the present authors with his coworkers [2.101].

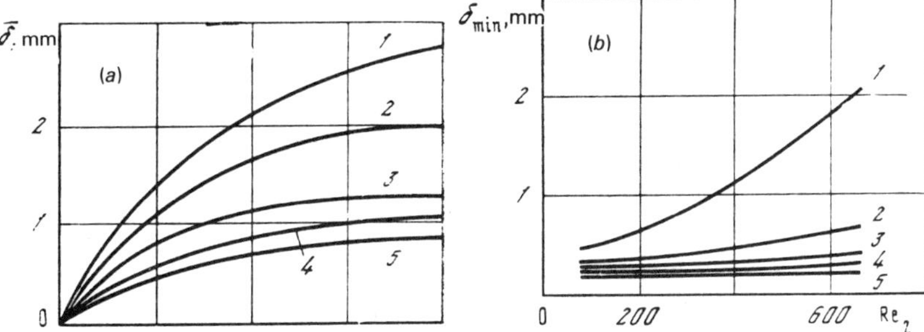

Figure 2.32 Mean film thickness. (*a*) Mean film thickness; (*b*) minimum film thickness; 1–5 w_1 equal to 7.2, 10.8, 14.4, 18 and 21.5 m/sec, respectively.

over the gas velocity range from 6 to 12 m/sec at Re_2 between 100 and 800; in addition, these waves have a typically high frequency and disorderly structure. When the gas velocity is increased further there form large-amplitude (rolling) waves, with minimum height several times the mean film thickness. The frequency of the roll waves is significantly lower than that of the two-dimensional waves, and, according to measurements, the phase velocity as high as 1.5 m/sec. In flows with roll waves there is a typical difference between the maximum film thickness (up to 5 mm) and the minimum thickness (0.2–0.35 mm), i.e., the waves are high.

Increasing the gas velocity, in the presence of roll waves, to values in excess of 18 m/sec results in inception of droplet entrainment from the film surface. Studies were not performed in the region of high entrainment rate (i.e, for $w_1 > 22$ m/sec).

Statistical analysis of the oscillograms allowed the construction of curves of the mean film thickness $\bar{\delta}$ vs. Re_2 at different Re_1. The oscillograms were obtained by means of a sensor placed at 950 mm from the liquid-film producing slot. It is seen from Fig. 2.32, which shows curves of $\bar{\delta}(Re_2)$ (a) and δ_{min} (b) vs. Re_2 for three-dimensional waves ($w_1 = 7.7$ and 10.8 m/sec), the mean film thickness $\bar{\delta}$ increases over the entire range of Re_2 from 100 to 800. In the cases where roll waves existed ($\bar{w}_1 = 14.4$, 17.6 and 20.8 m/sec) the value of $\bar{\delta}$ first increases with the increase in the liquid-film flowrate, and then remains virtually constant. The minimum film thickness decreases with increasing air velocity.

Figure 2.33 shows experimental data on the mean velocities $\bar{w}_2$ of the liquid in the film (a) and the phase velocity c_{ph} (b) as a function of Re_2 and with Re_1 as the parameter. It is seen that $\bar{w}_2$ increases with the film flowrate; since $\bar{\delta}$ decreases upon transition from three-dimensional to roll waves, the mean liquid velocity in the film rises steeply in this region. An increase in the gas velocity also results in a rise in the phase velocity of the waves, which are driven by the kinetic energy of the gas at the phase interface; here the relative rise in the phase velocity is initially greater, and then (at $Re_2 > 200$) smaller, than the relative increase in w_2.

Figure 2.34 shows the measured relative phase velocity c_{ph} as a function of Re_2 according to data of different investigators. Experimental data due to Maksimov et al. [2.94], Chu and Duker [2.95], and Hewitt with Wallis [2.96] for small waves in the surface of a freely falling liquid film are in rather satisfactory agreement. Of the

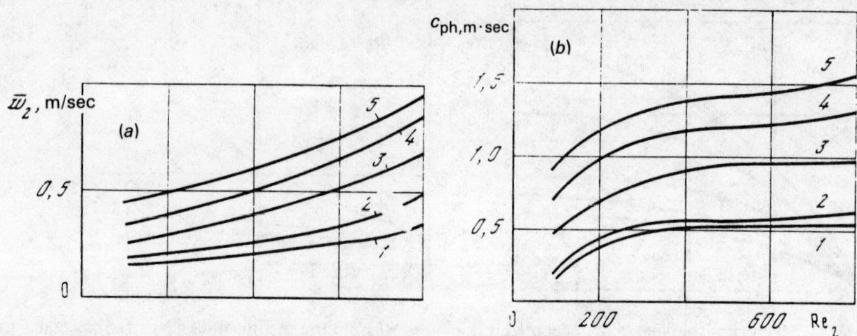

Figure 2.33 Mean (a) and phase (b) film velocities (w_1 is the same as in Fig. 2.32.

available analytic solutions for this case only the Krylov et al [2.97] and the Rushton-Davis [2.98] theories describe satisfactorily the experimental results over the range of Re_2 under study. Figure 2.35 also presents the experimental data of Miya et al. [2.99] for roll waves [2.95] on a freely-falling film. Also shown are the present authors data (wet steam) [2.100] and the results of measurements with an air-water mixture (in a rectangular duct) performed by the last of the present authors with his coworkers [2.101]. Comparison of these data shows that the nature of experimental curves of $c_{ph}/\bar{w}_2$ vs. Re_2 obtained for roll waves by Chu and Dukler [2.95] and by the present authors is qualitatively the same. The quantitative difference is due to the effect of gravity and absence of cocurrent gas flow in the Chu-Dukler experiments. Note the satisfactory agreement between our experimental points with those of Miya et al. [2.99]. It is seen from Fig. 2.35 that increasing the gas velocity results in a rise in the relative phase velocity.

The behavior of the frequency of large waves as a function of Re_2 is shown in Fig. 2.35, from where it follows that this frequency increases with Re_2 at $Re_1 = $ const and increases also with rising Re_1 at $Re_2 = $ const. Figure 2.36 also shows the

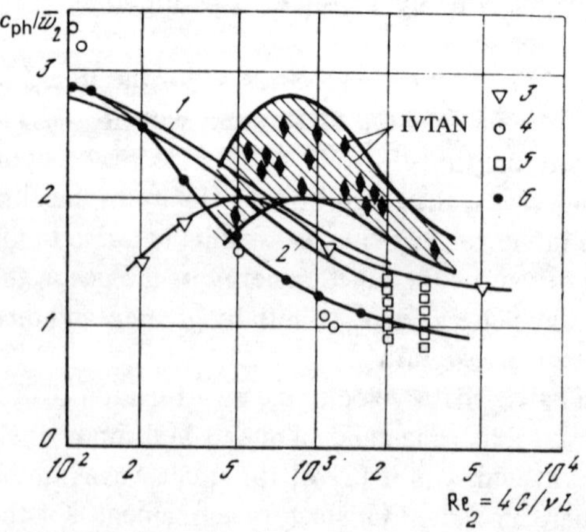

Figure 2.34 Variation in relative phase velocity. 1, Theory of Krylov et al. [2.97]; 2, theory of Ruston and Davis [2.98]; 3, data of Chu and Dukler [2.95]; 4, of Chu and Dukler [2.95]; 5, of Hewitt and Wallis [2.96]; 6, of Maksimov et al. [2.94].

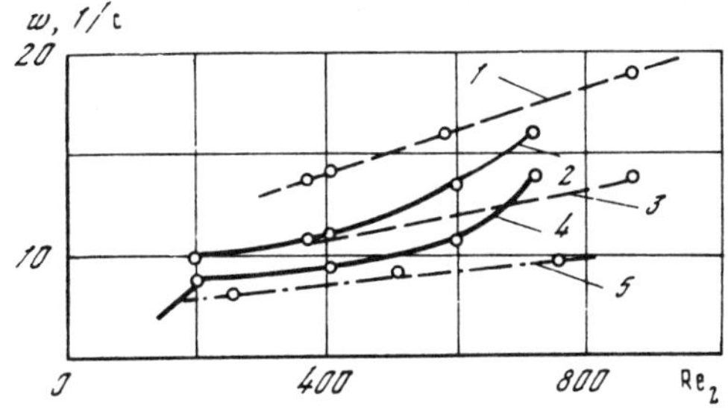

Figure 2.35 Frequency of large waves. 1, $Re_2 = 35 \cdot 10^3$; 2, $33 \cdot 10^3$; 3, $25 \cdot 10^3$; 4, $24 \cdot 10^3$; 5, $92.5 \cdot 10^3$; the solid curve represents data of the last of the present authors with his coworkers [2.101]; the dash-dotted line represents the data of Chu and Dukler [2.95]; the dashed line, of Maksimov et al. [2.94].

results of measurements of the frequency of large waves according to data of Chu with Dukler [2.95] and Maksimov et al. [2.94].

2.5.3 Friction at the Phase Interface

Let us consider the friction at the film-gas interface for the case of stabilized flow of both phases and absence of droplet entrainment from the film surface. The film flow is assumed to be laminar. For this case we can use an equation of motion in the form

$$dp/dx = \mu_2 \cdot dw_2^2/dy^2 \tag{2.102}$$

with the boundary conditions

$$y = 0, \quad w_2 = 0, \quad y = \bar{\delta}, \quad dw_2/dy = \tau_{12}/\mu_2 \tag{2.103}$$

The velocity profile in the liquid film at the specified boundary conditions is:

$$w_2 = \frac{1}{\mu_2} \frac{dp}{dx} \frac{y^2}{2} - \frac{1}{\mu_2} \frac{dp}{dx} \bar{\delta} y + \frac{1}{\mu_2} \tau_{12} y \tag{2.104}$$

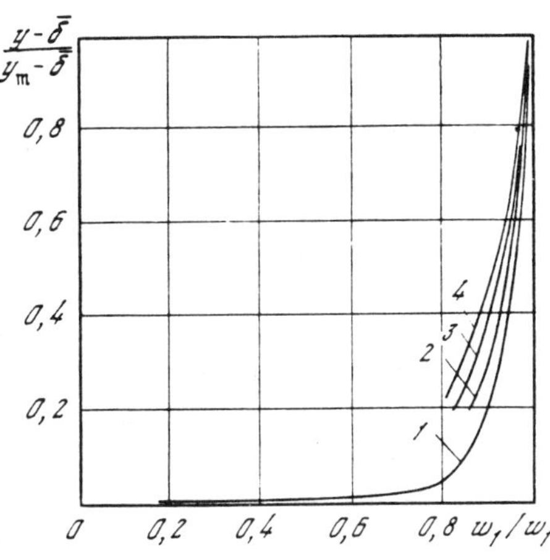

Figure 2.36 Effect of Re_2 on the air flow velocity ($w_1 = 10.8$ m/sec). 1, $Re_2 = 0$; 2, 400; 3, 600; 4, 800.

The mass flowrate of liquid through unit film cross section is found by integrating over its thickness from 0 to $\bar{\delta}$, i.e.,

$$j_2 = \frac{1}{\bar{\delta}} \int_0^{\delta} \rho_2 w_2 \, dy = \frac{\rho_2}{w_2} \frac{\tau_{12}}{2} \bar{\delta} - \frac{1}{3} \frac{\rho_2}{w_e} \frac{dp}{dx} \bar{\delta}^2 \qquad (2.105)$$

Equation (2.86) can be written as

$$\tau_{12} = 2 \frac{\mu_2^2}{\rho_2 \bar{\delta}^2} \, \mathrm{Re}_2 + \frac{2}{3} \frac{dp}{dx} \bar{\delta} \qquad (2.106)$$

The pressure drop dp/dx in a channel with a flowing film over its entire perimeter is expressed in terms of the tangential stress at the phase interface τ_{12} thus

$$dp/dx = \tau_{12} \Pi / F. \qquad (2.107)$$

Substitution of Eq. (2.107) into (2.106) yields

$$\tau_{12} = \left(2 \frac{\mu_2^2}{\rho_2 \bar{\delta}^2} \, \mathrm{Re}_2 \right) \Big/ 1 - \tfrac{2}{3} \frac{\pi}{F} \bar{\delta} \qquad (2.108)$$

Equation (2.108) makes it possible to calculate the shear stress τ_{12}, arising on the surface of a smooth film, moving in cocurrent gas flow.

For horizontal stratified flow in rectangular channels (the film moves along the bottom wall) if one assumes equality of shear stresses on the upper and side walls and their constancy, for a given Re, in the presence as well as in the absence of the liquid film (Re_1 = idem); the interphase shear stress can be represented in the form

$$\tau_{12} = \frac{2\mu_2^2/\rho_2\bar{\delta}^2 \cdot \mathrm{Re}_2 + \tfrac{2}{3}\,(bh/\Pi)\,\bar{\delta}\,(dp/dx)_0\,(\tfrac{1}{6} - 1\,(h - \bar{\delta}))}{1 - \tfrac{2}{3} \cdot (\bar{\delta}/h - \bar{\delta}))} \qquad (2.109)$$

where $(dp/dx)_0$ is the pressure drop in the channel in the absence of film at the given value of Re_1.

Equation (2.109) is used for comparing experimental and analytic values of the friction drag coefficients at the phase interface.

In the course of our experimental studies of friction at the phase interface we measured the velocity profiles in the gas flow above a dry plate and over wavy surface of the film. The measurements in our Institute were performed using a DISA hot-wire anemometer with 5 μm diameter wire. Figure 2.36 shows in coordinates $w_1/w_{1\mathrm{max}} = f[(y - \bar{\delta})/(y_m - \bar{\delta})]$ the velocity profiles at different Re_2. It is seen that the velocity profile of the gas flow decreases steep with increasing Re_2, as in the case of flow of gas above a rough surface.

Figure 2.37 presents experimental curves of $\Delta p_{\mathrm{t.ph}}/\Delta p_0$ vs. Re_2 at different air flow velocities. This ratio was determined by pertinent corrections of the measured

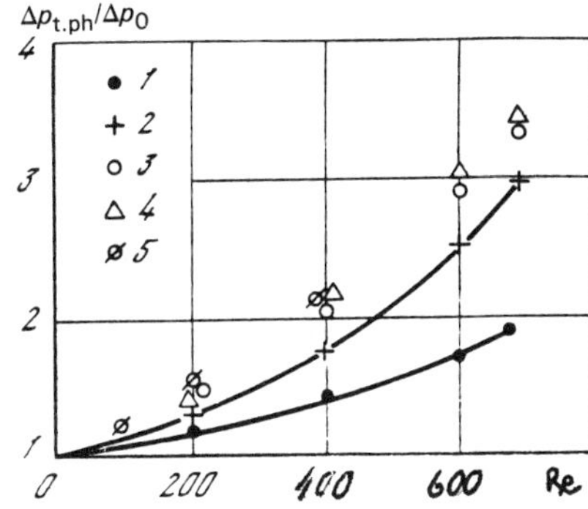

Figure 2.37 Relative rise in friction drag of a plate wetted by a liquid film. 1, $w_1 = 7.2$ m/sec; 2, 10.8; 3, 14; 4, 18; 5, 21.5.

pressure drops Δp and Δp_0 on the assumption of equality of tangential stresses on all the walls of the dry channel. It is seen from the plots of $\Delta p_{\text{t.ph}}/\Delta p_0 = f(\text{Re}_2)$ that increasing Re_2 at a given air flow velocity significantly increases the hydraulic drag, referred to the wetted channel surface as compared with a dry (smooth-walled) channel; here the hydraulic drag increases at a higher rate with the air velocity.

The last of the present authors with his coworkers [2.101] measured the pressure drops in a dry channel and in the same channel with a liquid film on the wall (at a given Re_1) and used these data to determine the frictional drag coefficient at the phase interface C_{fi}, using the familiar expression

$$C_{fi} = \frac{2\tau_{1\circ}}{\rho_1 \, (\overline{w}_1 - c_{\text{ph}})^2} \tag{2.110}$$

The shear stress τ_{12} can be determined from the balance equation

$$\int_{-b/2}^{b/2} \tau_t \, dz + 2 \int_{\bar{\delta}}^{h} \tau_s \, dy + \tau_{12} b = \frac{\Delta p}{\Delta z} \, (h - \bar{\delta}) \, b \tag{2.111}$$

In the flow of air in a dry channel ($\text{Re}_2 = 0$)

$$2 \int_{-b/2}^{b/2} \tau_{t_0} \, dz + 2 \int_{0}^{h} \tau_{s_0} \, dy = \frac{\Delta p}{\Delta z} \, hb \tag{2.112}$$

If it is assumed, in accordance with the data of Davis [2.98], that $\tau_s = \tau_t = \bar{\tau} = \tau_{s_0} = \tau_{t_0} = \bar{\tau}$ ($\text{Re}_1 = $ idem), then in this case we can obtain from Eqs. (2.110)–(2.112) an expression for calculating C_{fi} from the measured Δp and Δp_0, using the experimental values of $\bar{\delta}$:

$$C_{ji} = 2 \left[\frac{\Delta p}{\Delta z} (h - \bar{\delta}) - \frac{\Delta p_0}{\Delta z} \frac{F}{\Pi} \left(\frac{b - 2(h - \bar{\delta})}{b} \right) \right] \Big/ [\rho_1 (\bar{w}_1 - c_{ph})^2].$$

$$(2.113)$$

Figure 2.38 is a plot of C_{fi} vs. Re_2, calculated from the analytically obtained Eq. (2.110), and the results of data reduction using Eq. (2.113). The friction coefficients obtained from Eq. (2.113) are in satisfactory agreement with the experimental data of Cohen and Hanratty [2.102] and Miya et al. [2.99] plotted in the same figure. At low air velocities (to $w_1 \cong 10$ m/sec) the curves of $C_{fi} = f(Re_2)$ are highly stratified. As shown by measurements the amplitude and frequency of the waves undergo significant changes. However, as the mean gas velocity is raised further to 22 m/sec, the interfacial friction does not change further. It was possible to correlate experimental C_{fi} data for w_1 from 10 to 22 m/sec and Re_2 from 100 to 800, by a single correlation in the form

$$C_{fi} = (5.5 + 0.026 \, Re_2) \cdot 10^{-3} \qquad (2.114)$$

As should have been expected, the values of C_{fi} calculated from Eq. (2.113) obtained from theoretical analysis of a smooth film, lie below the test data; the difference between the experimental data and analytic results increases with Re_2 because this corresponds to increasing contribution of the wave "roughness" of the film surface to the interfacial shear stress. Note that it is highly advantageous to correlate the available experimental data by equations such as (2.113) in the region of existence of three-dimensional and roll waves, since the existing methods of determining C_{fi} are based on empirical equations given in the books by Wallis [2.104] and Hewitt and Hall-Taylor [2.105], which relate C_{fi} to the local thicknesses of the liquid film which, in its turn, is unknown and depends on the flow mode.

The behavior of C_{fi} at the vapor-liquid film interface in dispersed-annular pipe flows was investigated by Moeck [2.103], Wallis [2.104], and Subbotin et al. [2.71]. Figure 2.39 [sic] represents the experimental data of [2.71], and also the Moeck equation (curve 1)

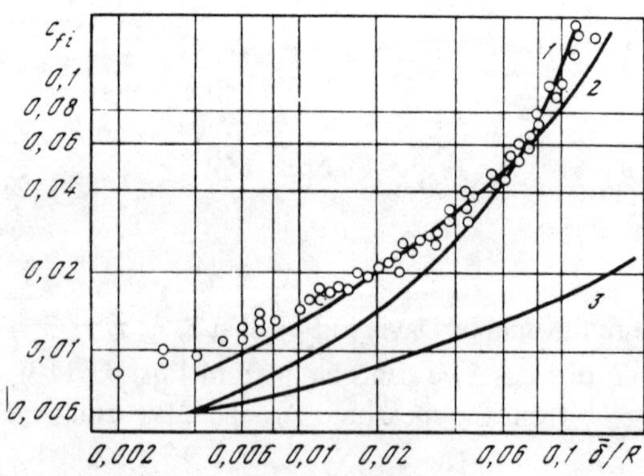

Figure 2.38 Interphase friction coefficient. 1 and 2, Calculated from Eq. (2.90) at $w_1 = 7.7$ and 14.5 m/sec, respectively; 3, data of Cohen and Hanratty [2.102]; the points correspond to data in [2.99, 2.101, 2.102].

$$C_{fi} = 0.005\left[1 + 545\left(\frac{\bar{\delta}}{R}\right)^{1,42}\right]$$ (2.115)

and the experimental correlation due to Wallis (curve 2)

$$C_{fi} = 0.005\left[1 + 150\left(\frac{\bar{\delta}}{R}\right)\right]$$ (2.116)

Comparison of the above experimental data with the equation for C_{fi} of tubes with sand roughness (curve 3) shows that the mechanism of friction between the gas core and the liquid film cannot be adequately represented by the surface roughness.

2.5.4 Entrainment of Liquid from the Film

Under adiabatic conditions droplets from the crests of disturbance waves at cocurrent flow velocities of $w \gtrsim 20$ m/sec are entrained by the gas flow. In spite of the fact that the entrainment mechanism was investigated in detail in Harwell and is described, among others, in the familiar book by Hewitt and Hall-Taylor [2.105], many problems remain insufficiently explored.

Let us consider dispersed-annular flow of wet steam in a long pipe. Experiments show that, irrespective of the film generation technique, equality of mass fluxes sets on between the film and the vapor-droplet core, i.e., the rate of liquid entrainment becomes equal to the rate of droplet deposition on the film. This state is known as hydrodynamic equilibrium. Data on this equilibrium, obtained for isothermal flow conditions, are useful for comparison with experimental data on entrainment in heated channels.

Figure 2.40 shows the experimental results obtained in Harwell by Bennet and Hewitt [2.110]. The dash-dotted curve corresponds to equilibrium entrainment under isothermal conditions, the solid line represents the variation in the total liquid flowrate through the tube as a function of the vapor quality. At point $x = 1$ the total liquid flowrate is equal to zero. When the experimental points are represented in this

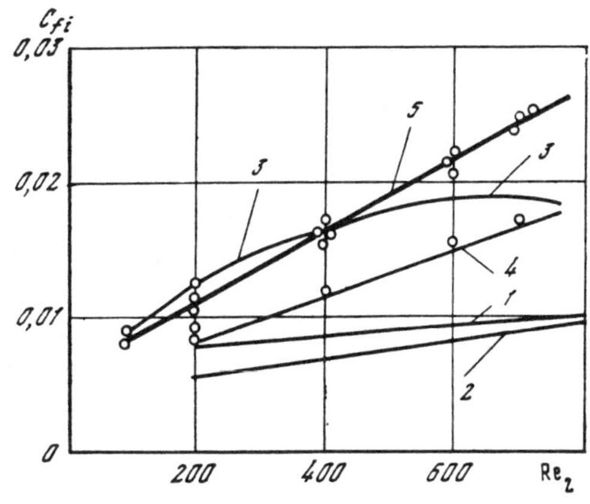

Figure 2.39 Friction coefficient C_{fi} at the vapor-liquid film interface. 1, The Moeck equation [2.103]; 2, the Wallis correlation [2.104]; 3, friction coefficient for tubes with sand roughness, $p = 3$–10 MPa; $\rho w = 500$–1500 kg/m^2·sec; 4, data of Subbotin with Sorokin [2.71]; 5, data of the last of the present authors with his coworkers [2.101].

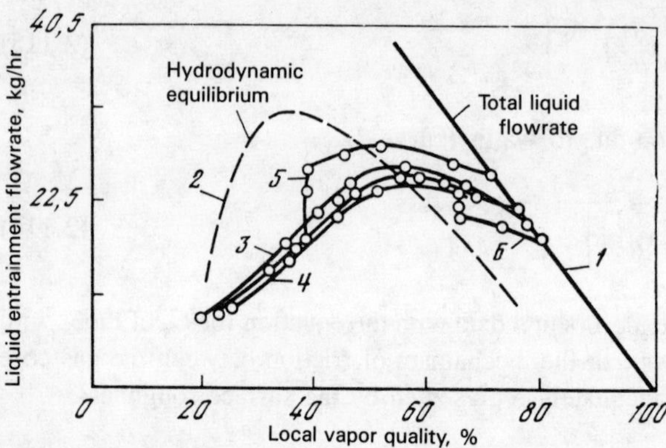

Figure 2.40 Entrainment curves in the presence of a cold section. 1, Total liquid flowrate; 2, hydrodynamic equilibrium; 3, uniform heating (heat flux 65.2 W/cm^2); 4, uniform heating (heat flux 40 W/cm^2); 5, cold patch (heat flux 61.7 W/cm^2); 6, cold patch (heat flux 69 W/cm^2).

manner the boiling crisis corresponds to the point of intersection of the curve of entrainment vs. x with the total flowrate line. The graph shows that the experimental data on entrainment for a heated tube differs significantly from the hydrodynamic equilibrium curve: at low local vapor qualities the experimental points for a heated tube lie much below the equilibrium curve, then as the quality increases, the entrainment under heating conditions exceeds the equilibrium entrainment. The shape of curves for entrainment with heating of the tube is the same as for isothermal conditions, however, the quantitative difference is rather significant.

Bennet and Hewitt [2.110] performed their entrainment experiments both with a uniformly heated tube, and with tubes some parts of which were not heated (cold patches). The entrainment curve along the cold patch (curve 5 in Fig. 2.40) exhibits a jumpwise rise in entrainment, since the quality remains constant along the nonheated tube. The subsequent behavior of the curve is analogous to the case of uniformly heated tube.

It is interesting to note that also when the cold patch is situated in the exit part of the tube (see Fig. 2.40, curve 6), the deposition on the nonheated section increases steeply (vertical drop in entrainment). Again, this is due to the system relaxing towards the equilibrium, the entrainment in this region being higher than equilibrium. The critical vapor quality in this case is found to be greater than for a uniformly heated tube. Consequently, for a given overall length, the total heat flux prior to crisis for a tube with a cold spot is greater than for a uniformly heating tube.

The main difference between entrainment conditions in heated channels and isothermal conditions is the change in the characteristics of the wavy surface of the film due to the heat flux. The entrainment increases in the initial length of the tube since the local quality and velocity of the cocurrent vapor flow rise there, which results in the appearance of large-amplitude waves and in rapid entrainment. A significant contribution in this part of the tube is also made by bubble entrainment. The entrainment flowrate at the exit part of the steam generating tube becomes lower than the deposition flowrate, since the reduction in film thickness brought about by evaporation and bubble entrainment results in vanishing of the waves.

Thus, it becomes obvious that theoretical description of entrainment must be based on knowledge of the wave structure on the film surface at given local conditions (vapor quality) and also statistical analysis of entrainment of droplets from a solitary wave.

Entrainment is usually described by the nondimensional entrainment rate E^*, equal to the ratio of the entrained liquid mass flowrate E to the total liquid flowrate $m(1 - x)$. Several techniques are currently used for reducing experimental data on entrainment. Paleyev with Filippovich [2.106] correlate the experimental points by means of the product $(\mu_2 w_1/\sigma_2)^2 \rho_0/\rho_2 \cdot 10^4$, where ρ_0 is the homogeneous density of the flow core.

Figure 2.41 presents the results of a detailed analysis of the experimental data of Paleyev with Filippovich [2.106] for an air-water mixture, of Nigmatullin et al. [2.71] in a high-pressure wet steam flow, and the data of Quandt [2.107] and the data from experiments performed in our Institute [2.100] for low-pressure wet-steam flows. There is little agreement between the data, most likely due to differences in experimental conditions. The experiments described by Paleyev and Filippovich [2.106] were performed in a narrow rectangular horizontal duct, in which the deposition and entrainment mass fluxes were not in equilibrium. In the experiments by Subbotin et al. [2.71], performed in small-diameter tubes, equilibrium apparently did prevail. In experiments performed in our Institute we measured "pure" entrainment

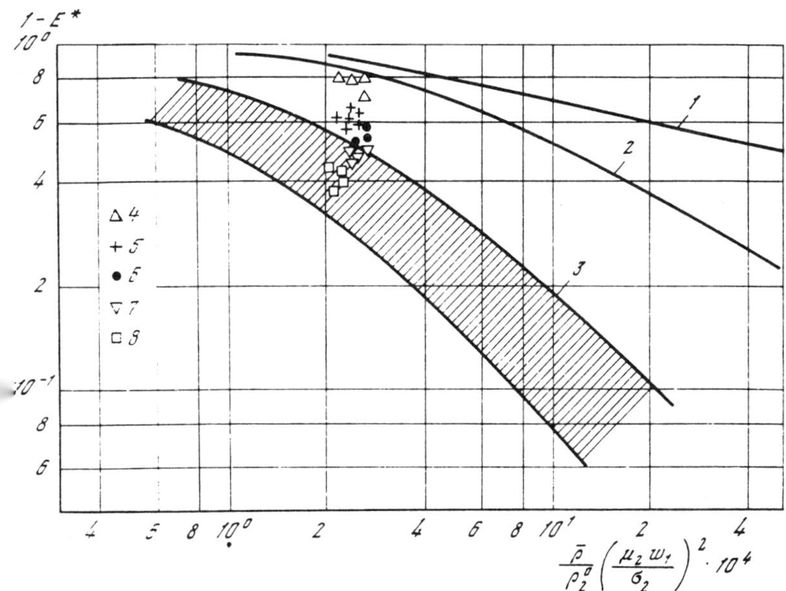

Figure 2.41 Entrainment of liquid in Paleyev-Filippovich coordinates. 1, Data of Quand E [2.107]; 2, of Paleyev and Filippovich [2.106]; 3, of Subbotin et al. [2.71]; 4, our Institute [2.100]; 5, $Re_2 = 209$; 6, 410; 7, 860; 8, 1010.

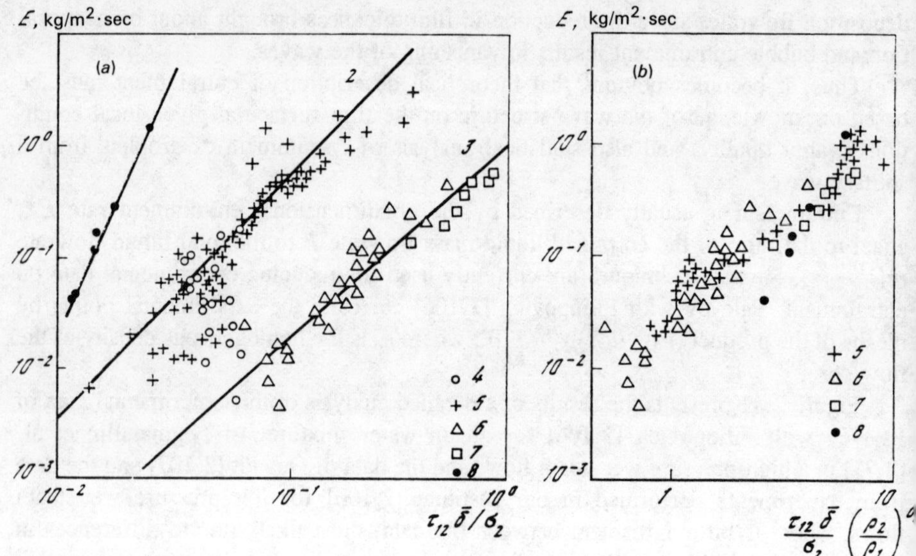

Figure 2.42 Entrainment E vs. $\tau_{12}\,\delta/\sigma_2$ (a) and $\tau_{12}\,\delta/\sigma_2(\rho_2/\rho_1)^{0.7}$ (b). 1, Data obtained in our Institute [2.100]; 2 and 3, data of Hewitt and Hall-Taylor [2.105], 4 and 5, air-water at 6.9 MPa; 6 and 7, steam-water at 6.9 MPa; 7, experimental results obtained in our Institute [2.100].

flux in a 60 mm high channel, when there was no deposition of moisture from the cocurrent flow. The experiments performed in our Institute point to very significant stratification with respect to Re_f, which cannot be attributed to the data reduction technique.

The effect of the film flowrate on the magnitude of the mechanical entrainment was taken into account by the technique developed in Harwell on the basis of extensive studies performed in tubes with gas-liquid flows. The experimental data are correlated by the ratio $(\tau_{12}\bar{\delta})/\sigma_2$. Hewitt with Hall-Taylor [2.105] suggest that friction at the phase interface be calculated from the Wallis formula (2.96).

The measured values of dropwise entrainment, processed according to the above technique are plotted in Fig. 2.42a. The experimental data exhibit a significant ascatter; however, similar variation in entrainment with the correlating group is exhibited by data of the different investigators at high and low pressures. It appears possible to unify all the available data on entrainment by multiplying $(\tau_{12}\bar{\delta})/\sigma_2$ by $(\rho_1/\rho_2)^{0.7}$. The results of this are plotted in Fig. 2.43b.

V. I. Rachkov [2.18] recommends the following expressions for the relative entrainment rate E^*:

$$E^* = 0{,}048\psi\,(N), \qquad N = 0{,}001\,(w_1 - v_1)^2\varphi\,(p/p_{cr}) \tag{2.117}$$

$$\psi\,(N) = N^{1/3}, \; N \leqslant 1 \quad \psi\,(N) = N^{1/2}, N \geqslant 1 \tag{2.118}$$

$$\varphi\,(p/p_{\text{cr}}) = 1 + 3.1\,(p - 5/p_{\text{кр}})(p/p_{\text{cr}})^{-0,75}, \quad p \leqslant 5 \ \text{MPa}$$
$$\varphi\,(p/p_{\text{cr}}) = 1 + 7.5\,(p - 5/p_{\text{кр}})^{1,36}(p/p_{\text{cr}}), \quad p > 5 \ \text{MPa} \tag{2.119}$$

These relationships describe experimental data of Rachkov over a pressure range of 0.96 to 11.77 MPa, steam velocities between 5 and 125 m/sec and initial flowrates of liquid in the film of 0.01–0.15 kg/sec to within 25%.

Recently Mayinger [2.108] performed an interesting experimental and analytic study of droplet entrainment. The working fluid in these experiments was Freon-12 at pressures between 0.8 and 1.2 MPa, specific flowrates from 300 to 1000 $kg/m^2 \cdot sec$ in a 5 m long vertical tube. These experiments showed that the principal parameters controlling entrainment are the local quality x and the specific flowrate j. Figure 2.43 shows results of experiments on entrainment for j = 500 and 1000 $kg/m^2 \cdot sec$ and pressures of 0.8, 1.28 and 1.4 MPa as a function of quality x. It is seen that the experimental points for the three pressures are very similar. Consequently, the relative entrainment E^* is shown in Fig. 2.43 as a function of x with j as a parameter.

In addition to experimental data, Fig. 2.43 shows correlating curves obtained from the equation

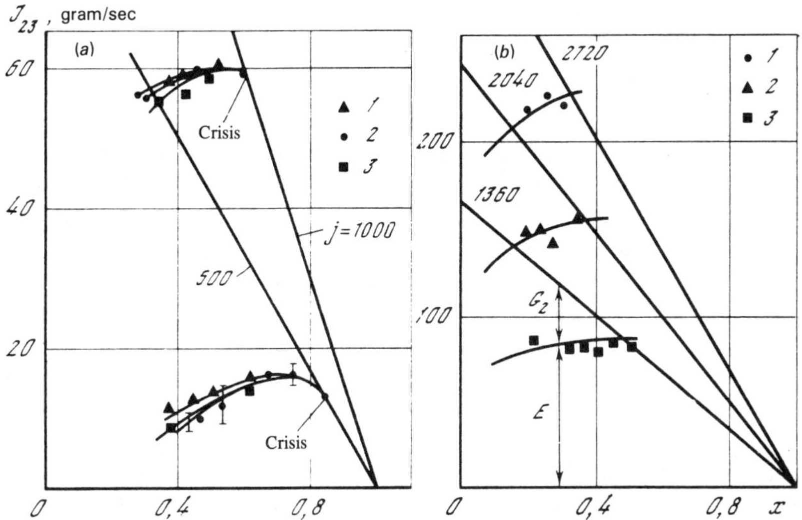

Figure 2.43 Comparison of calculations using Eq. (2.111) with experiments performed with Freon (*a*) and water (*b*). (*a*) 1, for p = 1.4 MPa; 2, 1.28; 3, 0.8 MPa; (*b*) 1, j = 2720 $kg/m^2 \cdot sec$, q = 133.8 W/cm^2; 2, j = 2040 $kg/m^2 \cdot sec$, q = 112.2 $kg/m^2 \cdot sec$; 3, j = 1360 $kg/m^2 \cdot sec$, q = 98.4 W/cm^2.

$$\frac{dE}{dz} = -\left\{ [2\tau_{12}/\mu_2 - dp/dz + x + E(1-x)/x + E(1-x)\rho_1/\rho_2 g\rho_1] \times \right.$$
$$\times \left[j^2 \left(\frac{r_0}{r_{1,2}}\right)^4 \left[\left(x + E(1-x)\right)\left(\frac{1-x}{\rho_2}\right) + \left(\frac{x}{\rho_1} - \frac{E(1-x)}{\rho_2}\right)(1-x) \right) \right]^{-1} +$$
$$\left. + dx/dz \, T_1/T_2 \right\}, \tag{2.120}$$

$$T_1 = [x + E(1-x)]\left[\frac{1}{\rho_1} - \frac{E}{\rho_2}\right] + \left(\frac{x}{\rho_1} + \frac{E(1-x)}{\rho_2}\right)(1-E);$$

$$T_2 = [x + E(1-x)]\left[\frac{1-x}{\rho_2}\right] + \left(\frac{x}{\rho_1} + \frac{E(1-x)}{\rho_2}\right)(1-x).$$

Equation (2.120) was derived from the laws of conservation of mass and momentum for the film and the vapor-droplet core using a one-dimensional approximation. The velocity of droplets in the flow core was assumed to be equal to the velocity of the steam; it was assumed that the droplets are uniformly distributed in the gas core. Mayinger determined the interphase stress σ_{12} and dp/dz from the Levy theory. The differential equation (2.111), together with equations for the shear stress and the balance equation for the vapor quality vs. heating power was solved numerically by the Runge-Kutta method. The results of calculations for freon at p = 0.8–1.4 MPa are shown by curves in Fig. 2.43a, whereas the data on Fig. 2.43b are for wet steam at 7.0 MPa. Satisfactory agreement is observed in both cases.

2.5.5 Bubble Entrainment on a Heated Wall

It is known that the emergence of a bubble on the film surface is accompanied by additional ejection of liquid into the cocurrent flow. Hewitt and Hall-Taylor [2.105] measured the flowrate of the film in the end of the heated part of the tube by suction of film through a porous wall. They showed that the effect of nucleate boiling on the entrainment E of the liquid and on the critical heat flux is insignificant at high vapor qualities and mass velocity. Figure 2.45a shows the reduction in the liquid film flowrate at the end of the heated rod in an annular experimental channel with rising heating power. The calculated variation in the film flowrate, with correction for evaporation, and on the assumption that hydrodynamic entrainment on the heated plate remains precisely the same as under isothermal conditions, is shown in the graph by the dash-dotted curve. It is seen that at high cocurrent vapor flow velocity and high qualities (lower curve in Fig. 2.44a) bubble entrainment, according to data of Hewitt and others, is insignificant, since the film thins out due to evaporation, waves on its surface decay and, as a result, the flowrate of the entrained liquid is smaller than under isothermal conditions (dash-dotted curve lies below experimental points).

Flow with low vapor quality and low vapor flowrate is characterized by another type of curve (upper curve in Fig. 2.44a). Here the experimental points lie significantly below the dash-dotted curve and the magnitude of bubble entrainment is determined rather easily if we assume that hydrodynamic entrainment on a heated plate remains the same as under isothermal conditions.

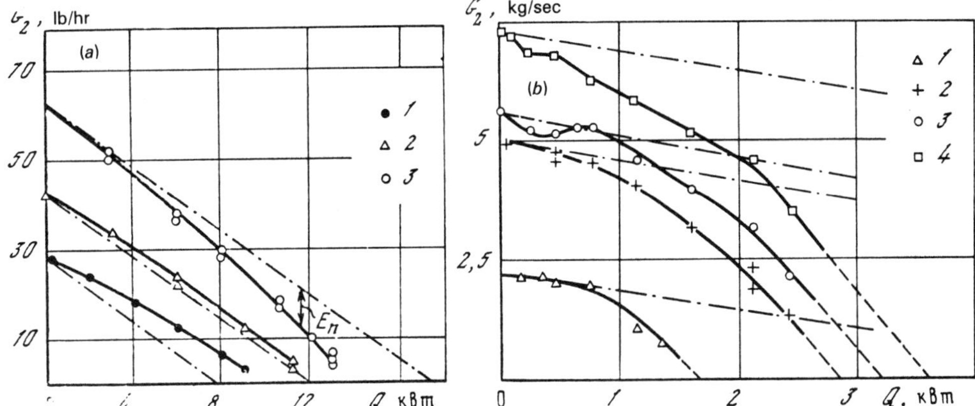

Figure 2.44 Film flowrate G_2 vs. the heating power Q. (*a*) Data of Hewitt with Hall-Taylor (Harwell) [2.105]: 1, x_{in} = 0.72; 2, 0.56; 3, 0.4; (*b*) data obtained in our Institute [2.100]: 1, experiments at Re_2 = 200; 2, 710; 3, 860; 4, 1010.

Experiments performed in our Institute [2.100] on a flat heated plate under conditions close to those in experiments by Hewitt with Hall-Taylor [2.105] showed that boiling in the film was observed virtually over the entire range of Reynolds numbers for the film (Re_2 = 80–2500) and for the vapor (Re_1 = 800–150,000).

The visually observed inception of boiling occurred at $qw \simeq 1.1 \cdot 10^5$ W/m^2. Figure 2.44 shows the variation in the film flowrate as a function of the heating power. It is seen that the analytic curve of evaporation and entrainment under isothermal conditions (dash-dotted curve) for all the conditions lies above the experimental points, i.e., the effect of bubble entrainment is perceptible.

We observed visually the ejection of large droplets, whose path and dimensions clearly differed from those which are torn off from the wave crests under isothermal conditions. At low cocurrent flow velocities the droplet penetrates virtually vertically the boundary layer above the film to a height of ~20 mm (some moving-picture frames show motion of droplets oppositely to the flow of the vapor). At high vapor velocities (to 60 m/sec) the droplet moves in the vapor-droplet layer near the film. The effect of bubble entrainment manifests itself the more perceptible, the higher the heat flux density. The critical heat flux, as in the book by Hewitt with Taylor [2.105], was determined by extrapolating the test curves to zero. As is known the film breaks at lower than critical values of q_w.

At high heat fluxes the onset of crisis with boiling of a thin film is preceded by formation of unsteady dry patches which are unstable and are rewetted by the liquid. A further rise in q_{cr} results in the formation of permanent dry patches, the film breaks up into rivulets and the wall temperature rises steeply. The last points in Fig. 2.44b correspond to the heat flux at which the film breaks.

Moving-picture photography and visual observations showed that the dry patches forming at close to critical heat fluxes are periodically rewetted liquid, which flashes explosively, ejecting a large number of droplets into the vapor, i.e., bubble entrainment increases. The increase in bubble entrainment is apparently one of the reasons for the significant difference between the calculated and experimental curves. The change in the wave structure on the film surface can be regarded as the second reason.

The last of the present authors with his coworkers [2.100] correlated the experimental points on bubble entrainment in the form

$$E_b = a \left(\frac{q_w - q_w^{i.b}}{r} \right)^m \frac{\rho_2}{\rho_1} \tag{2.121}$$

where a is a coefficient, equal to $(0.98–1.25) \cdot 10^3$, $q_w^{i.b}$ is the heat flux corresponding to inception of boiling in the film, and m is a power exponent, equal to 2.5.

Figure 2.45 shows the bubble entrainment E_b vs. $(q_w - q_w^{i.b})/r \cdot \rho_2/\rho_1$ according to data of our Institute [2.100] and of Petrovichev [2.109]. Note that the experimental points for different film Reynolds numbers cluster about a single curve, i.e., are controlled primarily by the heat flux and fluid variables.

Dolinin [2.117] measured bubble entrainment in a 13 mm diameter tube at 9.8 MPa. Figure 2.45b shows a plot of the rate of bubble entrainment E_b vs. the heat flux density, pressure and velocity of vapor in a tube, at a given film flowrate at the inlet to the heated section. The vapor velocity ranged from 3 to 18 m/sec. It is seen that the rate of bubble entrainment increases with the pressure and heat flux. This behavior is attributed by Dolinin to the drop in surface tension of the liquid in the film with increase in the pressure and reduction in the latent heat of vaporization. It is seen that the velocity of the vapor has no effect on the rate of bubble entrainment.

It is natural to assume that bubble entrainment is controlled by the number of active nucleation sites on the heated surface, which, according to data of Gaertner

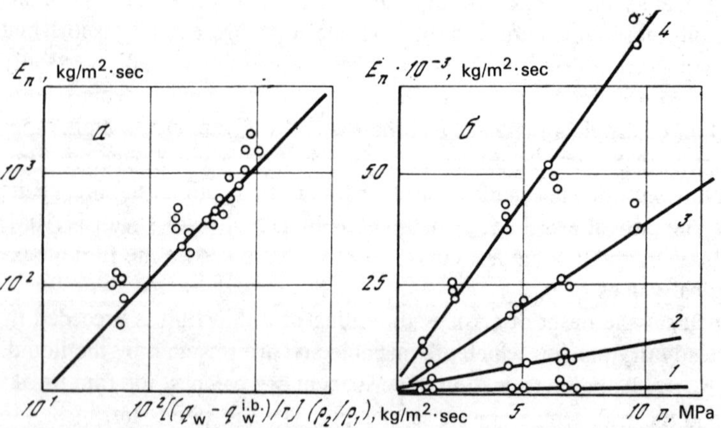

Figure 2.45 Bubble entrainment E_b vs. $(q_w - q_w^{i.b})/r)^m \cdot \rho/\rho_1$ (a) [sic] and pressure with heat flux as the parameter (b). (a) 1–6, data of our Institute at $Re_2 = 200–1700$; 7, data of Petrovichev et al. [2.109]; (b) data of Dolinin [2.117]: 1, $qw = 0.474$ MW/m^2 [sic]; 2, 1.77; 3, 1.222; 4, 2.42 MW/m^2; vapor velocity $w_1 = 2–24$ m/sec.

ith Westwater [2.118], is proportional to $j^{2.1}$. In addition, the ejection of droplets ith the surfacing bubble will be the greater, the thicker the liquid film. As a result of ata reduction, Nigmatulin and Dolinin suggested the following formula for bubble ntrainment:

$$E_b = 8 \cdot 10^{-5} \left(\frac{G_2}{\pi (\widetilde{D} - 2\overline{\delta}) \mu_2} \right) \left(\frac{\mu_1 w^{s.e}}{\sigma_2} \right) (q_w - q_w^{i.b})^{2.5} \tag{2.122}$$

here G_2 is the film flowrate; $w^{s.e}$ is the critical velocity of vapor, corresponding to e start of entrainment of liquid from the film surface in adiabatic flow, equal to $w^{s.e} = 9.4(0.71 - p/p_{cr})$.

It was noted both in experiments in AERE (Harwell) and in our Institute that the ilm flowrate curves tend to zero at a heat flux corresponding to the boiling crisis, as etermined by a steep rise in temperature at the end of the test section. Thus it has een demonstrated experimentally that the boiling crisis occurs at film thicknesses ending to zero. Due to hydrodynamic and bubble entrainment, and also due to evapo- ation, the flowrate in the film tends to zero at the critical heat flux position (it should e remembered that a large quantity of droplets are present in flow core at the crisis ocation).

Analogous conclusions were drawn by Bergles et al. [2.28], who performed their xperiments in 10 mm diameter tubes at pressures to 7 MPa. They measured the hickness of the film at the end of the test section as a function of the heating power, sing the contact-needle technique. A zero value of film thickness, obtained by ex- rapolating this curve, coincides, to a first approximation, with the boiling crisis.

Thus, it has been established experimentally that the boiling crisis in dispersed- nnular flow occurs due to a smooth approach of the flowrate in the liquid film to ero as a result of hydrodynamic and bubble entrainment.

FORCED–CONVECTION HEAT TRANSFER AND BOILING CRISIS

3.1 HEAT TRANSFER IN STEAM–GENERATING CHANNELS

Heat transfer in steam-generating channels is the subject of a very large number of experimental and analytic studies [3.1–3.3]. However, at present no qualitative theory of boiling heat transfer is yet available. The available correlations are based on approximate physical models and experimentally observed relationships. The results of calculations performed over a wide range of variables, differ significantly from one another. This is due to the complexity of the phenomenon as such, as well as to difficulties in investigating it. This becomes particularly obvious at high pressures, when wall-to-fluid temperature differences are small. The determination of the effect of a given factor (for example, mass velocity or quality) requires measurement of variations in temperature difference which are commensurable with the measuring error. It is hence not surprising that, up to present, there is no concensus between investigators on the effect of various parameters on the heat transfer coefficient.

Heat transfer occurs at the highest rates in surface and developed boiling. These boiling modes are, as a rule, the ones aimed for in designing and operating boiling-type equipment. Heat transfer under post-crisis (or post-dryout) conditions is encountered much less frequently. Due to the nature of the latter conditions, designers do not readily employ post-critical boiling in designing steam generators, and it is usually

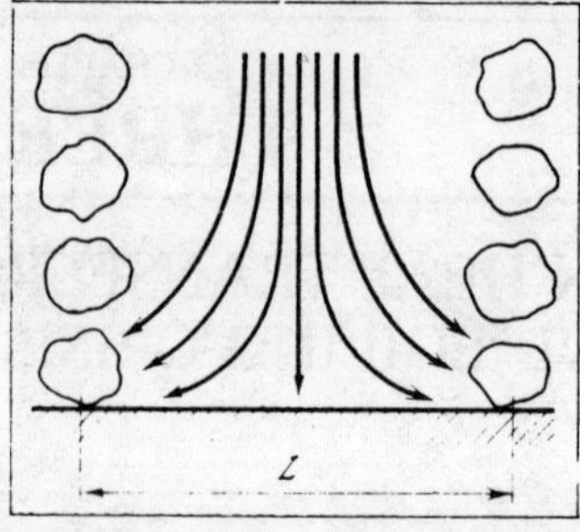

Figure 3.1 Schematic of nucleate boiling heat transfer.

found when, due to the considerations, it cannot be avoided. Post-crisis boiling is not considered in this section.

In boiling the heat transfer rate increases due to the appearance of microconvective flows in the vicinity of the wall. Microconvection, due to vapor bubbles, is then superimposed upon the principal macroconvection of liquid, which originates from its turbulence. The bubbles act as reciprocating pumps, displacing the liquid at the time of their growth and sucking it towards the wall after bubble breakoff (Fig. 3.1). The higher the heat flux, the higher the bubble departure frequency and the higher the bubble population per unit surface, and, hence, the greater the amount of heat removed by the liquid phase.

The microconvection model is based on the assumption that the bulk of heat is removed from the wall by the liquid phase. It correctly reflects the effect of factors such as the heat flux and pressure. However, new experimental findings, which became available recently, do not support the model. There is reduction in wall temperature at the time of bubble *growth* (not at time of departure) and the heat transfer coefficient increases with the void fraction.

In accordance with current concepts, the heat transfer coefficient is directly proportional to the quantity of heat carried from the surface by the generated vapor. The more the rate of vapor generation at the wall, the lower the effective temperature difference. This model explains both the previously known facts on the effect of p and q, and the effect of x. In fact, a rise in the vapor content of the flow is accompanied by a rise in void fraction at the wall. Also, in rather long channels the vapor quality at the wall is always greater than the cross-section average. This means that the rate of vapor generation rises with mean quality $\bar{x}$ (for more detail see Chapter 5) and the quantity of heat transmitted by the vapor increases. The coefficient of heat transfer (see below) also increases with the quality.

The region of high boiling heat transfer rate occurs over a quality range bounded by the quality for inception of surface boiling ($x_{\text{i.s.b}}$) and by the critical quality x_{cr}. Reliable methods are currently available for calculating $x_{\text{i.s.b}}$ and x_{cr}. Recommendations on x_{cr} were given previously. As to $x_{\text{i.s.b}}$, apparently the most extensively used formula in the USSR is that by Tarasova [3.4]

$$\Delta h_{\text{sub}} = 135 \frac{q^{1,1} D_{\text{э}}^{0,2}}{(\rho w)^{0,9}} \left(\frac{\rho''}{\rho'}\right)^{0,3} \tag{3.1}$$

where Δh_{sub}, the subcooling of water, is equal to $\Delta h_{sub} = h_{i.s.b} - h'$, and D_{eq} is the equivalent diameter. This expression describes satisfactorily experimental results over the range of variables employed in power engineering.

It is assumed by most of the investigators that the main parameters controlling heat transfer are the heat flux, pressure, mass velocity, and vapor quality of the flow. The determination of the effect of any parameter in forced convection boiling heat transfer requires solving a more general problem—that of the effect of induced motion and steam generation on the mechanism of boiling heat transfer. It was established by studies in the USSR and the West that there exist three heat transfer modes in two-phase flows. In the first region the rise in quality and speed of circulation have little effect on the heat transfer rate. The heat transfer coefficient is controlled by the heat flux and pressure of the boiling fluid. In the second region, the heat transfer coefficient is controlled primarily by the forced circulation and is virtually independent of the heat flux. These two heat transfer modes interface via a transition region, where both the heat flux and the circulation rate are controlling parameters.

Let us examine the variation in the heat transfer coefficient in steam-generating channels as a function of the respective parameters.

Effect of heat flux density. All the studies in which the velocity and vapor quality were not found to significantly affect the heat transfer rate show that the heat transfer coefficient is a function of the heat flux to a power of 0.6 to 0.7, i.e., $\alpha \sim q^{0.6-0.7}$. In the transition region, according to the results of most investigators, α behaves as a function of the heat flux in the same manner as in pool boiling. The degree to which the heat flux affects the heat transfer coefficient depends on the flow variables. However, the majority of formulas recommended for these conditions employ a constant power index of the heat flux, meaning that the index is independent of factors such as pressure and vapor quality.

The effect of pressure on the heat transfer coefficient is significantly more complicated. It is different according to the different authors. As a whole it can be claimed that the heat transfer coefficient is a direct function of the pressure. Usually $\alpha \sim p^{0.43}$ [3.3] or $\alpha \sim p^{0.14} + 1.83 \cdot 10^{-4} p^2$ [3.5].

Effect of flow velocity. It was noted above that there exists a range of variables over which a rise in velocity induces an increase in the heat transfer coefficient. Figure 3.2 shows the effect of vapor quality on the nondimensional heat transfer coefficient. The scaling factor in this figure is the pool boiling heat transfer coefficient α_p. In addition, it was found [3.6] that the circulation rate may also have a negative effect. The reduction in the heat transfer rate with rising circulation velocity is observed primarily at high heat fluxes. Reduction in heat flux or increase in quality, leads to the disappearance of this effect. At high circulation rates, the effect of the circulation rate itself becomes dominant. Here the values of α increasingly approach those characteristic of convective heat transfer without boiling. The effect of heat flux on the heat transfer rate weakens with rising circulation velocity. It follows from this that, under these conditions, the rate of heat transfer is affected primarily by flow turbulence. The velocity range over which the heat transfer rate is controlled primarily by vapor generation increases with the heat flux.

Effect of vapor quality. The greatest uncertainty in investigating boiling heat transfer is that concerning the effect of vapor quality. Some investigators assume that the quality (at least in the fully developed boiling region) does not affect the heat transfer coefficient. However, experimental data are available which do not support this assumption.

With rising flow enthalpy there arise, along the steam-generating channel several zones which have somewhat different heat transfer behaviors. These zones differ in both hydrodynamic and thermal aspects. A change in enthalpy changes the liquid content of the flow, which controls the diffusion of liquid to the wall. An increase in vapor quality results in a rise in the velocity of the vapor-liquid mixture, which in itself may cause a rise in the heat transfer coefficient. Some investigators assume that it is only the increase in velocity which is responsible for the rise in the heat transfer coefficient [3.7]. Andriyevskiy et al. [3.7] investigated separately the effect of vapor quality and velocity of the steam-water mixture on heat transfer. They found that α is independent of x at values of the latter between 0.1 and 0.9, i.e., over the range of variables under study the liquid content of the flow does not affect heat transfer conditions.

Figure 3.3 presents data obtained by Tarasova et al. [3.8] for water boiling in a tube; here, $D_{in} = 8$ mm, $p = 16.7$ MPa, ρw ranged from 1245 to 2220 kg/m$^2 \cdot$sec, q varied between 233 and 930 kW/m^2 and x varied from -0.5 to 1.0. The heat transfer coefficient in the single phase convective heat transfer region is constant independent of v. From the start of surface boiling the value of α increases steeply, and the curve of $\alpha = \alpha(x)$ becomes increasingly steep with increasing q. An inflection occurs at $x \simeq 0$. Then the value of α increases smoothly up to the point where $x = x_{cr}$. The rate of rise in α in this region is significantly smaller. The value of α is at maximum in the pre-critical cross section.

A variation in the heat transfer rate as a function of the vapor quality, similar to that described below, was also found to exist by Shvetsov [3.6], Andriyevskiy et al. [3.7], Tarasova et al. [3.8], Dzarasov [3.9], Collier [3.10], TsKTI instructions

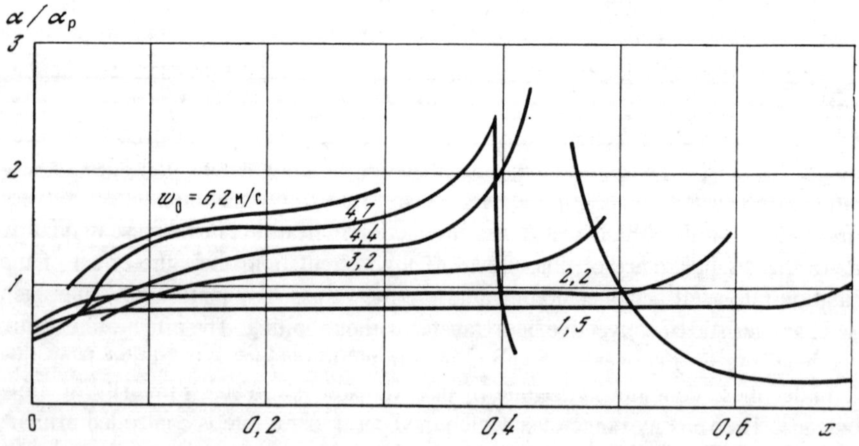

Figure 3.2 Forced-convection boiling heat transfer rate [3.6] for water at p between 6.9 and 7.8 MPa, $q = 696$ kW/m^2 and D_{eq} between 5 and 8.1 mm.

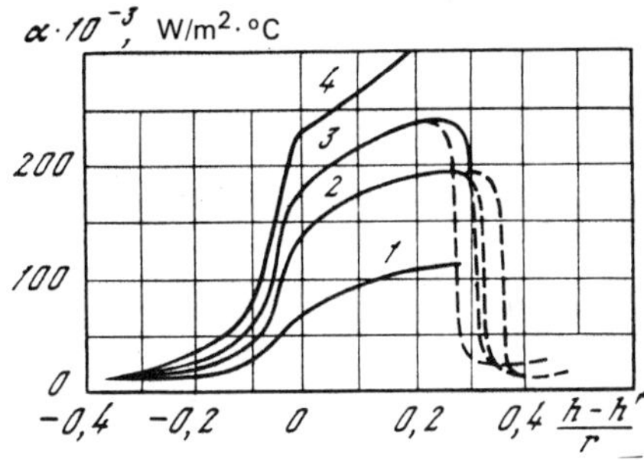

Figure 3.3 Heat transfer coefficient vs. vapor quality [3.8]. 1, q = kW/m²; 2, 522; 3, 697; 4, 929 kW/m².

[3.11], Przhiyalkovskiy and Petrova [3.12], Brantov [3.13], and Doroshchuk [3.14].

The above features of forced-convection boiling heat transfer are difficult to incorporate in mathematical models of the process. As a rule, the individual models describe various effects of the process, and do not represent the effect as a whole. During the past several years extensive use is made in the USSR of the TsKTI formula [3.11]

$$\alpha = \alpha^* \sqrt{1 + 7\cdot10^{-9} \left(\frac{w_{\text{mix}} r \rho' 3600}{q}\right)^{3/2} \left(\frac{0.7\alpha_p}{\alpha^*}\right)^2} \tag{3.2}$$

where $\alpha^* = [\alpha_t^2 + (0.7\,\alpha_p)^2]^{1/2}$, α_t is the coefficient of heat transfer in the flow of water in a tube, and α_p is the coefficient of heat transfer in pool boiling. This relationship incorporates most correctly the effect of the principal parameters of the process. It should be used at: $0.2 < p < 16.9$ MPa, $81 \le q \le 58\cdot10^2$ kW/m², $1.0 \le w_{\text{mix}} \le 300$ m/sec. At

$$\left(\frac{w_{\text{mix}} r \rho' 3600}{q}\right)\left(\frac{0.7\alpha_p}{\alpha^*}\right)^{4/3} < 5\cdot10^4$$

one should set $\alpha = \alpha^*$.

3.2 HEAT TRANSFER IN A FILM ON A STEAM–GENERATING SURFACE

Several forms of film are in principal possible in dispersed-annular flow, occurring in steam-generating channels of reactors or steam generators.

1. Thick films, which form after breakup of the slug flow mode. They form in transition flow between slug and dispersed-annular flow modes. The vapor released by the slug upon breakup of the liquid partitions, accelerates the film and produces a complex wave structure on it. Boiling occurs in such thick films, so that heat transfer through the film occurs by forced convection, evaporation from the surface and nucleate boiling. In this flow region there is no developed entrainment of droplets,

since the vapor released by the slug has a relative velocity lower than needed for entraining droplets from wave surfaces.

2. Dispersed annular flow with developed wave structure on the film surface. This flow mode is characterized by the formation of large disturbance waves (solitary roll and squall-type waves), whose height can be by an order of magnitude greater than the mean film thickness (a detailed description of waves characteristic of this flow region is given in Chapter 2). Nucleate boiling may also take place in the film.

3. At relatively low heat fluxes, characteristic, for instance, of the operation of boiling-water reactors [for example, the Duanne Arnold reactor (USA) where the mean heat flux density is $q_w \simeq 510{,}000$ W/m^2 (see Table 1.4)], the flow in long channels is dispersed-annular with a thin film, where the wave motion on the surface vanishes upon falling to some critical flowrate. There is no droplet entrainment in this zone, although the inverse process, i.e., deposition of droplets from the vapor core, apparently does take place. Transition from a dispersed-annular flow with a wavy film to one with a thin waveless film has an imperceptible effect on the thermal parameters: the wall temperature T_w and heat transfer coefficient α change monotonically in the transition zone.

When heat flux q_w at the wall increases, the zone of thermokinetic crisis (crisis of the second kind) is shifted upstream from the region of dispersed-annular flow with a thin film to a zone of annular flow with developed wave structure on the film surface, and this has, of course, a significant effect on heat transfer.

3.2.1 Visual and Moving-Picture Investigation

Let us consider the detail the mechanism of heat transfer in thick films with waves on the surface, characteristic of the first and second sections of dispersed-annular flow.

Visual observations and filming of wave generation on liquid films in heated horizontal plates, with lengths $L = 250$ and 450 mm, were performed at the Institute of High Temperatures of the USSR Academy of Sciences [3.15]; the plates were mounted in a rectangular channel with cocurrent vapor flow. Heat fluxes q_w ranged from 0 to the critical value of $q_{cr} \simeq 1.3 \cdot 10^6$ W/cm^2. Figure 3.4a shows moving-picture frames of the wave structure in the evaporating film, corresponding to constant cocurrent vapor flow velocity ($w_1 = 37$ m/sec, Re$_1 = 4.7 \cdot 10^4$ and Re$_2 = 130$); the moving-picture frames in Fig. 3.4b correspond to data for higher vapor-flow velocities (Re$_1 = 6.2 \cdot 10^4$) and higher film Reynolds number (Re$_2 = 449$), whereas the frames in Fig. 3.4c correspond to the same cocurrent flow velocity as in Fig. 3.4b, but for a higher film flowrate (Re$_2 = 1100$). The first frame in Fig. 3.4a corresponds to a film with $q_w = 0$; the subsequent frames show the effect of the heat flux. There are no waves on the film surface; there are only ripples (frames 1 and 2). At $q_w \leqslant 10^5$ W/m^2 nucleate boiling in the film is suppressed, the film is superheated. With increasing heat flux, as seen from Fig. 3.4a (frame 2) (Re$_2 = 130$ and $\bar{\delta}_2 = 0.6$ at $q_w = 2.1 \cdot 10^5$ W/m^2), semispherical bubbles appear with a very large diameter, which deflect the film, and break it when the bubble diameter d_1 reaches 4 to 5 mm, ejecting large droplets into the vapor flow. It is interesting to note that these macro-bubbles are unsteady, they originate and grow in different points on the plate; here the

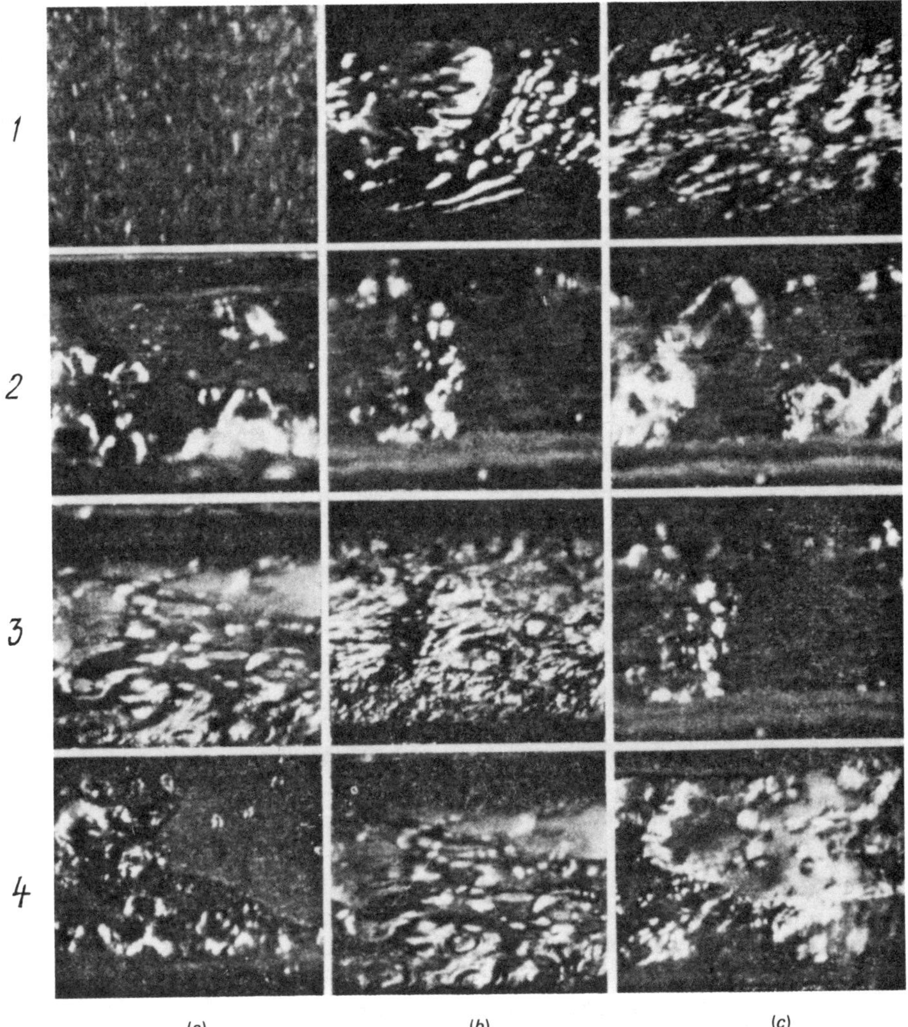

Figure 3.4 Moving-picture frames of boiling in a liquid film.

frequency of macrobubble generation is found to be smaller than the frequency of macrobubbles in pool boiling.

At high cocurrent vapor flow velocities ($Re_1 = 62{,}000$) the macrobubble (frame 3 in Fig. 3.4b and frame 4 in Fig. 3.4c) distorts at the leading stagnation point, and then the shell of the bubble breaks up in this region.

The presence of macrobubbles in thin films at moderate heat fluxes ($q_w \simeq 10^5$ W/m^2), at which microbubble generation due to roughness of the plate surface is impossible, points to the existence of another nucleate-boiling mechanism. Apparently the bubble operates as a vapor clot, periodically uncovering the wall. When the film flows onto the vapor clot, part of the vapor remains beneath the film and serves

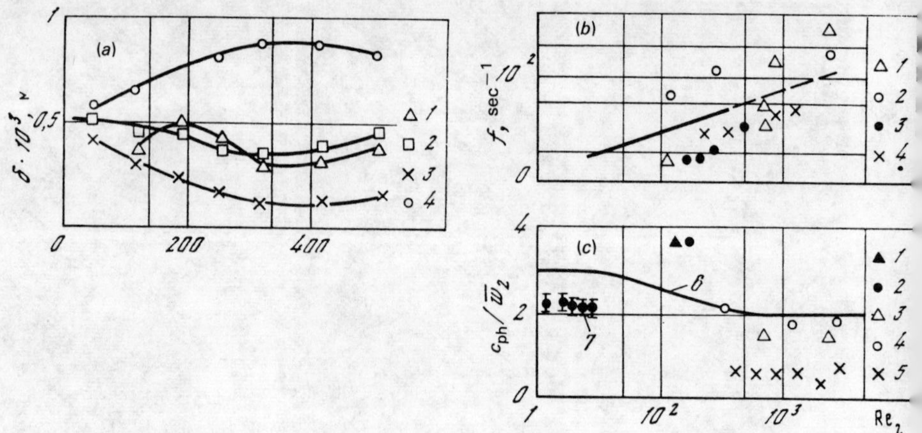

Figure 3.5 Data on film thicknesses (a), frequencies (b) and phase velocities of solitary waves (c). (a) At $Re_1 = 1.03 \cdot 10^5$: 1 and 2, δ—mean film thickness; 3, δ_{min} (minimum film thickness); 4, δ_{max} (maximum film thickness). (b) 1, $Re_1 = 0.62 \cdot 10^5$; 2, $Re_1 = 1.03 \cdot 10^5$, $q_w = 1.36 \cdot 10$ W/m²; 3, $Re_1 = 1.03 \cdot 10^5$; 4, Dukler [3.22]. (c) 1, $Re_1 = 0.62 \cdot 10^5$; 2, $Re_1 = 1.03 \cdot 10^5$, $q_w = 1.36 \cdot 10^5$ W/m²; 3, $Re_1 = 0.62 \cdot 10^5$; 4, $Re_1 = 1.03 \cdot 10^5$; 5, Dukler [3.22]; 6, Ishigai [2.89]; 7, Nakoryakov [3.29].

as a nucleation site for a new bubble. In this boiling mechanism, the diameter of the nucleation site is not controlled by the surface microroughness, but depends on the nature of the film flow and the heat flux.

A further rise in heat flux significantly reduces the film thickness $\bar{\delta} \ll d_1$ due to evaporation and bubble entrainment. It is seen from the cine frames (Fig. 3.4b, frame 3) that nucleation sites in the thin film operate as vapor channels (craters), periodically flooded with water. Droplets are entrained from the edges of such a crater. An increase in q_w is followed by formation of steady-state dry patches and zones (Fig. 3.4a, frame 3), and at $q_w = q_w^{d.o}$ (at $Re_2 = 130$, $q_w^{d.o} = 3.84 \cdot 10^5$ W/m²) the film breaks up into individual streams ($q_w^{d.o}$ is the heat flux at which wall dryout starts). Frame 4 shows the case of a crisis of the second kind for a laminar film without waves. The frame clearly shows droplets ejected by bubbles, however, there is no entrainment of liquid from the edge of the film, i.e., the film in the transition zone evaporates entirely.

The region of Re_2 between 300 and 1000, characterized by the appearance of large waves on the film surface are of great interest. Frame 1 (Fig. 3.4b) shows a large solitary wave at $Re_2 = 400$, $Re_1 = 62,000$ and $q_w = 0$. The leading front of the wave is steep, swell is seen upstream of the wave, whereas the film surface after the wave is smooth. As seen from Fig. 3.5a, the minimum film thickness for this case is 120 μm at a wave amplitude of 730 μm and frequency $f \simeq 40$ Hz, the ratio of phase velocity $c_{ph}/\bar{w} = 2.2$ (see Figs. 3.5b and c).

Boiling in the wave structure starts somewhat earlier than in a laminar film, at $q_w^n = 0.98 \cdot 10^5$ W/m², however, it occurs primarily within the wave, where the film thickness is at maximum. Frame 2 (see Fig. 3.4b) shows a boiling two-dimensional roll wave, whereas boiling between the waves is suppressed. At high heat fluxes and thin films dry patches appear between the waves (frame 3, see Fig. 3.4b). The wave frequency f, as its phase velocity, increases with q_w (see Fig. 3.5b). Under these flow conditions the bulk of the liquid drops are torn off the wave crests due to mechanical and bubble entrainment mechanisms.

At $Re_2 \geqslant 1000$ the two-dimensional wave structure breaks up and three-dimensional waves form (Fig. 3.4c). The process of wave generation and boiling in three-dimensional waves are shown in frames 1, 3 and 4, with the boiling crisis represented by frame 2 in Fig. 3.4c.

3.2.2 Suppression and Inception of Boiling in the Film

Analysis of experimental data on forced-convection heat transfer shows that, in order for boiling to start, the liquid in wall layer of thickness δ should be superheated to some value over the saturation temperature.

An analogous situation is observed also in thin liquid films. Thus, in experiments performed at the Institute of High Temperatures of the USSR Academy of Sciences [3.15] showed that for $p = 0.06$ MPa, boiling in the film starts with some value of q_w^n, ensuring the required superheat in the liquid for the formation of a vapor nucleus and for its growth.

Labuntsov [3.16] suggested a formula for the work of formation of a vapor embryo of arbitrary shape on a surface with any roughness

$$L_*^w = v_* \left[p'' - p' \right] + \sigma F_* \left[1 - \frac{F_w}{F_*} (1 - \cos \theta) \right] \tag{3.3}$$

where V_* is the volume of the vapor embryo, F_* is the latter's surface area, p' is the pressure in the superheated liquid, p'' is the pressure within the bubble, and F_w is the vapor-wall interface area.

Equation (3.3) shows that the value of L_*^w, all other conditions remaining constant, decreases upon transition from projections to crevices. Bankoff [3.17] also investigated this process in detail and arrived at the conclusion that the probability of formation of embryos in a crevice is greater than on a flat surface. Hsu [3.18] analyzed the formation of semispherical embryos in tapered crevices (Fig. 3.6) and concluded that a vapor-bubble embryo, once formed, may reside on the wall for a rather long time, unless the required superheat occurs in a liquid layer with characteristic dimension δ about the bubble; this superheat needs be

$$\Delta T_1 = T_1 - T_c = \frac{2\sigma T_{sat}}{c_2 r' r \rho''} \tag{3.4}$$

where r' is the radius of the active crevice, equal to the radius of the embryo. In Hsu's opinion, evaporation into the vaporization embryo occurs from superheated layer of liquid of thickness δ, in which the temperature changes linearly. Hsu's analysis was performed for pool boiling. It can be assumed that the characteristic superheated layer of liquid in the case of forced convection is equal to the thickness δ_{lam} of the laminar sublayer, in which the laminar sublayer, in which

$$\theta = \frac{T_2 - T_\infty}{T_w - T_\infty} = 1 - \frac{y}{\delta_{lam}} \tag{3.5}$$

Hewitt [3.19] suggests that the thickness of the superheated layer for a film be defined by the expression $\delta^+ = 7\mu_2/[(\tau_{12}/\rho_2)\rho_2]^{1/2}$.

Let us consider which embryos of radius r' in laminar sublayer δ_{lam} will grow, and which will vanish.

The dashed curve in Fig. 3.6 shows the variation in temperature in a bubble obtained from Eq. (3.4) and the linear variation in the liquid temperature, given by Eq. (3.5). It is seen from the figure that only embryos with $T_2 > T_1$ are capable of growth, and this occurs at $r_{min} < r' < r_{max}$; for $r' > r_{max}$ and $r' < r_{min}$ we have $T_2 < T_1$, i.e., boiling is suppressed. The embryo radius for the superheat curve of the liquid, tangent to the temperature curve for vapor $T_1(y)$, is equal to its critical value at $T_2 = T_1$. The latter circumstance makes it possible to determine analytically the critical embryo size by simultaneous solution of equations in the form [3.18]

$$r_{cr} = \frac{\delta(T_w - T_{sat})}{2C_1(T_w - T_\infty)}\left[1 \pm \left(1 - \frac{8C_1(T_w - T_\infty)T_{sat}\sigma}{C_2(T_w - T_{sat})^2 \delta\rho_1 r}\right)^{0,5}\right] \tag{3.6}$$

where C_1 and C_2 are constants, which are a function of the contact angle. For example, for $\theta = 63°$, $C_1 = 2$ and $C_2 = 1.25$ T_∞ for the film case is equal to T_{sat}.

Analysis of Eq. (3.6) shows that the wall superheat as a function of radius r_{cr} has a maximum which is obtained by differentiating

$$\frac{d\Delta T_w}{dr_{cr}} = \frac{C_1\delta^*\left[(T_w - T_\infty) + 2\sigma T_{sat}/C_2 r_{cr}r\rho_1\right] - (\delta^* - C_1 r_{cr})\delta^*}{(\delta^* - C_1 r_{cr})^2 (2\sigma T_{sat}/C_2 r\rho_1 r_{cr}^2)} = 0 \tag{3.7}$$

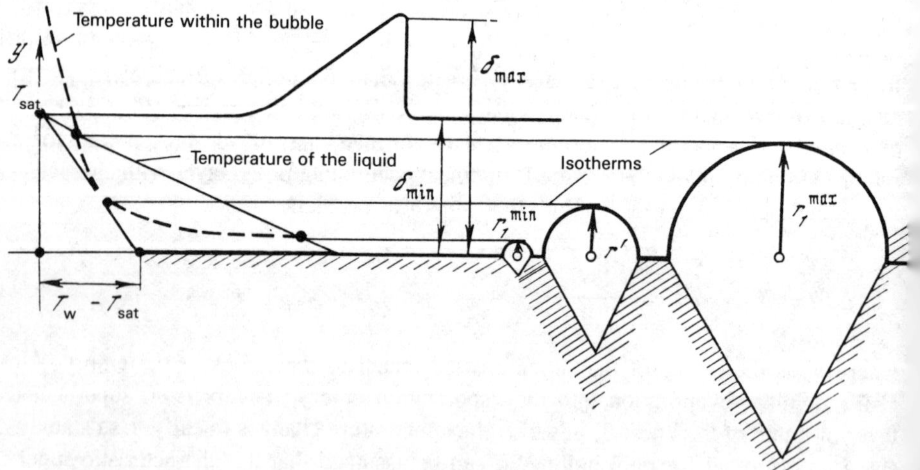

Figure 3.6 Variation of temperature in the boundary layer and within the bubble.

Hsu suggests that this value of the critical embryo be taken as a criterion for the inception of boiling.

In experiments performed at the Institute of High Temperatures [3.15] boiling inception was investigated at low pressures (p between 0.06 and 0.1 MPa), but at variable film Reynolds numbers. It was found that heat flux $q_w^{i.b}$ corresponding to onset of surface boiling in films with a developed viscous structure was almost independent of Re_2 and of the velocity of the cocurrent vapor flow. Boiling started within the large waves at $q_w^{i.b.} \simeq 1.1–1.3) \cdot 10^5$ W/m^2.

Bergles and Rohsenow [3.20] investigated inception of boiling in pipes under forced convection conditions for pressures from 0.1 to 14 MPa. They suggested that, in turbulent flow, the characteristic layer of superheated liquid at the wall, proportional to the thickness of the turbulent boundary layer, is smaller than in the case of pool boiling. They suggested that $q_w^{i.b}$ be expressed by the empirical equation

$$q_w^{i.b} = 443,7\,(T_{cr} - T_{sat})^{2,3/p^{0,234}}\,p^{1,156} \tag{3.8}$$

Davis and Anderson [3.21] analytically obtained expressions for inception of boiling in the form

$$q_w^{i.b} = \frac{\lambda_2}{4b}\,(\Delta T_w)^2$$

where

$$b = \left[\frac{2\sigma T_{sat}(v_1 - v_2)}{r} \right] \tag{3.8a}$$

Brantov [3.13] showed, as a result of an experimental study of the point of boiling inception for a range mass fluxes, that boiling under conditions of forced convection in tubes is suppressed in direct proportion to the mass velocity and subcooling of liquid. At $1 < p < 18$ MPa and mass velocities $500 < \rho w < 2000$ he suggested a formula which describes the experimental points to within 8.2 percent:

$$q_w^{i.b} = -0,0296 \cdot 10^6 \frac{(\rho w)^{0,812} x}{0,612 + 1,252 p/p_{cr}} \tag{3.9}$$

Unlike Eq. (3.8), Eqs. (3.8a) and (3.9) yield lower values of the incipient-boiling heat transfer with rising pressure, which appears to be physically more correct. Equation (3.9) has the additional advantage of being suitable over wide limits and yields satisfactory accuracy.

Hewitt [3.19] also measured experimentally the heat fluxes $q_w^{i.b}$ in dispersed-annular flow on a stainless-steel tube. Hewitt points to the existence of significant scatter of experimental data; most of the values for $q_w^{i.b}$ were found to be greater than from the Davis-Anderson equation. In Hewitt's opinion, this happened due to differences in the quality of surfaces (roughness, surface finish, etc.), which significantly affects the size of the active embryo.

An interesting fact was observed in experiments at the Institute of High Tempera-

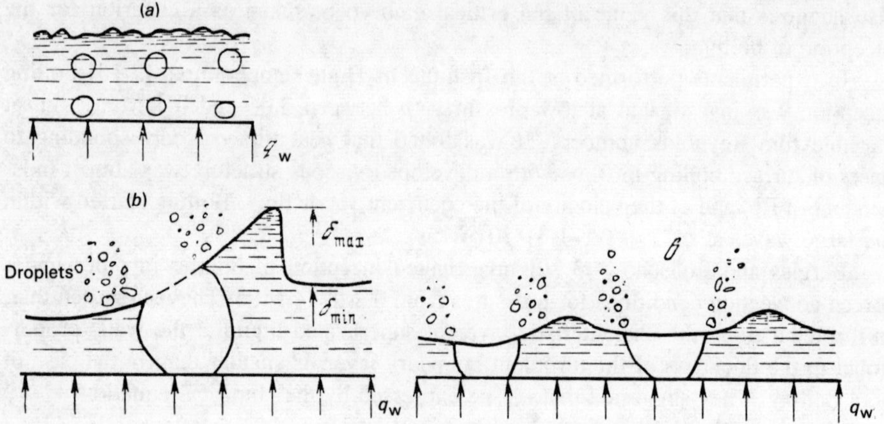

Figure 3.7 Examples of various boiling types. (*a*) Laminar thick film, $w_1 < w_1^d$, $\delta > d_{10}$; (*b*) wave motion $\delta \geq d_{10}$; (*c*) formation of "dry" patches $\delta \geq d_{10}$.

tures. In the suppressed-boiling region at $q_w < q_w^{i.b}$ the structure of waves on the film surface changes, the frequency and phase velocity of the waves increases. In the case of smooth laminar films ripples appear on the surface. This fact has not yet been sufficiently explored; however, it can be assumed that bubbles formed on devices with $r_1 < r_{cr}$ perturb the film and cause the effect.

3.2.3 Heat Transfer in the Film
in the Presence of Nucleate Boiling

Hewitt and Hall-Taylor [3.19] and Dukler [3.22] showed that the controlling factor in the transfer of heat in dispersed-annular flow is the thermal resistance of the film. Visual observations performed at the Institute of High Temperatures show that in thin laminar films with $\text{Re}_1 < 50$ and smooth surfaces without large waves and droplet entrainment at $d_1 < \delta_{\min}$ the bubble rises vertically in the film, as in pool boiling (Fig. 3.7*a*), without perceptible downstream drift and then leaves the film, carrying a part of the liquid with it.

With increasing liquid flowrate in the film and increasing velocity of the cocurrent vapor flow, and under conditions of laminar-wavy ($50 < \text{Re}_2 < 400$) and turbulent-wave film flow ($\text{Re}_2 > 400$) and at constant vapor velocity ($\text{Re}_1 = (15–150) \cdot 10^4$), the minimum film thickness decreases steeply as was shown in the preceding section, and, under certain conditions, particularly at low pressures, the departure diameter of the bubble is $d_{10} > \delta_{\min}$. The experimental data of Nishikawa et al. [3.23] for $p = 0.06$ MPa show that the departure diameter d_{10} lies within the limits of 1.8 to 3 mm, i.e., the breakoff diameter of the bubble is greater than the maximum film thickness $\delta_{\max}$. As shown by visual observation and measurement of liquid flowrate in the film, bubble entrainment of droplets into the vapor flow increases steeply, as if the film is "sprayed apart by the bubble" (see Fig. 3.7*b*).

A further rise in heat flux in the range $q_w^{i.b} < q_w \gtrsim q_w^{cr}$ and the corresponding reduction in the liquid flowrate in the film bring about a further thinning out of the liquid film and δ_{min} decreases. At $d_{10} > \delta_{min}$ the bubble acts as a steady-state vapor channel ("crater"), with droplets breaking off from its edges (see Fig. 3.7c) and evaporation occurs at a high rate.

Some investigators report thinning of the film between successive large waves due to the operation of surface tension, which stretches the film from the center to the wave crests due to the Marangoni force. Note that the film-vapor interface in single-component systems is virtually at constant temperature, equal to T_{sat}, and raising the temperature of the liquid in the troughs between the waves do not induce a change in the liquid temperature at the interface and the interface temperature $T_{12} = T_{sat}$; the Marangoni force does not operate under these conditions.

Another very important effect was detected in the experiments of the Institute of High Temperatures. When a wave passes over an operating vapor channel, the liquid covers the growing bubble (see Fig. 3.7b). Under these conditions the bubble beneath the wave tends to grow to the value of d_{10}. According to Treshchev [3.24], the bubble growth period in surface boiling under forced convection conditions is 3 to 5 m/sec; according to data of Hosler [2.17] for approximately the same conditions, the bubble lifetime lies between the limits of 10 and 20 m/sec. The bubble lifetime (t_1) is therefore of the order of 10 m/sec. A large wave, moving at a phase velocity between 1 and 1.5 m/sec (see data in the preceding section) traverses during time t_1 a distance of $\sim 10^{-2}$ m. Measurements of the wave structure from the cine frames showed that the mean width of the front of a large solitary wave ranges from 3 to 5 mm. This means that a bubble grows to $\sim d_{10}/2$ over the time of passage of a large wave over it, and then breaks up in the trailing part of the wave. This is responsible for the high-rate of nucleate boiling immediately after the wave front, as observed in the course of the cine study.

Upon further increase in the heat flux to q_w^b, corresponding to breakup of the film, the number of steady-state vapor channels increases and dry patches and zones appear.

Let us analyze in more detail experimental data on local values of the superheat $\Delta T_w = (T_w - T_{sat})$ and heat transfer coefficient α in the boiling film. Figure 3.8 shows the variation in ΔT_w and Fig. 3.9a the variation in α along a heated plate as found in experiments performed at the Institute of High Temperatures. These experiments were for laminar-wave flow at $Re_2 = 209$ and $Re_1 = 7.27 \times 10^4$ and for turbulent flow at $Re_2 = 1400$; the heat fluxes ranged between 3.63×10^4 and 3.14×10^5 W/m^2. The experiments were performed with saturated water (the subcooling of the water in the supply slot was minimal, ranging from 1 to 2 °C).

Boiling at the wall started at $\Delta T_w \simeq 16$ °C. Curves of $\alpha(r/R)$ and the wall superheat $\Delta T_w(z/L)$ under conditions without boiling are oscillatory. Under conditions of developed nucleate boiling in the film an increase in q_w also results in a rise in ΔT_w, but at a significantly lower rate.

Turbulence in the film enhances heat transfer for a given film thickness. Simultaneously with transfer of heat by turbulent eddies a change occurs in the nature of wave motion on the film surface, and entrainment increases. Figure 3.9b shows the variation in the local heat transfer coefficient α at $Re_2 = 1400$. It is seen that the

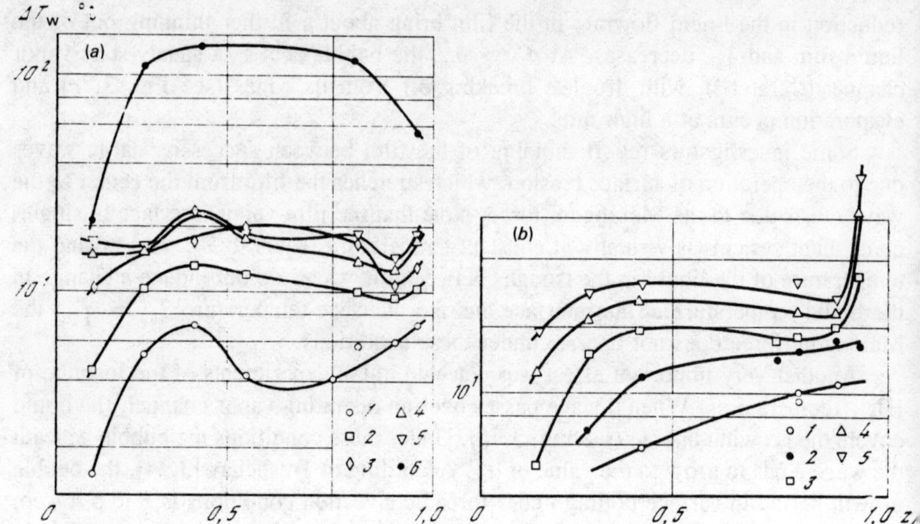

Figure 3.8 Variation in superheat ΔT_w along a plate. (*a*) For $Re_2 = 209$: 1, $q_w \cdot 10^5 = 0.36$; 2, 0.81; 3, 1.44; 4, 2.26; 5, 2.74; 6, 3.14 W/m^2. (*b*) For $Re_2 = 1400$: 1, $q_w \cdot 10^5 = 2.26$ W/m^2; 2, 3.27; 3, 4.43; 4, 5.69; 5, 7.29 W/m^2.

turbulent film is significantly more rupture resistant; film rupture in these experiments occurred only at the exit part of the plate. The wall temperature in the zone of stable dry patches is significantly greater in the turbulent than in the laminar film (this was deduced from the reddening of the patches at $T_w > 450\,°C$).

Figure 3.10 shows boiling curves for two plate cross sections, namely those at z/L of 0.306 and 0.78. At the inlet section ($z/L = 0.189$), where the wave motion and the hydrodynamic entrainment from the film surface are not stabilized, there is a significant effect of Re_2 on the boiling curves. This figure also shows clearly the transition from convective heat transfer to nucleate boiling. In section $z/L \geq 0.78$ the inflection of curves upon transition from convective heat transfer to nucleate boiling is less pronounced.

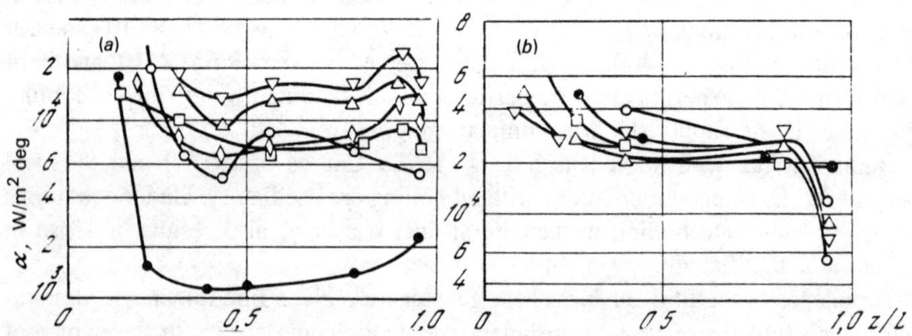

Figure 3.9 Variation in heat transfer coefficient α (*a*) for $Re_2 = 209$ and (*b*) for $Re_2 = 1400$ (for legend see Fig. 3.8).

Let us analyze qualitatively the transfer of heat in a boiling film. The overall heat flux to the wall in the region of forced convection in a film can be defined as the sum of forced convective heat transfer and transfer of heat by bubbles in the course of boiling:

$$q_w A = q^{ev} S_{12} + q_{conv} A_2$$

where s_{12} is the overall interphase vaporization surface; for spherical bubbles $s_{12} \simeq N\pi d_1^2$, the later equality is valid at $\delta_{min} \gtrsim d_{10}$. Then for spherical bubbles

$$q_w = q^{ev}\ \frac{N}{A}\ \pi d_1^2 + q_{cond}\ \frac{N}{A}\ \frac{\pi d_1^2}{4}\ (1 - \cos\theta)^2 \tag{3.11}$$

or

$$\frac{q_w}{q_{cond}} = n\ \frac{\pi d^2}{4}\left[4\ \frac{q^{ev}}{q_{cond}} + (1 - \cos\theta)^2\right] \tag{3.12}$$

where $n = N/A$ is the number of nucleation sites per m^2, and θ is the contact angle.

For vapor-generating channels

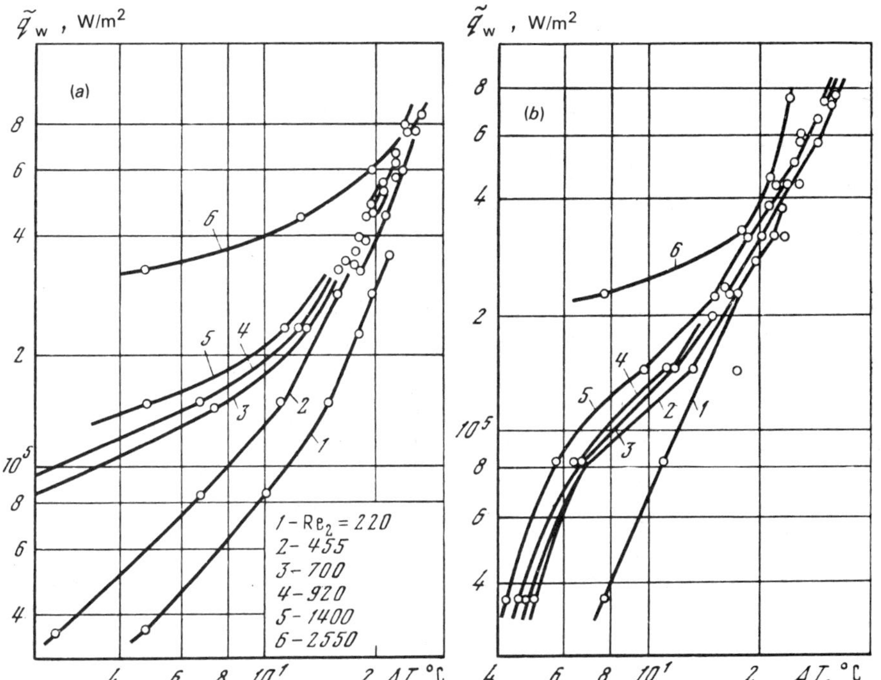

Figure 3.10 Boiling for two sections $z/L = 0.306$ (*a*) and 0.78 (*b*) [sic].

Table 3.1

Liquid	β	Liquid	β
Water	0.040	Methanol	0.045
Gasoline	0.9	Chloroform	0.16

$$\frac{q_w}{q_{cond}} = 1 + n\,\frac{\pi d_1^2}{2}\left[\frac{\angle\delta_{min}}{d_1}\,\frac{q^{ev}}{q_{cond}} - 1\right] \tag{3.13}$$

The heat flux corresponding to the total vaporization rate of a superheated liquid into a vacuum (i.e., with no condensation) is

$$q^{ev} = j^{ev}\,r \tag{3.14}$$

where j_{ev} is the specific mass flowrate of the steam being generated from unit surface per unit time.

In accordance with the kinetic theory of phase transition, Bankoff [3.25] expressed j_{ev} as

$$j_{ev} = \beta\left(\frac{2\pi R g T_{sat}}{gM}\right)^{-1/2}(p_2 - p_{sat}) \tag{3.15}$$

The evaporation coefficient β, measured for the case of evaporation of liquid into a vacuum, is given in Table 3.1 which is based on the data of Willie [3.26].

Equation (3.15) was employed by Hsu for calculating the heat fluxes in the evaporation of alcohols, liquid nitrogen and hydrogen into a perfect vacuum. These data are listed in Table 3.2 for atmospheric pressure.

The data in Table 3.2 cannot be used directly for calculating nucleate boiling heat transfer, since they are based on experimental data of [3.26] obtained for a flat surface with evaporation into a vacuum, whereas under real conditions of nucleate boiling evaporation occurs into a bubble at pressure $p_{sat}(T)$. Nevertheless these data

Table 3.2

Liquid	Boiling temperature, °C	Molecular weight, kg/mole	Evaporation coefficient β	Heat fluxes W/m$^2\cdot 10^6$
Mercury	357	200.6	1	140
Water	100	18	0.04	8.8
Methyl alcohol	77	46.1	0.02	2.7
Butyl alcohol	64	37.0	0.045	6.6
Nitrogen	−196	28.0	1	14.8
Hydrogen	−253	2	1	64.8

give the upper boundary for evaporative heat transfer. It is seen that q_{ev} is threefold higher than convective heat transfer into a liquid moving along a heated plate, which is responsible for the high coefficients of heat transfer in nucleate boiling.

Lately a theory of high-rate evaporation was developed by Labuntsov [3.27] on the basis of rigorous kinetic description of evaporation in a Knudsen layer. In this model (in addition to Maxwellian flux of vapor molecules from the phase interface, and flow of molecules from the latter into the Knudsen layer, which yields the conservation equation for mass) use was made also of conservation equations for the normal component of momentum and energy. For practical applications, interpolation formulas for high-rate evaporation were suggested on the basis of the Labuntsov theory

$$j_{ev} = 0{,}6 \sqrt{2RT_{sat}} \sqrt{\rho_\infty \Big/ \left(\rho_{sat}[1 - 2\sqrt{\pi}] \frac{i_{ev}}{\rho_{sat}\sqrt{2RTL}} \frac{1-\beta}{\beta} \right)} (\rho_{sat} - \rho_\infty)$$

$$(3.16)$$

where for evaporation into a bubble we can assume that $\rho_\infty \simeq \rho_{sat}(T_1)$, $\rho_{sat} = \rho_{sat} = \rho_{sat}(T)$ and β is the evaporation coefficient.

Using the ideal gas equation for vapor at $\bar{T}_1$ and $\bar{T}_2$ and assuming that for bubbles which develop on the surface $p_1 = p_2 = p$, we find

$$j_{ev} = 0{,}6 \sqrt{\frac{2R}{\bar{T}_2}} \frac{p}{R} \left[\bar{T}_2/T_{sat}[1 - 2\sqrt{\pi}] \frac{i_{ev}}{\rho_{sat}(T_2)\sqrt{2\pi T_2}} \frac{1-\beta}{\beta} \right]^{0{,}5} \times$$
$$\times \left[\frac{\bar{T}_2}{T_1} - 1 \right]$$

$$(3.17)$$

The mean temperature of the liquid surrounding the bubble is determined by the temperature profile in the laminar sublayer, for which one can assume a linear temperature distribution

$$T_2 = T_w - (T_w - T)/2$$

Assuming that $T' \simeq T_{sat}$, where T' is the temperature at the edge of the boundary layer, we obtain

$$j_{ev} = 0{,}6 \sqrt{\frac{2R}{T_{sat} + \frac{\Delta T_w}{2}}} \frac{p}{R} \left[1/[1 + 2\sqrt{\pi}] \frac{i_{ev}}{\rho_{sat}[T_2]\sqrt{2\pi T_2}} \frac{1-\beta}{\beta} + \right.$$
$$\left. + (T_w - T_{sat})/T_{sat}[1 - 2\sqrt{\pi}] \frac{i_{ev}}{\rho_{sat}[T_2]\sqrt{2\pi T_2}} \frac{1-\beta}{\beta} \right] \left[1 + \frac{T_w - T_1}{T_{sat}} \right]$$

$$(3.18)$$

Substitution of the above equation into Eqs. (3.12) and (3.13) makes it possible to calculate the heat flux for evaporation from a film at a given temperature difference.

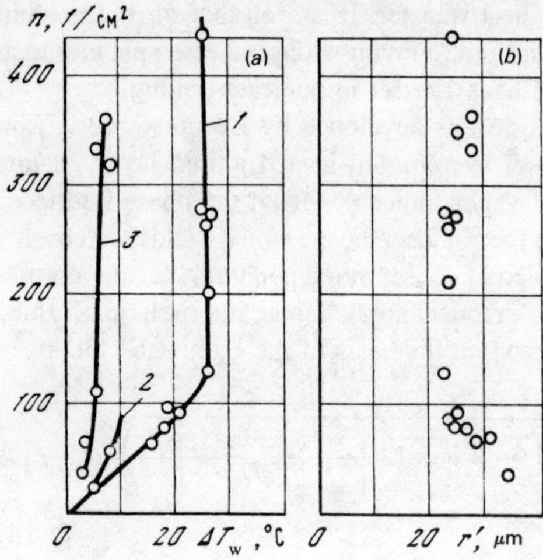

Figure 3.11 Variation in the population of nucleation sites vs. the superheat.

The density of active vaporization sites per unit area (or vapor channels) n is a function of wall superheat ΔT_w (or heat flux) for a given surface quality (number of nucleation sites). Note that the number of nucleation sites cannot be determined from measurement of the surface finish quality (height of asperities and their distribution). Hsu [3.18] clarifies this fact by an example from geography: it is impossible to judge the number and distribution of ponds and lakes from the mean height of mountains and hills in a given region and their distribution. Nevertheless it is possible to determine indirectly the size of the crevice serving as an active vaporization site. Figure 3.11 shows curves of the population of nucleation sites n vs. the superheat according to the experimental data of Griffith and Wallis [3.28] for a polished copper surface (see curves 2 and 3 in Fig. 3.11a). From Eq. (3.4) we may find the sizes of the active crevices (see curve 1 in Fig. 3.11a).

Analysis of points in Fig. 3.11b shows that experimental data for three liquids lie about a single value of the crevice radius of ~ 30 to $35\ \mu$m, i.e., upon attaining a certain superheat (according to the Griffith and Wallis data this is the transcritical superheat) the radius of an active crevice is a constant, characterizing the given surface. For a different surface roughness the radius of the active crevice will, naturally, be different. In other words calibration experiments for determining the dependence of the crevice radius r' on the superheat ΔT_w must be performed for each surface with given roughness parameters. The reader is referred to the book by Hsu and Graham [3.18] for more detailed familiarization with this problem.

3.2.4 Forced Convection Heat Transfer in a Film

Let us consider transfer of heat in a moving single-phase liquid film without nucleate boiling. As pointed out above, there exist several modes of laminar and turbulent flow of a film with a complex wave system. In the present section, analysis of heat transfer in a film will be performed without consideration of waves on its surface (we assume a smooth surface), $\delta \neq \delta_{min} \neq \delta_{max}$.

Laminar film. We shall analyze the motion of a film on a plate or in an annular tube at $\delta_{min}/D \ll 1$. Then the energy equation for a steady-state laminar film and smooth heat release over the channel perimeter (across the plate) can be written

$$\rho_2 c_{p_2} w \frac{dT}{dz} = \lambda_2 \frac{\partial^2 T}{\partial y^2} \tag{3.19}$$

with the boundary conditions:

$$y = 0;\ T = T_w;\ \lambda_2 \frac{\partial T}{\partial y}\bigg|_{y=0} = -q_w \tag{3.20a}$$

$$y = \delta;\ T = T_{sat};\ \frac{\partial T}{\partial y}\bigg|_{y=\delta} = 0. \tag{3.20b}$$

In addition, for constant heat flux at the wall (q_w = const) and stabilized flow (fully developed velocity and temperature profiles), we can write from the heat balance

$$\frac{dT}{dz} = \frac{d\bar{T}}{dz} = \frac{q_w}{Gc_{p_2}} \tag{3.21}$$

where $G(Q)$ is the mass (volumetric) flowrate in a film per unit wetted perimeter.

The use of Eq. (3.21) in Eq. (3.19) with boundary conditions (3.20a) yields for the temperature profile

$$\theta = \frac{T - T_w}{T_0 - T_w} = \frac{q_w \delta}{\lambda_2 (T_0 - T_w)} [0{,}5\tilde{y}^3 - 0{,}125\tilde{y}^4 - \tilde{y}] \tag{3.22}$$

where T_0 is the film temperature in the inlet cross section $z = 0$, $\tilde{y} = y/\delta$.

To find the mean temperature $\bar{T}$, the value of w should be substituted into the integral

$$\bar{T} = \frac{1}{G} \int_0^{\delta} wz\rho_2 T \, dy \tag{3.23}$$

To be able to integrate Eq. (3.19) it is also necessary to specify the velocity profile for the laminar film, which is found from the tangential stress distribution $\tau(y)$. The elementary force balance in the film for gradient less flow in a pipe or on a plate yields the expression

$$\tau = \tau_{12} \pm \rho_2 g (m - y) \tag{3.24}$$

The plus sign pertains to falling and the minus sign—to climbing flow. The velocity profile for Newtonian fluids is found by integrating Eq. (3.24)

$$w_z = \frac{\tau_{12}}{\mu_2} y \pm \left[\delta^2 y - \frac{y^2}{2} \right] \frac{\rho_2 g}{\mu_2}$$

or in nondimensional form

$$w_z = \frac{\tau_{12}\delta}{\mu_2} \tilde{y} \pm \frac{\delta^2 \rho_2 g}{\mu_2} \left[\tilde{y} - \frac{\tilde{y}^2}{2} \right] \tag{3.25}$$

The temperature profile is found by integrating Eq. (3.19) in the form

$$T - T_w = \frac{\rho_2 q_{w1}}{G \lambda_2 \mu_2} \frac{y^3}{6} \left\{ \left[\frac{\tau_{12}}{2} \pm \delta \rho g \right] \pm \frac{y}{2} \rho g \right] \right\} \tag{3.26}$$

and the coefficient of heat transfer in the laminar film is defined as

$$\alpha = \frac{q_w}{(T_w - T_{12})}$$

$$\alpha_{lam} = \frac{6 G \lambda_2 \mu_2}{\rho_2 \delta^3} \left[\frac{\delta^3}{2} \rho g \pm \frac{\tau_{12}}{2} \right] \tag{3.27}$$

Equation (3.27) is valid both for uptake and downtake flows.

Hewitt [3.19] obtained the following approximate expression for a falling film

$$\alpha = \left(\frac{\rho_2 g \lambda_2^3}{3 \mu_2 Q} \right)^{1/3} \tag{3.28}$$

$$\alpha_{lam} = \left(\frac{3 Q \mu_2}{\rho_2 g} \right)^{1/3} \tag{3.29}$$

For upwards flow at high shear stresses at the interface Hewitt suggested a linear velocity profile

$$w_z = 2G/\rho_2 \delta \tilde{y} \tag{3.30}$$

Then the temperature profile

$$\theta = \frac{q_w \delta}{2 (T_0 - T_w)} [0,33 \tilde{y}^3 - \tilde{y}] \tag{3.31}$$

and the heat transfer coefficient for a superheated evaporating film

$$\alpha \simeq 1,49 \lambda_2 / \delta_{lam} \tag{3.32}$$

The laminar film thickness in upwards flow is given by

$$\delta_{\text{lam}} = \frac{2G}{\rho_2 \tau_{12}} \,. \tag{3.33}$$

It is seen by comparing heat transfer coefficients from the Nusselt theory (λ_2/δ) with heat transfer coefficients calculated taking account of the shear stress at the interface [Eq. (3.28)], that τ_{12} increases the local heat transfer. The same conclusion can be drawn from experiments with film condensation of vapor.

Turbulent film. The theory of the turbulent film was developed by Hewitt [3.19] on the basis of the approximate theory of the turbulent boundary layer, according to which shear stress τ and heat flux q can be written as

$$\tau = - (\mu + \varepsilon\rho)\, dw/dy \tag{3.34a}$$

$$q = - (\lambda + \varepsilon_{\text{T}}\rho c_p)\, dT/dy \tag{3.34b}$$

Let us subdivide the wall, as is done in the von-Karman theory of the turbulent boundary layer, into three principal parts: laminar sublayer, buffer layer and the turbulent film. Then the velocity profile in the film in turbulent flow can be expressed by a universal profile which, as is known, has the form

$$y^+ < 5 - \text{laminar sublayer; } u^+ = w/w_* = y^+ \tag{3.35a}$$

$$5 < y^+ < 30 - \text{buffer layer; } u^+ = - 3{,}05 + 5\ln y^+ \tag{3.35a}$$

$$y^+ > 30 - \text{turbulent core; } u^+ = 5{,}5 + 2{,}5\ln y^+. \tag{3.35a}$$

where

$$w_* = \sqrt{\frac{\tau_w}{\rho}} \,; \qquad y^+ = \frac{w_* y \rho}{\mu}$$

It is usually assumed that the coefficient of turbulent momentum transfer ϵ_m and of turbulent heat transfer ϵ_{T} are equal. Hewitt calculated the turbulent viscosity coefficient ϵ from the Deissler equation for $0 < y^+ < 20$

$$\varepsilon = n^2 wy\, (1 - e^{-\rho n^2 wy}\,\mu) \tag{3.36}$$

where $n = 0.20$ and the von-Karman formula for $y^+ > 20$

$$\varepsilon = 0{,}36\, \frac{(dw/dy)^3}{(d^2w/dy^2)^2} \tag{3.37}$$

The nondimensionalized equation (3.34b)

$$\frac{dT^+}{dy^+} = \left[\frac{1}{\text{Pr}} + n^2 u^+ y^+ \left(1 - l^{-n^2 u^+ y^+}\right) \right] \tag{3.38}$$

$$T^+ = \frac{c_{p_2}\rho_2 w^*}{q}\,(T_w - T).$$

was solved by numerical integration for $y^+ < 20$. For $y^+ > 20$ temperature T^+ was determined by numerical integration of the von-Karman equation

$$\frac{dT^+}{dy^+} = \frac{(d^2w/dy^2)^2}{(dw/dy)^3}\,K$$

using profiles in accordance with Eqs. (3.35). The coefficient of heat transfer α in the turbulent liquid film is found in the form

$$\alpha = \frac{q}{T_w - T_{sat}} = \frac{c_p\rho_2 w^*}{T_{12}^+} \tag{3.39}$$

and the Nusselt number

$$Nu_\delta = \frac{\alpha\delta_{thr}}{K} = \frac{Pr\delta^+}{T_{12}^+} \tag{3.40}$$

where $\delta^+ = \delta w^*/\nu_2$.

In practice one usually investigates film moving under a high shear stress at the interface, when the film is sufficiently thin and the gravity force in it can be neglected. As seen from Eq. (3.24), in this case $\tau_{12} \gg \rho g(\delta - y)$ and it can be assumed that $\tau_{12} \simeq \tau_w = \tau$.

The mass flowrate through the film for this case is easily found by integrating the universal velocity profile, provided that a nondimensional liquid flowrate is introduced as [3.19]

$$G^+ = \frac{G}{2\pi R_0\mu_2}$$

we find

$$
\begin{aligned}
G^+ &= \frac{1}{2}\,y^{+2}; \quad y^+ < 5,\\
G^+ &= -8{,}05y^+ + 5y^+\ln y^+ + 12{,}45, \quad 5 \leqslant y^+ < 30\\
G^+ &= 8{,}0\,y^+ + 2{,}5\,y^+\ln y^+ - 214, \quad y^+ \geqslant 30
\end{aligned}
\tag{3.41}
$$

At low film Reynolds number the heat transfer coefficient tends to its value for laminar heat transfer from the Nusselt theory ($\alpha = \lambda/\delta$) for all the Pr. Note that departure of Nu from unity occurs at higher values of Re_2, the higher the value of Pr. For temperatures from 100 to 300 °C the Prandtl number for water changes from 0.85 to 1.7. This means that values of Re_2, corresponding to the inception of transition from laminar to eddy heat transfer, lies between 5 and 12. In experimental work the boundary between laminar and turbulent flow is usually found at higher Re_2 values. Thus, Nakoryakov et al. [3.29] show on the basis of measured instantaneous velocity profiles in a liquid film with waves at $Re_2 \leq 20$ that the velocity profile in this range obeys the parabolic relationship $w/w_{12} = 2(\tilde{y} - 1/2\tilde{y}^2$, i.e., it follows the rigorous laminar distribution.

As noted above, Hewitt's solution is valid for constant shear stress in the film, when gravity forces can be neglected. Condition $\tau_{12} = \tau_w$ is not satisfied for falling film flows in condensation systems.

Let us consider the case of variable τ. We assume for a tube [3.30]

$$\tau = \frac{R}{R_0}\tau_0 \pm (R_{12}/R_0)^2 \rho_2 g \tag{3.42}$$

where the minus sign pertains to a falling and the plus sign to a climbing film.

We write the equation in the more general form:

$$\tau = \tau_w \pm by; \quad b = \left(1 - 2\frac{\delta}{R_0}\right)\rho_2 g - \frac{\tau_w}{R_0} \tag{3.43}$$

The heat flux, as the shear stress, is written as a linear function of $\tilde{y}$

$$q = q_w [1 - \tilde{y}] \tag{3.44}$$

We now consider the case of a turbulent film, when the bulk of the thermal resistance is concentrated in the laminar sublayer. It then follows from Eq. (3.44) that

$$\varepsilon \simeq 0{,}4y \sqrt{\frac{\tau_w}{\rho_2}}; \quad \varepsilon \gg \nu_2 \tag{3.45}$$

Integrating Eq. (3.34) upon substitution of Eq. (3.45), we obtain the following expression for the film and the laminar sublayer

$$w \simeq \frac{2{\cdot}5}{\sqrt{\tau_w/\rho_2}} [\tau_w \ln \tilde{y} \pm b\delta (\tilde{y} - 1)] \tag{3.46}$$

$$T \simeq T_w - \frac{2{,}5q_w}{\rho_2 g c_p \sqrt{\tau_w/\rho_2}} [\ln \tilde{y} - (\tilde{y} - 1)] \tag{3.47}$$

Dukler [3.22] integrated numerically equations of heat flux q and shear stress of a tube for linearly varying τ with the turbulent transfer coefficient expressed by the von-Karman and Deissler laws (3.34) and (3.36). He suggested a nondimensional ratio for the heat transfer coefficient in the form

$$\tilde{\alpha} = \alpha \left(\frac{\mu_2^2}{\rho_2 g \lambda_2}\right)^{1/3}$$

The results of shear stress τ_{12} at the phase interface are incorporated by Dukler using the parameter

$$\beta = \left(\frac{\tau_{12}}{\rho_2 g^2 \mu L^2}\right)^{1/3}.$$

The results of numerical calculations obtained by the Dukler technique are given in the book by Hewitt and Hall-Taylor [3.19].

Figure 3.12 shows experimental data of the Institute of High Temperatures for a horizontal plate compared with the analytic equation obtained by Hewitt for $\mathrm{Nu}_\delta = f(\mathrm{Re}_2)$. It is seen by comparing the experimental and analytic results that for conditions with suppressed boiling at $q_w < q_w^{sup}$ for the entire range of Re_2 under study (209 to 2550) the experimental points lie 15 to 20% below the analytic values. Hewitt noted additionally, that in spite of the large scatter of results, most experimental data lies, as a rule, 30% below the analytic curve. This most probably occurs because the extension of laws of eddy transfer in a single-phase liquid in a channel to the case of a film is poorly validated. In fact, Hewitt [3.19] points out that turbulence is very rapidly suppressed at the phase interface, i.e., universal temperature and velocity profiles do not apply in this region. The second reason for disagreement between analytic and experimental results are waves on the film surface and entrainment. The latter manifests itself in cases of longitudinal temperature gradient, when cooler droplets are mixed into a film with a different temperature.

As should have been expected, for a boiling film, at $q_w > q_w^{i,b}$ the experimental points lie higher above the analytic curve, the higher the heat flux. The physical explanation for the enhancement of heat transfer with inception of nucleate boiling was given above.

3.3 INTRODUCTION TO THE BOILING CRISIS IN STEAM–GENERATING CHANNELS

Progress in power generation is significantly controlled by the detailed processes in power plant equipment. Particular attention is given to improving the effectiveness of

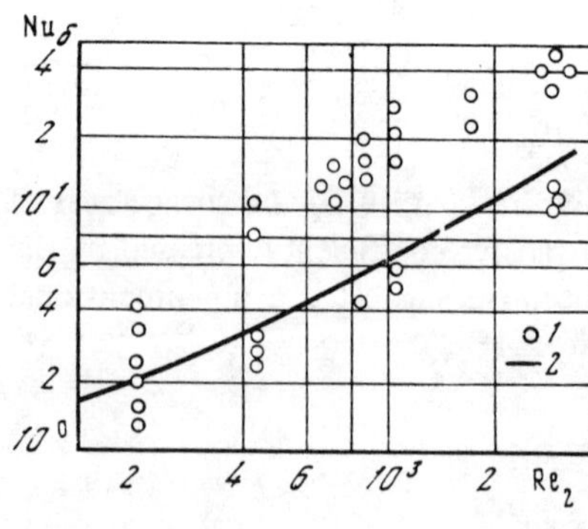

Figure 3.12 Plot of Nu_δ for a horizontal plate according to data of the present authors [3.15] and Hewitt and Hall-Taylor [3.19] as a function of Re_2. 1, Data of [3.15]; 2, [3.19].

operation of the major pieces of equipment, in particular reactors and steam generators. Steam generation at power plants occurs under surface or nucleate boiling conditions. It is natural for investigators and designers to increase heat fluxes, which means also reducing the size of heating surfaces and the unit as a whole also under these conditions. However, increase in heat fluxes in boiling-type equipment is limited by the boiling crisis.

At a certain level of heat flux nucleate boiling is replaced by film boiling. Heat transfer rates deteriorate significantly in film boiling, and at q = const this is accompanied by a significant rise of the wall temperature, and in certain cases in failure of steam-generating channels. Consequently, in order to ensure reliable operation it is necessary to know the principal relationships governing the boiling crisis, in order to select the safety margin on a rational basis. This, in part, is responsible for the continuous attention of scientists all over the world to this complex, and unfortunately, not as yet sufficiently explored phenomenon.

Studies of the boiling crisis have been continuing now for several decades. A large number of such studies were performed, but the problem still cannot be regarded as solved. Extensive work on the problem was performed over the last 10–15 years—experimental data were systematized, and the most reliable of these were selected to serve as a basis in developing practical recommendations. Experiments were performed which provided a deeper insight into the mechanism of the problem, validated models were suggested and engineering design methods were based on them.

At the same time the phenomenon is so complex that even the optimists do not dare to consider the possibility of a rapid solution of the problem. In spite of the fact that it is possible at present to list several tens of significant factors of the boiling crisis, its nature continues to present surprises. For this reason many leading organizations concerned with design of new power generating equipment (with different design and technological features, operating conditions, etc.) perform full-scale tests of individual components. This appears to be more rapid and reliable, and, what is most important, provides data which cannot be obtained analytically.

The results of studies of the boiling crisis were lately systematized by a large number of investigators [3.11, 3.14, 3.18, 3.31–3.35].

In this section we present the principal concepts of the boiling crisis, needed for analyzing the subsequent information.

It is generally assumed now that crisis inception conditions are significantly affected by the flow structure. At significant subcoolings and moderate vapor qualities, the boiling crisis is hydrodynamic in nature [3.1, 3.32]. At the high heat fluxes characteristic of these regions, transition to film boiling occurs due to instability of the two-phase boundary layer. The countercurrent motion of the vapor and liquid streams, between the flow core and the surface, is broken up, and a continuous vapor film arises at the wall. Recently this kind of crisis has been also termed "crisis of the first kind" [3.14].

With increasing quality the dryout or burnout transition (or "crisis") becomes governed by the transfer of liquid between the flow core and the film. Under certain conditions the wall film dries out and transition occurs to the post-dryout region. This

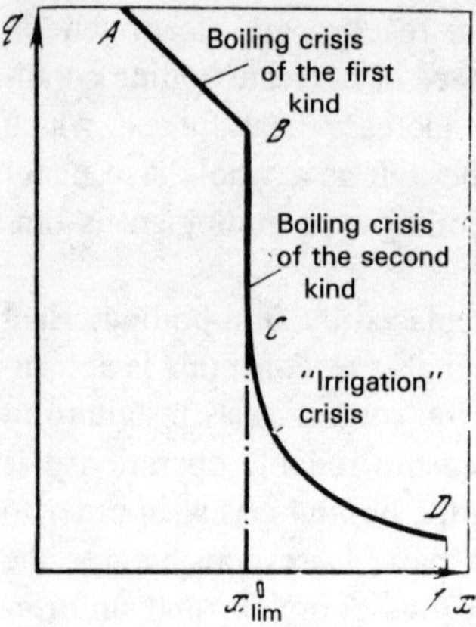

Figure 3.13 Boiling crisis.

kind of dryout crisis is not related to the hydrodynamic stability of the two-phase boundary layer and it is termed "thermokinetic crisis," or "crisis of the second kind."

Figure 3.13 illustrates schematically the two forms of the boiling crisis. Region *AB* has corresponding to the hydrodynamic crisis. A remarkable feature of the thermokinetic crisis in region *BC*, in Doroshchuk's opinion, is the absence of transfer of the liquid phase between the film and the flow core [3.14]. Droplet exchange exists in region *CD* and crisis in this part of the flow is sometimes termed the "irrigation" crisis.

At significantly flow qualities and two-phase flow velocities the liquid film may be rather thin. Heat is transferred through it by conduction. To be able to transmit high heat fluxes the film should be superheated relative to the saturation temperature. However, its temperature T_2 is limited by the limiting temperature for existence of liquid in the metastable state (T^*). At the time when T_2 becomes equal to T^* spontaneous flashing occurs. The loss of stability of a metastable liquid layer is related to the thermodynamic boiling crisis (Fig. 3.14). Calculation of critical heat fluxes from this theory yields the upper limit of critical heat fluxes.

Let us consider the effect of the principal variable—pressure, mass velocity and quality for the previously mentioned forms of the boiling crisis. A change in pressure causes a change in the phase velocity, surface tension force, viscosity, etc., which affects the vaporization parameters and the thickness of the boiling boundary layer. Different flow velocities ensures differences in the velocity gradient in the layer. This affects the dimensions of the departing vapor bubbles and the rate of their release into the flow core. Turbulent fluctuations, which also depend on the mean flow velocity, control the rate of diffusion of droplets from the core and entrainment of liquid from the film. A change in enthalpy changes the velocity, percentage liquid and rate of liquid transfer between the flow core and the wall layer.

Let us follow the variation of these parameters under the most extensively explored conditions (upwards flow in a vertical cylindrical channel). In the region of the crisis of the first kind, the value of q_{cr} decreases with pressure (Fig. 3.15). Physically this occurs because a rise in p reduces the latent heat of vaporization, which decreases

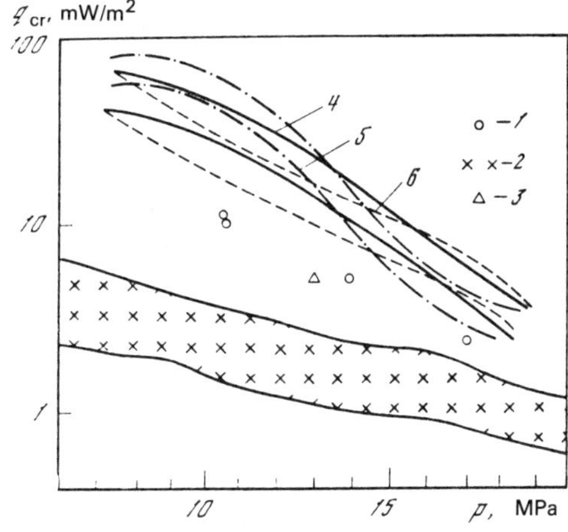

Figure 3.14 The thermodynamic boiling crisis. 1, Data of Zenkewicz; 2, of Doroshchuk (range of variation in q_{cr} in the boiling of water in tubes); 3, data of Khasanov-Ageyev, estimates of limiting values of q_{cr}; 4, estimate based on the concept of the minimum thickness of the liquid film and the limiting superheat; 5, estimate based on the concept of the limiting superheat; 6, estimate based on the limits on the thickness of the liquid film.

the heat removal rate from the wall for a given mass flux of vapor. In addition, the difference between the Leidenfrost and saturation temperatures becomes smaller, which also results in reduction in the critical heat flux.

The mass flux dependence of the critical heat flux [3.36] is shown in Fig. 3.16 as a function of quality. Over the range of parameters encountered in practical applications the heat flux drops with an increase in velocity, rising only at velocities which are relatively large for reactors and steam generators. For example, at $p = 19.6$ MPa the heat flux increases monotonically with mass velocity over the range of qualities under study. At low pressures the effect of pressure is negative, whereas at high pressures it is positive. Over a certain range the pressure has virtually no effect. The limits of this range depend on the relative enthalpy.

The critical heat flux drops with rising quality; the rate of this change can be

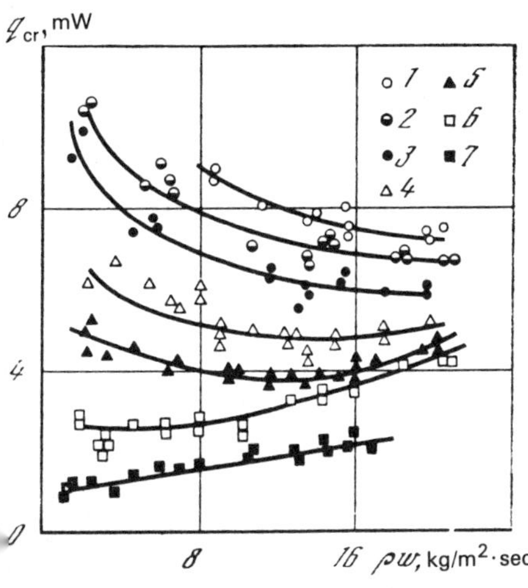

Figure 3.15 Effect of pressure on the critical heat flux. 1, $p = 2$ MPa; 2, 2.9; 3, 3.9; 4, 7.8; 5, 9.8; 6, 13.7; 7, 19.6 MPa.

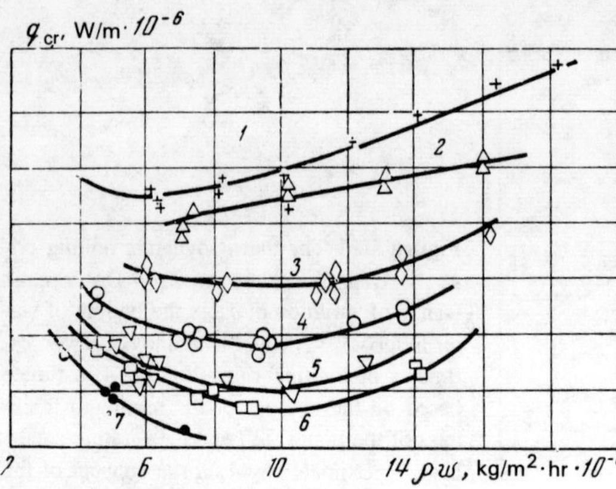

Figure 3.16 Critical heat flux vs. mass velocity at p = 13.7 MPa. 1, x = 0; 2, 0.05; 3, 0.1; 4, 0.15; 5, 0.2; 6, 0.25; 7, 0.3.

different over varied ranges of relative enthalpy, mass velocity, and pressure [3.36–3.40].

The effect of variables p, ρw and x is not monotonic. It should also be noted that the region of existence of crisis of the second kind without irrigation is rather limited (Fig. 3.17). However, it occurs over a range of parameters of importance for practical utilization, which is responsible for the close scrutiny afforded it in the various studies.

At a certain quality $x_{\lim 0}$ the heat flux ceases affecting the quality at which the boiling crisis occurs (see Fig. 3.13). Parameter $x_{\lim 0}$ is a complicated function of pressure (Fig. 3.18). First it increases with rising pressure, and then drops. The region where $x_{\lim 0}$ has a maximum corresponds approximately to the maximum of critical heat fluxes as a function of pressure under natural convection conditions. The relationship between $x_{\lim 0}$ and $x(p)$ is approximated thus

$$x_{\lim 0} = \left[0{,}39 + 1{,}57 \left(\frac{p}{98}\right) - 304 \left(\frac{p}{98}\right)^2 + 0{,}68 \left(\frac{p}{98}\right)^3 \right] \left(\frac{\rho w}{1000}\right)^{-0{,}6} \qquad (3.48)$$

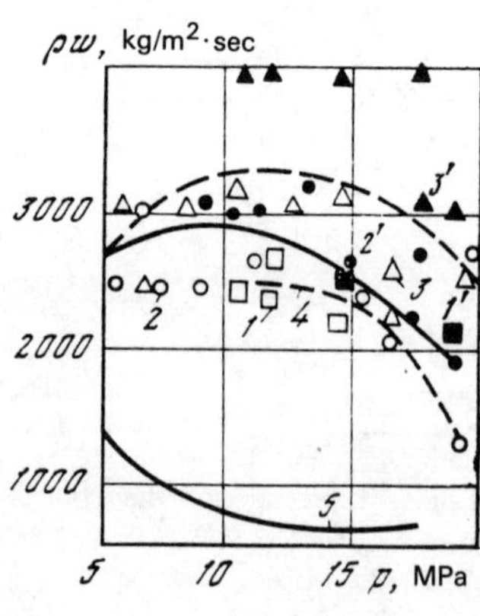

Figure 3.17 Region of existence of the crisis of the second kind without irrigation. 1, 2, 3, $x_{\lim 0}$ = const; 1', 2', 3', q = $f(x_{\lim 0})$; 1, 1', q = 3·10^5 W/m²; 2, 2', q = 5·10^5 W/m²; 3, 3', q = 8·10^5 W/m²; 4, upper limit; 5, lower limit.

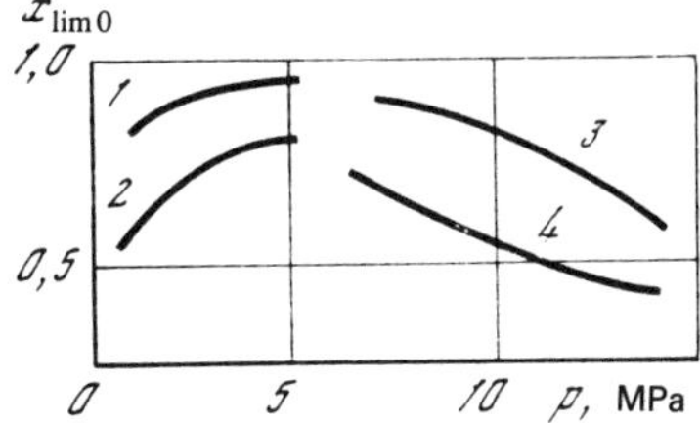

Figure 3.18 Plot of $x_{\lim 0}$ vs. pressure. 1, $\rho w = 500$; 2, 500; 3, 1000; 4, 1000 kg/m$^2 \cdot$sec.

The value of $x_{\lim 0}$ decreases with increasing mass velocity (Fig. 3.19). The law governing the effect of ρw on $x_{\lim 0}$ follows from the above formula.

The explanation put forward by Doroshchuk [3.14] for the independence of $x_{\lim 0}$ of the heat flux, consisting in lack of transfer of liquid between the flow core and the wall has been questioned by a number of investigators. Naturally, confusion stems also from the fact that the upper limit for the existence of $x_{\lim 0}$ (see Fig. 3.17) was found to be a function of the heat flux. Hence, some investigators assume that mass transfer does indeed exist in this region. The peculiar behavior of the critical quality is attributed to other factors, for example, attaining the critical velocity. Other investigators assume that in this region also q affects the critical vapor quality.

In sufficiently long channels, following the section with $x > x_{\lim 0}$ (or in short pipe runs, when $x_I > x_{\Delta p}$) the critical quality again rises with reduction in heat flux (thermokinetic crisis with "irrigation"). Figure 3.20 shows the effect of velocity in this range of variables. It is seen from the figure that over the range under study the critical heat flux rises with the mass velocity. The effect of pressure at $x > x_{\lim 0}$ is opposite, i.e., q_{cr} decreases with rising pressure.

3.4 MECHANISM OF THE BOILING CRISIS IN DISPERSED–ANNULAR FLOW

Let us consider the mechanism of the boiling crisis on the basis of a model proposed in Harwell [3.19]. We shall assume that in long tubes (or channels of arbitrary geometric shape), the coolant flow can be subdivided into two principal components: liquid film in the wall layer and the vapor-droplet core. We designate the film

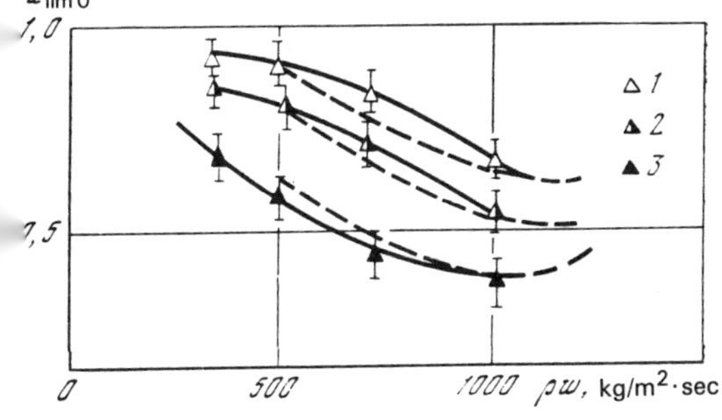

Figure 3.19 Effect of mass velocity on $x_{\lim 0}$ [3.33]. 1, $p = 6.9$ MPa; 2, 9.8; 3, 13.7 MPa.

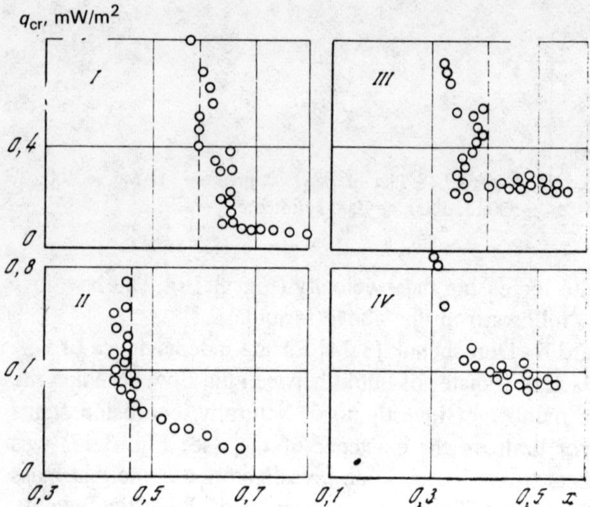

Figure 3.20 Plot of q_{cr} vs. the vapor quality at p = 9.8 MPa. I, $\rho w \cdot 10^{-6}$ = 3.6; II, 5.4; III, 7.2; IV, 10 kg/m²·hr.

flowrate by G_2; the mean film thickness is $\bar{\delta}$. Under conditions close to the boiling crisis, one cannot in general assume the existence of hydrodynamic equilibrium between the core and the film, when the mass fluxes of entrainment and deposition are equal. The entrainment flux from an elementary element of a tube with diameter D and length dz is equal to $dm_E = E\pi D\, dz$, the deposition flux is respectively $dm_D = D\pi D\, dz$. We write the mass balance for the liquid film in the form:

$$G_2 + dm_D = G_2 + \frac{\partial G_2}{dz} \tag{3.49}$$

or, after simple mathematical manipulations

$$\rho_2 \frac{\partial \bar{\delta}}{\partial t} + \frac{\partial G_2}{\partial z} = \pi D \left[\tilde{D} - E - \frac{q_w}{r} \right] \tag{3.50}$$

where G_2 is the liquid flowrate in the film, kg/sec.

We write similarly the mass balance for droplets in the flow core

$$\frac{\partial G_3}{\partial t} = \pi D\, [E - \tilde{D}] - \frac{\pi D^2}{4} \frac{\partial C_3}{\partial t} \tag{3.51}$$

and for the vapor

$$\frac{\partial \rho_1}{\partial \tau} + \frac{\partial G_1}{\partial z} = \pi D \frac{q_w}{r} \tag{3.52}$$

where G_3 is the flowrate of droplets in the flow core, kg/sec; C_3 is the mass concentration of droplets per unit volume, kg/m³ and G_1 is the vapor-phase flowrate, kg/sec.

To be able to integrate Eqs. (3.50) and (3.51) one must know the relationships for the entrainment and deposition mass fluxes. As shown in Chapter 2, the specific droplet deposition flux is

$$\tilde{D} = kC_3 \tag{3.53}$$

where k is the rate of deposition of droplets on the film.

The mass concentration of droplets in the core is expressed as

$$C_3 = \frac{4G_3}{w_3\pi D^2} = \frac{j(1-x) - \dot{j}_2}{w_3} . \tag{3.54}$$

It was shown in Chapter 2 that, as the crisis location is approached, the deposition flux exceeds the entrainment flux. However, in the first approximation Hewitt and Hall-Taylor [3.19] assume, the deflection from hydrodynamic equilibrium is insignificant, i.e.,

$$E = \tilde{D} = kC_3^{\text{eq}} \tag{3.55}$$

On the basis of a large number of experimental data (see Chapter 2), the equilibrium droplet population C_3 in the flow core is expressed as a function of the nondimensional ratio

$$C_3 = f\left(\frac{\tau_{12}\bar{\delta}}{\sigma_2}\right) \tag{3.56}$$

where σ_2 is the coefficient of the surface tension of the liquid.

Figure 3.21 shows experimental points according to Harwell data [3.10] for air-water, water-alcohol and steam-water mixtures. It is seen that the experimental data cluster satisfactorily about a curve with an rms deviation of about 15%. This validates the use of Eq. (3.56).

The mean film thickness δ is determined from the Armand equation

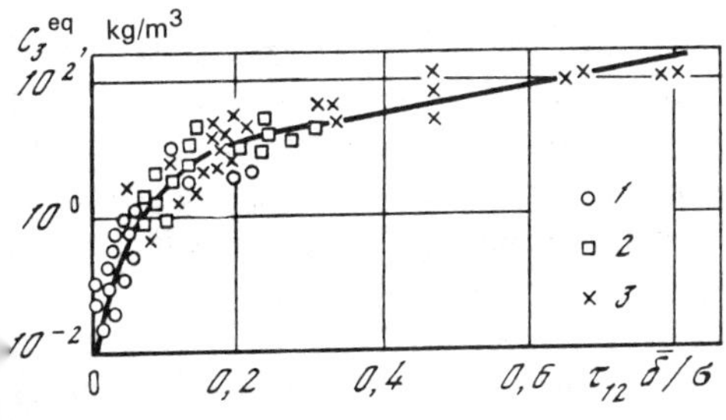

Figure 3.21 Equilibrium droplet concentration in the flow core C_3 vs. $\tau_{12}\delta/\sigma$. 1, Air-water; 2, air-alcohol; 3, steam-water.

$$\frac{4\bar{\delta}}{D} = \sqrt{\frac{(dP/dz)_2}{(dp/dz)_{\text{t.ph}}}} \tag{3.57}$$

Here $(dp/dz)_2$ is the pressure gradient in a single-phase liquid for the case when the liquid flowrate in the film moves over the entire tube cross section, $(dp/dz)_{\text{t.ph}}$ is the pressure gradient in the two-phase flow. Hewitt [3.5] suggested the following expression for $(dp/dz)_2$

$$\left(\frac{dp}{dz}\right)_2 = C_{f2}\frac{2i_2^2}{\rho_2 D} \tag{3.58}$$

where f_2 is the specific flowrate of liquid in the film, $kg/m^2 \cdot sec$; C_{f2} is the friction factor in the liquid.

Shear stress τ_{12} is calculated from the mean velocity of gas in the core w_1:

$$\tau_{12} = C_{12}\frac{\rho w_1^2}{2} \tag{3.59}$$

with the friction coefficient at the interphase found from hydrodynamic studies (see Chapter 2). In particular, one can use the Wallis formula (2.109)

$$C_{12} = C_{f_1}\left(1 + \frac{300\bar{\delta}}{D}\right) \tag{3.60}$$

The gas phase drag coefficient C_{f1} is found from the expression

$$C_{f_1} = 0{,}079\widetilde{\text{Re}}^{-1/4} \tag{3.61}$$

$$\widetilde{\text{Re}} = \frac{(i_1 + i_3)\widetilde{D}}{\mu_1}$$

The pressure gradient in the two-phase flow is expressed approximately as

$$(dp/dz)_{\text{t.ph}} \simeq -4\tau_{12}/D \tag{3.62}$$

The above means that the previously cited set of differential equations is integrated at specified boundary and initial conditions from the section of transition from slug to dispersed-annular flow ($x \simeq 0.01$) on the standard assumption that about 1% of the overall liquid flow moves in the initial section of the film. Integration is continued to the location where the film flowrate becomes equal to zero, which is the principal characteristic of the boiling crisis.

Under steady flow conditions Eqs. (3.50)–(3.52) become identical to the standard differential equations for continuity of the film, droplets and the vapor core, namely

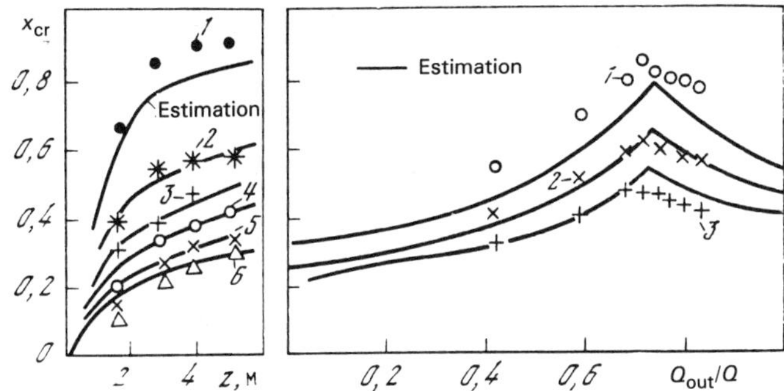

Figure 3.22 Critical quality vs. length (*a*) and the fraction of power at the outer surface (*b*). (*a*) 1, ρw = 678 kg/m$^2 \cdot$sec; 2, 2712; 3, 1356; 4, 4068; 5, 2034; 6, 5424 kg/m$^2 \cdot$sec. (*b*) 1, 600; 2, 1000; 3, 1400 kg/m$^2 \cdot$sec.

$$\frac{dG_2}{dz} = \pi D \left(\tilde{D} - E - q_w / r \right) \tag{3.63}$$

$$\frac{dG_3}{dz} = \pi D \left(E - \tilde{D} \right) \tag{3.64}$$

$$\frac{dG_1}{dz} = \pi D \left(\frac{q_w}{r} \right) \tag{3.65}$$

Equation (3.65) is a function of the two preceding, since the overall flowrate $G = (G_1 + G_3 + G_2) = $ const, however, it is convenient to use it in order to check the accuracy of the numerical method. For constant heat flux Eq. (3.65) integrates to the form

$$G_1 = \pi D L \cdot q_w / r. \tag{3.66}$$

In this case only one of the differential equations (3.63)–(3.65) is solved numerically. The results of calculations yield rather satisfactory agreement with experimental data for different geometries of steam-generating channels from a single tube to a fuel-element assembly.

Figure 3.22*a* compares the results of calculations of critical quality x_{cr} with experimental data of Bennet [3.41] for a wide range of mass flowrates (between 673 and 5424 kg/m$^2 \cdot$sec). The experiments were performed in a 12.6 mm diameter and 5 m long tube at p = 7 MPa. It is seen from analysis of Fig. 3.22*a* that the analytic curves are in both qualitative and quantitative agreement with experimental data.

Data for annular flows are plotted in Fig. 3.22*b*. Use was made of the above analytic method [see Eqs. (3.63–3.65)], with equations for the conservation of mass for the film employed both for the outer and inner wall. For comparison the graph shows experimental data due to Jensen and Manno [3.42] for an annular channel with both the inner and outer walls heated. It is seen that the curve of the critical quality vs. the fraction of power supplied to the outer tube has a maximum. The crisis on the

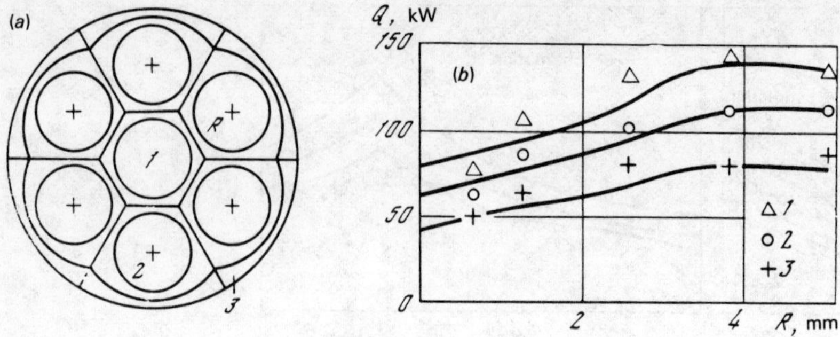

Figure 3.23 Comparison of calculations by Willie (3.45) with experiments using F-12 and a seven-rod assembly. (a) Schematic of assembly; (b) heat release over the assembly's radius; 1, ρw = 2609 kg/m^2·sec; 2, 1784; 3, 883 kg/m^2·sec; p = 1.07 MPa.

ascending branch of the curve starts on the inner surface, whereas, on the descending branch, it starts on the outer tube. At maximum heat flux crisis arises simultaneously at both surfaces of the annulus. It is seen that the technique suggested in Harwell describes experimental data with sufficient accuracy also for this case.

The boiling crisis in fuel element assemblies was investigated by Janssen [3.43], Kinneir et al. [3.44] and Whalley et al. [3.45]. The fuel-element assembly was also described by Eqs. (3.63)–(3.65), written for the three regions designated in Fig. 3.23a by numerals 1 through 3. An equivalent tube was selected for each region. The continuity equations (3.63)–(3.65) were supplemented by terms accounting for turbulent mixing between the channels for both the vapor phase and the droplets. Terms were also included for interchannel transfer of droplets and vapor in order to equalize the longitudinal pressure gradient. The boundary conditions, as in the case of steam-generating tubes, were specified in the form of x_0 = 0.01 and E_0 = 0.99, the film flowrate was taken to be proportional to the wetted nonpermeable surfaces.

Comparison of analytic results of Whalley [3.45] with experimental data for evaporation of freon in a seven-rod assembly at p = 1.07 MPa [3.44] is given in Fig 3.23a. The shape of the analytic curves is in satisfactory agreement with experimental data, the only exception being points for small spaces between the central and peripheral rods. Calculations, as well as experiments, show that for small spacings crisis arises on the central rod, whereas for large spacings, it occurs on the outer rods. Similar results were obtained for a boiling-reactor assembly of 16 rods with heated length of 1.83 m and p = 7 MPa [3.43].

3.5 METHODS FOR CALCULATING THE BOILING CRISIS

A large body of experimental data has been accumulated up to now on the forced convection boiling crisis. This phenomenon has been explored in most detail in upwards flow of water in uniformly heated tubes. The results of these experiments can serve as standard conditions for determining the conditions for inception of crisis in channels of other geometries, for different heat flux distribution, under dynamic conditions, etc.

In correlating experimental data, the boiling crisis is usually treated as one of the following phenomena: (1) local, i.e., the process is described by local values of quantities $x = x(q,p,\rho w)$; (2) integral, i.e., the behavior of the process is affected by its history. For example, in the case of non-uniform axial heat flux, the conditions of crisis inception are affected by the velocity, enthalpy and concentration fields which form under the specific flux profile; (3) global, i.e., the crisis is affected by all of the system's parameters. Since this statement of the problem makes it difficult to determine the role of each factor, the data are worked up in the form of a relationship between the overall supplied power and the controlling parameters. In addition, design equations can be constructed using physical models or on the basis of formal statistical data processing. Naturally, the former relationships are preferable, but, as a rule, their range of application is limited.

We shall now list certain relationships [3.2, 3.14, 3.46–3.61], which have come into comparatively extensive use for uniformly heated tubes.*

Hydrodynamic crisis (crisis of the first kind). The Kutaeladze formula [3.2]

$$q_{ev} = 3{,}22 \cdot 10^{-3} r \sqrt{g\gamma''} \sqrt[4]{\sigma(\gamma' - \gamma'')} \left(w_0 \sqrt[4]{\frac{\gamma' - \gamma''}{g^2 \sigma}} \right)^{0,5} \times$$

$$\times \left[1 + 0{,}065 \left(\frac{\gamma'}{\gamma''} \right)^{0,8} \frac{c_p \Delta T_{sat}}{r} \right], \quad \text{kcal/m}^2 \cdot \text{hr} \tag{3.67}$$

The limits of applicability are: p between 10 and 200 atm, $\Delta T_{sat} \leq 100\,°C$, w_0 between 0.7 and 7 m/sec.

The Labuntsov formula [3.46]

$$q_{cr} = 1{,}25 \cdot 10^6 \, \varphi(p) \sqrt[4]{1 + \frac{2{,}5}{\varphi(p)} w^2 \theta(\Delta T_{sat})}, \quad \text{kcal/m}^2 \cdot \text{hr} \tag{3.68}$$

$$\varphi(p) = p^{1/3}(1 - p/p_{cr})^{4/3}; \qquad \theta(\Delta T_{sat}) = 1 + \frac{15}{\sqrt{p}} \frac{c_p \Delta T_{sat}}{r}$$

Limits of applicability: p between 1 and 198 atm, ρw between 700 and 45,000 kg/m$^2 \cdot$sec, ΔT_{sat} between 0 and 240 °C.

The Miropol'skiy-Shitsman formula [3.47]

$$\frac{q_{ev}\mu'}{\sigma\gamma' r} = 0{,}174 \left(\frac{C_p' T_s}{r} \right)^{0,8} K_w^{0,4} \left[1 - 0{,}45 \frac{\Delta h_{sat}}{r} \left(\frac{\gamma'}{\gamma''} \right)^{0,85} \right], \quad \text{kcal/m}^2 \cdot \text{hr} \tag{3.69}$$

$$K_w = \frac{\gamma w \mu'}{\sigma \gamma'} \left(\frac{\gamma'}{\gamma''} \right)^{0,2}$$

---

*The formulas are given in that system of units in which they appear in the original studies.

Table 3.3

Pressure	560	1000	1550	2000	Pressure	560	1000	1550	2000
y_0	237	114	36,0	65,5	y_6	19,3	127	41,7	17,1
y_1	1,20	0,811	0,509	1,19	y_7	0,959	1,32	0,953	1,1
y_2	0,425	0,221	−0,109	0,376	y_8	0,831	0,411	0,0109	−0,4
y_3	−0,940	−0,128	−0,190	−0,577	y_9	2,61	−0,274	0,231	−1,5
y_4	−0,0324	0,0274	0,0240	0,220	y_{10}	−0,0578	−0,0397	0,0767	2,7
y_5	0,111	−0,0667	0,463	−0,373	y_{11}	0,124	−0,0221	0,117	2,2

limits of applicability: p between 36 and 200 atm, $\Delta h_{sat}/r$ between -0.5 and 0, ρw between 400 and 10,000 kg/m$^2 \cdot$ sec, D_{in} between 5 and 8 mm, $L/D \geq 100$.

The Kirillov formula [3.48]

$$q_{cr} = A_0 + A_1 \Delta T_{sat} + A_2 (\gamma w), \quad \text{kcal/m}^2 \cdot \text{hr} \tag{3.70}$$

$$A_0 = 11{,}25 - 7{,}08 \left(\tfrac{p}{100}\right) + 0{,}257 \left(\tfrac{p}{100}\right)^4 - 2{,}25 \exp\left[-4\left(\tfrac{n}{100}\right)\right]$$

$$A_1 = \left[1{,}44\gamma w - 1 + 4{,}5 \left(\tfrac{p}{100}\right) - 2{,}25 \left(\tfrac{p}{100}\right)^2 + 0{,}05 \left(\tfrac{p}{100}\right)^4\right] \cdot 10^{-2}$$

$$A_2 = 0{,}12 + 0{,}33 \sin 0{,}9\pi \left(\tfrac{p}{100} - 1\right)$$

Limits of applicability: p between 20 and 200 atm, ΔT_{sat} between 0 and 100 °C, ρw between 500 and 5000 kg/m$^2 \cdot$ sec, $L/D_{in} > 20$.

The Tong formula [3.49]

$$q_{cr} = (0{,}23 \cdot 10^6 + 0{,}094\gamma w) [1{,}7 - 1{,}4\exp(-a)] \times$$
$$\times [0{,}435 + 1{,}23\exp(-0{,}0093L/D)], \quad \text{BTU/ft}^2 \cdot \text{hr} \tag{3.71}$$

$$a = 0{,}532 \left(\frac{h' - h_{in}}{r^{3/4}}\right) \left(\frac{\rho'}{\rho''}\right)^{1/3}$$

Limits of applicability: p between 56 and 193 atm, ΔT_{sat} between 0 and 228 °C, ρw between $0.2 \cdot 10^6$ and $8 \cdot 10^6$ lb/ft$^2 \cdot$ sec, D_{in} between 0.1 and 0.54 inch, $L/D_{in} \geq 20$.

The Thompson-Macbeth formula [3.50]

$$\frac{q_{cr}}{10^6} = \frac{A' - 0{,}25D\,(\rho w \cdot 10^6)\,xr}{C'}, \quad \text{BTU/ft}^2 \cdot \text{hr} \tag{3.72}$$

$$A' = y_0 D^{y_1} (\rho w \cdot 10^{-6})^{y_2} [1 + y_3 D + y_4 (\rho w \cdot 10^{-6}) + y_5 D (\rho w \cdot 10^{-6})]$$
$$C' = y_6 D^{y_7} (\rho w \cdot 10^{-6})^{y_8} [1 + y_9 D + y_{10}'(\rho w \cdot 10^{-6}) + y_{11} D (\rho w \cdot 10^{-6})]$$

The values of y_0 through y_{11} are found from Table 3.3.

Limits of applicability: p between 15 and 2000 psia, ΔH_{sub} between 28 and 602 Btu/lb, ρw between 0.02 and 0.62 lb/ft^2·hr, D between 0.12 and 0.391 inch, L between 6 and 123 inch.

The Hewitt formula [3.51]

$$q_{cr} = K_1 (1 - K_2 x) \tag{3.73}$$

where K_1 and K_2 are coefficients, obtained from tables as a function of p and ρw.

Limits of applicability: p between 50 and 100 kg/cm^2, ρw between 500 and 4000 kg/m^2·sec, D_{in} between 5.6 and 11.5 mm.

Tong's formula [3.52]

$$q_{cr} = \frac{0,115 D^{-3,5} (\rho w)^{1/3} r}{\beta (p)} \left[\frac{2,5 \alpha (p)}{1 + \rho w / 135,6} - x_{cr} \right], \quad \text{W/cm}^2 \tag{3.74}$$

$$\alpha (p) = - 0,3126 \cdot 10^8 p^4 + 0,133 \cdot 10^{-2} p + 0,7113$$

$$\frac{\alpha (p)}{\beta (p)} = 0,34 + 9,4/(p - 7,3)$$

Limits of applicability: p between 49 and 112 kg/cm^2, ρw between 680 and 4100 kg/m^2·sec, D between 5.5 and 11.3 mm, $L_{max} = 2050$ mm.

In addition, recommendations were published in the form of skeletal tables, for example, in [3.54, 3.55].

Thermokinetic crisis (crisis of the second kind). Smolin's formula [3.56] for dispersed flow

$$q_{2cr} = 0,18 r \left(\frac{\mu''}{\mu'} \right) \tag{3.75}$$

The boiling crisis sets on when

$$K_w = K_{w_{cr}}, \qquad K_w = (\rho w) x \sqrt{\frac{D}{\rho'' \sigma}}$$

The transition between annular and dispersed flow is obtained from the expression

$$\frac{x_{lim}}{(1 - x_{lim})^{0,25}} = 1,36 \frac{(\rho' \sigma)^{0,5} (\mu' \mu'')^{0,0625}}{(\rho w)^{0,75} (D \mu')^{0,25} \exp [25, \angle q/[r(\rho' - \rho'')](\rho''/\sigma)^{0,25}]}$$

Limits of applicability: p between 49 and 196 bar, $0.75\, \rho''/(\rho' + \rho'') \leq x \leq 0.8$, ρw between 500 and 7500 kg/m^2·sec, D_{in} between 6 and 16 mm, L between 700 and 5000 mm.

Doroshchuk's formula [3.14]

$$x_{cr}^0 = (0{,}92 - 0{,}344p) \sqrt{\frac{1000}{\rho w}} \tag{3.76}$$

Limits of applicability: $50 < p < 160$ atm, $500 < \rho w < 2000$ kg/m$^2\cdot$sec, $D_{in} = 8$ mm.

Formula of the Physical Energetics Institute [3.57]

$$q_{cr} = 40\,(\rho w)^n\,(1-x)^m \left(\frac{\rho'}{\rho''}\right)^{2,2} \left[1 + \frac{8\cdot 10^9}{(\rho w)^k}\right], \text{ kcal/m}^2\cdot\text{hr} \tag{3.77}$$

$$n = 0{,}56 - 0{,}0189\,\frac{\rho'}{\rho''}; \qquad m = 0{,}7\,\frac{\rho'}{\rho''} - 0{,}4; \qquad K = 1{,}3 + 3{,}6\,\frac{\rho''}{\rho'} - 0{,}45$$

Limits of applicability: p between 100 and 200 atm, x between 0 and 0.4, ρw between 110 and 5000 kg/m$^2\cdot$sec, D_{in} between 4 and 12 mm, $L > 200$ mm.

The Campolunghi et al. formula [3.58]

$$x_{cr} = K q^{\alpha}\,(\rho w)^{\beta} p^{\gamma} \tag{3.78}$$
$$50 < p < 90 \text{ kg/cm}^2; \quad K = 3{,}75848; \quad \alpha = 0{,}381948; \quad \beta = -0{,}6039$$
$$\gamma = -0{,}005692; \quad 90 < p < 160 \text{ kg/cm}^2; \quad K = 2{,}92867; \ \alpha = 0{,}8850$$
$$\beta = -0{,}823414; \quad \gamma = -0{,}142230; \quad D = 11{,}8 \text{ mm}, \quad L = 22 \text{ mm}$$

Limits of applicability: x between 0.5 and 0.8 and between 0.25 and 0.5, respectively.

Formula of Bertoletti et al. [3.59]

$$\frac{Q}{\rho w A_f r} = \frac{1 - p/p_{cr}}{(\rho w/100)^{1/3}} \cdot \frac{L_s}{L_s + 0{,}315\,(p_{cr}/p - 1)^{0,4} D^{1,4} \rho w}, \quad \text{kW} \tag{3.79}$$

where L_s is the boiling length, cm; A_f is the cross sectional area, cm^2; p_{cr} is the critical pressure, kg/cm^2. Limits of applicability: $p > 45$ kg/cm^2, $\rho w > 100\,(1 - p/p_{cr})^3$ kg/m$^2\cdot$sec, $x > 0$, $x_{in} < 0.5(1 - p/p_{cr})/[(\rho w/100)^{1/3}]$.

Note that the separation of formulas for regions of hydrodynamic and thermokinetic boiling crises is rather arbitrary. In a number of cases the formulas somewhat overlap the regions with different types of crisis.

Relationships for other channel shapes, heat release distributions over the perimeter and length, channels with various enhancement devices and other structural elements are much more complex than in the case of tubes. This is particularly perceptible in the case of boiling crisis in rod assemblies. In the preceding section we already presented examples of utilization of equations of conservation of mass for calculating the boiling crisis in rod assemblies. Below we present in addition two correlation formulas for q_{cr} in rod bundle assemblies.

The Gaspari formula [3.60]

$$\frac{\bar{q}_{\text{cr.b}}}{r\rho w} = \left(\frac{D_{\text{eq}}}{4L_{\text{B}}}\right) \frac{1 - p_{\text{n}}}{\left(\frac{\rho w}{100}\right)^{1/3}} \frac{L_{\text{B}}}{L_{\text{B}} + 0{,}20 \left(\frac{1}{p_{\text{n}}} - 1\right)^{0{,}4} \rho w D_{\text{eq}}^{1{,}4}} \qquad (3.80)$$

where D_{eq} and L_{B} are parameters of the hot cell of the channel: the equivalent diameter and the boiling length; p_{n} is the reduced pressure.

Limits of applicability: p between 50 and 70 bar, ρw from 1000 to 4000 kg/m$^2\cdot$sec, number of rods in assembly between 7 and 37, rod diameter between 10 and 20 mm, length of L from 0.8 to 3.7 m.

The Gellerstedt formula [3.61]

$$\frac{q_{\text{cr}}}{10^6} = 3{,}16 \left\{ (1{,}55 - 16{,}0 D_{\text{eq}}) \left[0{,}37 \cdot 10^8 \left(0{,}435 \frac{\rho w}{10^3} \right)^{0{,}83_i' + 0{,}683 \left(\frac{p}{70} - 2\right)} - \right.\right.$$

$$\left.\left. - 0{,}0485 \rho w r x_{\text{cr}} \right] \right\} \Big/ \left[12{,}71 \left(2{,}25 \frac{\rho w}{10^3} \right)^{0{,}712 + 0{,}2073 \left(\frac{p}{70} - 2\right)} \right] \qquad (3.81)$$

Limits of applicability: p between 140 and 168 bar, ρw from 1000 to 5400 kg/m$^2\cdot$sec, D_{eq} between 5 and 12.5 mm, $L = 1.82$ m, and x_{cr} from 0.03 to 0.20.

It is seen from these examples that these relationships are quite complex. The form of the correlation is controlled to a large extent by the effect of structural elements. In the global approach to description of the above phenomenon, and when using formula statistical data processing methods, the coefficients and power exponents are mutually correlated and do not reflect the actual affect of the parameters. Obviously, such a description is rather convenient for practical purposes for a given assembly geometry; however, these expressions are of no special significance in determining the optimal parameters and prediction. In this sense it is better to use expressions such as

$$q_{\text{cr}} / q_{\text{cr}\,0} = \Psi_1 \Psi_2 \Psi_3 \dots \Psi_n \qquad (3.82)$$

where q_{cr0} is the critical heat flux under standard conditions, whereas Ψ_i is the effect of individual perturbing factors (spacing and kinds of grid, spaces between the rods, etc.). These formulas make it possible to compare results for different geometries and to determine the best assembly designs. A good survey of the literature on critical heat fluxes in rod assemblies is given by Tong [3.34].

3.6 EFFECT OF STRUCTURAL ELEMENTS OF STEAM–GENERATING CHANNELS UNDER CONDITIONS OF INCIPIENT BOILING CRISIS

Previously we considered basically the results of study of q_{cr} in circular tubes with upwards coolant flows (under standard conditions). The specific needs of reactor fuel element and steam-generator tube arrangements has resulted in a large number of

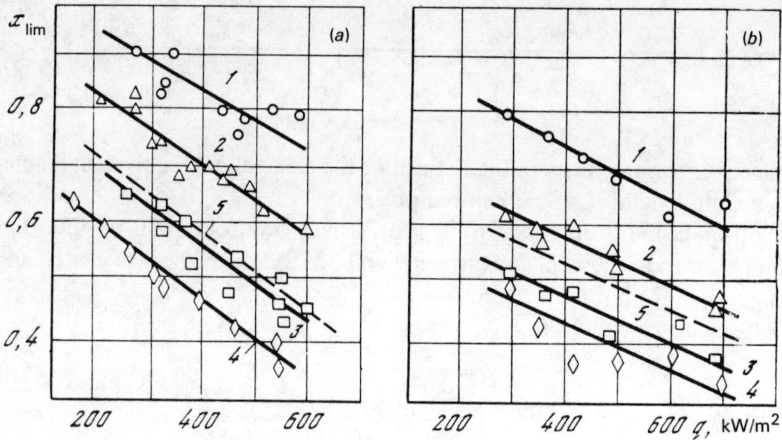

Figure 3.24 Effect of relative distance to a bend on the threshold values of parameters. (a) $\rho w = 700$ kg/ $m^2 \cdot$ sec, relative distance L/D to the bend: 1, 15; 2, 45 (past the bend); 3, 45; 4, 15 (ahead of the bend); 5, 90 to 95 mm (straight tube). (b) $\rho w = 1100$ kg/$m^2 \cdot$ sec (for legend see part a of the figure).

designs for these components. Without laying claim to completeness in analyzing the existing designs, we wish to note features specific to the effect of certain components of steam-generating channels on the critical parameters.

Current steam-generator design practice makes extensive use of coiled heating surfaces. A significant part of the surface in such designs is represented by bent parts of the tubes (bends). Clarification of the inception of the boiling crisis in bends and in the tube segments adjoining them is of great significance, since it allows practical recommendations to be made for designing heating surface with a large number of bends.

It was established by the second of the present authors [3.62] in experiments on channels with bends, that at any relative distances l/d from the bend the relationship between the limiting vapor quality on the one hand and the heat flux and mass velocity on the other, remains qualitatively the same. It is seen from Fig. 3.24 that, all other conditions remaining equal, the limiting vapor quality decreases monotonically with increasing heat flux. The value of the limiting quality decreases with increasing mass velocity. The slope of the curves of $x_{\text{lim}} = x(q)$ decreases with rising mass velocity.

The effect of the relative distance to the bend on the limiting values of parameters is shown in Fig. 3.25. A feature specific to the part preceding the bend is reduction in the limiting quality as a bend is approached. The effect of the bend on x_{lim} of the length past the bend is opposite. The maximum permissible value of x_{lim} is observed in the immediate vicinity of the bend, and the lowest far from it, the latter approaching the straight tube value of $x_{\text{lim.st}}$. The change in the relative limiting or threshold quality increases with the heat flux and mass velocity. The extent of change in the relative quality depends on the value of $x_{\text{lim.st}}$.

This effect of the bend is apparently associated with the manifestation of centrifugal forces. With respect to the upstream part of the tube the bend acts as an additional resistance, decelerating the steam-wave flow. The fraction of pipe cross section occupied by the steam increases, which results in reduction of the quantity of water in the film and interferes with the mass transfer between the flow core and the wall layer and thus reduces the critical heat flux. After the bend, the steam accelerates, which is accompanied by a reduction in the cross section occupied by the steam and an increase in the wall layer thickness. The transverse circulation fluxes arising in the bend continue to exist at some distance from the latter, enhancing mass transfer between the flow core and liquid film. As a result, the critical heat flux increases.

Upflow-downflow loops can have bends of two types. In a bend with upwardly facing convexity (top bend) the direction of motion changes from upflow to downflow. In the bottom bend the situation is opposite. Conditions of heat and mass transfer in such bends and in the straight pipe segments adjoining them may be different.

In the bottom bend the centrifugal and gravity forces are in the same direction. This may induce stratification of flow at low mass velocity and reduce the rate of mass transfer between the flow core and the wall layer on the inner generatrix. In the top bend these forces act in opposing directions. The effect of the gravity force could be perceptible in the case when it is greater than or commensurable with the centrifugal force. In these experiments only one design of a bend with a relatively small bend radius ($R_{bend}/D = 4$) was investigated. The rather high vapor quality resulted in high linear velocities. Under such conditions centrifugal forces greatly exceed the forces of gravity.

The above is confirmed by results of experiments performed with a top bend. Figure 3.26 compares the threshold qualities obtained with a top (x_{bend}^t) and a bottom (x_{bend}^b) bend. The comparison was performed at the same mass velocities, heat fluxes and relative distances from the bend. As seen from the figure, the deviations of the threshold qualities from the mean value are nonsystematic and do not exceed $\pm 15\%$ (with the exception of three points). This means that under the conditions studied, the

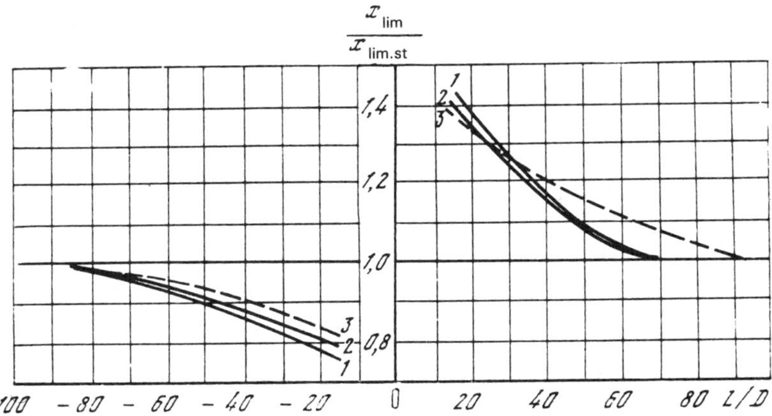

Figure 3.25 Effect of relative distance to bend on the threshold values of parameters. At $\rho w = 1100$ kg/m²·sec: 1, $q = 696$ kW/m²; 2, 348 at $\rho w = 7$ kg/m²·sec; 3, $q = 348$ kW/m².

orientation of the bend does not significantly affect the parameters responsible for the critical heat flux at the segments adjoining the bend. The observed reduction in critical heat flux in the tube segment upstream of the bend requires introduction of pertinent correction factors in designing such channels.

Pikus [3.63] and Campolunghi et al. [3.64] investigated critical heat flux in coils. Pikus investigated the region of negative and low positive qualities in rather extensive detail. He showed that the critical heat fluxes at $x \leq 0$ are lower than in straight tubes. At rather high positive values of x, the values of q_{cr} can be higher than in straight tubes. The effect of tube curvature decreases with increasing pressure. At $x < 0$ a reduction in D_{in}/R_{bend} first results in a reduction in q_{cr}, and then q_{cr} rises. The minimum values of q_{cr} were observed at D_{in}/R_{bend} between 0.05 and 0.15. The maximum reduction in q_{cr} in coils as compared with straight tubes was about fivefold. Pikus recommended that the critical heat fluxes in coils be calculated using the formula

$$q^c = q_0^{str} K_1 \left[1 - \left(1{,}5 - 4{,}7\frac{D_{int}}{R_b} \right) \left(5{,}5 - \frac{2{,}3\rho w}{1000} \right) x \right] \tag{3.83}$$

where

$$q_0^{str} = 2{,}7 \cdot 10^5 \left(\frac{p_{cr}}{p} \right)^{0,7} \frac{\rho w^{0,4}}{D_{int}^{0,5}}$$

here q_0^{str} is the critical heat flux in a straight tube, ρw is in kg/m$^2 \cdot$ sec, q is in W/m^2, D_{in} and R_{bend} are in mm, ratio $K_1 = q_0^c/q_0^{str}$ is the ratio of critical heat fluxes at $x = 0$. A special nomogram was constructed by Pikus for this parameter.

As noted above, under conditions characteristic of power generating equipment, the boiling crisis in channels occurs at rather high liquid contents in the flow. In the immediately pre-crisis zone the liquid is concentrated in the flow core. Thus, if the liquid could be moved to the wall, there is a real possibility of raising the critical power. The problem consists of designing various devices to direct this liquid to the channel walls.

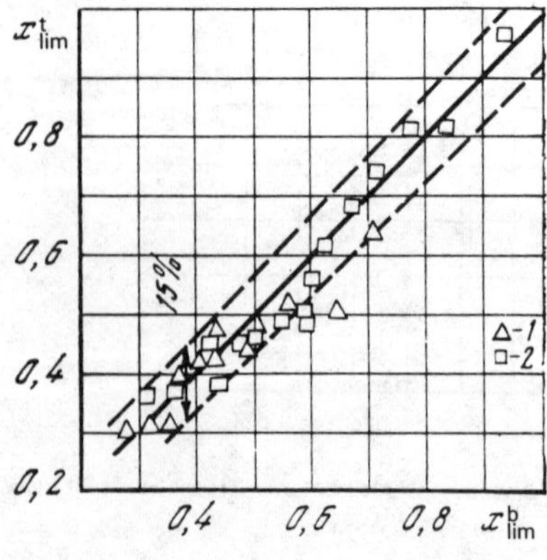

Figure 3.26 Comparison of experimental data in experiments with top and bottom bends. 1, $\rho w = 1100$ kg/ m$^2 \cdot$ sec; 2, 700 kg/m$^2 \cdot$ sec.

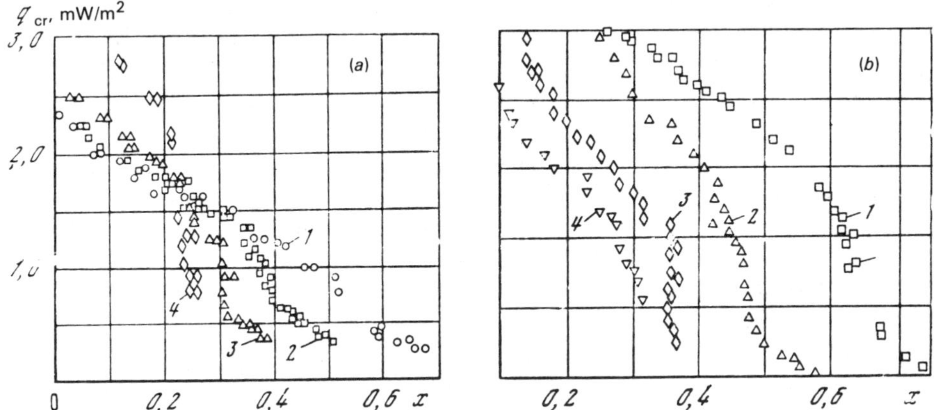

Figure 3.27 Critical heat flux vs. quality for rod assemblies at $p = 7.4$ MPa (a) and for at tube with $D_{in} = 8$ mm and $p = 7.8$ MPa (b). (a) 1, $\rho w = 600$ kg/m²·sec; 2, 1000; 3, 1500; 4, 2500 kg/m²·sec. (b) 1, $\rho w = 1000$; 2, 1600; 3, 2700; 4, 4000 kg/m²·sec.

The experience of using various enhancing devices in single-phase flows is rather extensive, and it is natural to use it for increasing the critical power in two-phase flows. One of the first studies in this direction is apparently that by Gambill and Greene [3.65]. Subsequently, these efforts have been continued by other investigators [3.66–3.85]. At least two methods are available for raising the critical power: (1) swirling of the flow, i.e., producing a radial field of mass forces, which would induce a flux of droplets to the wall; (2) turbulization of the flow, i.e., increasing the pulsatile component of the velocity, and thus also of the droplet flow to the wall. Up to now, flow swirling has been the more extensively used technique. Variations of this technique include swirling in twisted tubes of noncircular cross section, in spirally wound tubes, and also by twisted tapes (augers), local turbulence promoters, wire spirals, helical fins, tangential injection of liquid. Induced flow turbulence was attained by all kinds of projections, fins, etc.

At present rod-type assemblies are extensively used in reactor construction practice. The critical heat fluxes in such assemblies depend on the geometry of the rods, their spacing, type of spacing grids, etc. Usually, q_{cr} in assemblies is lower than in cylindrical tubes (Fig. 3.27). Increasing the critical power in such devices is of particular importance. The study by Smolin and Polyayev [3.83] can serve as an example of using local swirl-producing devices for increasing q_{cr} in rod-type assemblies. The assembly, with honeycomb-type grids, consisted of seven 12 mm diameter fuel pins with a wetted length of 2200 mm. It is seen from Fig. 3.28 that the use of local swirl generators significantly widens the region of noncritical operation.

A large cycle of studies was performed at the Physical Energetics Institute by Blagovestova et al. [3.82] and Subbotin et al. [3.84]. They investigated the effect of various flow swirling devices: twisted tapes, wire spirals, internal helical fins, and elliptical-cross-section twisted tubes. the tubes used in the studies were 10.8 to 11 mm in diameter and from 1 to 6 m long. It was found that swirling makes it possible not only to increase the critical heat flux (Fig. 3.29), but also, in a number of cases to prevent the occurrence of the boiling crisis as such. At constant inlet flow enthalpy the swirling effect is a function of the tube length (Fig. 3.30). In short tubes ($L < 0.7$ m) swirling does not result in heat transfer rate enhancement [3.84]. The swirling effectiveness increases with the flow quality. Up to a certain limit (as long as the pitch of the flow swirl is identical to the pitch of the swirling device and there is no slip). The value of q_{cr} can also be raised by reducing the swirling pitch.

Qualitatively similar results were also obtained with corrugated tubes [3.84]. Transverse corrugations were produced to a depth of 0.5 to 1.5 mm and a pitch between 10 and 400 mm. A typical feature in the temperature behavior of these tubes is the fact that the initial rise in temperature occurs directly upstream of the last corrugation. This is due to reduction in flow turbulization in the space between the corrugations and reduction of the droplet flux to the wall. Another important feature of corrugated tubes is the self similar behavior of the critical heat flux relative to the flowrate. In particular, for tubes with corrugation depth of 1 mm and a 100 mm pitch the value of q_{cr} remained unchanged while the flowrate was raised from 550 to 1250 kg/hr.

The value of q_{cr} in fuel-pin assemblies is significantly affected by the spacing grids. Depending on their type and spacing, the critical heat flux may both decrease and increase. It is noted by Klochkova and Sapankevich [3.80] that reduction in the distance between honeycomb-type grids from 700 to 175 mm resulted in a 25–30% increase in q_{cr}. In designing spacing grids it is desirable to select a spacing such that each successive grid is situated in the field of turbulization produced by the preceding grid.

Another geometry-produced perturbation which deserves mention is the effect of various flow constrictions. These may be various projections resulting from the fabrication technology, etc.

Let us consider the effect of flow constriction on q_{cr} for the case of an annular fin. Osmachkin et al. [3.86] performed experiments in an annulus with only the inner tube heated. An annular fin was placed on the inner surface of the outer tube, which produced a flow constriction and acceleration in the immediate proximity of the heated surface. It was found (Fig. 3.31) that a local constriction reduces the value of q_{cr}. The reduction in q_{cr} is the greater, the smaller the mass velocity of the flow and the higher the pressure. An analogous effect of flow constriction on the threshold

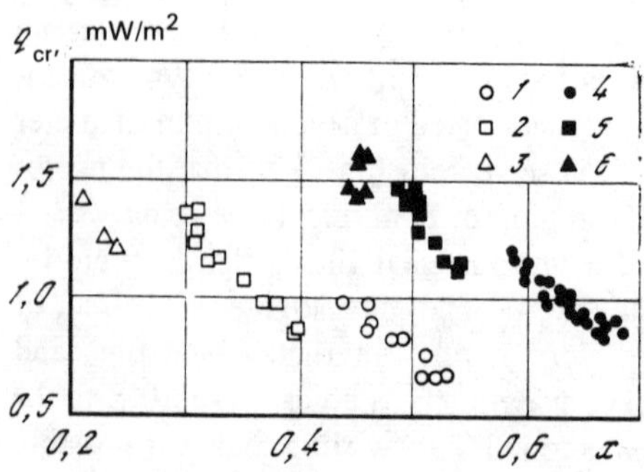

Figure 3.28 Critical heat flux vs. quality in rod assemblies at $p = 9.8$ MPa. In assemblies with honeycomb kind grids: 1, $\rho w = 600$ kg/m$^2 \cdot$sec; 2, 1000; 3, 1500; in assemblies with turbulence promoters: 4, $\rho w = 600$ kg/m$^2 \cdot$sec; 5, 1000; 6, 1500.

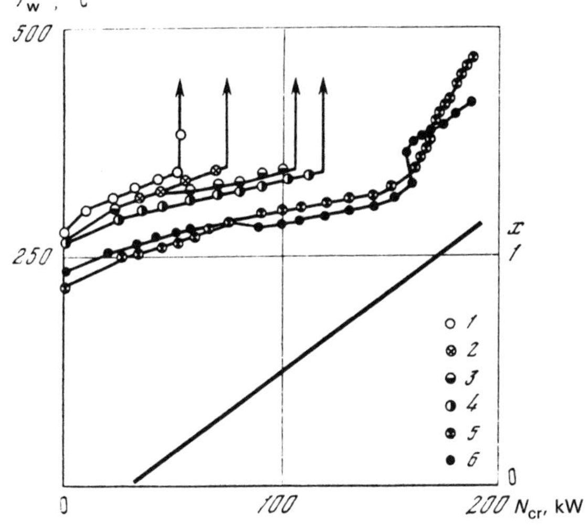

Figure 3.29 Temperature of the outer wall of tubes with three inner helical fins as a function of the supplied power at $p = 13.7$ MPa, flowrate of 500 kg/hr ($\rho w = 1500$ kg/m$^2 \cdot$ sec, inlet water temperature 320°C. Tube lengths: 1, 1m; 2, 1.5; 3, 2; 4, 2.26; 5, 4; 6, 6 m; the bottom line represents the output quality).

quality was noted also by the present authors in one of their previous papers [3.87].

It should be noted that the rate of deposition of impurities from circulating water increases significantly in enhanced channels, which may reduce the benefits due to flow turbulization. This circumstance must be taken into account both in designing and operating boiling-type equipment.

3.7 EFFECT OF DEPOSITS OF IMPURITIES FROM CIRCULATION–LOOP WATER ON THE BOILING CRISIS

The presence of deposits, consisting of precipitated water impurities, on the walls of steam-generating channels gave rise to a large number of problems of scientific interest and practical importance. Deposits, for example, of products of corrosion of materials of construction are permeable and have a capillary-porous structure. The supply of liquid to a heated wall is controlled primarily by laws of capillary hydrody-

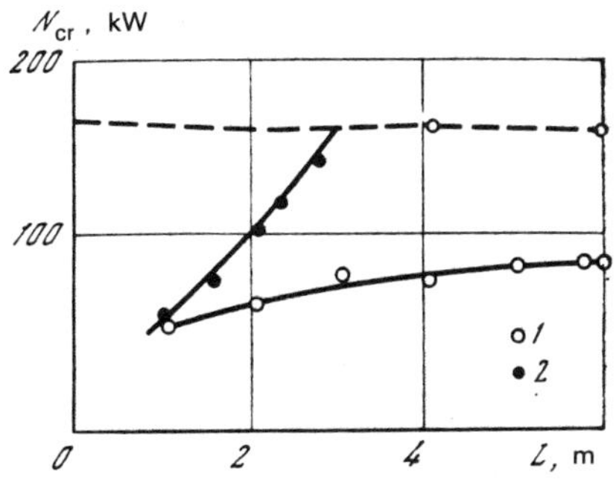

Figure 3.30 Effect of length on the critical power at $p = 13.7$ MPa, flowrate of 500 kg/hr, inlet water temperature 320°C. 1, Smooth tubes, data of the Physical Energetics Institute; 2, tube with three 1.2 mm high and 200 mm long inner fins; the dashed curve corresponds to the exit quality of 0.9.

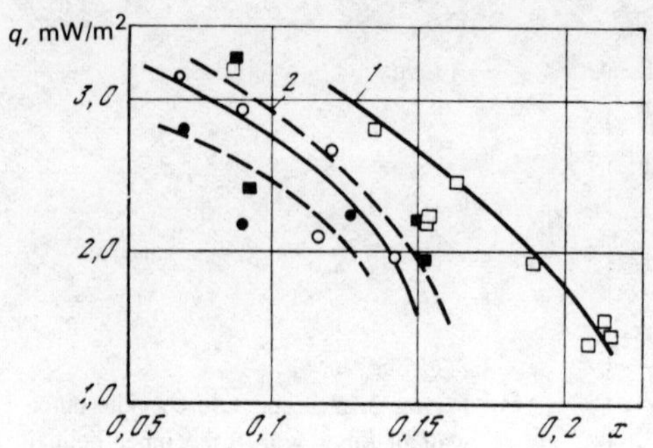

Figure 3.31 Effect of local channel constriction on critical heat fluxes; p = 9.8 MPa. 1, Smooth annulus, ρw = 900 and 1700 kg/m²·sec; 2, annulus with annular fin on the nonheated surface, ρw = 900 and 1700 kg/m²·sec.

namics. The nucleation and growth of a vapor bubble occur initially in the porous structure, and is not explicitly affected by the external coolant flow. The departure diameter of the bubble is affected by capillary pressure in addition to ordinary external-flow forces. These are the primary distinguishing features of boiling in a capillary-porous body as compared with boiling on an impermeable surface.

Deposits of products of corrosion affect the temperature distribution in steam-generating channels. Usually changes in the wall temperature are moderate. This is because the bulk of the heat is transmitted by means of the phase transition, analogous to the manner in which heat is transferred in heat pipes upon boiling within the wicks. However, the rise in wall temperature may also be significant, in the case when solidification of hard salts, SiO_2 and other low thermal conductivity compounds occurs within the pores.

It is of great importance to consider the inception of the boiling crisis in capillary-porous bodies and, in particular, in the layers of deposit corrosion products. This question has not been sufficiently explored. However, according to available data on the boiling crisis in heat pipes, the critical heat fluxes are several fold smaller than on impermeable walls. Boiling crisis in capillary-porous bodies with a matrix of high thermal conductivity occurs less suddenly and is accompanied by a smaller temperature jump. Under ordinary conditions, the boiling crisis manifests itself suddenly and can be clearly and easily detected since it is accompanied by a sudden rise in the wall temperature. Under operating conditions, the boiling crisis in capillary-porous bodies may remain unnoticed. However, when the porous matrix has a low thermal conductivity, the boiling crisis may manifest itself even more drastically than on an unpermeable wall. However, even in this case, it may be to detect the onset of the boiling crisis. The replacement of the ordinary boiling mode by film boiling is accompanied by a significant concentration of impurities (including aggressive agents), which may result in intensive corrosion.

3.7.1 Physico-Chemical Properties of Deposits and Coolant Impurities

Prior to considering details of boiling crisis inception in a porous layer, let us consider the main properties of the impurity materials from which the layer is formed. Analysis of a large body of published data shows that products of corrosion of the materials of construction comprise a significant part of nuclear power plant coolant impurities. As the water treatment arrangements and condenser designs are improved, increasingly less impurities enter the loop with the makeup water and intrusions of cooling water. At the same time, products of corrosion of materials of construction are continuously entering the working fluid. However, the chemical composition and concentration of these impurities are determined to a large extent by the size of the surfaces in contact with the coolant, by the properties of the materials and by the operating conditions.

The principal products of corrosion are various forms of iron oxides, and also oxides of Cr, Co, Ni, Mo, Mn, Zr, Cu, etc. As a rule, iron oxides comprise more than 70–80% of the products of corrosion.

The range of forms of iron oxide present in the water used in power plants is very wide. According to Homig [3.88], the following may, in principle, be present in the water: compounds of two-valent iron, which result in the formation of FeO; compounds containing in the molecule simultaneously two- and three-valent iron, which in the final analysis form Fe_3O_4; compounds of three-valent iron, which become transformed into α-Fe_2O_3 and γ-Fe_2O_3.

The main parameters controlling the probability of predominant existence of one or another forms of oxides are the pH index, the redox potential of the system and the temperature. In the iron oxide-water system the equilibrium between the ionic and molecular forms of iron with the same valency depend to a large degree on the value of pH. Magnetite is the thermodynamically stable oxide under normal operating conditions at the steam-generating surfaces of power plants (temperature $\leq$ 400 °C). Hematite or its hydrated forms can be stable oxides only at higher oxidation potential (for example, at high oxygen concentrations) and elevated temperatures.

Studies for determining the actual solubility of products of corrosion acquire particular importance in predicting the nature of the deposits. The published data on the solubility of iron oxides at high temperatures are highly contradictory, which is due to great difficulties in performing the experiments. On the whole it can be noted that the solubility of iron oxides under power plant operating conditions range from one to several tens of ppb parts per billion. This is significantly lower than their ordinary concentration in the coolant water of reactors and in nuclear steam generators. It suffices to say that during transition periods, for example, the concentration of iron oxides can amount to several ppm.

The products of corrosion in loop water may exist in the form of a real solution, colloidal solution or suspension: in a real solution the particle (molecule) size is up to 10 Å, in a colloidal solution it ranges from 10 to 1000 Å, and in a suspension, it is more than 1000 Å.

The spectrum of impurity particle size is usually analyzed either by microscopic methods, or by filtration of samples of water through a number of membrane-type ultrafilters with subsequent determination of the iron concentration in the filtrate, or of the amount of iron held back by the filter. The particle size is significantly affected by the water regime of the unit, the coolant temperature, the impurity concentration, the residence time of the suspension in the loop and other factors. The question of the particle size spectrum has not yet been sufficiently explored. However, it can be noted in general that the majority of particles encountered in power-generating equipment range in size from several hundreds of millimicrons to several tens of microns.

Deposits of coolant impurities on heated surfaces change the physical situation under which transfer processes occur. To clarify the general properties of deposits, let us consider their structure. According to Schoch [3.90] and Schuster [3.91], deposits consist of two layers, an inner dense oxide layer and an outer oxide layer with large monocrystals.

The inner layer arises as a result of oxidation and dissolution reactions. In the formation of this layer all the composite parts of the alloy have the same volumetric concentration as in the material of the wall. The only exception is Mn, whose concentration is somewhat lower, since it has a better solubility than other constituents.

Conversely, the composition of the outer layer differs significantly from that of the main alloy. In the first place it is determined by the constituents which dissolve well in water of the given chemical composition. These are primarily Fe and Mn. The presence of elements in the outer layer is a question of solubility of their ions. Consequently, this is a sedimentation layer. Both layers grow simultaneously in opposite directions from the initial surface. The inner layer adheres tightly to the metal surface and is an "intrinsic" protection layer.

The formation of the outer layer depends on the motion of soluble elements of the metal and hence its structure is significantly affected by the flow pattern and heat transfer conditions. According to data of Schoch [3.90] and Schuster [3.91], the inner layer consists of crystals with sizes between 1000 and 2000 Å. The outer layer consists of larger particles, their size being determined by the size of particles in the flow. The structure and properties of sedimentation deposits depends significantly on the size distribution of the flow impurities and the flow conditions under which the deposits form.

Deposits forming from colloidal and large-size particles are only weakly coupled to the surface. They have a porous structure and consequently may significantly affect the behavior of transfer processes. It is these deposits that are primarily considered here.

Macbeth [3.92] reported that the layers studied in his work had characteristic "chimneys" which carried the steam out of the porous layer. These had diameters of the order 5 μm, and there were about 5000 such chimneys per mm^2 of the layer. The liquid phase entered the layer through passages which were 0.1 to 0.5 μm with the mean distance between passages being about 14 μm. The passages are interconnected.

Rassokhin et al. [3.93] analyzed a depthwise section of a deposit layer and showed that the chimney diameters ranged between 5 and 10 μm and that their population may be from 3000 to 5000 per mm^2. The chimneys are directed in the direction of removal of vapor bubbles from the wall.

The variety of physico-chemical conditions under which the various power generating plants operate is responsible for a wide range of deposit densities. The latter depend to a large extent on the chemical composition, porosity and structure of the deposits. This explains the contradictions in data of various investigators. According to data of different investigators [3.93–3.97], the density ranges from 2.5 to 4.7

gram/cm^3. The top limit on the density of deposits of iron oxide, for example, is naturally limited by the theoretical density of $\rho = 5.2$ gram/cm^3.

The range of porosity of the deposits is rather wide. Macbeth [3.92] reported an effective porosity $\epsilon = 30\%$. Cohen [3.98] estimated the porosity of iron-oxide deposits at between 40 and 60%. Rassokhin et al. [3.93] note that the experimental heat transfer coefficient and that calculated from the Maxwell formula on the assumption of $\epsilon = 0.50$ are in satisfactory agreement.

One of the most important properties of porous deposits is thermal conductivity. The problem here is that it apparently cannot be regarded as a thermophysical constant of the porous material. Thermal conductivity here is a function not only of the thermal conductivity of the material of the basic matrix and of the filler, but also the processes which control the transport of heat through the deposit layer.

The question of thermal conductivity of iron-oxide deposits are considered in great deal by Rassokhin et al. [3.93]. Treating the deposits as a layer of porous material, and using the Maxwell equation for the conductivity of the two-phase (oxide-water or oxide-steam) layer, these authors constructed a graph for the thermal conductivity as a function of porosity and temperature. The thermal conductivity of water and steam was taken at $p = 9.8$ MPa. It is seen from the calculations that the thermal conductivity (λ) of magnetite deposits with water-filled pores ranges, depending on the porosity ϵ, between 1 and 3 W/m$\cdot$°C. In the case of vapor-filled pores λ ranges between 0.5 and 2.0 W/m$\cdot$°C. According to data of Kelen and Arvesen [3.97] the actual thermal conductivity of deposits ranged from 0.17 to 3.25 W/m$\cdot$°C.

The thermal conductivity of porous deposits increases significantly when boiling occurs within them. Thus, according to Pshemenskiy [3.96], λ is estimated at 35 W/m$\cdot$°C, according to Cohen [3.98], the value of λ at temperatures between 149 and 316°C ranged from 14.2 to 12.1 W/m$\cdot$°C. Under surface boiling conditions [3.93] the value of λ increased several fold and rose to 35 W/m$\cdot$°C. According to Kelen and Gustafsson [3.99], the thermal conductivity of deposits is 6.25 W/m$\cdot$°C.

The above shows that the thermal conductivity of deposits cannot be regarded as a thermophysical constant. It is more proper to use the concept of effect thermal conductivity, based on the physical nature of the working processes occurring in the layer and the physico-chemical properties of the deposits.

The thickness of deposits depends on a large number of factors: heat flux, mass flux, pressure, vapor quality, concentration and spectrum of particle sizes, roughness, time, etc. It is hence natural that this thickness varies within very large limits. According to limits established by various investigators it can be claimed that the minimum thickness is several microns whereas the maximum is several millimeters.

3.7.2 Boiling Crisis within Porous Deposits

A number of studies have been reported recently in which the effect of porous deposits on q_{cr} was investigated. Let us consider some of these in more detail.

It is known from experimental studies of boiling in heat pipes that the boiling crisis in porous structures arises at significantly lower heat fluxes than on an impermeable surface. Apparently, the first study of the effect of deposits on q_{cr} was that of

Table 3.4

Section	Mean size of droplets		Porosity, %	Chemical composition by weight, %							
	Thickness, mm	Quantity by weight, mg/mm^2		Fe	Cu	Ni	C	Si	Al	Cd	Mg
MS18/311CL	0,03	0,05	70	65	0,15	0,7	0,1	0,1	0,06	—	—
MS19/321CL	0,10	0,15	71	65	1,15	0,2	1,5	0,2	0,03	—	—
MS27/351CL	0,09	0,11	77	62	—	0,3	5,2	—	—	—	—
MS26A/331CL	0,15	0,18	77	66	0,68	0,4	5,0	1,0	0,06	2,0	0,1

Rice and Goldstein with their coworkers [3.100–3.103]. In their study of corrosion of metal in the deposit layer they added significant amounts of copper and iron oxides to the water, while monitoring the wall temperature of the steam-generating channel. Starting with addition of these impurities, deposits started forming and the wall temperature at the channel exit rose. This was regarded as the onset of crisis conditions, which disappeared after the deposits were washed off. This phenomenon as such is noted in [3.100–3.101], but is not investigated there in depth.

Macbeth et al. [3.104] performed a thorough study of critical heat flux in an annulus with deposits on the inner (heated) surface. They measured the temperature conditions and pressure differences. The deposits were produced on the outer surface of the heated tube before it was placed in the test section. The description of the deposits is given in Table 3.4.

The deposits on the first three sections consisted primarily of magnetite. The last section was also first coated by magnetite, but then it was permeated with impurities by boiling local tap water on its surface for 77 hrs. An item of importance in these experiments was to provide deposits with properties close to those actually encountered in industrial units. Chemical and structural analyses showed that these conditions were in fact met. On the last section, as under actual conditions, the inner layer was dense and without pores. As a whole Macbeth and his coworkers felt that they have been able to successfully solve the problem of producing representative deposits on the surface of the test section.

Investigations of the structure of the deposits used by Macbeth showed that the pores penetrate deeply into the deposits. A part of the pores has a relative large diameter. Macbeth et al. suggest that the large pores serve as steam channels and the small ones as water channels. The experiments were performed under the following conditions: $p = 6.9$ MPa, $\rho w = 1360, 2720$ and 3400 kg/m$^2 \cdot$ sec, inlet water subcooling ΔT_{sub} between 3 and 60 °C, heat flux density q between 120 and 320 W. cm^2.

The feedwater consisted of demineralized deaerated water with pH between 9 and 10. The value of pH was artificially lowered when performing experiments at length with deposits, in order to reduce the dissolution of the latter. The quality of the loop water deteriorated as the experiments progressed. This is apparently due to the fact that a part of the deposits was washed off and absorbed by the flow. In addition, the water quality deteriorated due to corrosion of structural parts of the test stand. Figure 3.32 shows the effect of the deposit on the critical heat flux as a function of subcooling of the liquid entering the section at mass velocities of 1360, 2720 and 3400 kg/m^2·sec. The thickness of deposits in this series of experiments was 100 μm. It is seen from the figure that q_{cr} in the sections with deposits falls by 5 to 10% compared to a clean test section. Varying the deposit thickness between 30 and 150 μm had virtually no effect on the level of q_{cr}. This is possibly because a part of the deposits was washed off from the tube surface, and hence the subsequent run was performed with a lower deposit thickness. Macbeth and his coworkers note that the deposits were washed off primarily in the course of cooling, startup and stopping of the unit. Individual measurements of deposit thickness after experiments showed that up to 50% of the deposits could wash off in the course of experiments. However, this reduction in deposit thickness did not affect the results, which was confirmed in special tests. A rod with deposits was removed from the test section after the first series of runs and was held in air for 14 days. Then it was again placed into the test section and the runs were repeated. The results of both series of runs on q_{cr} were identical.

A very thorough study of the effect of deposition of corrosion products on the boiling crisis was performed by Rikhter [3.105]. His experiments were performed in an annulus (δ = 2 mm) with inner heated tube of zirconium 1% niobium alloy. He investigated the effect of the chemical composition and concentration of the corrosion products of iron, copper, and calcium and the value of pH, structure and thickness of the layer of deposits on q_{cr}. The operating conditions of the experiments were: p = 7 MPa, ρw between 1250 and 5000 kg/m^2·sec, quality x between -0.309 and 0.168.

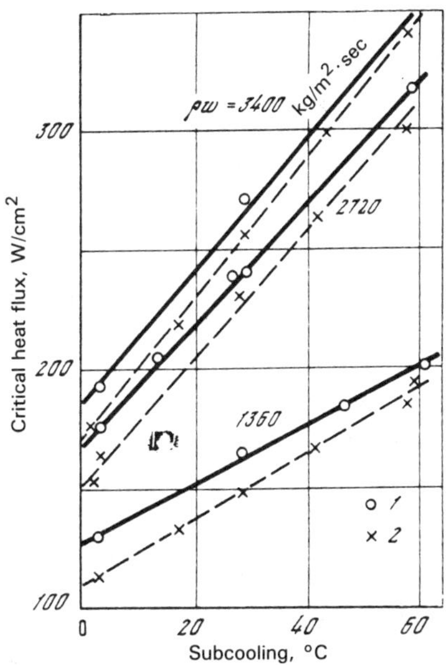

Figure 3.32 Critical heat flux vs. the subcooling of water [3.104]. 1, Clean tube; 2, tube with deposits.

The deposits were formed in the loop at pH between 4 and 9. For this purpose measured portions of iron and copper oxides, and also a calcium sulfate solution were added to the loop. He investigated basically two types of deposits: the basis in the first was magnetite (up to 80%) and the second was calcium and magnesium (up to 42%) with up to 18% copper. It was found that the deposits significantly reduce q_{cr}. Thus, for example, the difference in q_{cr} for δ_{dep} between 9 and 15 μm and q_{cr} for a clean surface at $x = 0$ is 79%. In the case of δ_{dep} between 48 and 90 μm the difference in q_{cr} is as high as 146%. The steepest change in q_{cr} occurs as the deposit thickness is increased to 5–15 μm. As δ_{dep} is increased further, the value of q_{cr} changes insignificantly.

The chemical composition of the deposits has a significant effect on the value of q_{cr}. The critical fluxes were the smallest in deposits consisting primarily of iron oxides and of calcium sulfate. Apparently, the deposit structure under these conditions forms in such a manner that the rate of coolant circulation in the pores decreases.

The following possible reasons for reduction in q_{cr} are listed in [3.105].

1. Significant increase in the boiling surface when steam is generated inside the pores of the deposit layer. This may result in premature appearance of a continuous vapor film as compared with an impermeable surface, which will induce the boiling crisis.
2. Possibility of interference with circulation in pores due to partial or complete plugging up of capillaries by coolant-carried impurities.
3. Inception of boiling crisis at lower values of q_{cr} due to the poorer thermophysical properties of the deposits as compared with a metal surface. In the case when local superheating of the wall occurs, the lower thermal conductivity of the layer as compared with the metal will not provide the necessary removal of heat and the crisis will extend over a rather large tube length.

Rikhter [3.105] presents an empirical correlation, which incorporates the effect of deposits by means of correction factors:

$$q_{cr} = C\tau\rho w K_1^{0,35} K_2^{0,8} K_3^{0,08} K_4^{0,6} K_5^{0,7} \tag{3.84}$$

where τ, K_1, K_2, K_3, K_4, and K_5 are parameters which are functions of the operating conditions and design; $C = f(\delta_{dep})$ which is the parameter representing the effect of deposits, ranges from 3.1 to 1.9.

The effect of deposits on q_{cr} was investigated also by others. For example, test section used by Tomas and Heitman [3.106], consisted of a 2500 mm long vertical tube with outer diameter of 12 and inner diameter of 5.6 mm. The tube wall and coolant temperatures were measured by Fe-K$_0$ thermocouples. Scintillators and a radiation counter were distributed over the height of the section for measuring the quantity of material being deposited. The pressure and pressure drops in the test section were measured by a standard pressure gauge and a Barton cell. Water with specified inlet parameters was first supplied to a mixing chamber where it was mixed

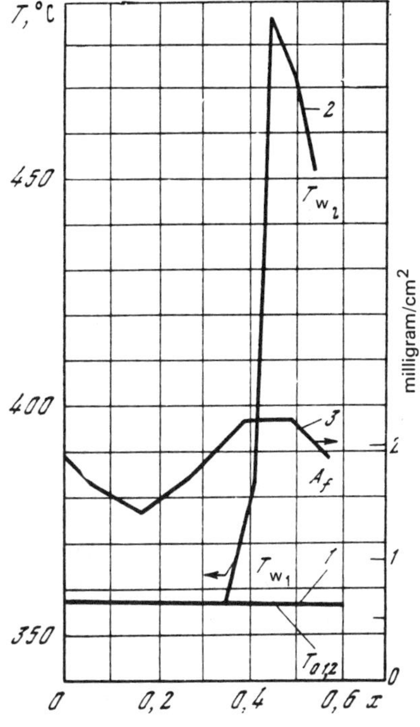

Figure 3.33 Deposits-induced boiling crisis [3.106] at p = 17.8 MPa, q = 371 kW/m^2 and ρw = 1100 kg<m^2·sec.

with a magnetite suspension delivered from a special vessel, and then was fed by a pump to the loop.

Figure 3.33 shows experimental results obtained at p = 17.8 MPa, q = 371 kW/m^2 and ρw = 1100 kg/m^2·sec. Curve 1 represents the temperature behavior of the experimental tube before scale deposition, and curve 2—after scaling. This temperature profile corresponds to the behavior of wall temperature along the channel upon inception of the boiling crisis. Curve 3 shows the size of deposits which induced the crisis. It is seen from the figure that the crisis sets on within the deposits at a lower quality of vapor than on an impermeable wall.

It is suggested by Tomas and Heitman [3.106] that heat transfer deteriorates as a result of the "smoothing" effect of deposits on the effective roughness of the surface. In our opinion, this point of view is incorrect. In fact, in the majority of experiments the ratio D/K ranged from 80 to 280, which corresponds to absolute roughness values K from 20 to 70 μm. As shown by the experiments of a large number of investigators, variation in the surface roughness within these limits has virtually no effect on the level of q_{cr}. The reason fro the effect is probably quite different. As previously noted, deposits forming under such conditions are porous. If the liquid encounters difficulties in reaching the heating surface through the porous layer, this could lead to premature overheating of the surface. A particularly important role in this phenomenon is played by pressure. The effect of deposits increases with rising pressure, which follows, in particular, from the results of [3.106]. Thus, whereas at p = 17.8 MPa the value of q_{cr} decreased by 23%, then at p = 20.7 MPa, it decreases by 45%.

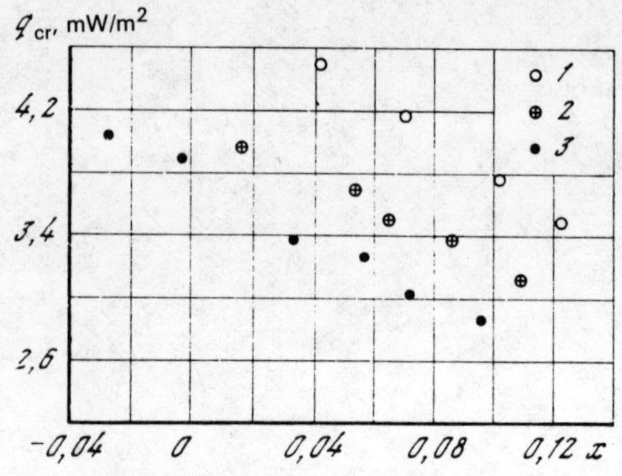

Figure 3.34 Plot of $q_{cr} = f(x)$ for different thicknesses of deposits. 1, $\delta_{dep} = 0$; 2, 20; 3, 55 μm.

Balabanov [3.107] continued studies started at the Moscow Energetics Institute [3.105] of the effect of deposits on q_{cr}. The experiments were performed in an annulus with inner heated tube. During the period when deposits collected on the tube, a suspension obtained by electrolytic dissolution of steel 20 was supplied to the loop. The iron concentration in the experiments ranged from 0.4 to 5.15 ppm. The deposits accumulated at a pressure of 7.0 MPa under surface boiling conditions (q between 1.41 and 1.7 MW/m^2, $\rho w = 3500$–4700 kg/m$^2 \cdot$sec, $x_{out} = -0.04$) at pH $= 5.5$–6.5 and pH $= 9.0$–9.5. The conditions of deposits accumulation were very close to experimental conditions. The accumulation period ranged from 10 to 106 hr. The thickness of deposits ranged between 6.5 and 55 μm. Chemical analysis showed that 97.7% by weight of deposits consisted of Fe_2O_3 (74.2%) and Fe_3O_4 (23.5%). The operating conditions were: $p = 7$ MPa, $\rho w = 1250$–5000 kg/m$^2 \cdot$sec, $q_{cr} = 2.04$–4.91 MW/m^2, $x_{cr} = -0.3$–0.16.

Figure 3.34 shows the effect of the thickness of deposits on the value of q_{cr} at mass velocity of 2500 kg/m$^2 \cdot$sec. It is seen that the effect of the layer is in direct proportion to its thickness. The variation in $q_{cr} = q(\delta_{dep})$ is shown in Fig. 3.35. The

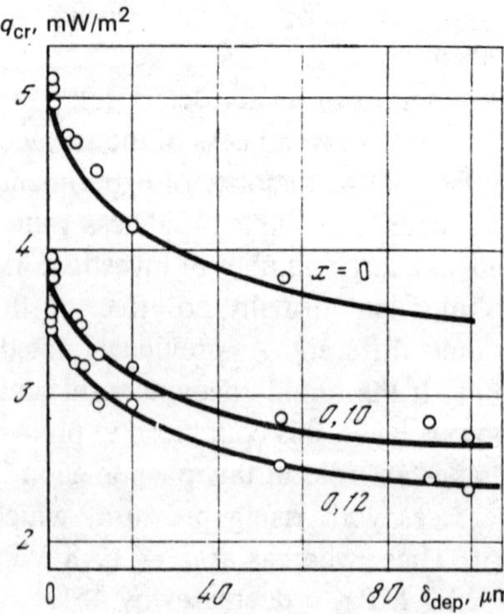

Figure 3.35 Plot of q_{cr} vs. thickness of deposits at different x [3.107].

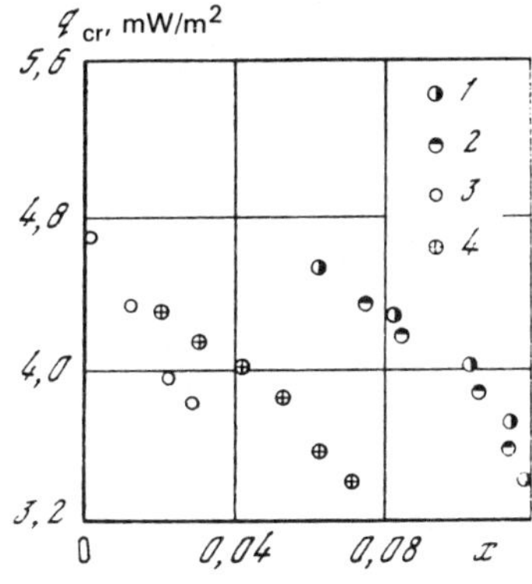

Figure 3.36 Plot of $q_{cr} = f(x)$ at $\rho w = 3500$ kg/m$^2\cdot$sec at different deposit thicknesses. 1, $\delta_{dep} = 12$; 2, 25; 3, 40; 4, 4 μm.

effect of the layer is greatest at moderate thicknesses (to 20–30 μm). As the layer thickness increases the absolute value of $\partial q_{cr}/\partial \delta_{dep}$ decreases, although it still remains perceptible.

The results of studies by Rikhter [3.105] and Balabanov [3.107] are compared in Fig. 3.36, from which it is seen that q_{cr} at $\delta_{dep} = 5$ μm [3.105] is approximately the same as q_{cr} at $\delta_{dep} = 40$ μm at low vapor qualities and is lower at $\delta_{dep} = 12$–25 μm [3.107] at somewhat higher qualities. This difference is attributed by Balabanov to different structures of the deposits. The working fluid in the study by Rikhter was turbine condensate and the deposits formed were Fe_3O_4 and $Fe(OH)_2$, whereas Balabanov used chemically purified water (its pH is lower than that of the turbine condensate, and the products of corrosion were obtained by electrolytic dissolution of steel 20).

Comparison of data of Macbeth et al. [3.104] and Balabanov [3.107] shows that their results are different. Balabanov ascribes this difference to the fact that the deposits in the study by Macbeth et al. have a structure with "steam-chimneys," whereas the deposits in the study by Balabanov had a porous structure without a clearly delineated vapor channel structure. This means that according to Balabanov, the conditions for crisis inception depend significantly on the structure of deposits.

Balabanov presents an empirical correlation for the critical heat flux in channels with porous deposits:

$$\frac{q_{cr}}{r\rho w} = 65{,}027 \left[\frac{\lambda}{\rho w c_p D_{eq}}\right]^{0,93} \left[1 - \frac{h_{cr} - h'}{r}\right] \left[\frac{h'\rho'}{r(\rho'-\rho'')}\right] [1 - 0{,}065\delta_{dep}^{0,323}]$$

$$(3.85)$$

As is seen from Eq. (3.85), correction for the effect of deposits is performed using the expression

$$q_{cr\ dep}/q_{cr\ 0} = 1 - 0{,}065\delta_{dep}^{0,323} \tag{3.86}$$

Rikhter [3.105] uses a correction in the form of Eq. (3.84). The difference in the

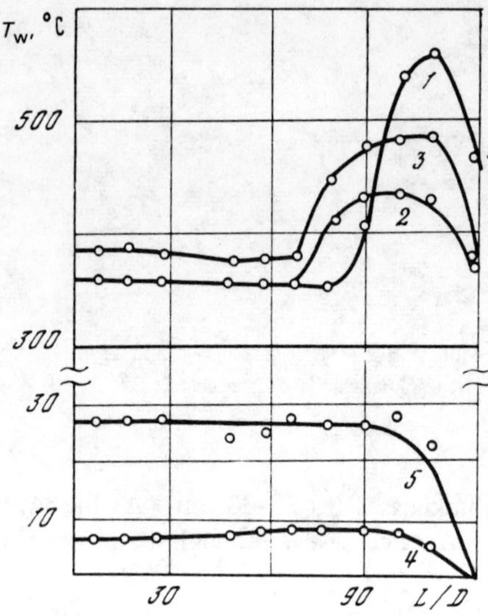

Figure 3.37 Effect of calcium sulfate deposits on the critical vapor quality [3.108] at p = 13.7 MPa, q = 600 kW/m^2, ρw = 1000 kg/m$^2 \cdot$ sec.

correction, as previously noted, is a result of the fact that the structure of deposits in experiments described in [3.105] and [3.107] was different. From this follows, in part, the conclusion that the process is affected not only by the layer's thickness, but also by its other properties.

It follows from the above that the effect of porous deposits may be significant, and must be taken into account both in designing and operating of steam-generating facilities. This must be particularly remembered in conjunction with the current tendency to use carbon steels in nuclear power loops.

In addition to porous deposits, one has to deal in practice with deposits of hard salts, which adhere tightly to the heating surface. In accordance with circulation-water standards, these compounds are present in the form of a true solution. The salt precipitates by crystallization, for which reason the deposits have a nonpermeable structure. The relationships governing the inception of the boiling crisis in these structures are the same as on a nonpermeable metal wall. The effect of impermeable deposits on the critical variables·has not yet been investigated thoroughly.

Figure 3.37 shows as an illustration the results of investigating the effect of calcium sulfate deposit son q_{cr}, based on data of the present authors [3.108]. Curve 1 represents the variation in the temperature of a clean wall along the channel Curve 2 gives the temperature of a clean wall following the accumulation of a certain amount of salt, corresponding to the increment of wall temperature in nucleate boiling (ΔT = 8–9 °C—curve 4). A further increment in the deposit layer (ΔT = 30 °C) results in a change in the temperature profile (curve 3), however, the value of the threshold vapor quality did not change. The value of q_{cr} is virtually unaffected by a variation in the depth of calcium sulfate deposits between 0 and 40 μm, which allows the claim that, to first approximation, thin impermeable deposits do not affect the critical parameters.

FOUR

HEAT TRANSFER IN THE POST–DRYOUT REGION

4.1 DETERIORATION OF HEAT TRANSFER IN THE POST–DRYOUT REGION IN STEAM–GENERATING CHANNELS

The deterioration of heat transfer in the post-dryout region of steam-generating channels was discovered a relatively long time ago. However, systematic studies were started relatively recently in the context of the design of once-through reactors and steam generators in the power-generating industry, cryogenic engineering and other fields.

A rather large volume of experimental analytical work has now been carried out on the tube wall temperature distribution under post-dryout conditions. These studies were carried out over a wide range of variables and under a large variety of experimental conditions: in tubes and annuli, in vertical, horizontal, upwards and downwards flows, and with various liquids. Post-dryout heat transfer has been most extensively explored in upwards steam-water flows in smooth vertical tubes.

One of the first experimental studies of the temperature profile under post-dryout conditions in the field of power generation was apparently that by the first of the present authors with his coworkers [4.1]. They established experimentally that, at a certain vapor quality (Fig. 4.1), the wall temperature of a steam-generating channel undergoes a steep change. The wall temperature first rises abruptly, attains a maximum, and then decreases. The maximum increment of wall temperature $T_{\mathrm{w}}^{\mathrm{max}} = T_{\mathrm{w}} - T_{\mathrm{sat}}$ is a function of the mass velocity. The higher the latter, the smaller the increment in wall temperature.

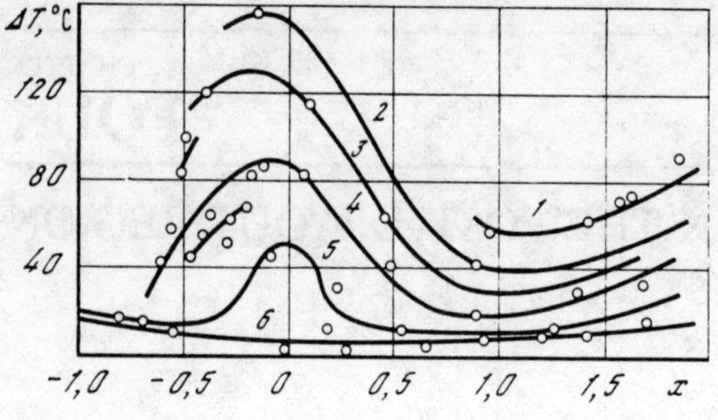

Figure 4.1 Temperature difference between the inner surface of a tube and the coolant as a function of the relative enthalpy [4.1]. q = 314 kW/m², p = 21.4–21.8 MPa; 1, ρw = 385 kg/m²·sec; 2, 440; 3, 480; 4, 550; 5, 700; 6, 1650 kg/m²·sec.

A very thorough study of heat transfer in post-dryout region performed by Miropol'skiy [4.2]. His experiments were performed over a wide range of variables (p between 3.92 and 21.6 MPa, ρw of 400 to 2000 kg/m²·sec and q from 232 to 1160 kW/m²). These studies verified the principal relationships (Fig. 4.2) governing film-boiling heat transfer. He additionally showed that the value of T_w^{max} increases significantly with the heat flux density. Upon attaining T_w^{max} the wall temperature falls off and a minimum in the curve of $T_w = T(x)$ occurs at $x \simeq 1$. Analogous results were obtained by other investigators such as Kon'kov [4.3], Smolin et al. [4.4], Schmidt [4.5] and Swenson et al. [4.6].

Even the earliest studies of heat transfer in the post-dryout region showed that under conditions typical of power plants the maximum wall temperature may not be such as to exceed permissible temperatures, from which it immediately follows that boiling equipment may operate under post-dryout conditions. However, one must know how T_w^{max} behaves as a function of the main operating variables, since it is precisely this temperature which controls the reliability of steam generating channel operation. In addition, since under real conditions the portion interface between the pre-dryout and post-dryout boiling modes fluctuate with time, it is also necessary to know the amplitude and frequency of these fluctuations, since they control the fatigue strength of the tube material.

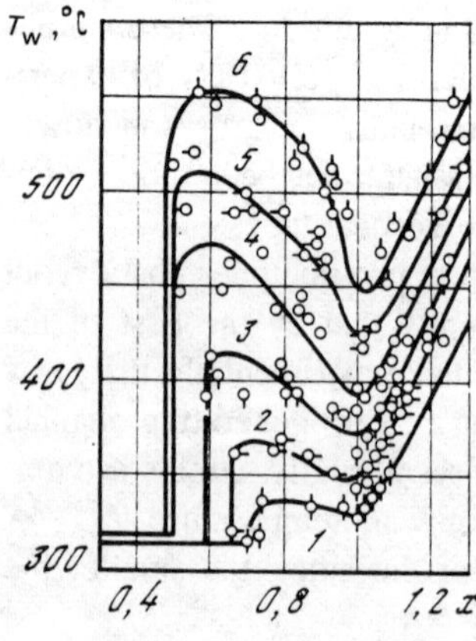

Figure 4.2 Wall temperature vs. vapor quality [4.2]. ρw = 1400 kg/m²·sec, p = 9.8 MPa; 1, q = 230 kW/m²; 2, 350; 3, 460; 4, 600; 5, 700; 6, 875 kW/m².

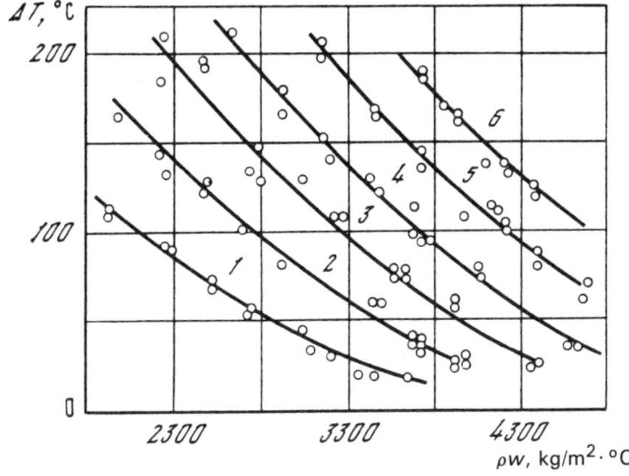

Figure 4.3 Wall temperature increase on entering the post-dryout region as a function of mass velocity at $p = 14.7$ MPa [4.4]. 1, $q = 470$ kW/m²·sec; 2, 580; 3, 700; 4, 810; 5, 930; 6, 1050 kW/m².

Figure 4.3 shows the effects of mass velocity on the maximum temperature rise. At constant pressure and vapor quality the heat transfer coefficient increases, and the maximum temperature decreases, with the mass flux. This is a result of the fact that an increase in the mass flux causes the droplets to break up into smaller units, which significantly increases their overall area. Thus, for example, for $p = 6.8$ MPa an increase in the mass flux from 1000 to 2000 kg/m²·sec caused the droplet diameter to decrease several fold (from 50–200 to 10–50 μm).

The increase in heat transfer coefficient increasing mixture Reynolds number ($Re_{mix} = \rho w D/\mu_{mix}$) causes a reduction in T_w. In addition, the increasing phase slip also amplifies the dynamic and thermal interaction between the phases and the rate of droplet evaporation. As a result, the thermal nonequilibrium under conditions of high mass velocities may be found smaller than at low mass velocities. When mass velocities are increased, even fine droplets acquire sufficient kinetic energy to penetrate steam boundary layer. Even if these droplets do not reach the heating surface, their evaporation in the superheated steam layer at the wall (which is more rapid than in the flow core) additionally turbulizes and cools the wall region (particularly at significant superheats and low vapor quality).

With increasing pressure the boundary of the post-dryout region is shifted in the direction of lower qualities, and the value of T_w^{max} falls off significantly (Fig. 4.4). This behavior of heat transfer as a function of pressure in post-dryout is due to lessening of the nonequilibrium of the flow due to the fact that the physical properties of both phases come closer together. The effect of pressure on the heat transfer coefficient is particularly significant at $p \geq 16.6$ MPa [4.7] where significant changes in the physical occur with increasing pressure.

Tobilevich and Yeremenko [4.8] investigated the effect of vapor quality on post-dryout heat transfer to water at atmospheric pressure. The following ranges of parameters were covered: w between 0 and 0.55 m/sec, q_w from 11.6 to 93 kW/m², and x from 0 to 1. The heat transfer coefficient rises with x, attains a maximum, and then decreases steeply (Fig. 4.5). Maximum heat transfer rate were attained at volumetric percentage moisture of ~2–3%. However, since the heat fluxes were low, the heat

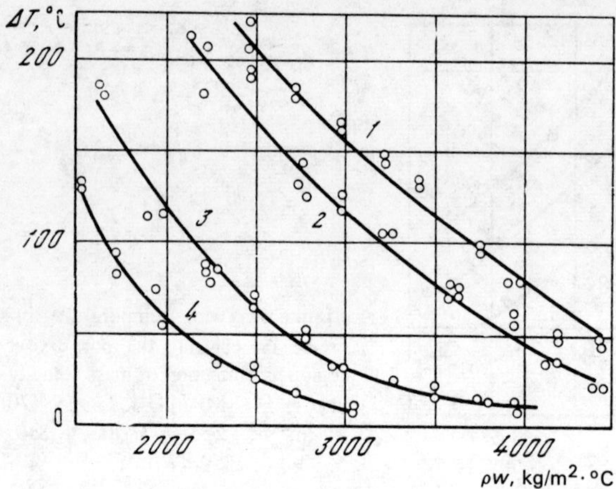

Figure 4.4 Effect of pressure on the magnitude of the wall temperature increase on entering the post-dryout region at $q = 700$ kW/m^2 [4.4]. 1, $p = 11.7$ MPa; 2, 14.7; 3, 17.6; 4, 19.6 MPa.

transfer coefficient was rather high even at low liquid concentrations. The ratio $\alpha^{max}/\alpha^{min}$ of the maximum to the minimum heat transfer coefficient was 50.

The surface temperature under post-dryout conditions is significantly affected by the flow quality in the crisis location. Figure 4.6 shows curves of wall temperature as a function of the transition quality. As x_{cr} increases, the absolute value of wall temperature (for a given value of dryout quality) decreases [4.9]. The tube length between the dryout point at the point at which T_w^{max} occurs increases.

There is no concensus concerning the effect of heat flux on the heat transfer coefficient. Cumo [4.10] notes that this coefficient falls off with increasing q_w.

In our opinion, an increase in q_w increases the degree of the flow's nonequilibrium, and this results in a deterioration in heat transfer. Belyakov et al. [4.11] and Subbotin et al. [4.12] did not detect any effect of the heat flux on post-dryout heat transfer.

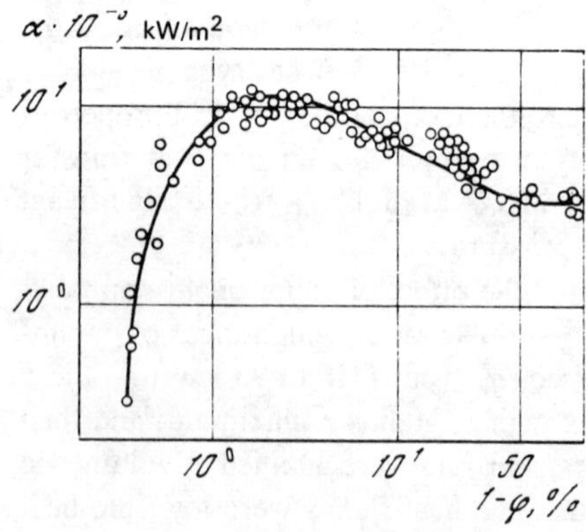

Figure 4.5 Heat transfer coefficient vs. volumetric percentage moisture at $p = 0.1$ MPa [4.8].

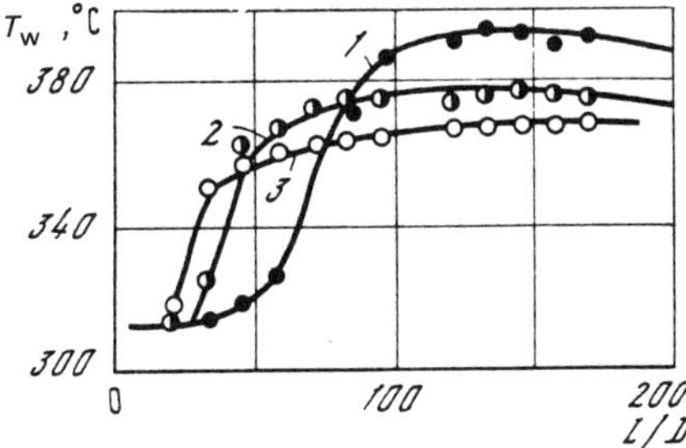

Figure 4.6 Distribution of inside wall temperature along the tube [4.9]. p = 9.8 MPa, ρw = 1750 kg/m^2·sec, q = 500 kW/m^2; 1, x_{in} = 0.59, x_{out} = 0.78; 2, 0.73 and 0.91; 3, 0.81 and 1.

The post-dryout region is seen to consist of two parts (Fig. 4.7), differing by conditions of heat transfer [4.13–4.15]: the transition boiling zone, situated immediately downstream of the dryout point, and the stable film boiling zone.

The region of stable film boiling is typified by absence of contact between the liquid and the wall. The existence of contact between the liquid and the wall in the transition region is confirmed by visual observations [4.14] and by studies using the "salt" method [4.15]. In the salt method, a salt is added to the inlet water. If, in the post-dryout region, the water contacts the wall, a layer of salt deposits grows with time and the wall temperature rises. Conversely, the wall temperature does not change in the region where there is no contact between the liquid and the wall. As compared with the zone of stable film boiling, the transition zone has lower values of wall superheat ($T_w - T_{sat}$) and higher values of the heat transfer coefficient. The heat transfer coefficient in this zone is a function of the heat flux and mass velocity.

Wall temperature fluctuations arise in the transition region due to local fluctuations of the liquid-film dryout boundary. The length of the zone with temperature fluctuations and the amplitude of the latter depend on the heat flux and mass velocity. It can be concluded from various data that the length of the zone with fluctuations is between 25 and 70 mm, and the amplitude of temperature fluctuations may be as high 100°C and more.

The bulk of the energy of temperature fluctuations lies within the range of 0–0.5 Hz (Fig. 4.8). An increase in heat flux and mass velocity decreases the length the zone of fluctuations, while increasing the fluctuation amplitude. This amplitude increases with reduction in x_{cr} (as a result of the large difference between the rate of

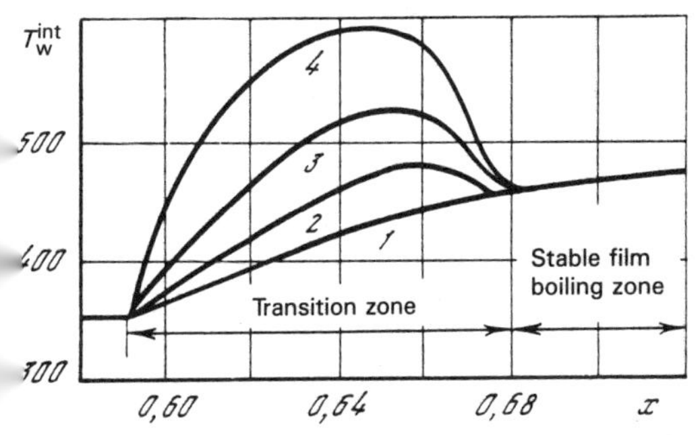

Figure 4.7 Schematic of zones in the post-dryout region and of the variation in wall temperature with time. p = 13.7 MPa, q = 400 kW/m^2, ρw = 715 kg/m^2·sec, $t_4 > t_3 > t_2 > t_1$ (t is the time between temperature recording cycles); 1–4, represent t_1 through t_4, respectively.

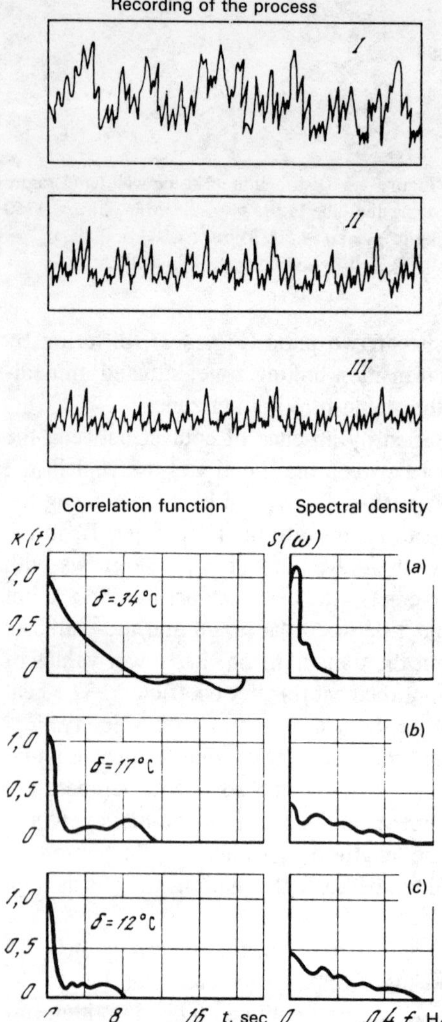

Figure 4.8 Fluctuations of wall temperature in the post-dryout zone [4.16]. $p = 13.7$ MPa, $\rho w = 350$ kg/m$^2 \cdot$ sec, $x_{in} = 0.18$, $x_{lim} = 0.61$, $q = 337$ kW/m^2; I through III designate tube cross sections, (a) through (c) correspond to sections I through III, respectively.

heat transfer in nucleate and film boiling at low vapor qualities). The amplitude of temperature fluctuations decreases with increasing pressure, which also causes a reduction of the wall superheat at which stable film boiling sets on. The temperature in this region ranges from its value in nucleate boiling to the minimum value in film boiling [4.16, 4.17].

4.2 THERMAL NONEQUILIBRIUM OF FLOW IN THE POST–DRYOUT REGION

The nonequilibrium of the flow in the post-dryout region is controlled by the rate of evaporation of droplets and the coefficient of heat transfer between the vapor and the

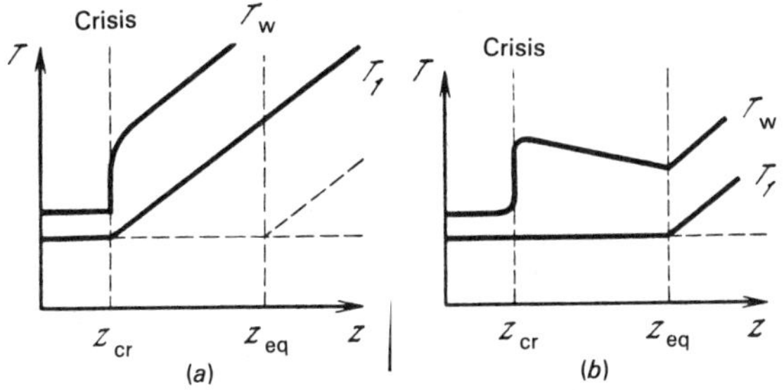

Figure 4.9 Variation in temperature along a tube with nonequilibrium (*a*) and equilibrium (*b*) heat transfer.

droplet. In the case of low masses velocities and pressure, the droplets have a large mean diameter and low overall interphase area. In addition, the velocity of the steam relative to the droplet is also not too high and, in view of the low value of the heat transfer coefficient, the evaporation rate is rather low. As a result, the heat which is supplied to the steam is expended to a large degree in superheating it. The curve of wall temperature along the tube length grows monotonically (Fig. 4.9*a*). At high pressures and mass velocities (for example, $\rho w = 3000$ kg/m$^2\cdot$sec) the interphase heat transfer at fine droplets occurs at such a high rate that there is an insignificant difference between the mean temperature of the steam and the saturation temperature. The wall temperature rises to T_{max} (Fig. 4.9*b*) and then starts decreasing at the start of the post-dryout region reaching a minimum at a point corresponding to $x = 1$, after which it rises again.

The above two cases of wall temperature variation are limiting cases: limiting nonequilibrium and complete thermal equilibrium between the phases. In practice one usually deals with intermediate cases, when the two-phase flow is not in equilibrium, but the heat and mass transfer between the steam and droplets cannot be neglected.

Let us consider the qualitative pattern of variation in the mean parameters in the flow and the tube wall under real conditions at $q_w = $ const (Fig. 4.10). The small thermal nonequilibrium in the flow at the dryout location (z_{cr}) due to subcooling of the droplets can be neglected, i.e., $x_{cr} = x_{cr}^{eq}$ (for an evaporating nonboiling film the quality at the wall upstream of the crisis location is close to zero, whereas for a boiling film the wall quality is significant).

Had there been thermodynamic equilibrium, the longitudinal variation in x would have consisted of a continuation of the straight line from the crisis location, i.e., at point Z_{eq}^{x-1} (see Fig. 4.10) all the droplets would have evaporated and $x(Z) = 1$. However, due to nonequilibrium of the flow (superheat of the vapor), and in some cases also due to subcooling of the droplets, the rise in quality downstream of the dryout point is slower, and complete evaporation occurs downstream of point Z_{eq}^{x-1}. The wall temperature at point Z^{x-1} rises more steeply than in the region with dispersed-droplet flow.

It is of interest to consider the behavior of the wall quality x_w. In the initial sections of the tube following the boiling crisis the droplet concentration at the wall significantly exceeds the mean concentration of liquid in the flow (i.e., $x_w < x$). Also, the superheat of the wall is not sufficiently high to inhibit heat transfer by direct

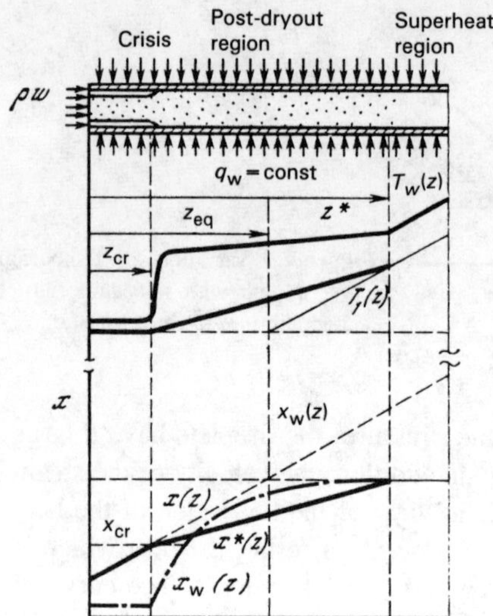

Figure 4.10 Qualitative pattern of variation in mean variables in the flow and the tube wall.

contact between the droplets and the wall. At significant irrigation rates the wall surface is partially wetted by droplets impinging on the wall. When the droplet passes into the vapor boundary layer under conditions in which the wall temperature is lower than the Leidenfrost temperature, T_L, a "puddle" of the evaporating film remains on the wall. At $T_w > T_L$ heat is transferred to the droplet by conduction, and part of the liquid in the droplet is superheated to its limiting value and vaporizes explosively.

Note that contact of droplets with the heating surface does not always make a significant contribution into the heat transfer, but is decisive for the transfer of mass. In fact, a supersaturated vapor solution forms in steam-generating channels immediately beyond the dryout zone. As a result of contact between liquid droplets containing corrosive salts and of their explosive vaporization, salt crystals precipitate on the wall. This question will be considered below in more detail.

We see from the above that the heat flux from the wall (without consideration of radiant heat flux) is expended in evaporating the droplets (at the wall and in the flow) and in superheating the vapor. We shall assume that the fraction of heat used for evaporating the liquid $\epsilon = q_3/q_w$ is independent of the channel length. Then the mean temperature of the flow and the quality will be linear functions of the tube length. This assertion is confirmed by the experimental data of Plummer [4.18], obtained with nitrogen (Fig. 4.11); the curves of measured qualities as a function of the equilibrium quality (or length) do not differ too significantly from straight lines, so that to a first approximation, ϵ can be regarded as independent of the channel length.

The thermodynamic quality for constant heat flux at the wall [4.18] is found from the expression

$$x_{eq}(z) = x_{cr} + \frac{4q_w}{D j r}(z - z_{cr}^{eq})$$

(4.1)

and the actual quality (ratio of vaporous mass flow to total mass flow)

$$x(z) = x_{cr} + \frac{4q_w\,\varepsilon}{Djr}\,(z - z_{cr})\qquad(4.2)$$

Then ϵ can be expressed as

$$\varepsilon = \frac{x(z) - x_{cr}}{x^{eq}(z) - x_{cr}} = \frac{z^{x=1} - z_{cr}}{z_{eq}^{x=1} - z_{cr}}\qquad(4.3)$$

The actual vapor temperature T_1 for $z < z^{x-1}$ is

$$T_1(z) = T_{sat} + \frac{4(1-\varepsilon)\,q_w\,(z - z_{cr})}{jC_{p1}D}$$

and for $z > z^{x-1}$

$$T_1(z) = T_{sat} + \frac{4q_w\,(z - z_{eq}^{x=1})}{Gc_{p1}D}\qquad(4.4)$$

Plummer [4.18] suggested experimental correlations for ϵ on the basis of experimental data obtained with freon and water. His expression for water is

$$\varepsilon = 0{,}402 + 0{,}0674 \ln\left[j\left(\frac{D}{\rho_1\sigma}\right)^{0,5}(1 - x_{cr})^5\right]\qquad(4.5)$$

and for freon

$$\varepsilon = 0{,}236 + 0{,}0811 \ln\left[j\left(\frac{D}{\rho_1\sigma}\right)^{0,5}(1 - x_{cr})^5\right]\qquad(4.6)$$

Figure 4.12 shows experimental points for water, freon and nitrogen and also the correlations (4.5) and (4.6). It is seen that the experimental points for nitrogen lie between (4.5) and (4.6): for high mass fluxes and low values of quality, the experi-

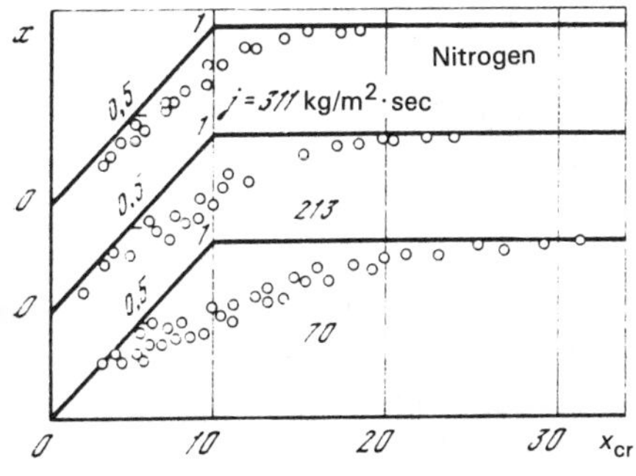

Figure 4.11 Deviation of the actual vapor quality from its equilibrium value according to Plummer [4.18].

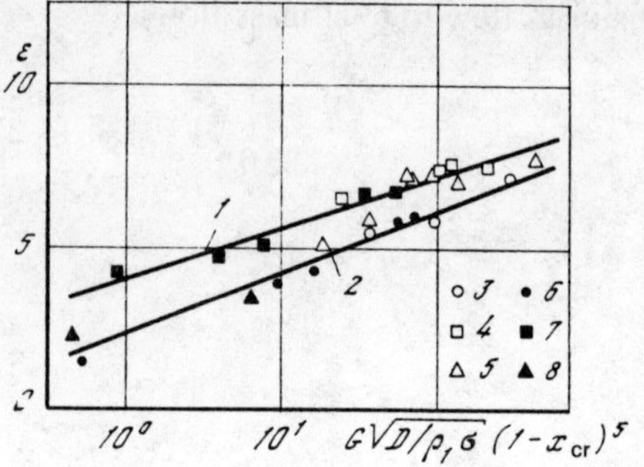

Figure 4.12 Variation in the relative quality as a function of $G\sqrt{D/\rho_1\sigma}\,(1 - x_{cr})^5$. 1, Equation (15) from [4.18]; 2, Eq. (16) from [4.18]; $x_{cr} = 2\%$; 3, Freon-12; 4, water; 5, nitrogen; 6–8, same as above at $x_{cr} \geq 10\%$.

mental points for ϵ lies near the plot of Eq. (4.5), whereas for low velocities and high qualities the nitrogen data cluster about curve (4.6) for freon.

4.3 MECHANISM OF INTERACTION BETWEEN MOVING DROPLETS AND A SUPERHEATED WALL

Most methods for calculating heat transfer under post-dryout conditions are based on the assumption of the formation of a stable vapor layer which separates the droplet from the wall; in these methods, calculation on heat transfer to the droplet moving in the boundary layer are here based on analytic or experimental data for a droplet suspended above a heated wall. The study of evaporation of a nonmoving droplet above a heated surface is one of the classical topics and has been extensively studied [4.19–4.29] and, for this reason, it will not be considered in this book.

The physical aspects of heat and mass transfer between a moving droplet and a superheated solid surface have insufficiently explored so far. Only two studies [4.30, 4.31] are known in which an attempt was made to measure the amount of heat absorbed by a single droplet. The results from the respective studies are contradictory and largely incompatible.

Cumo with Farello [4.21] performed a detailed visual study (by high speed photography) of a water droplet falling onto a surface from a steel needle. The droplets retained their spheroidal shape and performed oscillations about the heated surface. Measurements showed that the maximum wall to droplet distance was 1 mm. Optical measurements also showed that the droplets come into contact with the superheated surface, but no other methods of recording this contact were used.

In experiments with a droplet spray, Cumo and Farello recorded the time of residence of the droplet at the superheated surface; the latter was found to decrease steeply as the temperature was raised to 250–300 °C. Further increases in temperature gave only a very small reduction in residence time. Cumo and Farello suggested that the region of rapid residence time reduction corresponded to a change in heat transfer mechanism (transition to film boiling). The magnitude of the limiting wetting temperature was independent of the velocity over the range of 2 to 4 m/sec.

Wachters and Westerling [4.32] measured the amount of heat absorbed by a single two-millimeter droplet of water, falling on a polished gold surface. Their experiments were performed in a steam atmosphere at a pressure of ~ 0.1 MPa the droplet temperature being 100 °C. They measured the relative reduction in the volume of liquid caused by a single interaction of the droplet with the heating surface over a velocity range of 1.08 to 1.47 m/sec. The reduction in volume and consequently, in the droplet mass was between 1 to 1.5% at 200 °C and was not more than 0.2% at higher temperatures. The velocity range studied was selected such that the droplet would not break up on contact with the surface. Wachters and Westerling found a maximum limit of the Weber number, at which the droplet retains its shape after interacting with the heating surface

$$\mathrm{We}_{\mathrm{cr}} = \frac{2\rho_3 u^2 R}{\sigma} = 80 \tag{4.7}$$

The variation in the droplet radius with time was determined from photographs for three droplet velocities. Wachters and Westerling showed that with an initial droplet size of 2.3 mm and velocities of 0.63 to 1.09 m/sec (i.e., $\mathrm{We}_3 < 80$), and with $T_w = 400$ °C, the maximum contact time is 12×10^{-3} sec; under these conditions the droplet shows minimal flattening upon impact and its surface interacting with the wall changes smoothly with time. AT 2.2 m/sec ($\mathrm{We}_3 > 80$) the time of contact decreases approximately twofold and is equal to 6×10^{-3} sec; at this velocity, the droplet flattens significantly on impact, the maximum radius exceeding the initial radius by almost twofold and then decreasing steeply with time upon droplet rebound.

To describe the interaction between the "jumping" droplet and the superheated wall, Wachters and Westerling suggested analytic formulae for a droplet lying motionless on a vapor cushion. They introduce a correction for the hydraulic impact force, but do not explain the course of formation of the vapor layer. In addition, the interaction of a moving droplet with the hot wall is clearly unsteady from the point of view of heat transfer and droplet shape, whereas the expressions obtained for a motionless droplet are the results of a solution for a steady-state case.

Wachters and Smulders [4.33] investigated heat transfer to small droplets. Their experiments were performed in an 8 cm diameter horizontal glass tube with a spray nozzle placed at one of its ends. A 2.5 cm diameter and 1 mm thick metal disk was placed in the center of the tube. The disk temperature was measured by a thermocouple, placed on its downstream side. The disk was preheated to close to 500 °C. When stable spray nozzle operation was attained, the heating element was removed and the subsequent cooling of the disk of the vapor-droplet flow was recorded. The magnitude of the heat flux was determined from the heat balance

$$q_w = c_p \rho V \frac{dT}{dt}$$

where ρ, V and c_p are the density, volume and specific heat at constant pressure of the disk. It was then assumed that heat is used up for heating the saturated steam, flowing along the disk, and for partial evaporation of droplets interacting with the heated surface. The heat flux to the droplets was calculated by the equation:

$$q_{w,3} = \rho_w V_w c_{pl} \left\{ \left(\frac{dT_w}{dt} \right)_\Sigma - \left(\frac{dT}{dt} \right)_{w.s} \right\}$$

The number of liquid droplets interacting with the leading side of the disk was determined using an experimentally determined coefficient [4.33], which accounts for the fraction of all the droplets produced by the spray nozzle which are incident on the disk. It was assumed here that each droplet impinges on the disk and interacts with it only once; there is no interaction between the droplets and the downstream side of the disk. The droplet diameters were determined from the dimensions of the salt track of the dried-out droplets, to which nickel sulfate was first added. The droplet diameters ranged between 50 and 70 μm. The droplet velocity was measured using cine photography and ranged between 4.6 and 5 m/sec. The vapor flowrate was 4.8×10^{-3} kg/sec, that of the liquid ranged from 0.4×10^{-6} to 2×10^{-6} m^3/sec. The experiments were performed using disks of different materials: gold, platinum, stainless steel and copper. The experimental results were represented as the ratio $\epsilon = \Delta V/V$

$$\varepsilon = \frac{\Delta V}{V} = \frac{\rho_w c_{pw} V_w}{\rho_3 r^* E \Phi_{1,3}} \left\{ \left(\frac{dT_w}{dt} \right) - \left(\frac{dT}{dt} \right)_{sat} \right\}$$

where r^* is the heat vaporization, $\Phi_{1,3}$ is the volumetric flux of the liquid to the disk in the absence of vapor flow from the disk wall; E is an experimental coefficient, making allowance for the effective, liquid wetted surface of the disk.

The relative reduction in the volume of small-diameter droplets deposited on the wall was approximately 1%, which is smaller than that obtained with large single droplets. Wachters and Smulders also found that the Leidenfrost temperature was 400 °C, and estimated that the likely thickness of the vapor layer for droplets 60 μm in diameter was about 1 μm.

Pedersen [4.36] investigated heat and mass transfer in the impingement on a cylinder of uniformly sized droplets with diameters between 200 and 400 μm moving at 3 to 9 m/sec, respectively. The droplet temperature was 22 °C and the impingement frequency was 40 to 100 sec^{-1}. The horizontally moving particles were incident on a stainless-steel cylinder ($d = 6$ mm, length $= 6$ mm) preheated in an oven. In the course of the experiment Pedersen measured the change in the cylinder's temperature with time. The heat transferred to a single droplet was calculated by dividing the overall heat flux, determined from the heat balance, by the impingement rate; dividing the heat transferred by heat needed for complete vaporization of the droplet gave the value of ϵ. It was found that ϵ was directly proportional to the velocity. At low values of the latter (~ 3 m/sec) ϵ is about 5% and at 9 m/sec $\epsilon \cong 20\%$. ϵ did not vary significantly with droplet diameter. ϵ varied strongly with temperature; for instance, droplets with a diameter of 330 μm moving at 8 m/sec evaporate by 90% at 200 °C and about 10% at 500 °C. The values given differ on the average by an order of

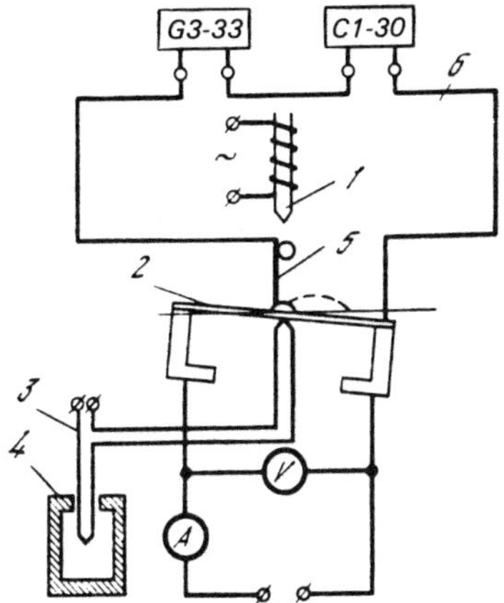

Figure 4.13 Schematic of arrangement for investigating transfer of heat between a droplet and superheated wall. 1, Droplet generator; 2, heated plate; 3, differential thermocouple; 4, calorimeter; 5, wire electrodes; 6, electric circuit for measuring the contact between a droplet and wall.

magnitude from those obtained by Wachters [4.33] at high temperatures, and almost 15-fold at temperatures close to the Leidenfrost point.

The difference in droplet diameters (in Wachters' experiments it was 60 μm whereas in the study by Pedersen it was 300 μm) should not result in such a difference, the more so since according to Moriyama [4.25] ϵ is entirely independent of the droplet diameter over the range 200 to 400 μm. Such a significant disagreement of data is apparently due to the difference in the medium in which the experiments were performed. Pedersen obtained his results in air at room temperature, whereas Wachters' experiments were performed in an atmosphere of wet steam.

Kutateladze [4.23] performed an experimental study of the heat transfer between an incident droplet and a hot surface. His experiments demonstrated the existence of contact between the droplet and the hot wall at temperatures up to 800 °C; above this temperature the sensitivity of instruments was apparently insufficient for measuring the short contact time.

The mechanism of interaction between an incident droplet and superheated surface was investigated experimentally in the Institute of High Temperatures [4.30] using an arrangement shown schematically in Fig. 4.13.

The diameter of the droplet was 2.2 mm, and the velocity range was 0.1 to 1.7 m/sec. To avoid errors which may appear in investigating unsteady processes, the wall temperature (T_w) beneath the droplet was investigated by using 30 μm diameter microthermocouple.

The interaction was photographed using high-speed cine-photography performed at a moving rate of 3000 frames/sec. This made it possible to observe dynamic features of the contact between a droplet and heated surface. A typical sequence of moving-picture frames is shown in Fig. 4.14.

Special experiments were performed to determine whether direct contact occurs between the droplet and the wall, in which a droplet was incident simultaneously on the tip of a 0.1 mm diameter needle and on the surface of superheated carefully polished stainless-steel plate. The tip of the needle was placed at ~0.4 mm from the surface in such a manner that its contact with the liquid would not cease upon flatten-

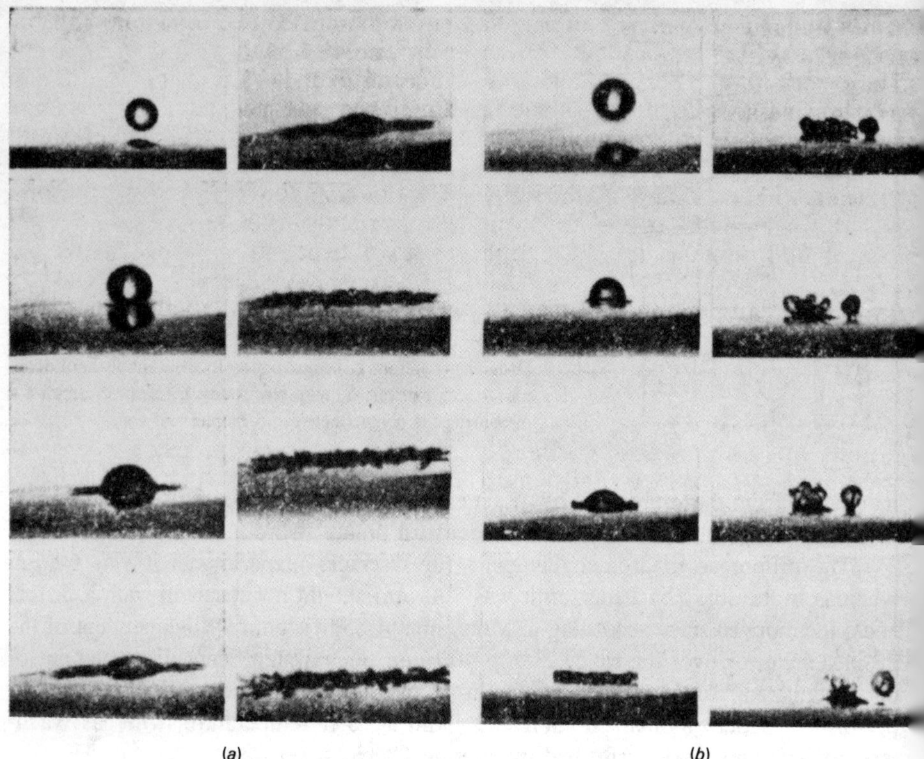

(a)   (b)

Figure 4.14 Moving-picture representation of the interaction of a droplet of distilled water. (a) Height of droplet fall 380 mm, $T_w = 613\,°C$; (b) 50 mm and 270 °C.

ing of the droplet. The existence of contact between the liquid and the superheated surface was determined from closing of an electric circuit (see Fig. 4.13). Contact was recorded by a Cl-30 electronic oscillograph, operating in the waiting scanning mode. The duration of the signal was determined by supplying a signal from a G3-33 oscillator with a specified frequency to the second beam. During each individual contact event a stable and continuous pulse was observed whose duration decreased with increasing temperature. There was no pulse interruption or secondary closing of the circuit during the contact event. The dependence of the circuit closing time on the plate superheat is shown in Fig. 4.15. The experiments were performed at droplet velocities $w_3 \simeq 1$ m/sec.

The heat transfer coefficient between the droplet and the surface was investigated only during the initial contact. The plate was inclined at a small angle and the droplet rolled down the plate following contact. The temperature of the distilled water droplet was close to saturation.

Figure 4.16 shows how the temperature of a thin film at the point of droplet incidence changes in time. The graph shows clearly two virtually linear segments. The duration of segment AB was 0.01 sec, which is commensurate with the time of

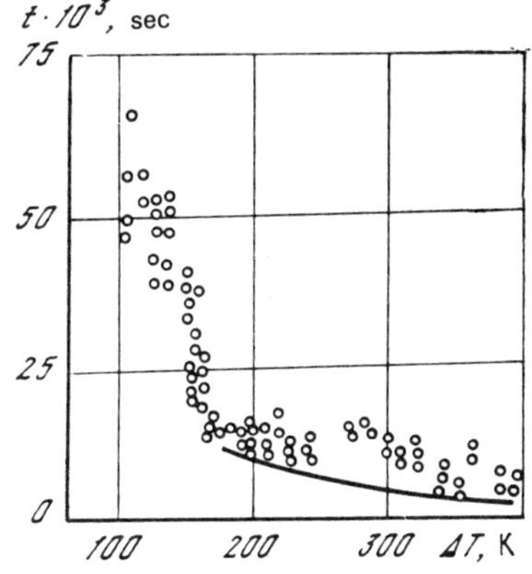

Figure 4.15 Time of droplet contact as a function of the wall temperature.

electrical contact and also with the time of complete detachment of a droplet from the surface as observed from the cine frames.

The heat transfer coefficient from the surface to the liquid can be estimated using Newton's cooling law for a thin wall, since the value of the Biot number for a 0.1 mm thick nichrome heater is small. Thus,

$$\alpha = \frac{C_p \rho}{T_w - T_{sat}}\left(\frac{dT}{d\tau}\right)\frac{V}{A} + \frac{q_j}{T_w - T_{sat}}$$

where A is the surface area, and V is the volume (for our case $V/A = \delta$ is the heater thickness). As noted, the heater thickness was 0.1 mm, and the maximum diameter of the liquid spheroid is ~4 mm. Hence, in this case, edge effects can be neglected. In addition, the thermal perturbation wave moved from the center of droplet incidence.

The measured coefficients of heat transfer during the time period AB (Fig. 4.16) are plotted in Fig. 4.17; their magnitudes exceed by more than an order of magnitude the predictions obtained for a droplet at rest on a superheated surface.

The existence of electrical contact between the liquid and superheated surface, and also the shape of the curve of wall temperature variation (Fig. 4.16), allows the postulation of the following mechanism of heat transfer to the droplet impinging on

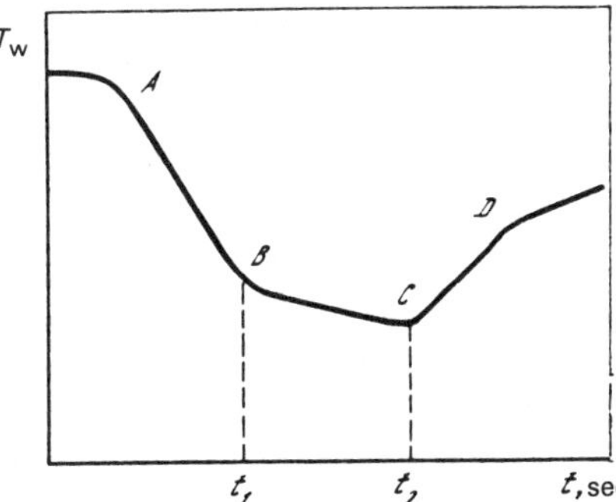

Figure 4.16 Variation of temperature with time.

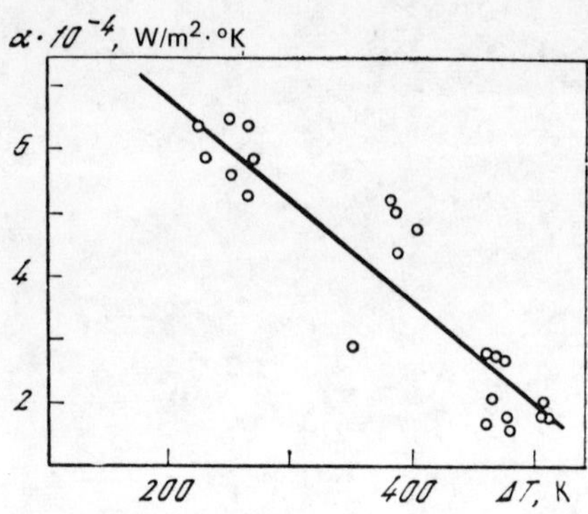

Figure 4.17 Heat transfer coefficient vs. the superheat [4.29].

the surface. Initially the liquid wets the wall even if T_w is significantly higher than the saturation temperature. The layers of the liquid adjacent to the wall then start heating up. The rate of this heating depends on the thermal resistance of the liquid-wall interface.

When the liquid temperature becomes commensurate with the limiting superheat temperature (for homogeneous nucleation, the superheated layers flash and the droplet is ejected from the surface. At significant wall superheats the main contributory factor in this flashing process is avalanche-like nucleation and growth of fluctuating vapor embryos. The growth rate and the number of the embryos is controlled by the magnitude and growth rate of the liquid superheat, and also by the properties of the liquid. At low wall superheats, when heat transfer occurs at a rather low rate and, accordingly, the contact time is large, a significant contribution can also be made by heterogeneous nucleation sites, such as individual roughness elements on the surface. In this case single large bubbles may develop within the liquid, which results in breakup of the droplet or ejection of large-sized spray.

One should point out the significant lack of agreement between the measured contact time, which is about 5 to 20 msec, and the time needed for the nucleation of a critical embryo and its growth, which is several orders of magnitude smaller than the contact time. A qualitative explanation for this disagreement may be as follows. Due to roughness and other statistical microirregularities, the spontaneous flashing occurring when a droplet impinges on a hot surface, occurs nonuniformly over the droplet-wall interface. The limiting superheats in this case are attained at different times at different points on the surface and the growth of the nascent bubbles causes cooling of the surrounding liquid. This cooling served as a heat sink, which prolongs the time of formation of a continuous vapor blanket. The contact time is thus assumed to consist of the statistical sum of times needed for the nucleation and, primarily, growth of individual microbubbles.

The above mechanism of heat transfer to a droplet incident on a superheated surface is qualitatively confirmed by mass transfer experiments with $CaSO_4$ solutions; large salt crystals remained on the wall following the droplet contact. The appearance of these crystals around the point of droplet incidence confirms the suggested model of contact. As a result of flashing of the thin superheated liquid layer there, a highly supersaturated vapor solution forms from which salt crystals precipitate.

4.4 METHODS FOR CALCULATING HEAT TRANSFER IN THE POST–DRYOUT REGIONS OF STEAM–GENERATING CHANNELS

Two main alternative approaches have been made to prediction of heat transfer and hydrodynamics in the post-dryout region. In the first (homogeneous mixture) approach, the two-phase mixture is treated as a single-phase system, whose parameters are determined as a function of the phase concentrations. In the second approach the behavior of each of the phases is determined separately, with the equations of conservation of momentum, heat and mass being written separately for each phase. Appropriate correlations make allowance for the interaction between phases at the interface. This approach is based on one-dimensional models. Intermediate models, employing various elements of the two above approaches are also used. In addition, a number of studies have been published in which a purely empirical correlation of the experimental results is given. Thus, we can subdivide the studies into three groups: (1) one-dimensional nonequilibrium models; (2) boundary layer approximations; (3) empirical correlations.

4.4.1 One-Dimensional Nonequilibrium Models

The one-dimensional model for heat transfer in post-dryout region was first employed by Laverty with Rohsenow [4.42], and was subsequently developed also by others [4.43–4.56]. Laverty and Rohsenow suggested that heat transfer occurs in two stages: first, heat is transmitted from the wall to the superheated vapor and the liquid droplets which bombard the heating surface (first stage), and then by convection from the superheated vapor to the bulk of liquid droplets (second stage). It is assumed in these models that all the parameters of the vapor and liquid change only along the channel and in time, but are constant over the cross section. This is done by introducing the concept of bulk velocities of vapor w_v and liquid w_3 and the bulk enthalpies h_v of the vapor and h_3 of the liquid. The phase interface temperature is usually assumed to be equal to the saturation temperature T_{sat}.

The equations for one-dimensional description of two-phase flows can be obtained as a generalization of an analogous set of equations for single-phase flow. Neglecting, in the energy equations, a number of terms such as longitudinal thermal conductivity, pressure energy and dissipating (which are usually small), Kalinin et al. [4.57] obtained the following set of equations:

$$\frac{\partial \rho_3}{\partial t} F_3 + \frac{\partial G_3}{\partial z} = - \frac{\Pi_{\text{intf}} q_{\text{ev}}}{r} \tag{4.8}$$

$$\rho_3 \frac{dw_3}{dt} = g_z \rho_3 - \frac{\partial P}{\partial z} + \mathrm{sign}\,(w_v - w_3)\,\frac{\tau_{\mathrm{intf}}\Pi_{\mathrm{intf}}}{F_3} \tag{4.9}$$

$$\frac{G_3}{w_3}\frac{\partial h_3}{\partial t} + G_3 \frac{\partial h_3}{\partial z} = q_3 \Pi_{\mathrm{intf}} \tag{4.10}$$

$$\frac{\partial \rho_v}{\partial t} F_v + \frac{\partial G_v}{\partial z} = \frac{q_{ev}\,\Pi_{\mathrm{intf}}}{r} \tag{4.11}$$

$$\rho_v \frac{\partial w_v}{dt} = g_z \rho_v - \frac{\partial p}{\partial z} - \mathrm{sign}\,(w_v - w_3)\,\frac{\tau_w\,\Pi_{\mathrm{intf}}}{F_v} - \frac{\tau_w\,\Pi_w}{F_v} \tag{4.12}$$

$$\frac{G_v}{w_v}\frac{\partial h_v}{\partial t} + G_v \frac{\partial h_\Pi}{\partial z} + (h_v - h_{v\,\mathrm{sat}})\frac{\partial G_v}{\partial z} = q_w \Pi_w - q_{ev}\Pi_{\mathrm{intf}} \tag{4.13}$$

$$G_3 = \rho_3 w_3 F_3 \tag{4.14}$$

$$G_v = \rho_v w_v F_v \tag{4.15}$$

$$q_w = \alpha\,(T_w - T_{\mathrm{sat}}) \tag{4.16}$$

The above equations were written in a general form taking account of the unsteady nature of the process. To close this set of equations one must specify initial and boundary conditions. Usually this set of equations is used for specific two-phase flow patterns, which makes it possible to obtain the closure equations. In the majority of investigations heat transfer in post-dryout region is treated as steady or quasisteady. This allows significant simplification of the starting set of equations.

Equation of phase transition

$$\frac{dx}{dz} = \frac{q_{eq}\,\Pi_{\mathrm{intf}}}{rG} \tag{4.17}$$

Equation of motion for the liquid

$$\rho_3 w_3 \frac{dw_3}{dz} = \rho_3 g_z - \frac{dp}{dz} + \mathrm{sign}\,(w_v - w_3)\frac{\tau_{\mathrm{intf}}\Pi_{\mathrm{intf}}}{(1 - \varphi)\,F} \tag{4.18}$$

Energy equation for the liquid

$$(1 - x)\,Gc_{p3}\frac{dT_3}{dz} = q_3 \Pi_{\mathrm{intf}} \tag{4.19}$$

Equation of motion for the vapor

$$\rho_v w_v \frac{dw_v}{dz} = \rho_v g_z - \frac{dp}{dz} - \mathrm{sign}\,(w_v - w_3)\frac{\tau_{\mathrm{intf}}\Pi_{\mathrm{intf}}}{\varphi F} - \frac{\tau_w\,\Pi_w}{\varphi F} \tag{4.20}$$

Energy equation for the vapor

$$xGc_{pv}\frac{dT_v}{dz} + Gc_{pv}(T_v - T_{sat})\frac{dx}{dz} = q_w\Pi_w - (q_3 + q_{ev})\Pi_{intf} \qquad (4.21)$$

Equation relating the mass and volumetric fraction of the liquid

$$(1 - x)G = \rho_3 w_3 (1 - \varphi)F \qquad (4.22)$$

Equation relating the mass and volumetric fraction of the vapor

$$xG = \rho_v w_v \varphi F \qquad (4.23)$$

The uniqueness conditions of the system are: channel geometry (for a tube the diameter D and the length L); physical properties of the vapor and liquid; mass acceleration force field g_z; boundary conditions at $Z = 0$ (p_0, φ_0, x_0, G, T_3 and T_v) and the temperature profile $T_w = T(z)$ or the heat flux distribution along the channel $q_w = q(z)$.

Equations (4.17)–(4.23) contain thirteen unknowns: p, w_v, w_3, T_v, T_3, x, φ, Π_b, τ_w, τ_b, q_3, q_{ev}, q_w or T_w. Additional conditions are needed for closure. Writing equations for coefficients which appear in the one-dimensional description of the process and relating friction τ_w at the wall and τ_{intf} at the interface, heat fluxes q_3, q_{ev}, q_w (or T_w) and geometry of the interface to the flow variables, Kalinin et al. [4.57] found that

$$\tau_{intf} = f(w_v - w_3, \; \ldots) \qquad (4.24)$$
$$\tau_w = f(w_v, \; \ldots) \qquad (4.25)$$
$$q_3 = f(T_{sat} - T_3, \; \ldots) \qquad (4.26)$$
$$q_{ev} = f(T_v - T_{sat}, \; \ldots) \qquad (4.27)$$
$$f(T_w - T_v, \; q_w, \; \ldots) = 0 \qquad (4.28)$$
$$f(\Pi_{intf}, \; \varphi, \; \ldots) = 0 \qquad (4.29)$$

The specific form of these equations is determined by the flow pattern. In particular, condition (4.29) has the following form for dispersed flow

$$\Pi_{intf} = \frac{1,5\pi(1 - \varphi)D^2}{d_3} \qquad (4.30)$$

$$\varphi = 1 - n\frac{\pi d_3^3}{6} \qquad (4.31)$$

$$d_3 = d(\sigma, \; w_v - w_3) \qquad (4.32)$$

$$n = n(\rho w, \; x, \; d_3, \ldots) \qquad (4.33)$$

Equations (4.17)–(4.29) represent the full set of equations. However, none of the analytic studies employed was set in this form, but rather in a simplified manner, depending on the purpose of the study. These simplifications become rather clear when the aforementioned studies are analyzed on the basis of Eqs. (4.17)–(4.29).

Let us now examine the best known analytic studies. Among the few early studies of heat transfer in dispersed flow at DNB, which are relatively complete with respect to the mathematical statement are those by Rohsenow and his coworkers [4.42, 4.43]. In the latter study the set of equations has been simplified as follows.

1. The equation of motion for the steam (4.20) is not considered. This makes expressions for τ_w and τ_{intf} unnecessary. Naturally, it is then impossible to determine the reduction in p along the channel.

2. It is assumed that $T_3 = T_{sat}$. Equations (4.19) and (4.26) then drop out. The heat flux is used in heating the liquid [Eq. (4.21)] then equal to zero.

3. The momentum equation (4.18) is written for upwards motion of a single droplet

$$\rho_3 \frac{\pi d_3^3}{6}\left(w_3\frac{dw_3}{dz}+g\right)=\Phi_{dr} \tag{4.34}$$

where Φ_{dr} is the drag of the droplet.

The closure equations.

1. The dimension of droplets at $We \leq We_{cr}$ is determined assuming the number of droplets is constant.

$$(1-x)/(1-x_0) = d_3^3/d_{30}^3 \tag{4.35}$$

It is assumed that at $We > We_{cr}$ each droplet divides into two.

2. The drag of a particle in the vapor flow is calculated from the formula

$$\Phi_{dr}=C_D\frac{\rho_v\,(w_v-w_3)^2}{2}\frac{\pi d_3^2}{4} \tag{4.36}$$

$$C_D=C_{D1}-\kappa\,(C_{D1}-C_{D2}) \tag{4.37}$$

where D_{D1} is the drag coefficient of a solid sphere, equal to

$$C_{D1}=\begin{cases}\dfrac{24}{Re_{d_3}}(1+0{,}142Re_{d_3}^{0,698}) & \text{at}\quad Re_{d_3}<200 \tag{4.38a}\\[2ex] 0{,}45 & \text{at}\quad Re_{d_3}>200 \tag{4.38b}\end{cases}$$

C_{D2} is the drag coefficient of evaporating droplets in the vapor flow, equal to

$$C_{D2}=27Re_{d_3}^{-0,84} \tag{4.39}$$

Coefficient κ in Eq. (4.37) is defined as a function of the gravitational loading $n_g = (w_3/g_z)(dw_3/dz)$

$$\kappa = \begin{cases} 0 & \text{at} \quad n_g \leqslant 15 \\[1mm] \dfrac{n_g - 15}{150} & \text{at} \quad 15 < n_g < 165 \\[1mm] 1 & \text{at} \quad n_g \geqslant 165 \end{cases} \qquad (4.40)$$

The droplet Reynolds number is determined from the slip velocity

$$\mathrm{Re}_{d_3} = \frac{\rho_v \, (w_v - w_3) \, d_3}{\mu_v} \qquad (4.41)$$

3. Equations for heat fluxes were assumed on the basis of the two-stage model. For the first stage

$$q_w = \alpha_{wv} \, (T_w - T_v) + \alpha_{w3} (T_w - T_{sat}) \, \omega_1 \qquad (4.42)$$

where the ratio ω_1 is to represent the total cross sectional area of the droplets in contact with the tube surface per unit tube surface area. For the second stage of heat transfer

$$q_{ev} = \alpha_{v3} \, (T_v - T_{sat}) + \alpha_{w3} (T_w - T_{sat}) \, \omega_2 \qquad (4.43)$$

where ω_2 is the ratio of the total cross sectional area of droplets in contact with the tube surface to the total surface area of droplets in the volume of unit tube length. In addition to determining the coefficient of heat transfer from the wall to the vapor $\alpha_{w\text{-}v}$ from the expression for single-phase flows

$$\frac{\alpha_{wv} D}{\lambda_v} = A \left(\frac{\rho_v W_v D}{\mu_v} \right)^a \mathrm{Pr}_v^{B} f \left(\frac{T_w}{T_v} \right) \qquad (4.44)$$

the heat transfer coefficient from the wall to the droplet was calculated from an expression for spheroidal boiling of droplets

$$\frac{\alpha_{w3} d_3}{\lambda_v} = \kappa_1 \left(\frac{6}{\pi} \right)^{1/12} \left[\frac{\rho_v^2 g d_3^3 \rho_3 r^*}{\mu_v^2 \rho_v c_{pv} (T_w - T_{sat})} \mathrm{Pr}_v \right] \qquad (4.45)$$

$$r^* = r \left[1 + \frac{7}{20} \frac{c_{pv} (T_w - T_{sat})}{r} \right]^{-3} \qquad (4.46)$$

the heat transfer coefficient from the vapor to the droplets α_{v3} was calculated from an empirical expression for the motion of a unit liquid sphere in superheated vapor

$$\frac{\alpha_{v3} d_3}{\lambda_v} = 2 + 0{,}55 \sqrt{ \frac{\rho_v \, (w_v - w_3) \, d_3}{\mu_v} \mathrm{Pr}_v } \qquad (4.47)$$

The relative cross sectional area of contacting droplets per unit tube surface is given by

$$\omega_1 = \pi d_3^2 N / 4 \tag{4.48}$$

where N is the number of droplets in the wall layer, per unit heating surface area. The value of N was estimated using the expression

$$N = \kappa_2 n^{2/3} \tag{4.49}$$

The ratio of the total cross sectional area of droplets contacting with the wall to the total surface area of droplets in the volume is

$$\omega_2 = \left(\frac{\pi d_3^2}{4} N \pi D \Delta z \right) \Big/ \left(\pi d_3^2 n \frac{\pi D^2}{4} \Delta z \right) = \frac{N}{nD} \tag{4.50}$$

4. It was assumed, in calculating variables for the cross section upstream of the dryout, that the flow is in equilibrium; the quantities T_v, φ and d_3 were taken as those for equilibrium flow.

Ganic and Rohsenow [4.56] further developed the Laverty and Rohsenow [4.42] physical concepts on the mechanism of post-dryout heat transfer. An item of particular interest in their study was the analysis of the effect of a spectrum of particle sizes on heat transfer. The particle size distribution given by Ganic and Rohsenow is:

$$P(d_3) = 4 \frac{(d_3)}{(\bar{d}_3)^2} \exp \left[-2 \left(\frac{d_3}{\bar{d}_3} \right)^2 \right] \tag{4.51}$$

It is assumed that the mean droplet diameter $\bar{d}_3$ is equal to

$$\bar{d}_3 = 2 d_3^* \tag{4.52}$$

where d_3^* is the most probable droplet diameter.

Ganic and Rohsenow showed that

$$\bar{d}_3 = \frac{1.83 C}{w_v - w_3} \sqrt{\frac{\sigma}{\gamma_3}} \tag{4.53}$$

In their opinion, only droplets with diameters in excess of d_{3c} make contact with a heated surface. This latter quantity is a function of the flow variables and is determined by solving a set of differential equations describing the motion of the droplets

$$\rho_3 \frac{\pi}{6} d_3^3 \frac{d^2 z}{dt^2} = \frac{\pi}{16} d_3^3 \rho_v \frac{dw}{dy}\left(\frac{dy}{dt}\right) - 6\pi\mu_v \frac{d_3}{2}\left(\frac{dz}{dt} - w\right) - (\rho_3 - \rho_v)\, g\, \frac{\pi}{6} d_3^3$$

$$(2.54)$$

$$\rho_3 \frac{\pi}{6} d_3^3 \frac{d^2 y}{dt^2} = -\frac{\pi}{16} d_3^3 \rho_v \frac{dw}{dy}\left(\frac{dz}{dt} - w\right) - 6\pi\mu_v \frac{d_3}{dt}\left(\frac{dy}{dt}\right) +$$

$$+ \frac{\pi d_3^2 \alpha^2}{4r^2 \rho_v}\,(T_w - T_{sat})\frac{d_3}{\delta}\left(1 - \frac{y}{\delta}\right)$$

$$(4.55)$$

The forces incorporated in the above equations are: lift and drag of the droplets, gravity and buoyancy forces, and also the reaction force arising due to nonuniformity of droplet evaporation in the boundary layer. This set of equations was solved for the wall flow region on the assumption that

$$w = w_0\,(y/\delta)$$

$$(4.56)$$

and under the following conditions:

$$t = 0;\ z = 0;\ dz/dt = w_0$$

$$(4.57)$$

$$t = 0,\ y = \delta,\ \frac{dy}{dt} = w_0^*$$

$$(4.58)$$

Solving the above equations one obtains the fraction of droplets in contact with the wall

$$f(d_3) = \frac{\displaystyle\int_{d_3^{\,w}}^{d_3^{\max}} d_3^v F(d_3)\, d(d_3)}{\displaystyle\int_{d_3}^{d_3^{\max}} d_3^v F(d_3)\, d(d_3)}$$

$$(4.59)$$

The amount of heat removed from the wall by a single droplet is:

$$Q = \frac{\pi}{6} d_3^3 \rho_3 r\,\exp\left[1 - \left(\frac{T_w}{T_{sat}}\right)^m\right]$$

$$(4.60)$$

The droplet irrigation flux along the wall is given by

$$M_w = W_0^*\,(1 - \varphi^*)\,\rho_3 f$$

$$(4.61)$$

The number of precipitating droplets is

$$N = M_w \left/ \left(\frac{\pi}{6} \rho_3 \bar{d}_3 \right) \right.$$

(4.62)

where $d_{3c} < \bar{d}_3 < d_3^{max}$ and is given by the equation

$$\bar{d}_3 = \left(\frac{\displaystyle\int_{d_3^w}^{d_3^{max}} d_3^3 F(d_3)\, d(d_3)}{\displaystyle\int_{d_3^w}^{d_3^{max}} F(d_3)\, d(d_3)} \right)^{1/3}$$

(4.63)

With reference to the above, the heat flux to the droplets was found from the expression:

$$q_3 = NQ = w^* (1 - \varphi^*)\, \rho_3 rf \exp [1 - (T_w / T_{sat})^2]$$

(4.64)

The heat flux to the vapor was written as

$$q_v = 0{,}023\, \frac{\lambda}{D}\, Re^{0,8}\, Pr^{0,4}\, (T_w - T_{sat})$$

(4.65)

Neglecting the radiant heat flux, we obtain

$$q_w = q_v + q_3$$

(4.66)

This model contains two constants: m and C. Their values were found by fitting the experimental results

$$m = 2; \quad C = 0{,}4/[(W_v - W_3)\alpha\rho_v/\mu_v]^{0,5}$$

(4.67)

The suggested model gave satisfactory agreement with experimental data. It should, however, be pointed out that the model makes no allowance for flow nonequilibrium, which significantly detracts from the usefulness of the study. This approach can apparently be used only in the case of low heat fluxes and high mass velocities, when in effect nonequilibrium is low. In addition, the fact that the values of m and C were obtained by fitting experimental data also limits the utilization of the model for ranges of variables other than those investigated by the authors.

The above approach has been extensively developed in the USSR [4.43–4.61]. Let us consider in more detail the study by Inkov [4.55]. Its most remarkable feature is the most complete utilization of Eqs. (4.17)–(4.29). Closure of this set of equations by relationships obtained by measuring the quantities being monitored (profile of wall temperature along the tube, pressure drop, etc.), makes it possible to analyze the effect of various factors (p, ρw, T_w, etc.) on the flow variables w_v, w_3, T_v and T_3.

This made it possible to gain a deeper insight into the relationships governing the process, to correlate the results and to recommend empirical equations for calculating the hydrodynamics and heat transfer in the post-dryout region.

The starting set of equations is

$$G \frac{dx}{dz} = \frac{\Pi_3}{r} q_3 \tag{4.68}$$

$$G_3 + G_v = G \tag{4.69}$$

$$G_3 = G(1-x) = \rho_3 w_3 (1-\varphi) F \tag{4.70}$$

$$G_v = \rho_v w_v \varphi \tag{4.71}$$

$$\rho_3 w_3 \frac{dw_3}{dz} = \rho_3 g_z - \frac{dp}{dz} + \mathrm{sign}\,(w_v - w_3) \frac{\Phi}{1-\varphi} - \frac{\beta\,(w_v - w_3)\,G}{(1-\varphi)} \frac{dx}{dz} \tag{4.72}$$

$$\rho_v w_v \frac{dw_v}{dz} = \rho_v g_z - \frac{dp}{dz} - \mathrm{sign}\,(w_v - w_3) \frac{\Phi}{\varphi} =$$

$$= \frac{(1-\beta)\,(w_v - w_3)\,G}{\varphi F} \frac{dx}{dz} - \tau_w \frac{\Pi_w}{F\varphi} \tag{4.73}$$

$$G_v \frac{dh_v}{dz} + (h_v - h'') \frac{dG_v}{dz} = q_w \Pi_w - q_3 \Pi_3 \tag{4.74}$$

$$q_w = q_v + q_3 \frac{\Pi_3}{\Pi_w} \tag{4.75}$$

The closure relationships are

$$x = \frac{G_v}{G_v + G_3} \qquad \varphi = \frac{F_v}{F_v + F_3} \qquad F_3 = F(1-\varphi)$$

$$\varphi = 1 - n \frac{\pi d_3}{6} \qquad \Pi_3 = \frac{6 F_3}{d_3} \qquad \frac{1-x}{1-x_0} = \left(\frac{d_3}{d_{30}}\right)^3$$

$$\Phi = \frac{3}{4} \frac{1-\varphi}{d_3} C_D \rho_v (w_v - w_3)^2$$

$$C_D = f(\mathrm{We}) \begin{cases} 12,3\, \mathrm{Re}_{d_3}^{-0,5} & \text{at} \quad 13 \leqslant \mathrm{Re}_{d_3} \leqslant 800 \\ 0,44 & \text{at} \quad \mathrm{Re}_{d_3} > 800 \end{cases}$$

$$f_1(\mathrm{We}) = 1 \qquad f_2(\mathrm{We}) = \varphi(\mathrm{We})$$

$$q_3 = \frac{\lambda_v}{d_3} (T_v - T_{\mathrm{sat}})(2 + 0,55\, \mathrm{Re}_{d_3}^{0,5}\, \mathrm{Pr}_v^{0,39}) f(\Theta)$$

$$\Theta = \frac{h_v - h''}{r} \qquad f_1(\Theta) = \frac{\ln(1+\Theta)}{\Theta} \qquad f_2(\Theta) = 1 \qquad \mathrm{We}_{\mathrm{cr}} = 5-10$$

where the interaction coefficient $\beta = 0, \ldots, 0.5$. The uniqueness conditions are as follows: $F = \pi D^2/4$, $D(z) = D$, $\Pi_w = \pi D$, $g_z = -9.81$ m/sec^2, $G(z) = $ const, $h_v = h(p, T_v)$, $\rho_v = \rho(p, T_v)$, $\rho_{sat} = \rho(T_{sat})$, $r = r(T_{sat})$, $T_{sat} = T(p)$. The system is completely specified by experimentally obtained relationships of $q_w = q(z)$ and $p = p(z)$ or dp/dz.

On the basis of workup of experimental data of In'kov [4.55], and Yarkho with his coworkers [4.59–4.61], the above set of equations was used to establish the following important features of two-phase flow in the post-dryout region.

1. Phase slip ($w_v - w_3$) depends to a large degree on the pressure, mass velocity, and wall temperature. Since phase slip controls their thermal and dynamic interaction, generation of vapor superheat and thermal nonequilibrium also depend on these parameters. The value of the slip factor $S = w_v/w_3$ increases with the mass velocity and wall temperature and decreases with rising pressure.

2. The thermal nonuniformity lessens with rising pressure.

3. The reduction of droplet size along the steam generating channel is due primarily to breakup of larger droplets. The effect of vaporization of droplets on their size is secondary.

4. The direction of flow (up or downflow) has a perceptible effect on the inlet parts of channels. Due to the greater slip of droplet in upwards flow, the heat transfer rate is higher.

5. The rate of heat transfer to vapor-liquid flow is higher than to vapor. The evaporating droplets turbulize the flow, particularly near the heating wall. In addition, vapor generation near the heating wall is basically equivalent to injection of cold vapor into regions of maximum heat fluxes. These effects are incorporated by introducing the parameters

$$\chi = D/x \cdot dx/dz \tag{4.76}$$

Naturally, the effect of this parameter is a function of droplet concentration $(1 - x)_w$ in the vicinity of the wall. The value of $(1 - x)_w$ depends on the velocity and temperature gradients, which control the aerodynamic and reaction forces acting on the droplet. As a whole, the effect of these two factors is defined by the parameter

$$\Theta = [c_p(T_w + T_{sat} - 2T_v)]/r \tag{4.77}$$

With allowance for the phase slip, temperature factor, parameters χ and Θ, the expression for Nu_v can be written in the generalized form [4.61]

$$Nu_v = f(Re_v, Pr_v, \chi, \Theta, S, T_w/T_v) \tag{4.78}$$

The specific form of this relationship was obtained from a graphical analysis of a very large body of experimental data. It was found that there exist two regions, which are a function of χ. At $\chi > \chi_{lim}$ the effect of χ is significant, whereas at $\chi < \chi_{lim}$ parameter χ does not affect Nu_v. We can therefore recommend two empirical formulas

$$\mathrm{Nu_v} = \frac{0{,}394\theta^{0{,}5(\mathrm{th}\,S-1)}}{3{,}42^{\mathrm{th}\,S}}\,\mathrm{Re_v^{0{,}8}}\,\mathrm{Pr_v^{0{,}4}}\chi^{0{,}61-(0{,}62-0{,}074\theta)^2\,\mathrm{th}\,S} \tag{4.79}$$

$$\mathrm{Nu_v} = 0{,}023\,(1 + 0{,}15x^{-1{,}5})\,\mathrm{Re_v^{0{,}8}}\,\mathrm{Pr_v^{0{,}4}}\,(T_w/T_v)^{-[0{,}55(1-x)]/(1+0{,}9x)} \tag{4.80}$$

The above expressions (or their modifications) serve as the basis for analysis and correlation of data for rough channels. For example, Khasayev [4.54] used the same set of equations with closure equations (4.17)–(4.29) for calculating the flow variables. The effect of roughnesses (annular turbulence promoters) was evaluated by comparing values of Nu for a smooth tube $\mathrm{Nu^{sm}}$ and a tube with turbulence promoters $\mathrm{Nu^r}$. Khasayev obtained the following expression for the heat transfer coefficient at the post-dryout region in tubes with turbulence promoters

$$\frac{\mathrm{Nu^r}}{\mathrm{Nu^{sm}}} = 1 + \left\{ \left[1{,}047\,\frac{\cos[1{,}74\,(\lg 10^3\chi)] + 1}{2} \right] \middle/ [(D_T/D)^{34{,}011}\,(t/D)^{1{,}628}\,\mathrm{Re_v^n}] \right\} \tag{4.81}$$

where $n = 0.1793/(D_T/D)^{3.203}\,(t/D)^{0.265}$, D_T is the diameter of the turbulence promoter, and t is the spacing of these promoters.

The above relationship is recommended for use within the limits:

$$D_T/D = 0{,}863 - 0{,}957 \quad t/D = 0{,}5-2; \quad \mathrm{Re_v} = 2\cdot 10^4 - 6\cdot 10^5$$

$$\chi = 0{,}05-1{,}85$$

$$\mathrm{Nu^{sm}} = 0{,}023\mathrm{Re_v^{0{,}8}}\,\mathrm{Pr_v^{0{,}4}}\,(1 + 0{,}15\chi)\,(T_w/T_v)^{-[0{,}55\,(1-\chi)]/(1+0{,}9\chi)} \tag{4.82}$$

or

$$\overline{\mathrm{Nu}}^{sm} = 2{,}3\,(1 + 0{,}15\chi)\,(T_w/T_v)^{-[0{,}55\,(1-\overline{\chi})]/(1+0{,}9\chi)} \tag{4.83}$$

at

$$\lg(10^3\chi) < 1{,}2; \quad \overline{\chi} = x/x_{\mathrm{lim}} < 1; \quad x_{\mathrm{lim}} = 0{,}016$$

$$\mathrm{Nu^{sm}} = 0{,}023\mathrm{Re_v^{0{,}8}}\,\mathrm{Pr_v^{0{,}4}}\,[1{,}67\lg(10^3\chi) - 1{,}17] \tag{4.84}$$

or

$$\mathrm{Nu^{sm}} = 2{,}3\,[1{,}67\lg(10^3\chi) - 1{,}17] \tag{4.85}$$

at $1{,}2 < \lg(10^3\chi) < 1{,}9; \; x > 1$

The temperature factor in these experiments (T_w/T_v) ranged between 1.08 and 3.807; the parameter typifying the rate of vapor generation had the range $\lg(10^3\chi) = (-0.335) - 0.977$.

The hydraulic drag coefficient was calculated from expressions for single-phase flow in tubes with turbulence promoters

$$\frac{c_f^{\mathrm{r}}}{c_f^{\mathrm{sm}}} = \left[1 + \frac{100\,(\lg \mathrm{Re}_v - 4{,}6)\,(1 - D_{\mathrm{T}}/D)^{1{,}65}}{\exp\,(t/D)^{0{,}3}} \right] \exp\left[\frac{25 - (1 - D_{\mathrm{T}}/D)^{1{,}32}}{(t/D)^{0{,}75}} \right] \tag{4.86}$$

for D_{T}/D = 0.9–0.97 and t/D = 0.5–1

$$\frac{c_f^{\mathrm{r}}}{c_f^{\mathrm{sm}}} \left[1 + \frac{\lg \mathrm{Re}_v - 4{,}6}{3/4(\mathrm{Re}_v/10^5) + 6} \right] (1{,}3 - \sqrt{|D_{\mathrm{T}}/D - 0{,}93|}\,\exp\,[20{,}9\,(1 - D_{\mathrm{T}}/D^{1{,}0} \tag{4.87}$$

for D_{T}/D = 0.88–0.98 and t/D = 0.5

$$c_f^{\mathrm{sm}} = \frac{0{,}316}{\mathrm{Re}_v^{0{,}25}} \tag{4.88}$$

In earlier studies [4.44, 4.52] the main features were analyzed using a significantly simplified set of equations: equations of motion, phase slip and liquid subcooling were omitted, and the following additional simplifications were introduced

$$\frac{dT_v}{dz} = 0 \quad \text{and} \quad T_3 = T_{\mathrm{sat}}$$

The heat fluxes to the vapor and droplets were written in the form

$$q_{\mathrm{ev}} = \alpha_{v3}\,(T_v - T_{\mathrm{sat}}) \tag{4.89}$$
$$q_w = \alpha_{wv}\,(T_w - T_v)$$

Two limiting cases were considered

$$\frac{\alpha_{wv}}{\alpha_{v3}}\,\frac{F_w}{F_3} \gg 1 \qquad \frac{\alpha_{wv}}{\alpha_{v3}}\,\frac{F_w}{F_3} \leqslant 1$$

In addition, it was assumed that

$$\frac{\alpha_{v3}\,d_3}{\lambda_v} = \text{const} \qquad \frac{F_w}{F_3} \sim \frac{d_3}{d_3\,(1 - \varphi)}$$

Setting d_3 = const, it was possible to obtain a set of dimensional variables for analysis of the experimental data

$$\frac{q_w}{\lambda_v (T_w - T_v)} = f \frac{1}{(1 - \varphi) D} \tag{4.90}$$

Using this coordinate system, the qualitative effects of principal variables are correctly predicted; however, the quantitative differences between the analytic and experimental results is quite significant.

Extensive studies of steam generating channels in the post-dryout region were performed by Italian scientists [4.46, 4.62]. Apparently, they were the first to use a one-dimensional two-phase two-velocity model for steam-water flows. Let us enumerate the principal features of their model.

1. The droplet drag coefficient is given by the formula

$$c_f = (0{,}63 + 4{,}8 \mathrm{Re}_{d_3}^{-0,5})^2 \tag{4.91}$$

2. The heat flux to the droplet is calculated from the expression

$$\mathrm{Nu} = \frac{\ln (1 + B')}{B'} [2 + 0{,}355 \mathrm{Pr}_v^{1/3} \mathrm{Re}_{d_3}^{0,59}] \tag{4.92}$$

where

$$B' = \frac{c_p (T_v - T_{sat})}{r} \qquad \mathrm{Pr}_v = \frac{c_p \mu_v}{\lambda_v} \qquad \mathrm{Re}_{d_3} = \frac{\rho_v (w_v - w_3) d_3}{\mu_v}$$

The set of equations for heat transfer were written as:

$$\frac{dT_v}{dz} = \frac{4}{\rho_v w_v D c_p} \left[q_w - \frac{n_0 w_{30}}{w_3} \frac{D}{4} \pi d_3^2 \alpha_3 (T_v - T_{sat})]; \right] \tag{4.93}$$

$$\frac{d (d_3)}{dz} = - \frac{2\alpha_3 (T_v - T_{sat})}{r \rho_3 w_3} \tag{4.94}$$

$$\frac{dw_v}{dz} = - \frac{n_0 w_{30} \pi \rho_3 d_3^2}{2\rho_v} \frac{d (d_3)}{dz} \tag{4.95}$$

$$\frac{dw_3}{dz} = \frac{1}{\rho_3 w_3 d_3} \left[\frac{3}{4} c_f \rho_v (w_v - w_3)^2 + d_3 g (\rho_v - \rho_3) - 3\rho_3 w_3^2 \frac{d (d_3)}{dz} \right] \tag{4.96}$$

The boundary conditions were

$$T_v (z = 0) = T_{sat} \qquad d_3 (z = 0) = d_{30} \qquad w_v (z = 0) = w_{v0}$$

$$w_3 (z = 0) = w_{30} \tag{4.97}$$

Calculations performed with these equations for different values of slip made it possible to determine the effect of flow variables (ρw, x_{cr}, q_w, etc.) on the velocities of the vapor and liquid, vapor temperature and droplet size. Analysis of calculations shows that: (1) liquid droplets in nonequilibrium flow vanish after traversing a channel length equal to several hundreds of equivalent diameters; the nonequilibrium in the steam-water flow may persist to equilibrium qualities as high as 200%; (2) the mean droplet diameter decreases with increasing heat flux and mass velocity; (3) over a wide range of variables the droplet velocity is independent of the droplet size.

A statement of the problem close to that of Cumo et al. [4.46, 4.62] is given by Subbotin et al. [4.50]. Its salient features are.

1. The heat transfer coefficient to the vapor-droplet flow is determined from Bishop's formula

$$\mathrm{Nu} = 0{,}0073\,\mathrm{Re}^{0{,}886}\,\mathrm{Pr}^{0{,}61}\,(1 + 2{,}76\,D/L) \tag{4.98}$$

2. The number of droplets

$$N = \frac{6\rho w\,(1 - x_{\mathrm{lim}})}{\pi d_3^2 \rho'} \tag{4.99}$$

remains constant along the steam-generating channel.

The constant in the equation for heat transfer from the vapor to the droplets was changed to

$$\mathrm{Nu} = 2 + 0{,}74 \left[\frac{(w_{\mathrm{v}} - w_3)\,d_3}{\nu_{\mathrm{v}}}\right]^{0{,}5} \mathrm{Pr}^{0{,}33} \tag{4.100}$$

The working equations were as follows

$$\frac{d\,(d_3)}{dz} = -\frac{2 d_3 \Delta T}{w_3} \tag{4.101}$$

$$\frac{dx}{dz} = \frac{N \rho' \pi d_3^2}{\rho w}\frac{d\,(d_3)}{dz} \tag{4.102}$$

$$\frac{dT_{\mathrm{v}}}{dz} = \left(q \pi D - \frac{\pi D^2}{4}\,r\rho w\,\frac{dx}{dz}\right) \Big/ \left(\frac{\pi D^2}{4}\,\rho w x \bar{c}_p\right) \tag{4.103}$$

$$\left(\frac{\rho'}{\rho_{\mathrm{v}}} + 0{,}5\right) w_3\,\frac{dw_3}{dz} = \frac{3\varphi}{4D}\,(w_{\mathrm{v}} - w_3)^2 + 1{,}5 w_3\,\frac{dw_{\mathrm{v}}}{dz} - g\left(\frac{\rho'}{\rho_{\mathrm{v}}} - 1\right) \tag{4.104}$$

$$w_{\mathrm{v}} = \frac{\rho w x}{\rho_{\mathrm{v}}}\left[1 - \frac{\rho w\,(1 - x)}{\rho' w_3}\right]^{-1} \tag{4.105}$$

which, with the appropriate closure expressions, were solved numerically by computer. Analysis of data on heat transfer in the post-dryout region of channels cooled by wet steam at $p = 13.7$ MPa showed that the numerical and experimental results were in satisfactory agreement, which was improved by selecting an appropriate

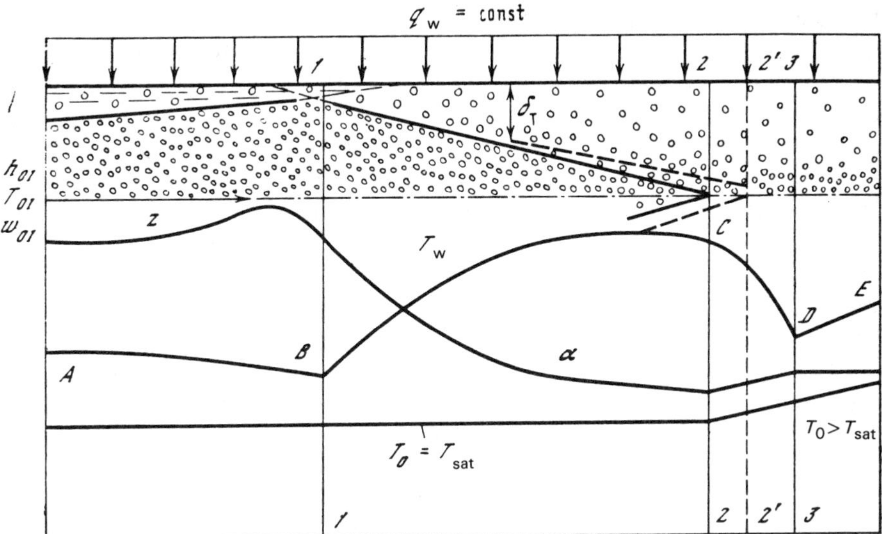

Figure 4.18 Physical model of heat transfer in the post-dryout region.

value of the droplet diameter. The droplet sizes needed to give the best data fit are much higher than those physically expected and this is a reflection of all the simplifications underlying the model. The best agreement was obtained when the diameter d_3 of the droplets at the dryout location was taken to be 1.2 mm.

4.4.2 The Boundary Layer Approximation

Analytic studies using one-dimensional models have made it possible to obtain deeper insight into the features of heat transfer in the post-dryout region, to correlate individual experimental results and to obtain working equations. At the same time, the inherent nature of the one-dimensional description makes it impossible to make allowance for the variation of parameters across the flow, which is necessary for an adequate description of the process. In previous studies by the present authors [4.63, 4.64], a two-dimensional model of heat transfer for the post-dryout region in steam-generating channels was developed. We shall examine this model in somewhat more detail.

The flow of coolant in steam-generating channels undergoes changes in structure along the channel. At high vapor qualities, the critical heat flux on dryout phenomenon is associated with transition from dispersed-annular to dispersed flow (the transition is known here as a "crisis of the second kind). In dispersed-annular flow the liquid phase is partially concentrated in the annular film and partially, as seen from Fig. 4.18, distributed in the form of droplets within the flow. In this flow region heat is transferred at a high rate between the wall and the two-phase flow (line AB). The wall temperature exceeds the saturation temperature only by several degrees.

The heat transfer conditions undergo a radical change at the dryout location. As dryout is approached, the liquid film thins out significantly, leading to it breaking down locally. Dry patches (islands of dry wall within the film) appear and vanish. Ultimately, the boiling liquid film disappears and the wall is in contact with the droplet-laden vapor, which has a significantly higher thermal resistance. The location in which a vapor boundary layer (averaged in time) appears over a significant part of the tube perimeter is usually taken as the dryout location. Starting with this location the wall temperature rises steeply (line BC) and then, depending on the flow variables (see Sec. 4.1) may continue to grow or fall. At high mass velocities and relatively low vapor qualities the heat transfer coefficient attains its minimum at point C, where the wall temperature attains its maximum value. The temperature falls to a minimum at point [D] and then rises again.

The flow in region AB is slightly nonequilibrium due to metastable superheating of the liquid phase at the wall by several degrees. In BC there may be much larger departures from thermodynamic equilibrium. Superheated vapor moves along the wall while liquid droplets exist in the flow core. The droplets that enter the vapor layer evaporate and increase the rate of heat transfer between the wall and the mixture. The physical situation which comes about due to onset of dryout is quite similar to that occurring with heat transfer in the inlet regions of tubes. Beginning with section 1-1 (see Fig. 4.18), there starts developing a thermal boundary layer of superheated vapor, whose formation obeys virtually the same laws as those in inlet region single phase heat transfer in a channel. The dryout location can be conditionally regarded as the leading edge of the tube, where the mixture temperature is virtually constant over the tube cross section and equal to the saturation temperature. At high heat fluxes, when the wall temperature in the post-dryout region exceeds the maximum liquid superheat temperature $T_{\lim}$, the surface of the tube is in contact with superheated vapor, whose temperature varies over the boundary-layer cross section from the wall temperature to the saturation temperature in the flow core. By analogy with the thermal boundary layer, we can postulate a diffusion boundary layer, in which the droplet concentration varies from a maximum in the flow core to zero at the wall or at some distance from it.

Until the boundary layer reaches the axis of the tube, liquid droplets can only evaporate within the thermal layer δ_T, and we can assume that the enthalpy of the mixture on the tube axis remains unchanged and equal to its value h_{0I} at the dryout location. The flow velocity at the tube axis increases due to growth of the boundary-layer thickness and evaporation of droplets. The increase in boundary layer thickness decreases the coefficient of heat transfer and causes a rise in wall temperature up to section 2.2. At point C, the wall temperature reaches its maximum $T_w^{\max}$. It is this temperature which controls the reliability of steam-generating channels and is of primary interest to designers. In this sense the initial part BC of the post-dryout region is of great practical interest, particularly to designers. It is precisely this part which is considered in the model analyzed here.

The liquid droplets are relatively small. It is usually assumed that

$$\mathrm{We}_{cr} = \frac{d_3 \rho'' \, (w_v - w_3)^2}{\sigma} = 7{,}5 \qquad (4.106)$$

When the Weber number attains this value the droplets break up. The mean dimensions of droplets under conditions characteristic of dryout in the crisis of the "second kind" region are between several tens and several hundreds of microns. In spite of such a small size, the droplets may continue to exist up to the very high thermodynamic qualities, i.e., in the flow of highly superheated vapor. This means that the rate of droplet evaporation is not very high.

Due to the small droplet size, the flow can often be regarded as approximately homogeneous. The homogeneous approximation becomes close with reducing volumetric fraction of the liquid phase $(1 - \beta)$ and with reducing droplet size. Since the rate of evaporation of the droplets is relatively low, their concentration at each point of the flow is controlled primarily by turbulent diffusion. In this sense heat transfer between the vapor flow and the evaporating droplets is analogous to heat transfer in dissociating or chemically reacting gases in a "frozen" boundary layer. Also, this process in the inlet region of the tube occurs under conditions of significantly variable fluid density.

The droplet evaporation rate is highest in the immediate vicinity of the wall. It can be assumed in the first approximation that the droplets evaporate in a thin layer adjoining the wall and that the distribution of droplet concentration over the boundary layer cross section is controlled by eddy diffusion and is similar to the distribution of mixture enthalpy and velocity. In accordance with our model, the rate of convective heat transfer between the tube surface and the wet vapor will be significantly affected by the variations of mixture's density across the boundary layer and the heat flux to the rapidly evaporating droplets. The effect of these two factors can be incorporated using asymptotic boundary layer theory [4.65].

The relative expression for the limiting heat transfer can in general be written as:

$$\Psi'_s = (St/St_0)_{Re_T^{**}} = \left[\int_0^1 \left(\tilde{\rho}\, \frac{\tilde{q}_0}{\tilde{q}}\, \frac{\partial \omega}{\partial \xi_T}\, \frac{\partial \vartheta}{\partial \xi_T} \right)^{1/2} d\xi_T \right]^2 \tag{4.107}$$

where $St = q_w/(\rho_0 w_0\, \Delta h_\Sigma)$ is the Stanton number under the given conditions; St_0 is the Stanton number under standard conditions; $Re^{**} = (\rho_0 w_0 \delta_T^{**})/\mu_0$ is the Reynolds number based on the energy-thickness; $\tilde{q} = q/q_w$ is the nondimensional heat flux under the given conditions; $\tilde{q}_0 = q_0/q_{w0}$ is the same under standard conditions; $\omega = w/w_0$ is the nondimensional velocity; $\vartheta = h/h_0$ is the nondimensional enthalpy; $\xi_T = y/\delta_T$ is the nondimensional transverse coordinate; $\rho_0 w_0$ is the mass velocity on the tube axis; μ_0 is the characteristic of viscosity on the tube axis; $\Delta h_\Sigma = h_w - h_0$ is the total-enthalpy difference, and $\tilde{\rho} = \rho/\rho_0$ is the relative flow density.

On the basis of the above assumptions one may postulate a nonequilibrium diffusion model for the vapor-droplet flow. The set of equations is written as:

$$\frac{h_w - h}{h_w - h_0} = \omega \tag{4.108}$$

$$\frac{1-x}{1-x_0} = \omega \tag{4.109}$$

$$v = v_v x + v'(1-x) \tag{4.110}$$

$$v_v = v'' + \frac{\partial v}{\partial h}(h_v - h'') \tag{4.111}$$

$$h = h_v x + h'(1-x) \tag{4.112}$$

where x and x_0 are the real vapor quality in an arbitrary flow location and on the tube axis; v, v_v, v' and v'' are the specific volumes of the mixture, superheated steam, water, and saturated steam respectively and h, h_v, h' and h'' are the respective enthalpies.

The laws of heat transfer for turbulent boundary layers (as stated by Kutateladze and Leont'yev [4.65]) were used to represent the combined effect of the various perturbing factors. Since $\partial\omega/\partial\xi_T = \partial\vartheta/\xi$ for the case of $\tilde{q}_0/q \simeq 1$ (there are no internal heat sinks), the expression for the limiting heat transfer simplified to

$$\Psi_{sv} = \left[\int_0^1 \frac{d\omega}{\sqrt{\tilde{v}}}\right]^2 \tag{4.113}$$

where $\tilde{v} = v/v = \rho_0/\rho$ is the nondimensional specific volume.

It follows from the solution of Eqs. (4.108)–(4.112) that

$$\tilde{v} = \frac{v_w}{v_0} - \left(\frac{v_w}{v_0} - 1\right)\omega \tag{4.114}$$

The final form of the limiting heat transfer equation is

$$\Psi_{sv} = \left(\frac{2}{\sqrt{v_w/v_0}+1}\right)^2 \tag{4.115}$$

Internal heat sinks (evaporation of droplets in the boundary layer of superheated steam) were incorporated into the heat transfer equation for the case of a mixture with constant density $\tilde{\rho} = 1$. The expression for the limiting relative heat transfer in this case is:

$$\Psi_{sq} = \left[\int_0^1 (\tilde{q}_0/\tilde{q})^{1/2}\,d\vartheta\right]^2 \tag{4.116}$$

Using the approximation for $\tilde{q}_0/\tilde{q}$ [4.65]

$$\tilde{q}_0/\tilde{q} = 1 + \frac{\mathrm{St}^{-1} \int_0^{\xi_T} \tilde{q}_v \, d\xi_T}{1 + 2\xi_T} \tag{4.117}$$

where $\tilde{q}_v = (q_v \delta_T)/(\rho_0 w_0 \, \Delta h_\Sigma)$, we obtain an expression for the volumetric heat sink q_v due to droplet vaporization:

$$q_v = 2\pi d_3 m \, (T_v - T_{sat}) \, \lambda_v \, (1 + 0{,}275 \, \mathrm{Re}_{d_3}^{0,5} \, \mathrm{Pr}_0^{1/3}) \tag{4.118}$$

where m is the volumetric concentration of droplets.

It was assumed in the expression for q_v that $\mathrm{Nu} = 2 + 0.55 \, \mathrm{Re}_{d3}^{0.5} \, \mathrm{Pr}_0^{1<3}$. The droplet concentration was obtained from the expression:

$$\left(m \, \frac{\pi d_3^3}{6} \, \gamma' \right) \Big/ \gamma_{mix} = 1 - x \tag{4.119}$$

Since

$$\gamma_{mix} = 1/v_{mix} = 1/[v_w - \omega \, (v_w - v_0)] \tag{4.120}$$

and $\omega = \xi_T^n$, we obtain

$$1 - x = (1 - x_0) \, \xi_T^n, \tag{4.121}$$

$$m = \frac{6 \, (1 - x) \, \xi_T^n}{\pi d_3^3 \gamma' \, [v_w - \xi_T^v \, (v_w - v_0)]} \tag{4.122}$$

It is found from Eqs. (4.108)–(4.112) that

$$T_v - T_{sat} = (T_w - T_{sat}) \, (1 - \xi_T^n) \tag{4.123}$$

Thus,

$$q_v = \frac{12\lambda_v \, (1 - x_0) \, (T_w - T_{sat}) \, (1 - \xi_T^n) \, \xi_T^n \, (1 + 0{,}275 \, \mathrm{Re}_{d_3}^{0,5} \, \mathrm{Pr}_0^{1/3})}{d_3^2 \gamma' \, [v_w - \xi_T^n \, (v_w - v_0)]} \tag{4.124}$$

Since the effect of q_0/q on the behavior of heat transfer is relatively small, the expression for q_v can be simplified to:

$$q_v = A \, (1 - \xi_T^n) \, \xi_T^n \tag{4.125}$$

where

$$A = \frac{12\lambda_v \, (1 - x_0) \, (T_w^{max} - T_{sat})}{d_3^2 \gamma' v_w} \tag{4.126}$$

The denominator of the fraction in Eq. (4.117) is rewritten to the form:

$$St^{-1} \int_0^{\xi_T} \tilde{q}_v \, d\xi_T = \frac{A\delta_T}{q_w} \left(\frac{\xi_T^{n+1}}{n+1} - \frac{\xi_T^{2n+1}}{2n+1} \right) \tag{4.127}$$

Approximating the expressions

$$\frac{\xi_T^{n+1}}{1 + 2\xi_T} = 0{,}41\xi_T \quad \text{and} \quad \frac{\xi_T^{2n+1}}{1 + 2\xi_T} = 0{,}375\xi_T$$

we find

$$\frac{\tilde{q}_0}{\tilde{q}} = 1 + \frac{A\delta_T}{q_w} \left(\frac{0{,}41\xi_T}{n+1} - \frac{0{,}375\xi_T}{2n+1} \right) \tag{4.128}$$

Then the expression for Ψ_{sq} becomes

$$\Psi_{sq} = \left[\int_0^1 \sqrt{1 + a\xi_T} \, d\vartheta \right]^2 \tag{4.129}$$

where

$$a = \frac{0{,}675 A \delta_T}{q_w} \tag{4.130}$$

Replacing ξ_T by ϑ, assuming $n = 1/7$ and approximating the expression ϑ^7 by 0.25ϑ, we finally obtain

$$\Psi_{sq} = \left\{ \frac{8}{3a} \left[\left(1 + \frac{a}{4} \right)^{3/2} - 1 \right] \right\}^2 \tag{4.131}$$

The effect of heating-surface roughness on the heat transfer was incorporated on the basis of the data of Kalinin et al. [4.58] and Gomelauri et al. [4.67]. These investigators showed that St rises in proportion to the rise in the "pure" friction factor, which comprises a part of the total drag coefficient, i.e.,

$$\Psi_{sr} = \left(\frac{c_f^t}{c_{f_0}^t} \right)_{Re} \tag{4.132}$$

where c_f^t is the friction coefficient under the conditions at hand (without correction for the roughness form drag); $c_{f_0}^t$ is the friction coefficient of a smooth wall.

The problem consequently reduces to separators of the coefficient of "pure" friction from the total drag coefficient. The coefficient c_f^t was found from the data of Gomelauri et al. [4.67]:

$$c_f^t = c_{f_0}^t \, \sqrt{c_f/c_{f_0}} \tag{4.133}$$

Consequently, the analytic value of the Stanton number $\mathrm{St_{an}}$ was determined from the expression

$$\mathrm{St}_{an} = \mathrm{St}_0 \Psi_{sv} \Psi_{sq} \Psi_{sr} \tag{4.134}$$

The heat transfer at the inlet region under standard conditions is described by the expression [4.65]:

$$\mathrm{St}_0 = \frac{B}{2 \, (\mathrm{Re_T^{**}})^{0,25} \, \mathrm{Pr}_0^{0,75}} = \frac{0,014}{(\mathrm{Re_T^{**}})^{0,25} \, \mathrm{Pr}_0^{0,75}} \tag{4.135}$$

In accordance with the above model we analyzed our own experimental data and that obtained by several other investigators [4.2, 4.3, 4.68, 4.69]. The experimentally measured Stanton number $\mathrm{St_{exp}}$ is determined from the equation:

$$\mathrm{St}_{exp} = \frac{q_w D}{(\mathrm{Re}_{DI} + 5,2\psi \, \mathrm{Re_T^{**}}) \, \mu_0 \Delta h} \tag{4.136}$$

where Re_{DI} is the Reynolds number in the crisis location and $\psi = h_w/h_0$ is the enthalpy factor.

The value of $\mathrm{Re_T^{**}}$ was found from the energy equation

$$\frac{d\,\mathrm{Re_T^{**}}}{dz} + \frac{\mathrm{Re_T^{**}}}{\Delta h} \frac{d\Delta h}{dz} + \frac{\mathrm{Re_T^{**}}}{\rho_0 w_1} \left(\frac{\delta_T}{\delta_T^{**}}\right) \frac{D}{\Delta h} \int_0^1 \tilde{q}_v \, d\xi_T = \mathrm{St} \, \mathrm{Re}_D \tag{4.137}$$

The distribution $\Delta h = h(\tilde{z})$ was determined experimentally. Setting

$$\left(\frac{\delta_T}{\delta_T^{**}}\right) \frac{D}{\rho_0 w_0 \Delta h} \int_0^1 \tilde{q}_v \, d\xi_T = c = \mathrm{const} \tag{4.138}$$

we find the solution of the nonhomogeneous linear equation

$$\mathrm{Re_T^{**}} = \frac{q_w D}{\mu \Delta h} \left[\frac{e^{c\tilde{z}} - 1}{ce^{c\tilde{z}}}\right] \tag{4.139}$$

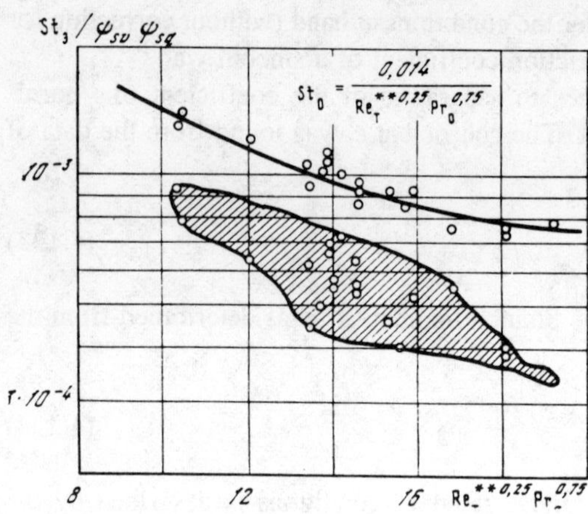

Figure 4.19 Plot of St_{exp} and $St_0/\Psi_{sv}\Psi_{sq}$ vs. $Re^{**0.25}Pr_0^{0.75}$.

The value of c is of the order of several hundredths or thousandths. For low values of $c(c \rightarrow 0)$ the last equation was approximated by the expression

$$Re_T^{**} = \frac{q_w D\tilde{z}}{\mu_0 \Delta h\,(1 + c\tilde{z})} \tag{4.140}$$

The parameters h_0, v_0, μ_0, w_0, etc. on the flow axis was determined from standard equations for two-phase flows, the parameters at the wall were based pressure p and temperature T_w. The results of this analysis are presented in Fig. 4.19. The shaded region contains results calculated from experimental values of the Stanton number (St_{exp}) as a function of the parameter $(Re_T^{**})^{0.25}Pr_0^{0.75}$. It is seen from the figure that the deviation from the curve of $St_0 = 0.014/(Re_T^{**0.25}Pr_0^{0.75})$ in individual cases is as much as 200 to 300%. The introduction of ratios Ψ_{sv} and Ψ_{sq} causes these data to cluster about the analytic curve.

Note that data obtained at low volumetric percentage moisture cluster satisfactorily about the theoretical curve. With increasing percentage moisture the experimental values of $St_{exp}/\Psi_{sv}\Psi_{sq}$ start deviating from the analytic values, which might be expected considering the limitations of the model with respect to flow homogeneity. Data obtained where the wall temperature in the post-dryout region is lower than the limiting superheat temperature of the liquid (T_{lim}) along the entire inlet length, or over a significant part of it, also deviate from the theoretical curve. This means that a two-phase boundary layer with unknown quality was developed in the inlet region. If h_w is calculated on the basis of the wall temperature and pressure, then its value will be perceptibly overestimated. This naturally resulted in a significant underestimation of $St_{exp}/\Psi_{sv}\Psi_{sq}$ as compared with St_0.

Figure 4.20 is a plot, in coordinates $St_{exp}/\Psi_{sv}\Psi_{sq}\Psi_{sr}$ vs. $(Re_T^{**})^{0.25}Pr_0^{0.75}$, of the results of analysis of data obtained on a rough surface. The rate of heat transfer on such surfaces is significantly higher than on commercially smooth surfaces. The value of Ψ_{sr} was obtained from calibrations of hydraulic drag of commercially smooth and rough tubes. When corrected for Ψ_{sr}, the data for rough surfaces cluster satisfactorily about the theoretical curve.

Reduction of experimental data confirmed the fact that heat transfer for the initial length of the post-dryout region is described by Eq. (4.134). The fact that it was possible to establish the actual form of $\mathrm{St} = f(\mathrm{Re}_T^{**}, \ldots)$, taking account of the energy and continuity equations, makes it possible to calculate analytically heat transfer along the inlet length. In the first approximation, use was made of the energy equation without the heat sink provided by vaporization of droplets, since the effect of the latter on St can be incorporated by using the factor Ψ_{sq}. The energy equation contains two unknowns: Re_T^{**} and Δh. Note that Ψ_{sv} is a function of h_w. Hence we need an additional condition allowing the flow density to be expressed in terms of enthalpy. The expression for Ψ_{sv} was approximated as:

$$\Psi_{sv}' = \left(\frac{2}{\sqrt{v_w/v_0}+1}\right)^2 = \frac{2,2}{1+v_w/v_0} \tag{4.141}$$

This expression describes rather satisfactorily the relative heat transfer at $0.5 \leq v_w/v_0 \leq 3$.

The specific volume of the mixture at the tube axis in the initial region was expressed in terms of the inlet enthalpy h_0:

$$v_0 = \hat{e} + \hat{d}h_0 \tag{4.142}$$

where

$$\hat{e} = v' - (v'' - v')\,h'/r; \quad \hat{d} = (v'' - v')/r \tag{4.143}$$
$$v_w = v'' + \partial v/\partial h\,(h_w - h'')$$

Here

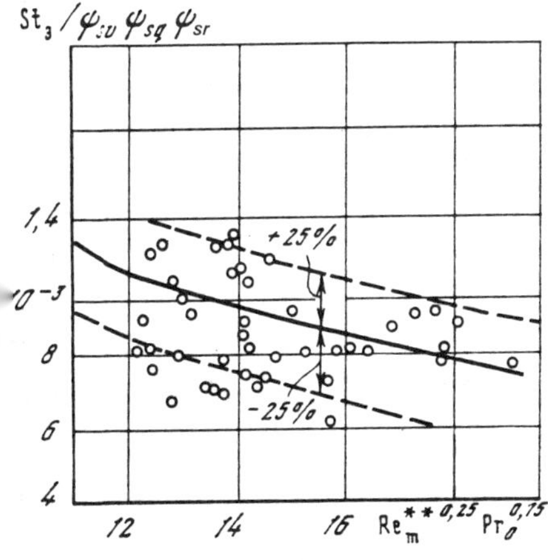

Figure 4.20 Comparison of experimental results for roughness of the first type with standard conditions.

$$\partial v/\partial h = \Delta v/\Delta h = (v_{\text{sup}} - v'')/(h_{\text{sup}} - h'')$$

Then

$$\Psi_{sv} = \frac{2,2}{\hat{c}_1 + \hat{d}_1\psi} \tag{4.144}$$

where

$$c_1 \cong \left[\left(v'' - h''\frac{\partial v}{\partial h}\right)\Big/(\hat{e} + \hat{d}h_0)\right] + 1; \qquad \hat{d}_1 = \left(h_0\frac{\partial v}{\partial h}\right)\Big/(\hat{e} + \hat{d}h_0)$$

and Eq. (4.134) becomes:

$$St = (B\cdot 2,2/2)\Psi_{sr}\ /[(Re_T^{**})^{0,25}\ Pr_0^{0,75}\ (\hat{c}_1 + \hat{d}_1\psi)] \tag{4.145}$$

For the case of q_w = const the solution of the energy equation (without correction for q_v) is written as:

$$\psi = 1 + \frac{N\tilde{z}}{Pe^{**}} \tag{4.146}$$

where

$$Pe^{**} = Re_T^{**}Pr_0, \qquad N = \frac{Dq_w c_p}{h_0\lambda} \tag{4.147}$$

Since

$$St = \frac{N}{Pe_D\,(\psi - 1)} = \frac{1,1B\Psi_{sr}}{(Pe^{**})^{0,25}\ Pr_0^{0,5}\ (\hat{c}_1 + \hat{d}_1\psi)} \tag{4.148}$$

$$Pe_D = Pe_{DI} + 5,2\psi\,Pe^{**}, \tag{4.149}$$

we obtain a set of three equations (4.147)–(4.149) with three unknowns: ψ, Pe^{**} and Pe_D. Solving these equations for the independent variable $\tilde{z}$, we find

$$\tilde{z} = 3,36N^{-1}\left[\frac{\hat{d}_1 N\ Pr_0^{0,5}\ Pe^{**0,25}}{\Psi_{sr}} - 0,0286\,Pe_{DI} - 0,149\,Pe^{**} + \right.$$
$$\left. + \sqrt{\left(\frac{\hat{d}_1 N\ Pr_0^{0,5}\ (Pe^{**})^{0,25}}{\Psi_{sr}} - 0,0286\,Pe_{DI} - 0,149\,Pe^{**}\right)^2 + } \right.$$
$$\left. \vphantom{\sqrt{X}} +\ \frac{0,595\,(\hat{c}_1 + \hat{d}_1)\ Pr_0^{0,5}\ (Pe^{**})^{1,25}}{\Psi_{sr}}\right] \tag{4.150}$$

Specifying a value of Pe**, we find $\tilde{z}$ and calculate ψ from Eq. (4.150). The entrance length was found from the following considerations. The relative velocity on the tube axis at the point where the boundary layer reaches the axis is, according to Kutateladze and Leont'yev [4.65]

$$\tilde{w}_{0\,\text{in}} = \frac{w_0}{w_{cI}} = \left(1 - 2{,}6\,\frac{\delta^{**}}{R_0}\,\psi\right)$$
(4.151)

Since

$$\delta^{**}/R_0 = 0{,}074\psi^{-0,318}$$
(4.152)

$$\tilde{w}_{0\,\text{in}} = \text{Pe}_D/\text{Pe}_{DI}$$
(4.153)

we find that

$$\text{Pe}_{DI} + 5{,}2\psi\,\text{Pe}^{**} = \frac{\text{Pe}_{DI}}{1 - 0{,}192\psi^{0,682}}$$
(4.154)

This means that the set of equations (4.146), (4.150) and (4.154) is available for calculating the initial length $\tilde{z}_{\text{in}}$. These equations cannot be solved in terms of $\tilde{z}_{\text{in}}$, Pe** and ψ. The value of $\tilde{z}_{\text{in}}$ is found graphically. Specifying values of Pe**, we find z from Eq. (4.150). We find ψ from Eq. (4.146) and construct a curve of $\psi = f(\tilde{z}/\text{Pe}^{**})$ we obtain

$$\text{Pe}^{**} = \frac{\text{Pe}_{DI}\psi^{0,632}}{5{,}2\psi\,(5{,}21 - \psi^{0,682})}$$
(4.155)

Then Eq. (4.155) is used for constructing a plot of $\text{Pe}^{**} = f_1(\psi)$. The point of intersection of curves $\psi = f(\tilde{z}/\text{Pe}^{**})$ and $\text{Pe}^{**} = f_1(\psi)$ yields the values of ψ_{in} and $\text{Pe}^{**}_{\text{in}}$ which makes it possible to obtain the value of z_{in} from Eq. (4.150). The value of z_{in} can also be determined by examining the deviation of predicted points from the curve of the heat transfer equation.

A method for calculating heat transfer under post-dryout conditions on the basis of a two-layered model (wall layer and flow core) was suggested by Miropol'skiy et al. [4.70]. Momentum, heat and mass transfer occurs between the two regions of the flow. It is assumed that the liquid phase within the wall layer is fully evaporated. The balance equations of conservation of mass and heat are written in the form:

$$g_{\text{rad}} = g_{\text{rad.3}} + g_{\text{v.0}} = g_{\text{v.0}} + g_{\text{w.l}}$$
(4.156)

$$q = g_{\text{v.0}}\Delta h_{\text{v.sup}} + g_{\text{w.l}}\,\Delta h_{\text{v.sup}} + g_{\text{r.3}}r$$
(4.157)

where $\Delta h^n_{\text{sup}} = \Delta T_{\text{sup}}c_p$.

Assuming that $g_{\text{v.0}}$ is proportional to the actual vapor quality x_{act}, Miropol'skiy et al. present the following expressions for $g_{\text{w.l}}$, $g_{\text{rad.3}}$ and $g_{\text{v.0}}$:

$$g_{v.l} = g_{rad.3} \tag{4.158}$$

$$g_{rad.3} = g_{rad}(1 - x_{act}) \tag{4.159}$$

$$g_{v.0} = g_{rad} x_{act} = g_{rad.3}\, x_{act}/(1 - x_{act}) \tag{4.160}$$

The first two terms in the right-hand side of Eq. (4.157) define the removal of heat from the wall by convection (q_{conv}) and the last term of this equation expresses the removal of heat due to phase transition (q_{ev}). It was assumed that q_{conv} can be calculated from the Fourier law

$$q_{conv} = \lambda_1^{ef}\, \Delta T_{sup}/\delta_{cl} \tag{4.161}$$

It was additionally assumed that

$$\delta_{cl} = \delta_{0\,s.ph}\, v_{cl}/v_0 \tag{4.162}$$

where $\delta_{s.ph} = D_{int}\lambda_1^{ef}/\mathrm{Nu}\lambda_v$ is the thickness of the layer in single-phase flow.
It was assumed that for the conditions under study

$$\mathrm{Nu} = 0{,}023\ \mathrm{Re}_0^{0,8}\ \mathrm{Pr}^{0,4} \tag{4.163}$$

$$\mathrm{Re}_0 = \mathrm{Re}_{sup}\left[\frac{v_{03}}{v_{sup.l}} + x_{act}\frac{v_{0v} - v_{03}}{v_{sup.l}}\right] \tag{4.164}$$

The vapor Reynolds number Re_{sup} is calculated at $x = 1$. The final expression for the heat transfer coefficient is

$$\alpha = \left\{\frac{1}{0{,}023\ \mathrm{Re}_{sup}^{0,8}\ \mathrm{Pr}_v^{0,4}}\frac{D_{BH}}{\lambda_v}\frac{1}{\left[\dfrac{v'}{v''} + x_{act}\left(\dfrac{v'' - v'}{v''}\right)\right]^{1,8}} - \frac{r}{c_p q}(1 - x_{act})\right\}^{-1} \tag{4.165}$$

where x_{act} is given by the expression:

$$x_{act} = x_{cr} + \sum_{i=1}^{n}\left\{\Delta x_5 \Big/ \left[1 + \frac{\Delta T_{sup}\, c_p^v}{(1 - x)\, r}\right]\right\} \tag{4.166}$$

Borishanskiy et al. [4.17] postulate the existence, in the post-dryout region, of two regions with different heat transfer mechanisms. In the first (transition) region, there is direct contact between the droplets and the wall; in the second region, there is no such contact. Heat from the wall is removed by the vapor and evaporating droplets

$$q = q_v + q_3 \tag{4.167}$$

The heat flux components q_v and q_3 were written as

$$q_3 = \alpha_3 \Delta T \omega \tag{4.168}$$

$$q_v = \alpha_v \Delta T (1 - \omega) \tag{4.169}$$

where ω is the part of the heated surface wetted by liquid droplets precipitating from the flow core. The total heat transfer coefficient is

$$\alpha = \alpha_v + (\alpha_3 - \alpha_v) \omega \tag{4.170}$$

Assuming that the droplets incident on the wall evaporate fully, Eq. (4.170) was written in terms of the deposition flux j:

$$\alpha = \alpha_v \Big/ \left[1 - \frac{\alpha_3 - \alpha_v}{\alpha_3} j \frac{r}{q} \right] \tag{4.171}$$

where

$$\alpha_n = \frac{\lambda''}{D} \left[0{,}023 \left(\frac{w_{mix} D}{v} \right)^{0,8} Pr^{0,4} \right] \tag{4.172}$$

$$\alpha_3 = 3 q^{0,7} [p^{0,14} + 1{,}83 \cdot 10^{-4} p^2] \tag{4.173}$$

It is seen from Eq. (4.173) that heat is supplied to the droplet under nucleate boiling conditions.

Another feature, distinguishing this study from, for example, that by Cumo et al. [4.62], is expressing the relative area of contact between the droplets and the wall by the deposition flux. Using data on droplet transfer between the flow core and the film, the following expression was obtained by Borishanskiy et al. [4.71]:

$$j = \rho w (1 - x) \left[\frac{\xi}{8} - k^2 \left(\frac{\lambda''}{r \rho'' v''} \Delta T \right)^2 \frac{\rho_{mix}}{\rho'} \right] \tag{4.174}$$

We see that a constant k has appeared in the above formula. The numerical value of this constant was obtained from the correlation of the experimental data of [4.71] in the coordinates

$$\left[\frac{\xi}{8} - \frac{j}{\rho w (1 - x)} \right] = f \left\{ \left[\frac{\lambda'' (T_w - T_{sat})}{r \rho'' v''} \right]^2 \frac{\rho_{mix}}{\rho'} \right\} \tag{4.175}$$

The value of k was found to be equal to $(17.5)^{1/2}$. After the constant is determined, the temperature regime of the channel under post-dryout conditions can be calculated.

The above homogeneous models are not free of a number of shortcomings. The first model [4.63, 4.64] considers in considerable detail the basic processes in the post-dryout region. However, this was done on the assumption that the enthalpy, concentration and velocity fields at the dryout location are two-dimensional. However, it is clear from physics of the phenomenon that the thermal boundary layer of superheated vapor starts developing under conditions of an already-formed momentum boundary layer. The assumption of similarity of the enthalpy, concentration and velocity fields has not been proven. As to other models [4.70, 4.71], they are relatively heuristic. In addition, the selection of thicknesses of layers δ_l [4.70] and δ [4.71], vapor temperature and values of constants in [4.71] have been poorly validated. At the same time it is clear that the homogeneous models have not yet exhausted all their capabilities. With certain refinements they may rather reliably describe heat transfer in the post-dryout region not withstanding their inherent limitations.

4.4.3 Empirical Correlations

Elementary correlations for the heat transfer coefficient usually have the form $\alpha = \alpha(p, \rho w, x, q, \ldots)$. The derivations of equations in papers by Collier [4.72], Lee [4.73], Mattson et al. [4.74], Remizov et al. [4.75, 4.76, 4.78] and Cumo and Urbani [4.77] are based on a formal analogy with single-phase flow. As an illustration we present several such formulas.

The Mattson formula [4.74]

$$\frac{\alpha D}{\lambda''} = 3{,}28 \cdot 10^{-4} \left[(\mathrm{Re}'')^{1,5} \left(x + \frac{\rho''}{\rho'\,(1-x)} \right) \right]^{0,777} \mathrm{Pr}_{\mathrm{w}}^{1,69} q^{0,18} \left(\frac{\lambda''}{\lambda_{\mathrm{cr}}} \right)^{-0,294}$$

$$(4.176)$$

has the limits of applicability: $p = 6.8{-}20.6$ MPa, $\rho w = 700{-}5300$ kg/m$^2 \cdot$sec, $x = 0.1{-}0.99$.

Subbotin's formula [4.75]

$$\alpha = \frac{12{,}5 + 0{,}025\rho w}{x - x_{\mathrm{lim}}} (4650 - 8\rho w)(x - x_{\mathrm{lim}}) + 1240 \qquad (4.177)$$

has the limits of applicability: $p = 10{-}14$ MPa, $\rho w = 300{-}700$ kg/m$^2 \cdot$sec, $x = x_{\mathrm{cr}} - 1$.

These equations reflect in the most direct form the effect of flow conditions on the post-dryout heat transfer coefficient. They are simple to use, but are applicable only over a narrow range of variables.

The majority of equations listed in studies [4.79–4.95] for calculating heat transfer in the post-dryout region are based on an equation for the nondimensional forced convection heat transfer coefficient with a single-phase liquid and have the form

$$\mathrm{Nu} = a\,\mathrm{Re}^m\,\mathrm{Pr}^n f(x, T_{\mathrm{w}}/T_{\mathrm{sat}}, \ldots) \qquad (4.178)$$

The following are the most familiar formulas of this type: (1) Miropol'skiy's formula [4.80]

$$\mathrm{Nu}_x = 0{,}023\ \mathrm{Pr}_w^{0,8}\ \mathrm{Re}''^{\,0,8}\left[x + \frac{\rho''}{\rho'\,(1-x)}\right]^{0,8} y \qquad (4.179)$$

$$\mathrm{Nu}_x = \frac{\alpha_x D}{\lambda''} \qquad \mathrm{Re}'' = \frac{\rho w D}{\mu'' g} \qquad y = 1 - 0{,}1\left(\frac{\rho'}{\rho''} - 1\right)^{0,4}(1-x)^{0,4}$$

which has the limits of applicability: p = 3.9–21.6 MPa, ρw = 400–2000 kg/$m^2 \cdot$ sec, x = 0 − 1; (2) Tong's formula [4.86]

$$\mathrm{Nu}_{fm} = 0{,}0193\,\mathrm{Re}_{fm}^{0,8}\,\mathrm{Pr}_{fm}^{1,29}\left(\frac{\rho''}{\rho'}\right)^{0,068}\left[x + \frac{\rho''}{\rho'\,(1-x)}\right]^{0,63} \qquad (4.180)$$

which has the limits of applicability: p = 4–21.5 MPa, ρw = 570–3400 kg/$m^2 \cdot$ sec, x = 0.07 − 1; (3) Groenewald's formula [4.91]

$$\mathrm{Nu} = 1{,}09 \cdot 10^{-3}\left\{\mathrm{Re}''\left[x + \frac{\rho''}{\rho'\,(1-x)}\right]\right\}^{0,89}\mathrm{Pr}_w^{1,41} y^{-1,15} \qquad (4.181)$$

$$y = 1 - 0{,}1\left(\frac{\rho'}{\rho''} - 1\right)^{0,4}(1-x)^{0,4}$$

which has the limits of applicability: p = 3.34–21 MPa, ρw = 700–5300 kg/$m^2 \cdot$ sec, x = 0.1 − 0.8; (4) Remizov's formula [4.76]

$$\mathrm{Nu}'' = 0{,}023\left(\frac{\rho w x D}{\mu''}\right)^{0,8}\mathrm{Pr}_w^{0,8} \qquad (4.182)$$

which has the limits of applicability: p = 6.8–13.7 MPa, ρw = 350–700 kg/$m^2 \cdot$ sec.

The following must be additionally noted. Equations in which it is assumed that the liquid is in equilibrium with the vapor, and the heating surface is cooled only by convection to saturated steam [the rate of the latter is calculated for homogeneous

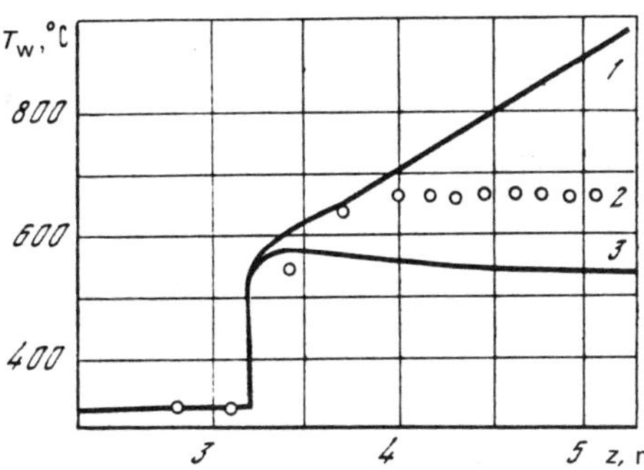

Figure 4.21 Effect of thermal nonequilibrium of the flow on the tube wall temperature in the post-dryout region [4.48]; p = 6.8 MPa, x_{cr} = 0.585, ρw = 1017 kg/$m^2 \cdot$ sec, q = 920 kW/m^2. 1, Analytic curve without allowance for droplet evaporation; 2, experimental data; 3, analytic curve under thermal equilibrium conditions.

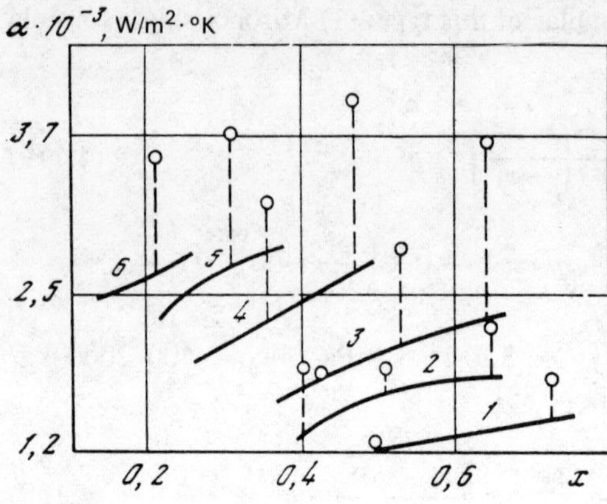

Figure 4.22 Effect of man-made roughness of the channel on the heat transfer coefficient at post-dryout [4.103]. 1, $\rho w = 680$ kg/m$^2 \cdot$ sec; 2, 950; 3, 1360; 4, 2040; 5, 2720; 6, 3800 kg/m$^2 \cdot$ sec.

flow ($s = 1$) or with allowance for the phase slip (Fig. 4.21)] yield the lower limit of temperature.

4.5 HEAT TRANSFER IN THE POST–DRYOUT REGION OF STEAM–GENERATING CHANNELS WITH ARTIFICIAL ROUGHNESS AND LOW THERMAL CONDUCTIVITY COATINGS

4.5.1 Effect of Channel Roughness on Post-Dryout Heat Transfer

Artificial roughness elements of various shapes have been used successfully for a long time now for enhancement of heat transfer in the wall region of single-phase flows, where the bulk of the thermal resistance is concentrated. For example, Kalinin and Yarkho [4.100] recommend that the height of the roughness elements should be made equal to the thickness of the wall layer, where 99% of the total temperature difference occurs. Naturally, wall roughness will also increase the heat transfer effectiveness in dispersed-annular two-phase flow. It can be expected here that the roughness elements will not only additionally turbulize the wall layer, but will aid in a deeper penetration of liquid droplets into it, which will also increase the heat transfer rate.

From this point of view it is of interest to analyze the relatively scarce recent literature. in the first studies on this subject [4.101–4.105] it was simply established that the heat transfer coefficient is enhanced by roughness elements. No special effort was made to determine the optimum form for these elements. The experimental conditions were apparently determined by the engineering problem to be solved. The experiments were performed in channels with roughness elements of small disturbance height (30 to 125 μm). It was shown that the heat transfer coefficients increased by 30–40% (Fig. 22) and even by 80% [4.102]. This was accompanied by a significant reduction in the maximum temperature. In particular, it is noted by Kutateladze

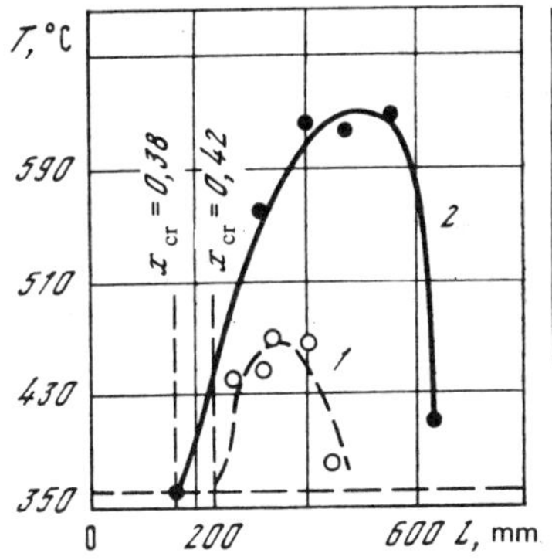

Figure 4.23 Distribution of wall temperature along the test section [4.105]; p = 13.7 MPa. 1, For a rough tube, ρw = 935 kg/m²·sec, q = kW/m²; 2, commercially smooth tube, ρw = 950 kg/m²·sec, q = 665 kW/m².

and Styrikovich [4.105] that a 100 to 150 °C reduction of the maximum temperature was achieved (see Fig. 4.23). The effect of roughness increases with the mass velocity and flow quality.

The subsequent studies [4.106–4.111] were more comprehensive, involving variation of the flow variables as well as the roughness-element geometry. The most thorough of these studies appears to be that by Nichigawa et al. [4.106], where a large variety of roughness-element geometries was used in post-dryout heat transfer measurements. Nichigawa et al. investigated tubes with different types of thread on the inner surface. They tested tubes with single- and twelve-start threads with pitches from 6.43 to 7.32 mm, and heights from 0.39 to 0.64 mm. Detailed information on the tubes used in their study is given in Table 4.1. The experiments were performed over a rather wide range of variables: p of 16.7, 18.6, and 20.6 MPa, q between 116 and 1050 kW/m² and ρw from 380 to 1270 kg/m²·sec.

The most perceptible changes in the temperature regime of the channel in the post-dryout region was produced by roughness type 3 (Fig. 4.24a). For example, at p

Table 4.1

Description of tube	Type 1 (tube with double helical thread)		Type 2 (tube with ordinary helical thread	Type 3 (the Babcock & Wilcox tube)	
Thread direction	Left	Right	Left		Right
Number of thread starts	12	12	12	1	
Thread pitch, mm	6.89		7.32	7.25	6.43
Groove width, mm	0.442		0.226	3.71	2.42
Groove depth, mm	0.64		0.47	0.39	0.39
Diameter, mm:					
Minimum	9.66		9.66	9.45	9.99
Maximum	10.75		10.75	10.22	10.77
Mean	10.16		10.18	9.77	10.38
Outer	20.02		20.02	20.06	19.98

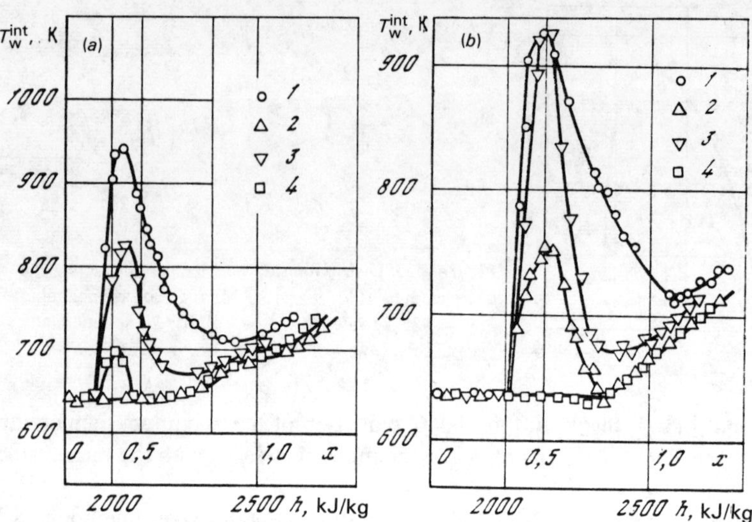

Figure 4.24 Temperature profiles of tubes with different kinds of roughness elements [4.106]. (a) ρw = 629–727 kg/m^2·sec; (b) 408 kg/m^2·sec; 1, smooth tube; 2, type 1; 3, type 2; 4, type 3.

= 20.6 MPa and q between 629 and 727 kW/m^2 the maximum temperature dropped from 940 °K for a smooth wall to 690 °K for a wall with type 3 roughness (flow temperature 640 °K), whereas for p = 18.6 MPa, ρw = 624–748 kg/m^2·sec and [sic] q = 462–470 kW/m$_2$ the maximum wall temperature fell from 930 to 750 °K (flow temperature 630 °K). The temperature difference in the first case changed 6-fold, whereas in the second, 2.5-fold. It was also shown that the use of various kinds of threads not only reduced T_w^{max}, but also perceptibly changed the dryout quality, x_{cr}. For example, for p = 18.6 MPa, ρw = 700 kg/m^2·sec, q = 233 kW/m^2 the value of x_{cr} changed from 0.5 to 0.9, whereas for p = 20.6 MPa, ρw = 408 kg/m^2·sec, q = 350 kW/m^2 x_{cr} changed from 0.35 to 0.8 (Fig. 4.24b). The significant rise in dryout quality x_{cr} indicates that heat transfer was enhanced by very large roughness elements, which protrude from the viscous and intermediate layers of the wall region and which turbulize not only the post-dryout regions with low eddy thermal conductivity, but the pre-dryout regions with high values of the transfer coefficient. A significant increase in the hydraulic drag also resulted from the introduction of the roughness elements.

The study by Rogers and Mesler [4.108], though performed at low pressures and low mass velocity, is of definite interest. Their experiments were performed at p = 0.24 MPa, ρw between 38 and 82 kg/m^2·sec and q between 7.3 and 38 kW/m^2. Though coolant parameters are significantly lower than those formed in operation, low mass flux and pressure may be found under transient conditions (for example, in loss of coolant accidents) and hence the Rogers and Mesler study is of interest. This study is distinguished by the fact that the authors were interested not only in the variations in heat transfer coefficient, but they also measured the hydraulic drag coefficient as a function of the type of roughness used. Five types of artificial rough-ness elements were investigated. The results of one of the series of experiments are

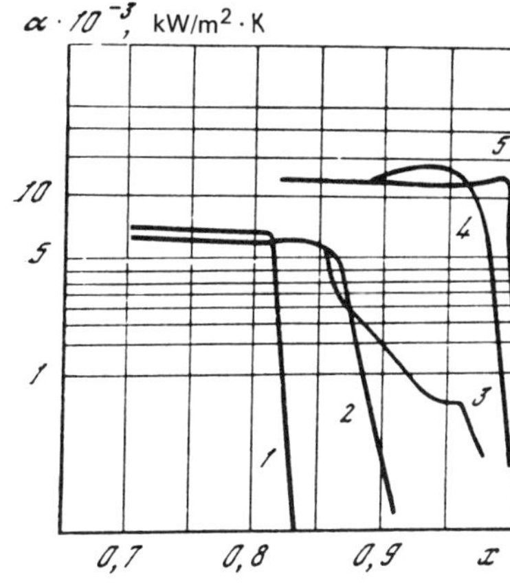

Figure 4.25 Local heat transfer coefficients as a function of the flow steam quality for the test sections in tube types 1 through 5 [4.108].

shown in Figs. 4.25 and 4.26. It is seen from the figures that roughness types 4 and 5 significantly increase the critical quality. The overall trend of the data is such that at a given vapor quality, the heat transfer coefficient increases with the hydraulic drag.

During the past several years a large series of experiments was performed at the Institute of High Temperatures [4.109-4.111] aimed at determining the effect of "small" roughness elements on post-dryout heat transfer. The roughness consisted of a single-start screw thread. The profilograms of the roughness elements under study are shown in Fig. 4.27. Type 1 roughness had a projection height $K = 4.5$ μm, pitch $S = 450$ μm, whereas for the second type these were 100 μm and $S = 2000$ μm, respectively. Figure 4.28 shows a plot of the coefficients of hydraulic resistance vs. the Reynolds number for rough and for commercially smooth tubes. These roughness elements did not greatly change the value of x_{cr} (see Fig. 4.29).

These roughness elements affected the wall temperature profile in the channel (see Fig. 4.30). The wall temperature with the rough surface of type 2 is lower than for surface type 1 and both are lower than the values for a smooth surface. In Fig.

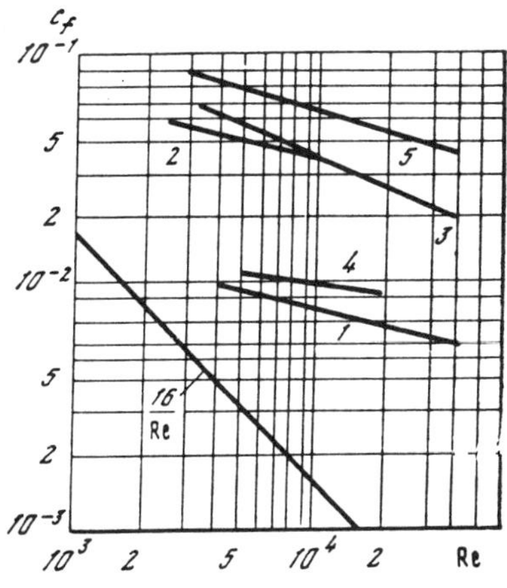

Figure 4.26 Hydraulic drag coefficient as a function of the Reynolds number for test section types 1 through 5 [4.108].

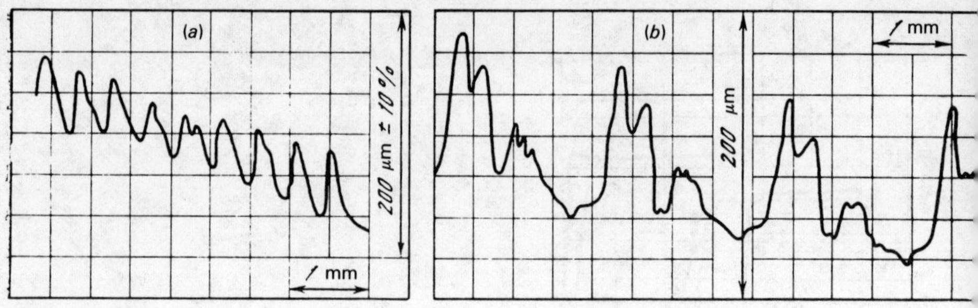

Figure 4.27 Roughness profilogram of some of the test sections. (*a*) Test section 1; (*b*) test section 2.

4.30, the critical quality (x_{cr}) was somewhat different for the three cases due to differences in inlet quality (and not due to the roughness as Fig. 4.29 demonstrates). However, the differences in x_{cr} are less than 0.1, which makes it rather obvious that the development of the temperature profile is affected primarily by the surface roughness. This becomes clear when the hydraulic drag coefficient is used as the parameter (Fig. 4.31). It is seen that the higher the hydraulic drag coefficient, the greater the effectiveness of the roughness elements, the effectiveness with increasing heat flux.

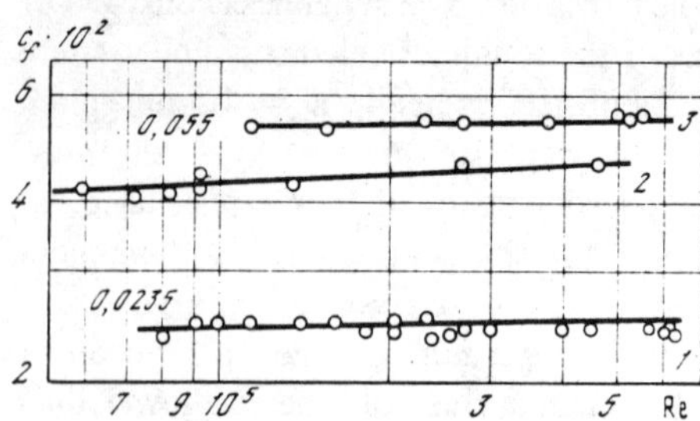

Figure 4.28 Hydraulic drag coefficients c_f vs. Re. 1, Commercially smooth tube; 2, test section No. 1; 3, test section No. 2.

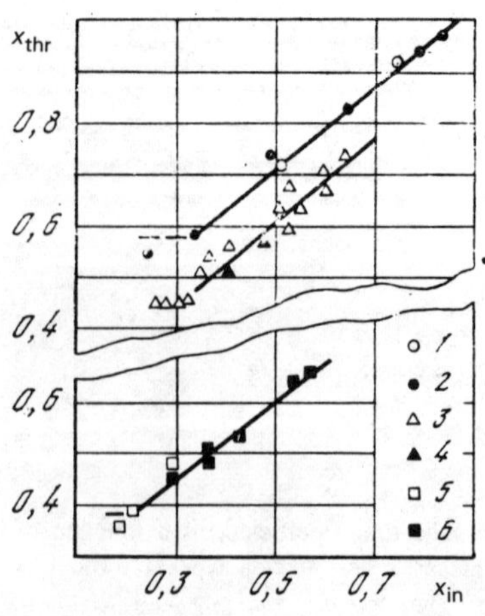

Figure 4.29 Plot of x_{cr} vs. x_{in} [sic]. 1,2, p = 9.8 MPa, ρw = 1000 kg/m^2·sec; 3,4, p = 13.7, ρw = 1000; 5, 6, p = 15.7, ρw = 700; 1,3,5, smooth tube; 2,4,6, rough tube.

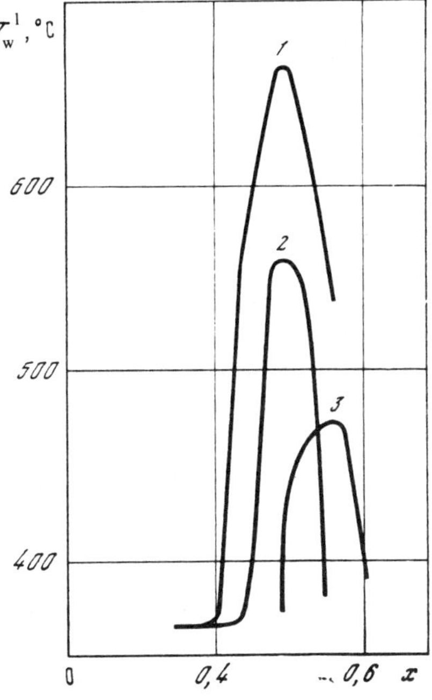

Figure 4.30 Wall temperature vs. vapor quality. 1, Smooth wall; 2, rough wall, section 1; 3, section 2.

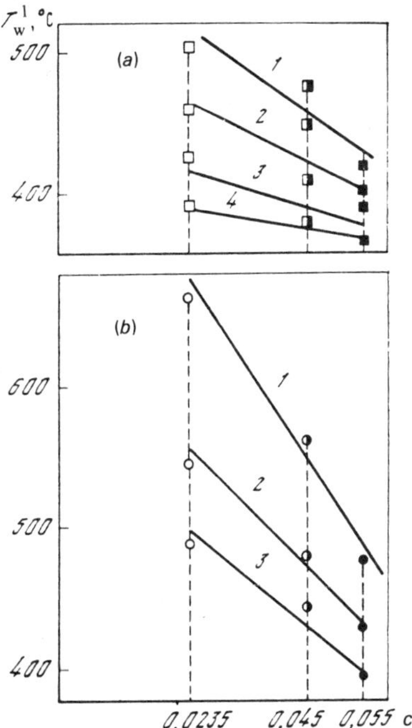

Figure 4.31 Wall temperature vs. roughness coefficient. Open symbols, smooth wall; half blackened symbols, type one roughness; fully blackened symbols, type two roughness; (a) $p = 13.7$ MPa, $\rho w = 1250$ kg/m$^2 \cdot$ sec; (b) $p = 16.6$ MPa, $\rho w = 750$ kg/m$^2 \cdot$ sec.

These experiments data were correlated using the technique described in Sec. 4.4.2. The results of this correlation are plotted in Fig. 4.32, from which it is seen that all the data are correlated by a single equation

$$St = St_0 \Psi'_{\varepsilon v} \Psi'_{sr} \Psi'_{sq}$$

where St_0, Ψ_{sv}, Ψ_{sr} and Ψ_{sq} are calculated from Eqs. (4.145), (4.131), (4.133) and (4.135), respectively.

The engineering method suggested above for calculating heat transfer in the inlet region of the post-dryout zone for both smooth and rough steam-generating channels, has been tested over a rather wide range of variables and is recommended for practical use.

Enhancement of heat transfer in the post-dryout region by means of artificial roughness elements has been investigated not only with water, but also with cryogenic fluids [4.112–4.114]. The working fluids were nitrogen and hydrogen. The geometry of the roughness elements and test conditions were varied over a relatively wide range.

It was found that the turbulence promoters significantly reduced the wall temperature in the post-dryout region. As seen from Figs. 4.33 and 4.34, the heat transfer coefficient increased 4- to 10-fold and its growth exceeded that of the hydraulic drag coefficient. Increasing the height of the disturbances did not result in reaching the limit characteristic of single-phase flows, i.e.,

$$NU_r/Nu_{sm} > c_{fr}/c_{fsm}$$

The following conclusions can be drawn from the above survey of experimental data on the effect of artificial roughness on heat transfer in the post-dryout region:

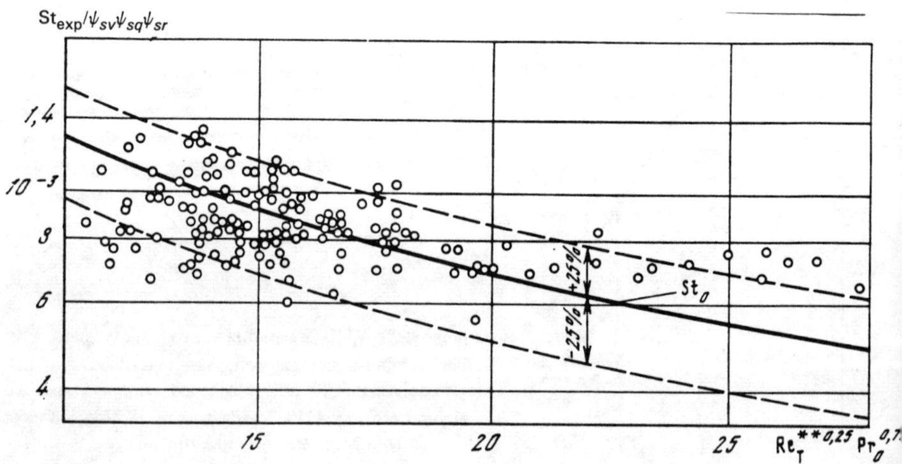

Figure 4.32 Correlation of experimental data.

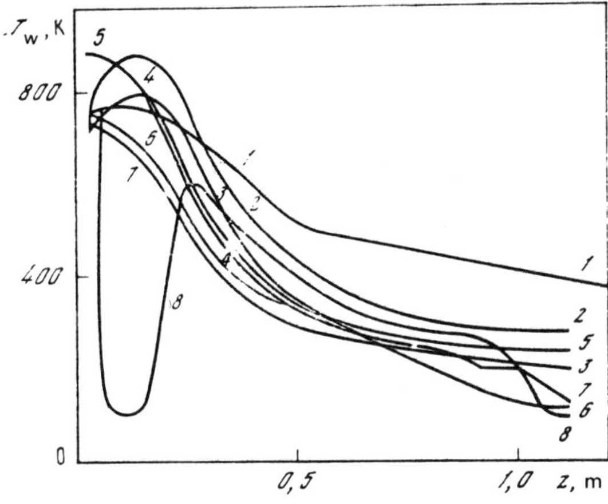

Figure 4.33 Variation in wall temperature along test sections of [4.13]. 1–8, different types of roughness.

1. Roughness decreases the wall temperature in the post-dryout region; in certain cases this reduction may be quite significant.
2. The roughness effect is similar for both tubes and annuli.
3. Roughness reduces the thermal in homogeneity of flow.
4. The increase in heat transfer coefficient in rough channels is greater at high pressures and mass velocities.
5. At constant mass velocity and pressure, the increase in heat transfer coefficient is greater at high qualities and heat flux densities.
6. Different kinds of roughness differ in their effect on post-dryout heat transfer. No general recommendations can yet be made on the selection of optimum roughness geometry.
7. Engineering methods have been developed for calculating heat transfer in the post-dryout region in rough-walled steam-generating tubes and for similar conditions with cryogenic fluids. However, they have only been tested with limited ranges of roughness elements and operating conditions.

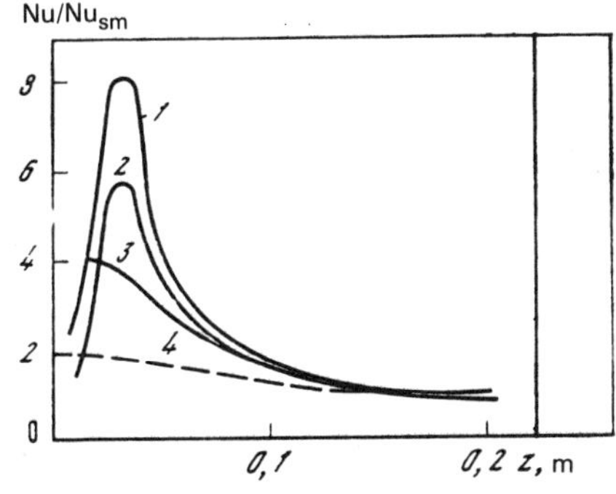

Figure 4.34 Enhancement of heat transfer along the tubes past orifices as compared with smooth tubes [4.114]. 1, $x = 0.77$, $q_w = 35$–49 kW/m^2; 2, 0.65 and 55–71; 3, 0.54 and 53–66; 4, 0.58 and 51–66 kW/m^2.

4.5.2 Effect of Low-Conductivity Coatings on Heat Transfer in the Post-Dryout Region

The question of the effect of low thermal conductivity coatings on post-dryout heat transfer has been poorly explored. Whereas many individual studies were performed in the case of pool boiling [4.104–4.111], little has apparently been done for forced convection conditions. To the authors knowledge, the only studies are those by Dreytsez et al [4.112], In'kov et al [4.113], Petukhov et al. [4.114], and Koshkin et al. [4.115]. However, this method for enhancing film boiling heat transfer appears to be highly promising. For example, according to Koshkin et al., coating surfaces with a ~ 170 μm layer of resin makes it possible to enhance the heat transfer 2- to 3-fold. The same results are obtained by depositing a 40–50 μm enamel coating.

Usually the effect of a coating as compared with a bare metal surface is expressed by the ratio $(\rho_c, c_{p.c}, \lambda_c/\rho_M c_{p.M} \lambda_M)^{1/2}$, which characterizes the ratio of densities, specific heat capacity and thermal conductivities of the coating and the metal. On the basis of this ratio, calcium sulfate exhibits positive properties. Hence the previous authors experiments discussed above were repeated with channels whose inner surface was artificially coated by a layer of $CaSO_4$ deposit.

The studies were performed according to techniques usually employed in investigating post-dryout heat transfer with one exception, which consisted in creating conditions for developed boiling before the start and at the end of experiments. Here all the variables $(p, q, \rho w)$, with the exception of the inlet enthalpy (h_{in}) was maintained at a specified level. Conditions were maintained steady over 40–60 min. This made it possible to obtain reliable statistical-average data for temperature profile during the different stages of the experiments and to compare them under identical conditions. The experiments were performed in several stages as follows.

First stage. It was found in investigating mass transfer in steam-generating channels with $CaSO_4$ as the indicator salt, that the deposition of calcium sulfate has a different effect, depending on the heat transfer region where it is present. Figure 4.35 shows the results of run 1.1. The temperature profile within the tube changes with time. in the nucleate boiling region the wall temperature increases due to the deposit layer, whereas in the post-dryout region the wall temperature decreases. The rate of temperature change in the post-dryout zone is significantly greater than in the nucleate boiling region; in this case the most rapid drop in temperature was observed at the exit from the test section.

The thickness of the deposit can be estimated by comparing boiling coefficients obtained under conditions of nucleate boiling over the whole of the tube, these coefficients being obtained at the start and at the end of the run respectively, by adjustment of the inlet enthalpy. The thickness of deposit varies along the channel, being smaller in the region which had been in the post-dryout condition during the main part of the experiment. The temperature variation can be followed more clearly in Fig. 4.36 (run 1.2). In this run the crisis location was about $x \simeq 0.38$–0.4. The quality at the start of the heated length was 0.23 and 0.26. It is seen that the maximum deposits lie between the inlet and the dryout point (the curves in Fig. 4.36 were joined arbitrarily).

In the experiments described above the feedwater consisted of a supersaturated

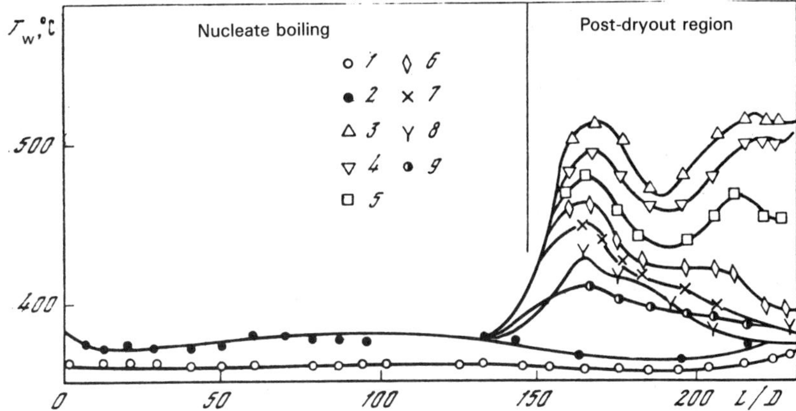

Figure 4.35 Variation in wall temperature along the channel vs. time when the loop is supplied by a supersaturated solution of $CaSO_4$. Run 1.1: $p = 13.7$ MPa, $q = 580$ kW/m^2, $\rho w = 1050$ kg/m$^2 \cdot$sec, $x_{in} = 0.08$, $x_{out} = 0.504$, $x_{cr} = 0.40$, $L/D = 31$, $C_{0I} = 3.33$–5.97 ppm; calibration in nucleate boiling: 1,2, at the start and end of run, respectively; temperature regime of tube: 3, at the start of run; 4, 2 hr after start; 5, 6 hr; 6, 0 hr; 7, 14 hr 30 min; 8, 18 hr; 9, 21 hr 30 min after start of run.

calcium sulfate solution. The $CaSo_4$ concentration at the inlet to the loop was 3.26–5.97 ppm, the inlet quality was $x_{in} = 0.04$–0.08, the exit quality $x_{out} = 0.464$–0.504 and the critical quality $x_{cr} = 0.38$–0.04. In run 1.3 (Fig. 4.37a), however, the $CaSO_4$ concentration was much lower ($C_{0I} = 0.42$ ppm, $x_{in} = 0.02$, $x_{out} = 0.44$, $x_{cr} = 0.38$–0.4). Whereas in the preceding runs deposits were observed along the entire tube, in this run they started only at $x = 0.30$–0.34. The maximum deposit thickness was at the dryout cross section, whereas in the nucleate boiling region the temperature was unchanged to within the accuracy of the thermocouple (Fig. 4.37b). The remaining parameters behave in the same manner as they did when supersaturated solution were used. In particular, over a 53 hr run, the wall temperature at the

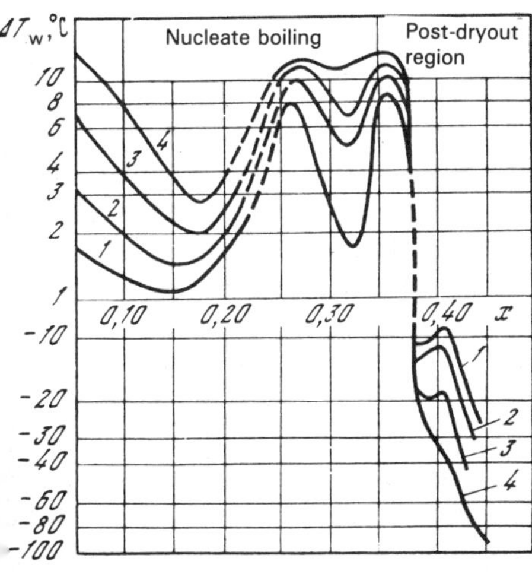

Figure 4.36 Change in wall temperature along the channel as a function of time when the loop is supplied by a supersaturation solution of $CaSO_4$. Run 1.2: $p = 13.7$ MPa, $q = 580$ kW/m^2, $\rho w = 1000$ kg/m$^2 \cdot$sec, $x_{in} = 0.464$, $x_{cr} = 0.4$, $(L/D)_{cr} = 71$; $C = 3.29$–4.76 ppm; 1–4, correspond to different values of temperature [4.112].

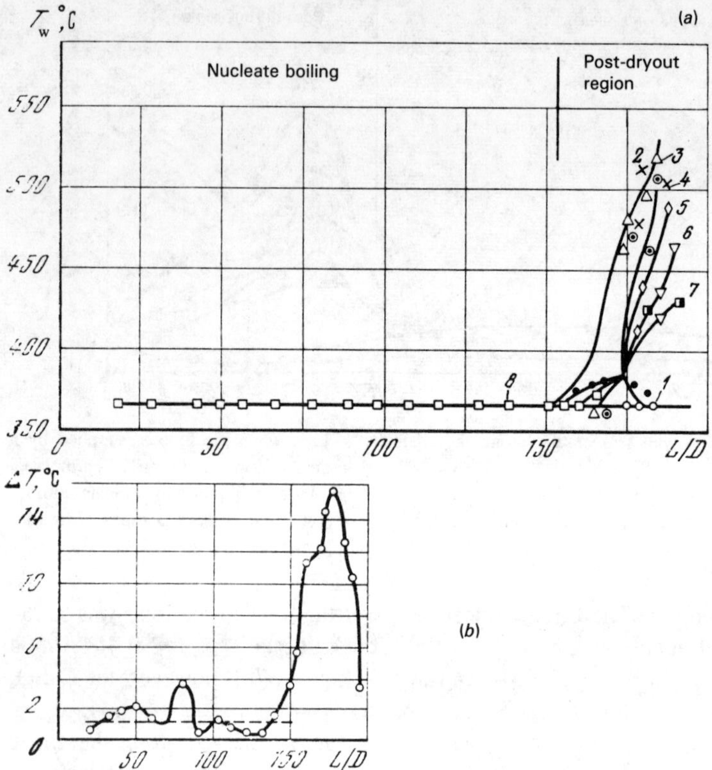

Figure 4.37 Variation in wall temperature (a) and its increment (b) along a channel as a function of time when the loop is supplied by a nonsaturated solution of $CaSO_4$. Run 1.3: $p = 13.7$ MPa, $q = 582$ kW/m^2, $\rho w = 1000$ kg/m$^2 \cdot$sec, $x_{in} = 0.02$, $x_{out} = 0.44$, $x_{cr} = 0.30$-0.34, $(L/D)_{cr} = 175$, $C_{0I} = 0.42$ ppm; calibration in nucleate boiling: 1, at the start of experiment; 2, at the end of experiment (after 53 hr), respectively; temperature regime of tube: 3, at the start of experiment; 4, after 12 hr; 5, after 24 hr; 6, 37 hr; 7, at the end of experiment (8 designates merged points for $\Delta T = 0$-3°C.

channel exit decreased from 527 to 439 °C. The salt content of the feedwater affected the rate of variation in heating surface temperature both in nucleate and film boiling. Whereas when the loop was supplied by a supersaturated solution of $CaSO_4$, the maximum rate of variation in wall temperature upon nucleate boiling was ~ 0.5-1.25 °C/hr, and in the film boiling region ~ 4-13.5 °C/hr, when nonsaturated solution was supplied it was ~ 0.3 °C/hr and 0.8-1.5 °C/hr. Note that the temperature difference in the post-dryout zone decreased in a number of cases from 150-170 to 10-30 °C.

Second stage. The first-stage experiments were performed with an "old" (ran-in) tube. It can be assumed that small deposits remained on its inner surface even after prolonged washing with condensate and the surface roughness was comparatively high. It was interesting, as a second stage, to perform experiments with a "new" tube, not previously used in experiments. The results of these experiments are shown in Fig. 4.38. These experiments were performed over several successive periods.

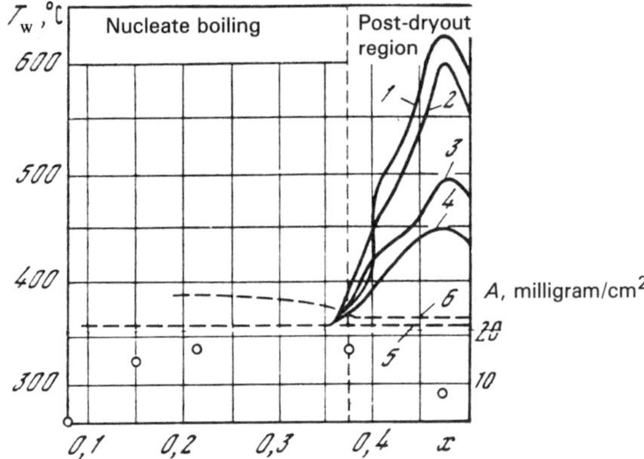

Figure 4.38 Variation in wall temperature and amount of deposits along the channel in time when the loop is supplied by a supersaturated $CaSO_4$ solution. $p = 13.7$ MPa, $q = 580$ kW/m^2, $\rho w = 1000$ kg/m^2·sec, $x_{in} = 0.08$, $x_{out} = 0.51$, $x_{cr} = 0.38$, $(L/D)_{cr} = 140$, $C_{0I} = 10$–13 ppm; temperature regime of new tube: 1, at the start of run (II-1.1); 2, at the end of run (after 33 hr); temperature regime of tube after prolonged boiling: 3, at the start of run (II-1.2); 4, at the end of run (after 34 hr); calibration in nucleate boiling: 5,6, at the start and end of run.

The duration of the first period was 33 hr (run II-1.1). The maximum temperature reduction at the highest-temperature location was 20 °C. Note that the initial temperature level in the post-dryout region, in spite of identical experimental conditions, was higher than in the preceding first stage experiments. Whereas previously the temperature level did not exceed 570 °C, in this experiment $T_{max} = 620$ °C.

Next, several individual short-duration (5–6 hr) runs were carried out using the "new" tube, with a resultant reduction in T_{max} to 500 °C. Then, a prolonged run (34 hr) was performed which terminated with the rig dry. Towards the end of this experiment the maximum temperature in the post-dryout region fell to 448 °C. This means that, after a "new" tube was ran in for a prolonged time period it was possible to attain a significant drop in wall temperature also with this tube.

Dry termination of the run was achieved by reducing the power supply from the entire tube and stopping the supply of coolant to the test section. This was done in order to prevent the washing off of the deposits from the tube, in order to allow their analysis.

For this purpose six specimens were cut out from the tube. Each of them was additionally slit into two parts. The first of these, 100 mm long was used for determining the amount of deposit; the second part, which was a 20 mm long half cylinder was used for measuring the scale topography. The amount of deposit was determined by the cathode method.

The results of these experiments are presented in Table 4.2 and Fig. 4.38.

It is seen that the deposits in the nucleate boiling region amount to 15.5–17.4 milligram/cm^2, increase somewhat along the channel and attain their maximum in the location preceding dryout. In the post-dryout region the deposits amounted to 8.33 milligram/cm^2. It should be remembered that these deposits accumulated in the course of several experiments. Consequently, the section which, under the specified test

Table 4.2

Specimen	Time of presence in solution	Initial weight, gram	Last weighting, gram	Amount of deposits, milligram/cm^2
1	3 hr 30 min	28.4090	28.3049	8.33
2	4 hr	29.0623	28.8450	17.4
3	2 hr 40 min	28.4030	28.1943	16.7
4	3 hr	27.6862	27.4916	15.5
5	3 hr 25 min	28.9298	28.9266	0.27
6	2 hr 35 min	31.5114	31.5119	

Note. Specimen 6 was cut out from a tube not previously used in experiments.

conditions had been subjected to post-dryout conditions had, under start-up and calibration conditions, been subjected to nucleate boiling or was entirely free of boiling. All this increased somewhat the amount of deposits in the post-crisis region. There were virtually no deposits on the nonheated part of the experimental tube.

The temperature profile of calibration runs with nucleate boiling at the start and end of experiments corresponds to the profile of the deposits. From these data it is possible to perform approximate calculation of the deposit thickness. If the specific weight of the deposits is taken to be 2.5 gram/cm^3, then in the nucleate boiling region $\delta_{dep} \simeq 62\text{–}70$ μm, whereas for the post-dryout region $\delta_{dep} \simeq 33$ μm. Assuming that the thermal conductivity of the scale is 1.16 W/m·°C we find for nucleate boiling (for example, $\delta_{dep} \simeq 60$ μm). Estimates of deposit thickness obtained by different methods are in satisfactory agreement.

The topography of the deposits was determined by a recording roughness indicator (model 201, produced by the *Kalibr* Plant). The height of protuberences on specimens of section 5, cut out from the nonheated part of the experimental tube, and on section 6, cut out from a "new" (not previously used) tube on which there were no deposits, did not exceed 5–8 μm (Fig. 4.39). The height of protuberances on specimens with deposits (section 1, taken from the post-dryout region, section 2 from the crisis location and sections 3 and 4 from the section with nucleate boiling) is significantly greater 30–35 μm, with the largest irregularities spaced relative far apart. No significant differences were noted between the nature of profilograms situated in different boiling heat transfer regions.

The last results support the assumption that the rise in heat transfer coefficient in the post-dryout region is associated with changes in the flow conditions at the surface. The increased roughness additionally turbulizes the wall layer, which results in a deeper penetration of droplets into this layer, which lowers its effective temperature. However, the deposits modify not only the roughness, but also the thermal conductivity, specific heat capacity and other properties of the surface film, which could also affect the heat transfer. Hence, in order to determine the manner in which the deposits affect heat transfer conditions, it appeared best to perform experiments with a rough surface, whose geometry would be close to that of a layer of calcium sulfate deposits.

Such an experiment is described below. These experiments were also performed with a supersaturated solution of calcium sulfate. The $CaSO_4$ concentration ranged from 1.36 to 86.5 ppm. The test section had both smooth and rough surfaces. The experiment were performed in several stages, each of which was characterized by accumulation of a certain amount of salt on the wall. The experimental results are plotted in Fig. 4.40.

It is seen from the figure that the wall temperature of a new tube does not exceed T_w^{max} = 510 °C on the rough part, and T_w^{max} = 560 °C on the smooth part, i.e., the temperature difference ΔT_{sm} = 225 °C is greater than ΔT_r + 175 °C by a factor of 1.28. The figure shows by solid curves conditions when post-dryout conditions occur simultaneously on the rough and smooth parts, whereas the dashed curves represent conditions with the post-dryout region on the smooth part only.

Data on the change of wall temperature ΔT_{dep} = $T(L)$ at stage 2 due to $CaSO_4$ deposits in the pre-dryout region are shown by a solid curve, whereas those following the crisis—by dash-dotted curve. The slight difference in values of ΔT_{dep} is due to washing off of a part of the deposits in the course of transition from nucleate boiling to post-dryout heat transfer and vice versa.

After a certain quantity of salt, causing a wall temperature change in nucleate boiling of ΔT_{dep} = 5 °C, has accumulated on the rough part (in the region of maximum wall temperature), T_w^{max} decreases to 465 °C. The effective temperature differ-

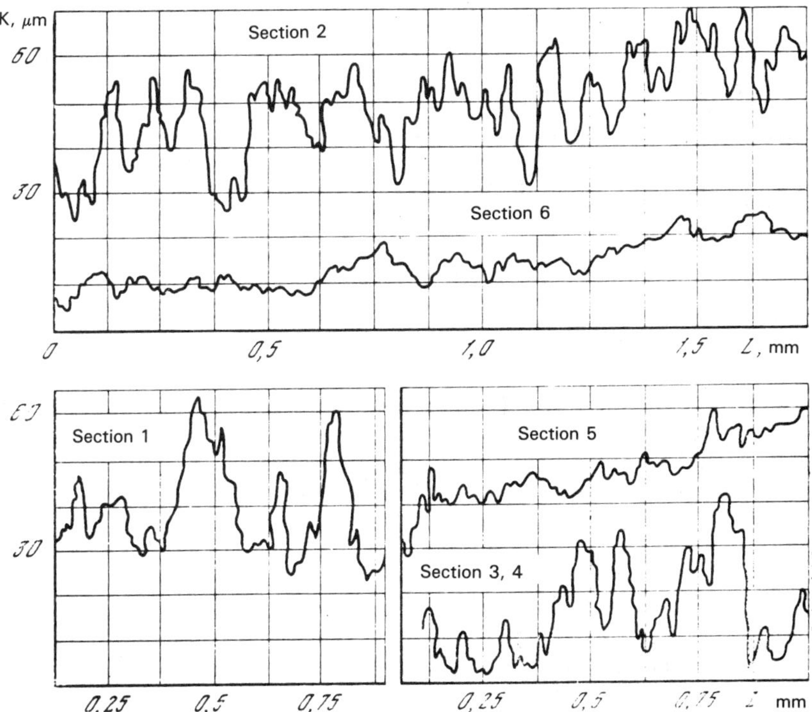

Figure 4.39 Profilograms of specimens cut out from experimental tube.

ence decreased (without consideration of the temperature difference in the deposit layer) by 50 °C. Increment $\Delta T_{dep} = 8\text{--}9\,°C$ on a smooth wall corresponds to a reduction in the maximum temperature to $T_w^{max} = 445\,°C$. The effective temperature difference decreased by 106 °C.

Assuming, as in the previous cases, that the thermal conductivity of calcium sulfate deposits $\lambda_{dep} = 1.16$ W/m·°C, it can be estimated that the thickness of the calcium sulfate scale at T_{dep}^{max} in the rough part will be $\delta_{dep} = 10\ \mu m$, whereas on the smooth part $\delta_{dep}^{max} = 17\ \mu m$. It was noted above that, for a $CaSO_4$ layer with thickness

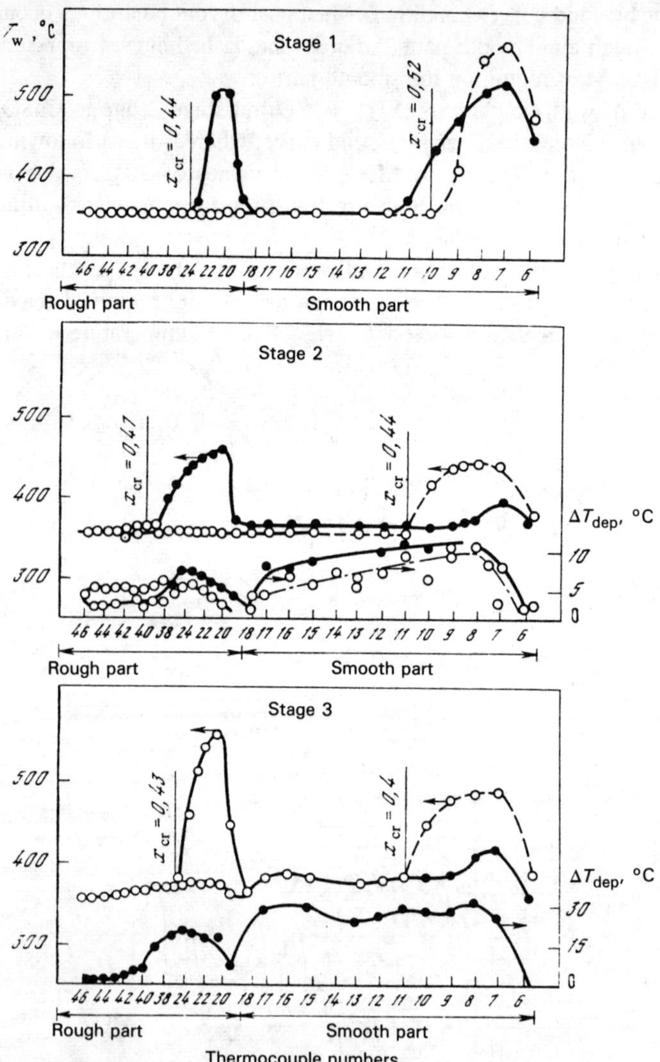

Figure 4.40 Effect of calcium sulfate deposits on the temperature regime of the channel at $p = 13.7$ MPa, $q = 600$ kW/m², $\rho w = 1000$ kg/m²·sec.

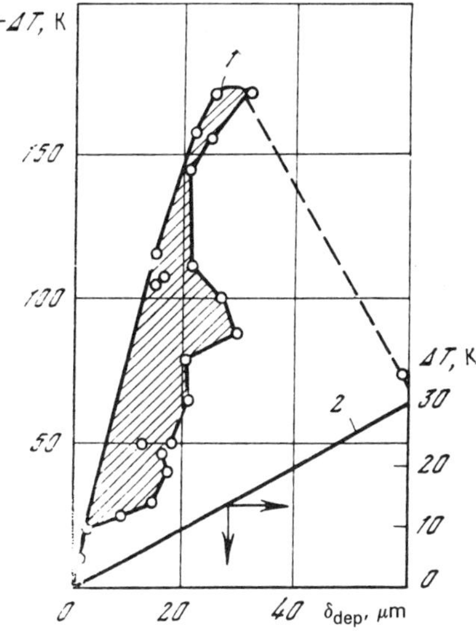

Figure 4.41 Variation in wall temperature and the temperature difference (ΔT) in the post-dryout region in the calcium-sulfate deposit layer as a function of its thickness.

Δ_{dep} = 33 μm, the spacing of the large roughness elements is $\sim$ 150–250 μm. This means that, if we assume that the roughness forming on the smooth part has approximately the same geometry as on the rough part, then this circumstance allows a reduction in T_w^{max} of 50 °C (in fact, the roughness effect should be smaller than this since the thickness of the layer of deposits was not 33, but 17 μm). From this follows the important conclusion that a layer of calcium sulfate deposit with a thickness of only $\sim$ 17 μm provided a reduction in T_w^{max} of at least 50–56 °C.

Further growth in the thickness of deposits on the rough part in the course of the third stage to δ_{dep}^r = 36 μm ($\Delta T_{\text{dep}}^r \simeq$ 18 °C) and on a smooth part to $\delta_{\text{dep}}^{\text{sm}}$ = 60 μm ($\Delta T_{\text{dep}}^{\text{sm}} \simeq$ 30 °C) resulted in a rise in wall temperature. The value of T_w^{max} on a rough surface was 565 °C, and on the smooth—490 °C. Consequently, there exists an optimum scale thickness at which the maximum effect is attained.

On the basis of all the experimental results obtained on a smooth wall, Fig. 4.41 shows an approximate pattern of the reduction in wall temperature in the post-dryout region as compared with a smooth surface. It is seen from the figure that first, as δ_{dep} starts growing, the value of ΔT increases (curve 1) then, after the deposits attain the thickness δ_{opt}, the value of ΔT starts falling off. The optimum thickness δ_{opt} of the calcium sulfate deposits for the conditions at hand can be approximately be assumed to be 20–40 μm. The same figure (curve 2) shows calculated data on the temperature difference across the deposit layer.

In summing up, we wish to emphasize the promise held by enhancement of post-dryout heat transfer by using low thermal conductivity coatings. Smooth coatings of low thermal conductivity materials do not change the hydraulic drag of the surface, whereas rough surfaces cause this drag to rise. In addition, since the loop water always contains products of corrosion of materials of construction, rough surfaces will be coated by a layer of deposits more rapidly than a smooth surface. Consequently, the effectiveness of their operation will be below its design value.

BOILING MASS TRANSFER ON AN IMPERMEABLE SURFACE

The study of mass transfer in a boiling boundary layer is of great practical importance and scientific interest. Knowledge of the fundamentals of mass transfer makes it possible to predict the likelihood of the formation, on the heated surface, of deposits of substances dissolved in the heat transfer fluid. The presence of deposits can in a number of cases significantly affect the overall coefficient of heat transfer through the wall, and consequently, the metal wall temperature. It is known that, at the high heat fluxes characteristic of a number of fields of new technology (including large modern power generating facilities), rather insignificant deposits, having a thickness of tenths or even hundredths of a millimeter, may raise the temperature of the heating surface above the permissible limits. Some water impurities characteristic of the steam cycles employed at conventional and nuclear power plants (in particular, products of corrosion of the materials of construction), have such a low solubility, that even the modern highly effective methods of water purification are not capable of ensuring (at reasonable cost) complete absence of precipitation of the solid phase.

In once-through high-pressure steam generators, the water purity requirements are very strict and such units are usually completely free of deposits of not only sodium salts, but also of silicic acid. The absolute solubility of these compounds in high-pressure steam is much higher than the concentration corresponding to the permitted impurity levels. In drum-type steam generators, however, high purity steam can be obtained from water with a high salt content, significantly exceeding the solubility of the majority of compounds in the steam. Hence in such generators it is possible to have, at high concentrations of impurities of water, the appearance of deposits not only from poorly soluble calcium salts, but also from highly soluble sodium salts.

However, knowledge of the real solubility of a compound in boiling water does not suffice in itself to establish the permissible concentrations ensuring the absence of deposits. In the course of boiling of liquid in the boundary layer, the local concentrations of impurities dissolved in the water, may rise to significant magnitudes. Practically they are limited by the ratio of the concentrations in water and in the steam in equilibrium with it, i.e., by transfer of impurities into the vapor as a result of their (usually small) solubility in the vapor. The distribution coefficient for a given compound between the liquid and vapor is a function of compound properties and pressure. For example, at pressures $p = 7$ MPa the distribution coefficient for different compounds ranges from 10^2 to 10^9 and above. As the critical point is approached, the value of the distribution coefficient drops. For the majority of compounds of interest in power generators, this limitation becomes significant only at very high pressures.

The rise in the boiling temperature of the solution with rise in its concentration may also limit the concentration but this effect is usually of little importance. Hence, in the majority of cases, the rise in concentration near the wall is limited only by mass transfer between the boiling layer at the wall and the flow core. In the case of highly suppressed mass transfer (for example, in the case when the wall is irrigated by droplets which completely dry out on it) deposition of impurities on the wall occurs even if the impurity concentration in the flow is several orders of magnitude lower than the saturation concentration. It is clear from this that the detailed knowledge of the phenomena governing heat and mass transfer between the flow core and the wall layer is very important from the point of view of preventing deposits on heated surfaces.

In a number of cases, mass transfer is significant also from the point of view preventing or limiting the corrosion of heating surfaces. The importance of this circumstance for large steam generators is seen if only from the fact that many cases of corrosion of steam-generating tubes are ascribed precisely to significant concentration [of impurities] at the wall. Data on concentration and deposition are also of great importance for evaluating the range alternatives physical models for nucleate boiling and critical heat flux (dryout). In fact, physical models have, up to now, been evaluated on the basis of conformance to the accumulated experimental data on integral characteristics (coefficients of heat transfer and critical parameters). Under such conditions the availability of any additional information on the process is quite useful. In particular, knowledge of the degree of concentration makes it possible in a number of cases to reject models which, though they are in satisfactory agreement with information on heat transfer, do not agree with mass transfer data.

5.1 PHYSICO–CHEMICAL PRINCIPLES OF THE "SALT" METHOD

The question of deposition of highly and poorly soluble compounds has been investigated extensively. As a rule, the studies were performed under industrial conditions and primarily for determining the rate of deposition. Much less attention was paid to determining the threshold of the start of deposition and the degree of concentration of the impurities. The term "degree of concentration" or "concentration ratio" is given to the ratio n of the concentrations of impurities in the liquid phase on the wall (C'_w) to that in the flow core (C'_0) (or within the bulk liquid in the case of natural convection)

$$n = C'_w / C'_0 \tag{5.1}$$

The threshold (limiting) concentration is the maximum concentration C^*_{0I} in single-phase flow entering the steam-generating channel at which there is no precipitation of impurities in solid form. If concentration C_{in} at the channel inlet is smaller than C^*_{0I}, there are no deposits. In the case of C_{in} greater than C^*_{0I}, deposits start forming. This means that concentration C^*_{0I} separates regions of presence and absence of deposits.

Usually, we will be dealing with impurities which have negative temperature coefficients of solubility. Thus, under conditions of single-phase flow in heated channels the wall temperature uniquely controls the threshold concentration which prevents precipitation of solids from the water. In two-phase flows one must know, in addition to the wall temperature, the mass fraction of one of the phases at the wall. The latter quantity is usually unknown, and hence, the threshold value of the impurity concentration at the channel inlet is usually determined experimentally. The most extensive studies of this kind have been performed at the Moscow Energetics Institute and at the High-Temperatures Institute of the USSR Academy of Sciences. The threshold concentration for inception of precipitation was determined by the salt method, whose fundamentals are presented below.

When solutions of substance boil, the local concentrations of impurities in the liquid phase rise at the base of the growing bubbles on the heating surface. This occurs because the solubility of substances in the vapor phase is significantly lower than in the liquid. When the concentration exceeds the solubility threshold, the impurities start to crystallize on the heated surface around the nucleation sites. During the subsequent period following the breakoff of bubbles (waiting time) the deposits may fully or partially dissolve. The formation on the heating surface of local, periodically arising supersaturation zones is a necessary, but insufficient condition for the start of continuous rise in the quantity of deposits on the surface. This deposition occurs only when the effective (averaged in time and over the surface) concentration of impurities in the liquid phase in the wall layer exceeds the limit of solubility C's.

The concentration of impurities in the liquid phase in the wall layer is a function of the rate of vapor generation and mass transfer between the flow core and the wall layer. Both these processes are controlled by a large number of thermal and hydraulic parameters, such as the heat flux, subcooling, velocity, pressure, overall salt content of the flow, etc. Here if the vaporization results in a rise of impurity concentration in the liquid phase in the wall layer, then this layer is subsequently supplied from the flow core by liquid with lower impurity concentration. The dynamic equilibrium between these processes results in the establishing of some effective impurity concentration in the wall layer. Crystallization and dissolution processes are to a large measure mutually compensated and have a relatively minor effect on the magnitude of the effective concentration of impurities in the liquid phase in the wall layer.

To be able to estimate experimentally the degree of concentration, one must know the concentration of impurity in the liquid phase of the flow core and of the wall layer. Direct determination of concentration in the wall layer for an arbitrary channel location is technically extremely difficult. The problem simplifies for the boundary cross section, in which (provided there is solid salt present) the concentration is equal to the solubility of the impurity at the effective temperature of the wall layer—the temperature which controls the course of crystallization of the salt on the heated surface.

The effective temperature T_{ef}, controlling the precipitation of impurities can, in principle, range from the bulk temperature T_0 of the flow core to the wall temperature T_w. Actually, however, vaporization into a bubble may occur only over the temperature range from T_w to T_{sat} and consequently these temperatures should be used in the calculations. The question of selection of the effective temperature of the wall layer at high pressures is not of particular significance, since by virtue of the high values of heat transfer coefficients the difference $T_w - T_{sat}$ is small. The first of the present authors with his coworkers [5.1] performed calculations on saturation concentrations taken at the wall temperature under high pressures. In the case of low pressures the solubility of impurities was taken at the saturation temperature. Special precision experiments performed during the past several years have shown that T_w is the controlling temperature. This means that it can at present be regarded as rigorously established that the effective temperature controlling the precipitations of impurities is the wall temperature ($T_{ef} = T_w$).

The selection of the type of additive (indicator) is of significance. The results obtained using the indicator should be sufficiently general that they can be extended to other additives. On the whole it is desirable to select as an indicator a salt with a negative, but as small as possible value of the temperature solubility coefficient. In addition, this salt should be sufficiently soluble so that deposits from the saturated solution would grow rapidly, if only to reduce the experiment time. For this reason the solubility should be not less than between several and several tens of ppm. At the same time the solubility should be sufficiently low so that the properties of the saturated solution would be virtually identical to those of pure water. This is rather satisfactorily provided by using salts with C'_{sat} smaller than 1 to 10 ppt (parts per thousand) and even somewhat higher if reference is had to properties of the liquid phase proper.

Cases may be encountered when the presence of a salt even with a significantly lower solubility significantly modifies the conditions at the steam-water interface and accordingly the entire two-phase flow pattern. As known from experiments in which steam is bubbled through water, the presence in the latter of even minute quantities of colloidal iron yields a salt concentration which is higher than the specified limit and steeply changes the properties of bubble shells. This results in a significant modification of the structure of the steam-water mixture, involving change in the mean bubble diameter, the rate of bubble rise, height of the layer of the foam which forms on the surface of the water, etc. This change occurs only over a concentration range from C'_{cr}

to C_{cr}'', which is specific for each type of salt. In the range between pure water and C_{cr}' concentration is not seen to have any effect, whereas above C_{cr}'' this effect is small. The value of C_{cr}' falls with rising steam pressure for all the salts; typically at 0.1 MPa C_{cr}' mistake several ppt, whereas, for example, at 14 MPa C_{cr}' might be only hundreds and even tens of ppm. This limits the selection of salts but at the same time makes it possible, as shall be shown below, to obtain a deeper insight into the analogy between boiling heat transfer and bubbling.

The formation of deposits is most frequently monitored by two methods: from the wall temperature profile of the steam-generating tube and from chemical analyses of the indicator salt concentration at the inlet and exit from the channel. If the saturation concentration is sufficiently high, then it is usually possible to detect reliably the rise ΔT in the wall temperature within several hours (Fig. 5.1b).* If the saturation concentration is moderate, then one only needs to analyze the concentration of the indicator salt at the inlet and exit from the channel and to deduce from any difference in concentrations whether deposits are absent or present (Fig. 5.1b). At very low saturation concentrations the accuracy of this method may be found insufficient. Under these conditions one may use a closed loop and to determine the limiting concentration C_{I}^* from the concentration "overshoot curve" (Fig. 5.1c). To improve the reliability of the results it is best to monitor the formation of deposits *both* on the basis of the wall temperature profile *and* from the chemical analyses of the indicator salt concentration.

The selection of the method for monitoring the formation of deposits is affected by a number of factors. For example, in the case of arbitrary distribution of the heat flux along the tube monitoring on the basis of chemical analyses is not suitable. In fact, whereas the limiting concentration for a uniformly heated tube is referred to the exit cross section of the tube, where conditions for mass transfer are the poorest, the referring of C_{0I}^* to exit parameters in the case when the heat load on the tube decreases toward the exit is not obvious (Fig. 5.1e). Under these conditions the deposits must be monitored on the basis of the wall temperature profile along the channel. However, in cases when the heat flux distribution along the channel obeys a complex relationship, this method may also be found ineffective, since deposits of the indicator salt may occur in several channel locations. Whereas the indicator salt concentration near the coolant flow inlet is known, its determination in other locations involves serious difficulties. In this case it is best to perform the measurements using a closed loop (Fig. 5.1e).

The threshold at which impurity deposition starts can be approached by various methods: changing the heat flux, pressure, mass velocity, enthalpy of the flow and concentration of the salt at the channel inlet. The last two methods have come into most extensive practical use. Figure 5.1a shows experiments at constant values of p, q_w, ρw and x_I with only the inlet concentration C_{0I} varied. Figure 5.1f) represents schematically experiments with changes in the inlet enthalpy x_I, with other variables

---

*In Figs. 5.1 through 5.7 the blackened points correspond to values situated to the right of the curve; the open points, to the left of the curve; and the half blackened points correspond to values on the curve proper.

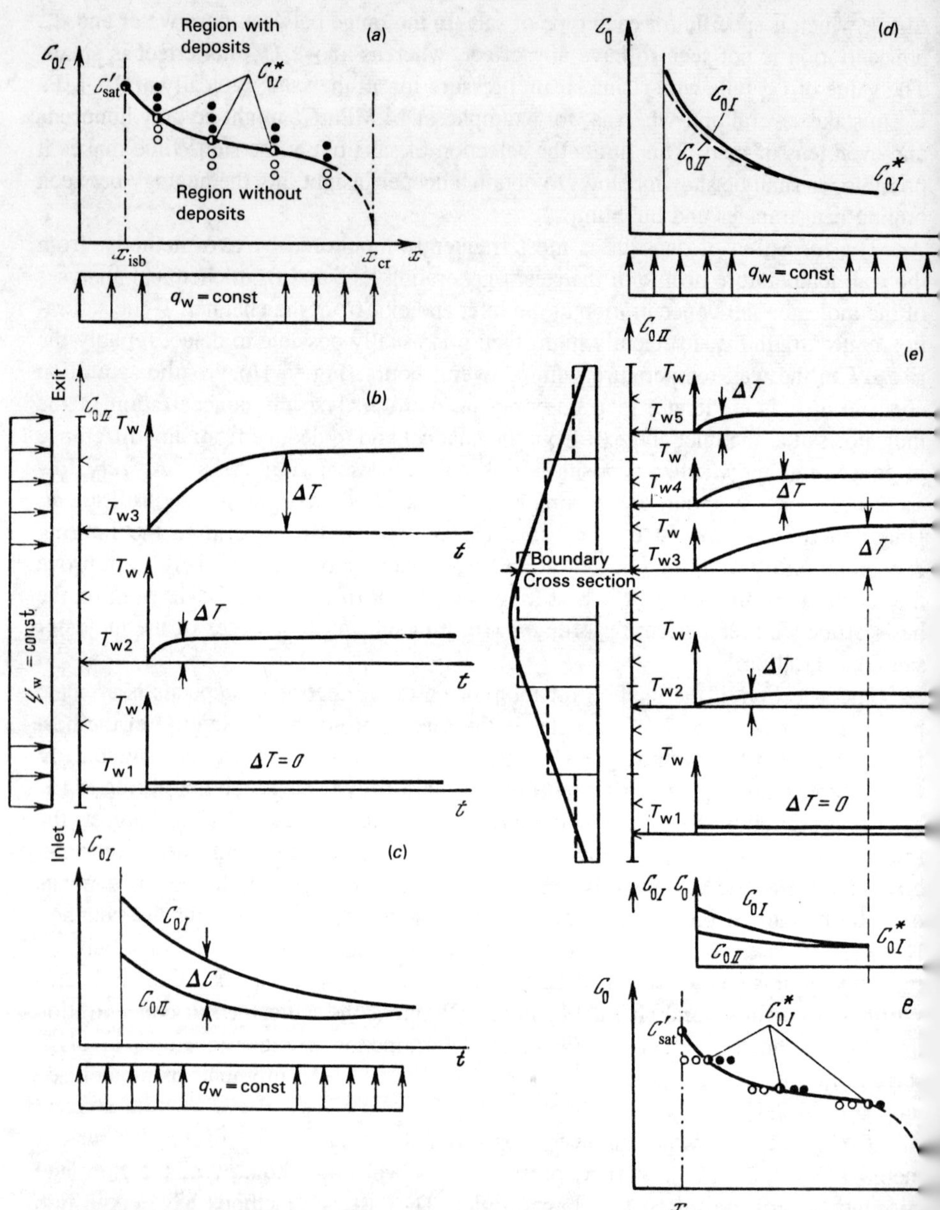

Figure 5.1 Interpretation of the principles of the "salt" method.

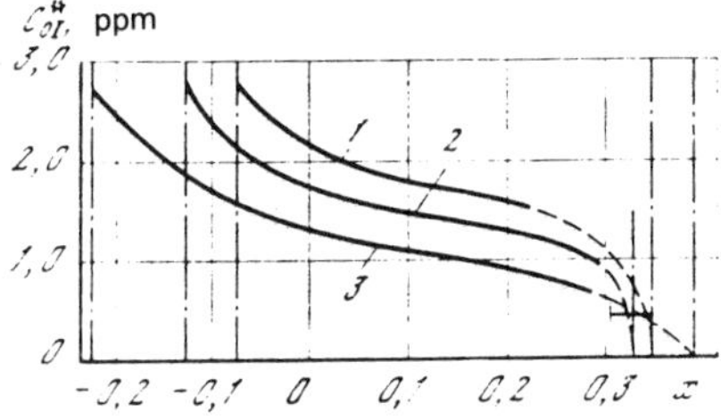

Figure 5.2 Effect of mass velocity on the threshold at which calcium sulfate deposition starts as a function of vapor quality; p = 13.7 MPa, q = 580 kW/m². 1, ρw = 3500 kg/m²·sec; 2, 2000; 3, 1000.

remaining unchanged. Such experiments can, in principle, be performed with changes in several variables.

Above we considered indicators which have high solubility in the liquid phase. Such indicators are best used in cases when investigating processes at the walls of steam-generating channels. If, however, processes within this flow cross section are to be investigated, special sensors must be placed into the channels. For small diameter channels usually employed in experimental practice the selection of such sensors involves serious difficulties. Under certain conditions it is simpler to sample the vapor rather than the liquid phase. In this case one must use indicators with preferential solubility in the gas (vapor) phase. Such indicators came into extensive use in investigating flow variables in the post-dryout region [5.2, 5.3]. As a rule the indicator used is helium, for which reason this version of the salt method has been termed the "helium indicator."

5.2 EXPERIMENTAL STUDY OF THE THRESHOLD IMPURITY CONCENTRATION AND MASS TRANSFER CHARACTERISTICS

The above experimental technique was used for performing tests over a wide range of parameters [5.4–5.9], with various modifications of calcium sulfate used as the indicator salt. The range of variables under study was basically typical of that prevalent in modern boiling-type facilities at conventional and nuclear power plants. Some of the experimental results are shown in Figs. 5.2 through 5.6.

To determine the deposit threshold concentration in the incipient surface boiling region x_{isb} a thorough study was performed at p = 13.7 MPa, q_w = 291 kW/m² and q_w = 1160 kW/m² at respective mass fluxes ρw of 1000 and 2000 kg/m²·sec. These experiments show that, at the location of incipient surface boiling, the threshold curve passes through the saturation concentration at the boiling temperature.

It is seen from Figs. 5.2 through 5.6 that, with increasing equilibrium vapor quality, deposits occur at lower C_{0I}^*. The point here is not only that the concentration in the flow core increases with the vapor quality. With increasing quality the mass transfer rate between the flow and the wall layer deteriorate, which results in a relatively premature precipitation of the salt on the heated surface. The curve of $C_{0I}^* = C(x)$ changes steeply in the subcooled (surface) boiling and the immediately

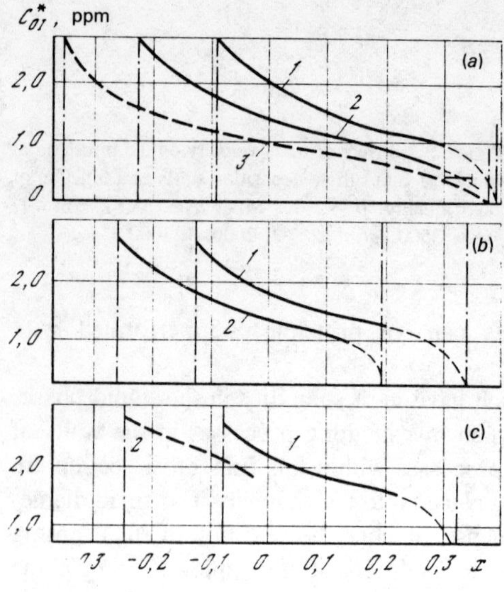

Figure 5.3 Effect of heat flux density on the threshold at which calcium sulfate deposition starts as a function of vapor quality. (a) ρw = 1000 kg/m$^2 \cdot$sec; 1, q_w = 290 kW/m^2; 2, 580; 3, 870. (b) ρw = 2000 kg/m$^2 \cdot$sec; 1, q_w = 638 kW/m^2; 2, 1160. (c) ρw = 3500 kg/m$^2 \cdot$sec; 1, q_w = 580 kW/m^2; 2, 1740.

pre-dryout regions, while being flat in nucleate boiling. The slope of theshold line in this region depends on the operating conditions.

Increasing the mass velocity results in enhancements of the transfer between the flow core and the wall layer (see Fig. 5.2). This reduces the radial concentration

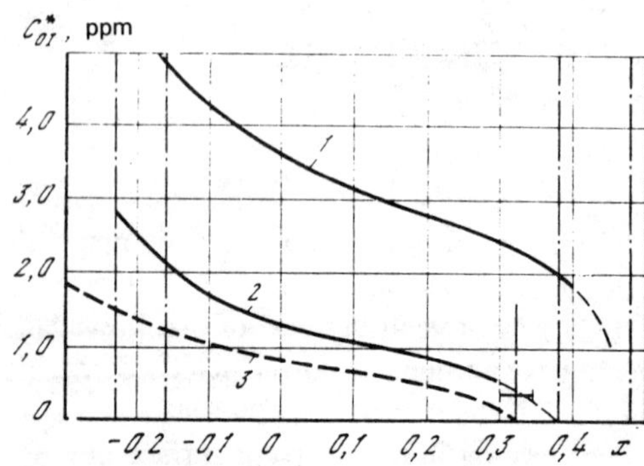

Figure 5.4 Effect of pressure on the threshold at which calcium sulfate deposition starts [as a function of vapor quality]. 1, p = 9.8 MPa; 2, 13.7; 3, 16.7.

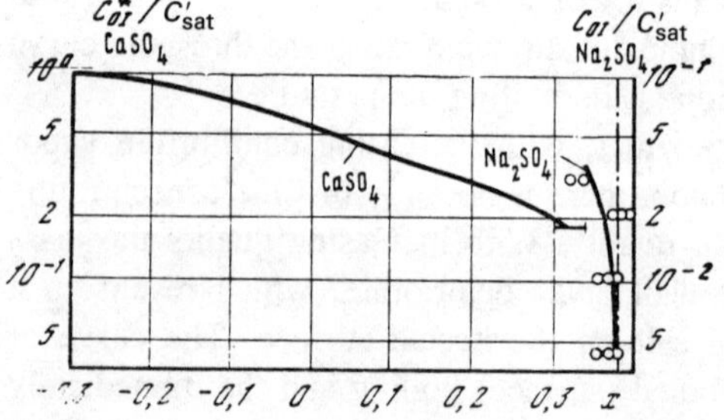

Figure 5.5 Comparison of results on the threshold at which deposition starts [as a function of vapor quality] obtained with solutions of CaSO$_4$ and Na$_2$SO$_4$ at p = 13.7 MPa, q_w = 580 kW/m^2 and ρw = 1000 kg/m$^2 \cdot$sec.

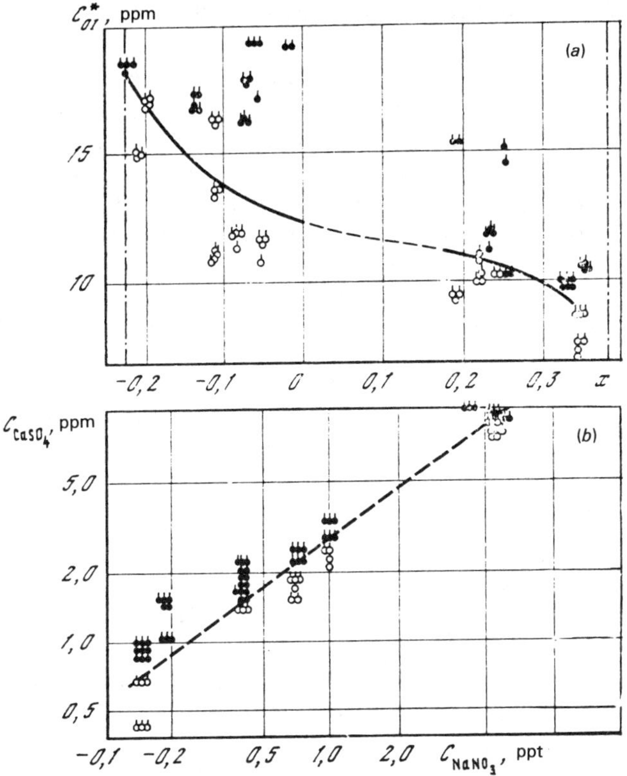

Figure 5.6 Effect of overall salt content of the threshold for inception of calcium sulfate deposition. (a) $C_{\mathrm{NaNO_3}} \cong 5$ ppt; (b) $x \cong 0.33$.

gradient and, all other conditions being equal, raises the curve at least in the region far from the boiling crisis. At higher heat fluxes the effect of velocity decreases.

A rise in the heat flux increases the rate of generation of steam by the heating surface, which results in a more rapid accumulation of salts in the wall layer. In addition, a rise in the vapor generation rate increases the resistance to the supply of liquid from the flow core to the wall layer. These factors aid the precipitation of salts at lower vapor qualities (see Fig. 5.3).

The relative concentration $C_{0I}^{*}/C_{\mathrm{sat}}'$ at which precipitation of salts on the heating surface starts is not a unique function of the pressure. First $C_{0I}^{*}/C_{\mathrm{sat}}'$ falls with pressure, and then increases. At high pressures ($p > 10$ MPa) the curve of $C_{0I}^{*} = C(x)$ is monotonic (see Fig. 5.4).

The effect of the overall salt content of the flow on the deposition inception threshold for calcium sulfate is shown in Fig. 5.5. In certain studies of mass transfer performed at the Moscow Energetics Institute the indicator salt was sodium sulfate. This salt has a high solubility and is convenient for determining the threshold of deposition inception on the basis of the wall temperature profile of the heating surface. Its advantages manifest themselves primarily in the pre-dryout region, where, for the conditions required for deposition, the salt concentration in the feedwater should be sufficiently low.

Let us compare the results at the threshold of salt deposition (see Fig. 5.5) obtained with aqueous solutions of $CaSO_4$ and Na_2SO_4 plotted in coordinates C_{0I}^*/C_{sat}'. This comparison shows that in locations with the same values of p, q_w, ρw and x there exists a break in the relative concentrations when the rig is supplied with various solutions. The relative concentration for inception of salt deposition when the rig was supplied by an Na_2SO_4 solution at $x = 0.3$–0.34 is significantly lower than when $CaSO_4$ was used as the salt. A series of special experiments was performed in which the overall concentration was modified by sodium nitrate (with a concentration of ~ 5 ppt). The experiments were performed at $p = 13.7$ MPa, $q = 580$ kW/m^2 and $\rho w = 1000$ kg/m$^2 \cdot$sec. The vapor quality ranges from x_{isb} to $x = 0.33$ (see Fig. 5.6a). Analysis of results shows that the maximum concentration C_{CaSO_4} at the deposition threshold is significantly higher, due to the increased solubility of $CaSO_4$ in the $NaNO_3$ solution than under the same conditions in pure water. However, the general relationships governing the deposition threshold as a function of vapor quality remain valid. Thus, the effect of the quality is again greater in the subcooled boiling and in the pre-dryout regions, which manifests itself in the steeper shape of the curve of $C_I^* = C(x)$. Again, this curve is flat in the nucleate boiling range.

The effect of the overall salt content was also investigated at different concentrations of $NaNO_3$. The experiments were performed at $p = 13.7$ MPa, $q_w = 580$ kW/m^2, $\rho w = 1000$ kg/m$^2 \cdot$sec and $x = 0.33$. The $NaNO_3$ concentration ranged from 0.135 to 5.27 ppt. The results of these experiments are plotted in Fig. 5.6b. The $CaSO_4$ deposition threshold at these concentrations of $NaNO_3$ is higher than that of $CaSO_4$ in water. This deviation increases with the concentration of $NaNO_3$.

The above experiments were performed with uniform heating ($q_w = $ const). Since under real conditions the heat flux distribution along the channel can be highly varied, there is sense in considering the question of the effect of the distribution of the heat flux on the relationships governing the transfer of mass.

Experiments were performed in which the heat flux varied cosinusoidally [5.9]. The heating load along the channel changes stepwise. The minimum heating load (sections I and VI) was maintained at the 330 kW/m^2 level, the maximum load (sections III and IV) was 582 kW/m^2. The heating load in sections II and V was 437 kW/m^2. The experiments were performed at $p = 13.7$ MPa and $\rho w = 3500$ kg/m$^2 \cdot$sec. The vapor quality changed from that for incipient surface boiling x_{isb} to the critical x_{cr}. Calcium sulfate was used as the indicator salt.

The presence of deposits was monitored on the basis of gain in the wall temperature. The value of the change ΔT wall temperature with time was measured over a time t between 6.5 and 13.5 hr. The wall temperature of sections I and II did not usually change significantly. The value of ΔT was usually 1 to 2 °C. The value of ΔT in the "hot" part of the experimental tube (sections III and IV) is significantly greater (ΔT between 3 and 7 °C). It should be emphasized that due to the insignificance of deposits of $CaSO_4$ in sections I and II the concentration of calcium sulfate at the inlet to sections III and IV was virtually equal to concentration C_{0I} in the feedwater. This circumstance made it possible to analyze experimental data at $q_w = 582$ kW/m^2 and

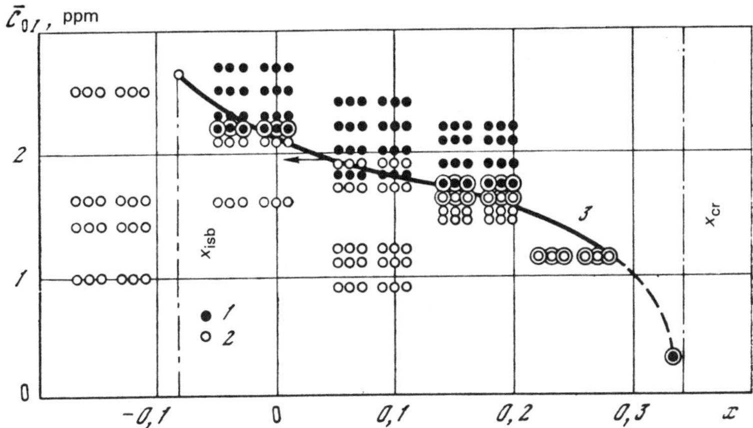

Figure 5.7 Plot of $\bar{C}_{0I} = C(x_0)$ in a channel with a cosinusoidal heat flux distribution at $p = 13.7$ MPa and $\rho w = 3500$ kg/m$^2 \cdot$ sec.

to compare them with analogous data [5.7] under conditions of uniform heating of the entire experimental tube.

Figure 5.7 shows experimental results as a plot of the flow concentration $\bar{C}_{0I}$ vs. the balance vapor quality x. Since deposition occurred over the entire "hot" part, the experimental points when plotted as $\bar{C}_{0I} = C(x)$ occupy a quality range $\Delta x = 0.058$. Hence in each such region of Δx there are six points corresponding to the tube locations in which the wall temperature was measured. Two experiments were performed with a CaSO$_4$ concentration such that deposits were not observed (see the open symbols in Fig. 5.7). Note that the experiments were performed both at constant CaSO$_4$ concentration in the feedwater and variable enthalpy at the channel inset, and with constant enthalpy at the inlet and variable concentration. The dashed curve (see Fig. 5.7) corresponds to data in [5.7]. This curve rather satisfactorily describes data obtained with cosinusoidal heating flux distribution.

The agreement between results obtained with different heat flux distributions along the channel points to the small effect of heat flux profile on parameter $\tilde{C}_I^* = C_{0I}^*/C_{int}'$. This conclusion pertains, naturally, only to that part of the channel in which q_w rises (i.e., sections I through IV); a similar observation may be made regarding single-phase flow heat transfer coefficients which are insensitive to flux profile in regions where the flux is rising slowly with distance [5.10].

The conclusion that mass transfer conditions are insensitive to the heat flux distribution along the channel is of important practical significance; it means that when the heat fluxes increase slowly along a channel, the value of C_{0I}^* can be determined from the local values of p, q_w, ρw and x. If the heat flux decreases along the channel, then, by analogy with single-phase flows the insensitivity of C_{0I}^* to flux profile may no longer hold. For this reason special experiments must be performed for these conditions.

We introduce the concept of liquid circulation ratio n_c in the wall boiling layer

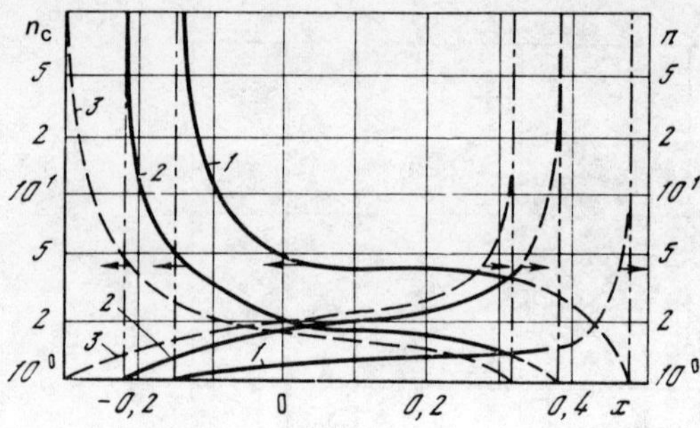

Figure 5.8 Parameters n and n_c vs. vapor quality; $q_w = 580$ kW/m^2; $\rho w = 1000$ kg/m$^2 \cdot$sec. 1, $P = 9.8$ MPa; 2, 13.7; 3, 16.7 MPa.

$$n_c = G_r/G_v \tag{5.2a}$$

where G_r is the mass velocity (radial) of influx of liquid to the steam-generating surface, whereas G_v is the mass rate of vaporization at the surface. On the basis of the two-layer model of mass transfer in steam-generating channels the circulation ratio and the degree of [impurity] concentration are related by the expression [5.8]:

$$n_c = n/(n - 1) \tag{5.2b}$$

Let us consider the effect of the principal operating variables on parameters n (Eq. 5.1) and n_c. In calculating these parameters the effective impurity concentration in the wall layer was assumed to be equal to the solubility based on the saturation temperature, rather than on the temperature of the wall. This introduces a rather insignificant error into the real values of mass transfer parameters, but provides a safety factor for practical calculations. The results of such calculations are shown in Figs. 5.8 through 5.13 where n_c is shown as a function of vapor quality x.

Effect of vapor quality. It is seen that in the region of incipient surface boiling (inception of surface boiling is designated by dash-dotted curves) the circulation multiplicity increases rapidly with decreasing quality, which is understandable from the physical point of view. The vapor generation rate in this region is low and is significantly smaller than the flux of liquid which is supplied to the heating surface. The circulation ratio falls with increasing vapor quality. This is because the vapor generation rate at the heating surface increases, since the quantity of heat needed for bringing the liquid to the saturation temperature becomes smaller (at least in subcooled boiling region). In addition, the flux liquid supplied to the surface decreases. The latter is due to reduction in the water content of the flow and increased resistance to the flow of liquid to the heated surface.

After its rapid fall in the region of the surface boiling inception, the circulation ratio then decreases smoothly with rising quality. The tangent of the curve in this region is a function of the operating variables. Near the crisis location the circulation ratio drops more rapidly than along the preceding length, tending to unity as $x \to x_{cr}$.

The degree of concentration n increases with the vapor quality. In the region of incipient surface boiling $n = 1$, which is due to the rather high rate of mass transfer between the flow and the wall layer. In a region removed sufficiently far from the

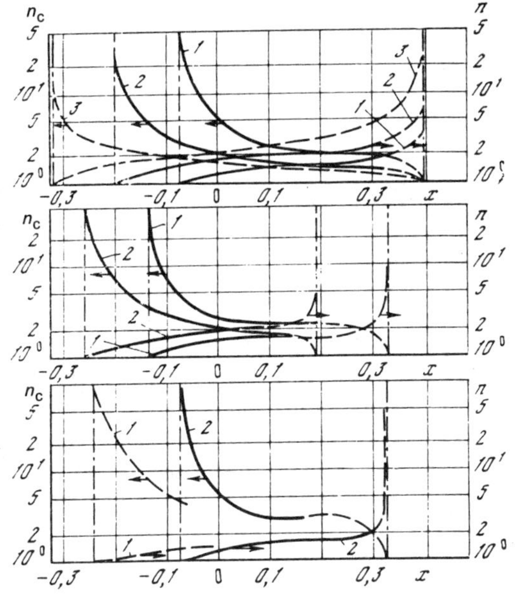

Figure 5.9 Effect of heat flux density on mass transfer parameters. (*a*) $\rho w = 1000$ kg/m$^2\cdot$sec; 1, $q_w = 290$ kW/m^2; 2, 580; 3, 870 kW/m^2. (*b*) $\rho w = 2000$ kg/m$^2\cdot$sec; 1, $q_w = 580$; 2, 1160 kW/m^2. (*c*) $\rho w = 3500$ kg/m$^2\cdot$sec; 1, $q_w = 1740$ kW/m^2; 2, 580 kW/m^2.

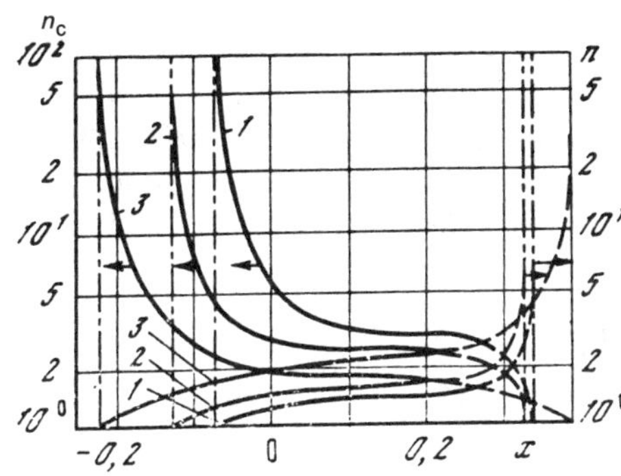

Figure 5.10 Effect of mass velocity on the mass transfer parameters at $p = 13.7$ MPa and $q = 580$ kW/m^2.

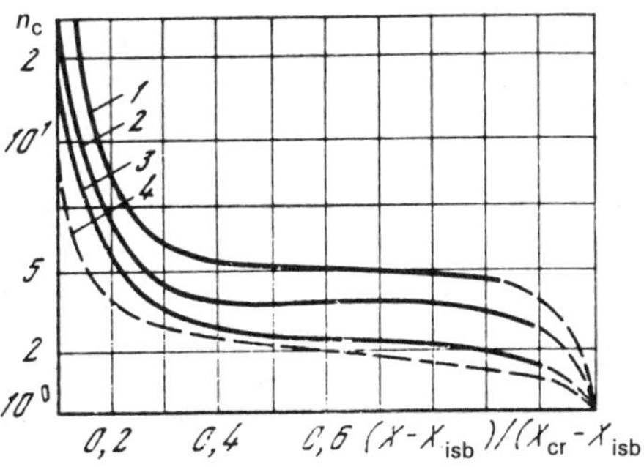

Figure 5.11 Dependence of circulation ratio on X under conditions of Fig. 5.10.

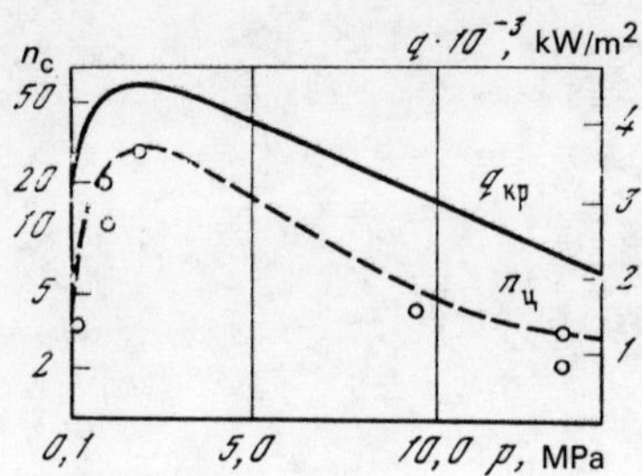

Figure 5.12 Effect of pressure on the circulation ratio and critical heat flux at $x = 0$ and $\rho w = 1500$ kg/m$^2 \cdot$sec.

dryout point the value of n rises smoothly with n but does not exceed values of 2–3. In the vicinity of the crisis location (designated by a dash dotted line in Figs. 5.8 through 5.12) the curve of $n = n(x)$ becomes very steep.

The effect of heat flux on the mass transfer rate can be established using Fig. 5.9; n is proportional to the heat flux density, whereas the circulation ratio n_c is inversely proportional to the flux. It was shown by Nevstruyeva with her coworkers [5.11, 5.12] that a rise in the heat flux is accompanied by a rise in vapor quality in the wall layer, i.e., the available cross section for passage of liquid to the heating surface decreases. The steam generating rate of the heating surface increases with the heat flux. All this raises the hydraulic resistance to the radial flux of liquid and, in the final analysis, reduces the amount of liquid flowing to the heating surface. The magnitude of the heat flux effect depends on the vapor quality; with increasing quality the effect heat flux becomes more pronounced, but decreases with increasing mass flux.

Effect of flow velocity. Analysis of the curves plotted in Fig. 5.10 shows that all other conditions remaining equal, the rate of mass transfer increases with the mass flux of the flow. The higher the mass flux, the higher the intensity of turbulent

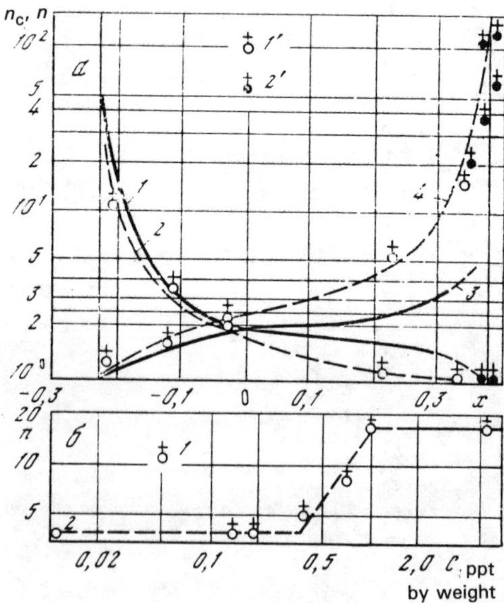

Figure 5.13 Effect of the overall salt content on mass transfer in steam-generating channels. (a) 1, n_c; 3, n; 1′, CaSO$_4$-NaNO$_3$-H$_2$O system; 2′, C $\approx$ 5 ppt and the Na$_2$SO$_4$-H$_2$O–NaNO$_3$ system; 2, n_c; 4, n. (b) $x = 0.33$; 1, CaSO$_4$-NaNO$_3$-H$_2$O; 2, CaSO$_4$-H$_2$O [sic]; point 2 has been plotted in arbitrary fashion.

fluctuations of the liquid phase in the flow core and the kinetic energy of droplets moving toward the wall layer. In addition, as noted by Nevstruyeva with her coworkers [5.11, 5.12], increase in flow velocity is accompanied by reduction in the wall-layer vapor quality and consequently to a reduction in the hydraulic resistance to the supply of liquid to the heating surface. All these factors are responsible for the increasing mass transfer rate with rising mass flux.

The extent to which velocity affects n and n_c is a function of operating variables. Thus, in particular, for $x = 0.1$ increasing the heat flux density from 279 to 872 kW/m^2 at $\rho w = 1000$ kg/m$^2 \cdot$sec reduces the circulation ratio approximately twofold, whereas at $\rho w = 2000$ kg/m$^2 \cdot$sec an increase in heat flux density from 580 to 1160 kW/m^2 decreases the circulation ratio only by a factor of 1.4. This response to velocity changes is due to the fact that, at high heat fluxes, the principal effect on mass transfer in nucleate boiling is exerted by the heat flux density itself, whereas at low values of the latter the flow velocity is the controlling parameter.

It is difficult to determine the effect of individual parameters using quality as coordinates, since the mass transfer coefficients at a given value of vapor quality are not the same. This is because, for a given set of conditions, the extent to which the system is removed from the critical vapor quality and inception of surface boiling is different. Because of time, other coordinates were used in Fig. 5.11, where the argument is the ratio $X = (x - x_{isb})/(x_{cr} - x_{isb})$.

It is seen that at $X = 0$ (i.e., when $x = x_{isb}$) the circulation ratio increases rapidly with decreasing x. The rate of change in n and n_c as a function of X is controlled by operating conditions. The nature of the effect of operating conditions in these coordinates is basically unchanged from that observed in terms of quality.

Effect of pressure. Figure 5.12 presents data on the circulation ratio for different pressures. It is seen that the effect of pressure is not monotonic. Figure 5.12 compares data on both heat transfer and mass transfer. The solid curve represents the dependence of the critical heat flux q_{cr} on the pressure, and the dash-dotted curve—the same for the circulation ratio. It is interesting to note that the shape of curves for n_c and q_{cr} is identical; the values of n_c and q_{cr} first increase at pressures from 0.1 to 7 MPa, and decrease again as the pressure is raised further towards the critical. This experimentally obtained fact is additional proof of the relationship between the critical heat flux or dryout condition with the maximum rate of mass transfer, it definitely deserves further study and confirms that the models of heat transfer deserving the greatest attention are those which are in one way or another based on the hydrodynamic nature of the critical (dryout).

Effect of overall salt content. It follows from experiments on the bubbling of steam through water that the presence of impurities may significantly modify the conditions at the steam-water interface, and consequently the entire two-phase flow pattern. Studies are known which confirm certain general relationships governing boiling and bubbling. It is natural to assume that water impurities may also affect the boiling mass transfer. The results of investigating the effect of the overall salt content on mass transfer are plotted in Fig. 5.13. The experiments were performed at $p = 13.7$ MPa, $q = 580$ kW/m^2 and $\rho w = 1000$ kg/m$^2 \cdot$sec. The vapor quality ranged from x_{isb} to x_{cr}. The left-hand side of the dashed curves (Fig. 5.13a) was obtained for the $CaSO_4$-$NaNO_3$-H_2O system. The overall salt content of feedwater was here *a priori* known to be post critical ($C_{0I} > C''_{cr}$) and ranged between 4.34 and 5.27 ppt. The presence of deposits was determined from chemical analysis of $CaSO_4$ concentrations entering and leaving the test section. The right-hand side of the dashed curves was obtained for the Na_2SO_4-H_2O system. The presence of deposits was determined from the tube surface temperature. The solid curves representing mass transfer for the $CaSo_4$-H_2O system were analyzed above.

It is seen from Fig. 5.13 that the curves obtained for the $CaSO_4$-$NaNO_3$-H_2O and Na_2So_4-H_2O systems differ from those obtained for the $CaSO_4$-H_2O system. Though the results for all the systems are similar in the region close to the inception of boiling, the systems with a high overall salt ($NaNO_3$ or $NaSO_4$) content demonstrate a significant deterioration in the rate of mass transfer as the critical vapor quality is approached.

The following conclusions can be drawn from the results mentioned above.

1. The nature of the $n = n(x)$ and $n_c = n_c(x)$ curves for the different salt content reflects the way in which mass transfer changes as a function of vapor quality.
2. The gradual divergence between mass transfer rates for the respective system indicates that the vapor generating capacity is variable along the channel, and increases as the critical quality is approach.

It is interesting to determine the critical salt content at which the differences between the salt-containing and the $CaSO_4$-H_2O systems begin to appear at given flow conditions. A special experiment was therefore performed at the operating conditions ($p = 13.7$ MPa, $q = 580$ kW/m^2 and $\rho w = 1000$ kg/m$^2 \cdot$ sec) mentioned above and at $x = 0.33$ for the $CaSO_4$-$NaNO_3$-H_2O system. The $NaNO_3$ concentration ranged between 0.135 and 5.27 ppt.

Until the overall salt content increases to some value C'_{cr} (see Fig. 5.13b) the concentration ratio n (Eq. 5.1) does not change. Then over the range between C'_{cr} and C''_{cr} the degree of concentration increases with the salt content. At $C_{0I} > C''_{cr}$ the concentration ratio n again remains constant but at a higher level. The values of C'_{cr} and C''_{cr} are of the same order of magnitude as in bubbling. Analogous results were also obtained at $p = 0.12$ MPa.

In calculating mass transfer in complex systems one must remember the following circumstance. As the overall salt content of the feedwater increases, the ionic strength of the solution, and consequently, the solubility of the calcium sulfate changes. For this reason all the calculations on the rate of mass transfer were performed by the method of successive approximations.

Analysis of results of this series of experiments confirms the analogy between mass transfer in steam-generating channels and bubbling of steam through liquids. Consequently, to a first approximation, mass transfer at different salt contents and operating conditions can be analyzed on the basis of corresponding data on bubbling.

The effect of salt content on the mass transfer rate in boiling can be explained as follows. Colloidal iron particles reinforce the strength of the bubble shell when the salt content is above some limit. This interferes with bubble coalescence and with release of steam from the wall layer. As a result there forms a swollen foamy layer which exerts a significant hydraulic resistance to the supply of liquid from the flow core. This results in a higher concentration of impurities in the wall layer compared with that obtained in the boiling of a solution with subcritical salt content.

These data must be taken into account both in selecting indicator salts in the study of mass transfer and in calculating mass transfer conditions in boiling equipment operating on feedwater with different impurity contents.

5.3 MODELS OF THE PROCESS

Several models have been suggested [5.13–5.17] physical processes associated with boiling. As a rule, these are based on interpretation of data on pressure drop and heat transfer. More recently, these models have incorporated results obtained by sampling the liquid phase by means of various devices. These models are rather widely known and are not considered here. We shall only consider in more detail those models which take account of data on the behavior of impurity deposition in the steam-generating channel.

5.3.1 The Two-Layer Model

Studies of the mechanism of mass transfer significantly widen our ideas about processes attending boiling. The hypothesis put forward as far back as by Prandtl and Taylor to the effect that the bulk of the resistance to heat and mass transfer between the wall and turbulent core of a single-phase flow is concentrated in the laminar sublayer, was subsequently extended to the flow of boiling liquid. This was the origin of the widely held concept that additional turbulization of the wall layer, and reduction of its thickness, in boiling was responsible for the high value of the heat transfer coefficient. Early studies of the nucleating boiling mechanism, based on visual analysis of still and cine photographs and on measurements of the departure diameter of the bubble and frequency of bubble generation, did not detect any specific contradictions between these concepts and experimental results. However, the above hypothesis did not always explain certain characteristic features in the behavior of the boiling heat transfer coefficient.

Studies determining the breakoff diameter of the bubble, the number of nucleation sites and the frequency of nucleation made it possible to determine heat and mass fluxes from the wall layer to the flow core. It was found that the heat flux transmitted by vapor bubbles in surface boiling is small compared with the overall heat flux. This is exemplified by the results obtained by Jacobs and Shade [5.18] who, in addition to making measurements of bubble dimensions and the frequency of their departure, also measured the temperature field about the rising bubble. The heat removed by bubbles due to the latent heat of vaporization was only 1–4% of the total amount of heat supplied. The heat removed by the superheated liquid, ejected by the bubble and entrained in its wake, is 13–34%. The mechanism of transport of the remaining 80–60% of the heat was unexplained.

If the remaining heat was removed from the regions not involved in vapor phase formation, then a reduction in heating surface temperature should be observed in these regions. However, measurements showed that, outside the bubble radius, the temperature of the heating surface remains constant. A steep temperature drop occurs only at the vaporization site. In addition, if rapid removal of heat is the result of the effect of cold water flowing to the surface, then the temperature at the nucleation site should decrease at the time of departure or breakup of the bubbles. However, it was established by Rogers and Mesler [5.19] and by Bankoff and Mason [5.20] that the temperature of the heating surface in the vicinity of the nucleation site actually increases in the course of bubble breakup. This means that neither the mixing of liquid in the wall layer, nor the flow of cold liquid to the nucleation site ensure sufficient rate of heat removal from the heating surface in the course of boiling.

In conjunction with this it is of particular interest to consider studies of the feasibility of heat removal at the nucleation site at the time of nucleation and growth of the bubble [5.20–5.22]. It was shown experimentally in these studies that vapor condensation occurs in the course of bubble growth in subcooled boiling. The condensation rate is a function of the subcooling and the velocity of the incoming liquid. In certain cases the quantity of steam generated during the formation of the bubble exceeded, by several orders of magnitude, the quantity of steam contained within a bubble of maximum size. The latent heat transported from the surface by the maximum size bubble amounts to between 1 and 10% of the heat expended for forming this bubble, depending on the flow velocity and subcooling. The bubble which grows at a nucleation site removes a significant amount of heat from the heated surface due to vaporization of liquid near its base. Simultaneously it transmits heat at a rapid rate to cooler layers of liquid in the form of heat of condensation. The condensed liquid is transported to the flow core by turbulent fluctuations.

The process of heat transfer between the heating surface and the flow core can be conditionally subdivided into two stages. Firstly, removal of heat from the heating surface by a coolant layer directly adjoining this surface, and secondly, transfer of the heat thus removed from the wall layer to the flow core. Studies of the boiling mechanism based on analysis of the heat and mass moved from the wall layer into the flow core involves difficulties in determining all the possible transport processes. Direct instantaneous measurement of the heat removal rate from the heating surface is rather difficult. However, independent examination of the simultaneous heat and mass transfer processes facilitates an understanding of the mechanism of nucleate boiling. To clarify this mechanism it is necessary to known the amount of liquid flowing from the flow core to the wall layer, the heat flux associated with vapor generating at the heating surface and the ratio of components of heat removal due to evaporation and convection of the liquid phase.

Given the above, we shall consider a specific two-layer model (the Reynolds flux model). We subdivide the channel cross section into two zones—the wall layer and the flow core (Fig. 5.14). It is assumed that the parameters of each zone remain unchanged over its thickness. Thus, for the wall layer $h_1 = h_w = $ const, $C_1 = C_w = $ const, $T_1 = T_w = $ const and for the two-phase flow core: $T_0 = $ const, $h_0 = $ const and $C_0 = $ const.

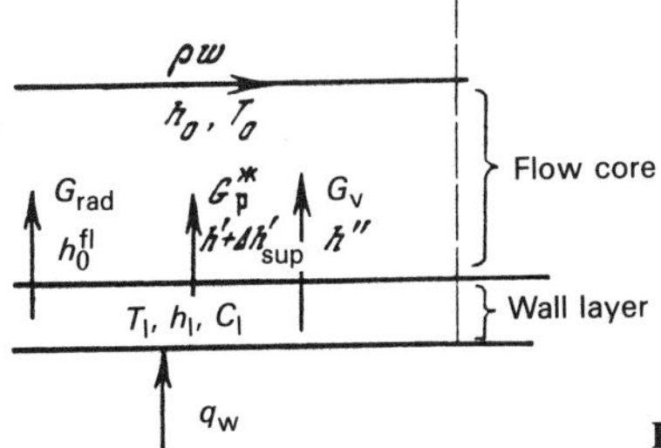

Figure 5.14 Model of transfer processes in boiling in a tube.

We shall assume that mass and heat transfer in a pipe in the course of boiling occur only by convection of the radial coolant flow. The radial flux G_{rad} to the heating surface consists only of the liquid phase with enthalpy h_0^{fl}. The radial flux from the heating surface consists of the liquid G_{rad}^{fl} and vapor G_v phases. The liquid phase is superheated relative to the saturation enthalpy by the amount Δh_{sup} and has an enthalpy equal to $h' + \Delta h_{sup}$. The vapor phase is assumed to have the enthalpy of dry saturated vapor h''. The heat transported to the wall layer is comprise of the external heat flux q and heat carried to the wall with the radial flow $G_{rad}h_0^{fl}$. The heat is removed from the wall layer by the liquid phase $G_{rad}^{fl}(h' + \Delta h_{sup})$ and by the vapor phase $G_v h''$.

The material and heat balance for the case under study can be written as

$$G_{rad}^{fl} = G_{rad} - G_v \tag{5.3}$$

$$q + G_p h_0^{fl} = (G_{rad} - G_v)(h' + \Delta h_{sup}) + G_v h'' \tag{5.4}$$

The superheat Δh_{sup} of the liquid phase can be determined if the coefficient α of heat transfer at the given location and the specific heat of the superheated liquid c_p^{sup} are known

$$\Delta h_{sup} = \left(\frac{q}{\alpha} - \Delta T_{sub}\right) c_p^{sup} \tag{5.5}$$

Since $G_{rad}/G_v = n_c$ and $(h' - h_0^{fl})/r = \Delta x_{sub}$, it is possible to obtain equations for the radial flow and the vapor generating rate at the heating surface

$$G_{rad} = \frac{q}{r} \frac{n_c}{1 + (n_c - 1)(q/\alpha - \Delta T_{sub}) c_p^{sup}/r + n_c \Delta x_{sub}} \tag{5.6}$$

$$G_v = \frac{q}{r} \frac{1}{1 + (n_c - 1)(q/\alpha - \Delta T_{sub}) c_p^{sup}/r + n_c \Delta x_{sub}} \tag{5.7}$$

Figure 5.15 shows the variation of G_{rad} and G_v along a uniformly heated tube at a pressure of 13.7 MPa for the three values of mass velocity and constant heat flux. The mass flux of liquid from the core to the wall layer increases with the axial mass velocity; however, even at high values of the latter (i.e., 3500 kg/m^2·sec) it does not

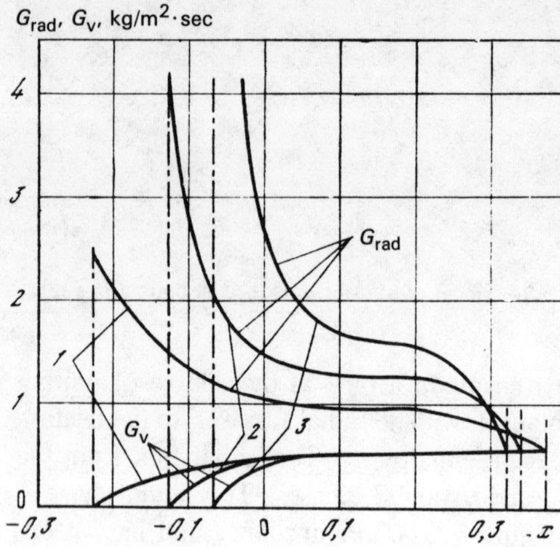

Figure 5.15 Radial fluid flow and steam-generating flux of heating surface in boiling in a tube vs. vapor quality at different mass velocities. p = 13.7 MPa, q = 580 kW/m^2; 1, ρw = 1000 kg/m^2·sec; 2, 2000; 3, 3500 kg/m^2·sec.

exceed by more than 10-fold the radial mass flux of vapor in the nucleate boiling region. It should be noted here that the values of circulation ratio n_c at point x_{isb} are very high, and their experimental determination is unreliable. It should be expected that actual value of the maximum G_{rad} will be somewhat higher than shown due to the rounding off effect.

Computational results for other conditions are plotted in Fig. 5.16. It is seen that the vapor generating capacity of the heating surface rises rapidly in the inlet region of the tube. However, a significant part of this vapor condenses when it enters the subcooled flow core. Knowing the amount of vapor in the flow and the vapor generating rate from the heating surface, we can define the amount of steam condensing as

$$G_{cond} = \frac{q_w}{r}\left[\frac{1}{1 + (n_c - 1)(q_w/\alpha - \Delta T_{sub})c_p^{sup}/r + n_c \bar{x}_{sub}} - \frac{\Delta x_{act}}{\Delta \bar{x}}\right] \qquad (5.8)$$

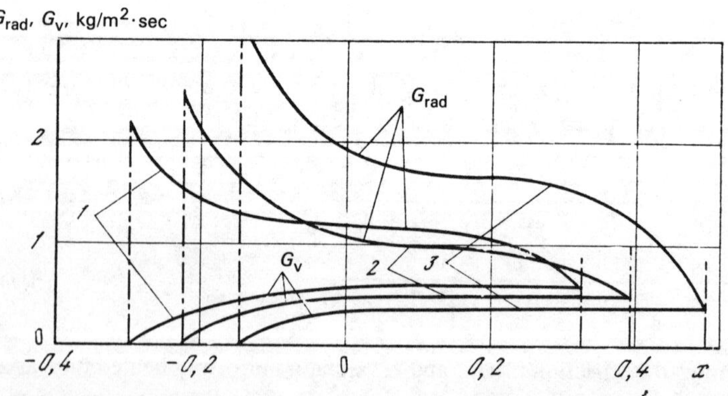

Figure 5.16 The radial flux to and the vapor generating capacity of the heating surface in boiling in a tube vs. the vapor quality at different pressures; q = 580 kW/m^2, ρw = 1000 kg/m^2·sec. 1, p = 16.7 MPa; 2, 13.7; 3, 9.8 MPa.

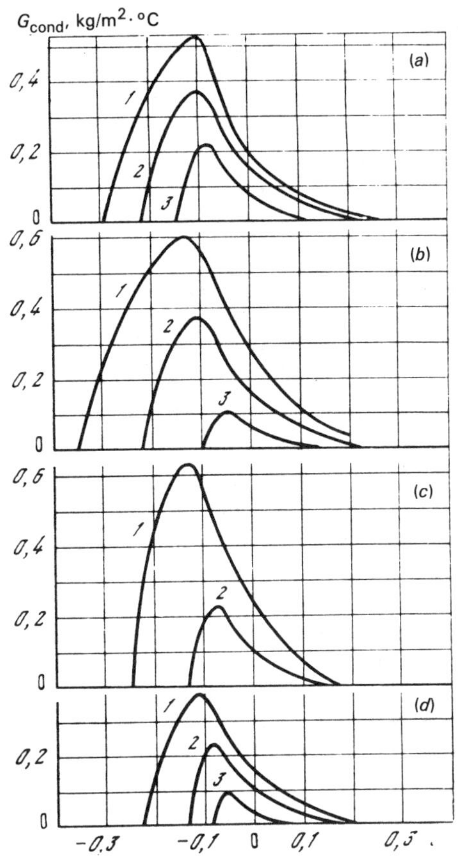

Figure 5.17 Effect of various variables on the condensation of vapor in a tube during boiling. (a) q_w = 580 kW/m², ρw = 1000 kg/m²·sec; 1, p = 16.7 MPa; 2, 13.7; 3, 9.8. (b) p = 13.7 MPa, ρw = 1000 kg/m²·sec; 1, q_w = 870 kW/m²·sec; 2, 580; 3, 290. (c) p = 13.7 MPa, ρw = 2000 kg/m²·sec, 1, q_w = 1160 kW/m²; 2, 580. (d) p = 13.7 MPa, q = 580 kW/m²; 1, ρw = 1000 kg/m²·sec; 2, 2000; 3, 3500.

where Δx_{act} is the gain of the actual value of vapor quality at the length under computation.

Figure 5.17 presents the results of calculating G_{cond}. It is seen that the vapor condensation rate first rises rapidly, "absorbing" a significant quantity of vapor flowing to the core from the wall layer. As a result of this, the fraction φ of the cross section occupied by the vapor (the "void fraction") increases very slowly in this zone. As the flow core becomes hotter and a shell of hot water forms about individual bubbles, the condensation rate drops and the absolute quantity of vapor condensing within the flow starts falling, in spite of the rise in the number of bubbles and in their overall surface.

It is important to note that the condensation zone moves far into the region of the positive values of thermodynamic quality (in the cases under study up to $\bar{x} \simeq 0.3$), with the rate of condensation at the end of this zone, naturally changing very slowly. Thus, the temperature of water in the flow core asymptotically approaches the saturation temperature, for which reason one can refer only conditionally to the existence of a given position where saturated nucleate boiling starts.

If it is assumed that the zone of nucleate boiling starts in the location where more

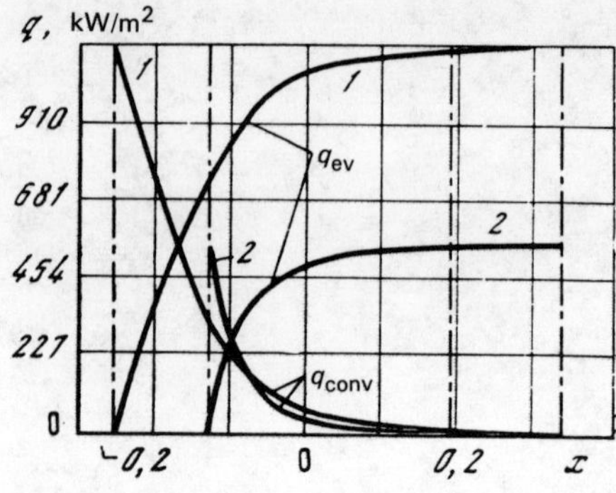

Figure 5.18 Plot of q_{conv} and q_{ev} vs. the vapor quality; p = 13.7 MPa, ρw = 1000 kg/m$^2\cdot$sec. 1, q_w = 568 kW/m^2; 2, 284 kW/m^2.

than 90% of the heat is transmitted from the wall layer to the vapor bubbles then, as seen from Figs. 5.16 and 5.17, this zone may start already in the region of slightly negative values of thermodynamic quality. If, on the other hand, the start of nucleate boiling is assumed to be the region in which the amount of vapor condensing comprises less than 10% of the vapor flowing to the flow core, then under conditions corresponding to those in Fig. 5.17, $\bar{x} \simeq$ 0.10–0.15.

Knowing the mass fluxes participating in transfer processes in the course of boiling in a tube, it is possible to determine the amount of heat removed from the wall by convection of liquid q_{conv} and due to change of phase q_{ev}

$$q_{conv} = q_w \frac{(n_c - 1)(q_w/\alpha - \Delta T_{sub})c_p^{sup}/r + n_c \bar{x}_{sub}}{1 + (n_c - 1)(q_w/\alpha - \Delta T_{sub})c_p^{sup}/r + n_c \bar{x}_{sub}} \tag{5.9}$$

$$q_{ev} = q_w \frac{1}{1 + (n_c - 1)(q_w/\alpha - \Delta T_{sub})c_p^{sup}/r + n_c \bar{x}_{sub}} \tag{5.10}$$

It is seen from Figs. 5.18 and 5.19 that most of the variation in q_{conv} and q_{ev} occurs in the subcooled boiling region, whereas the variation in these variables under the nucleate boiling conditions (defined above) is insignificant. The location of the start of subcooled boiling depends significantly on the wall heat flux q_w, but the defined location of the termination of this zone moves relatively little. It is obvious

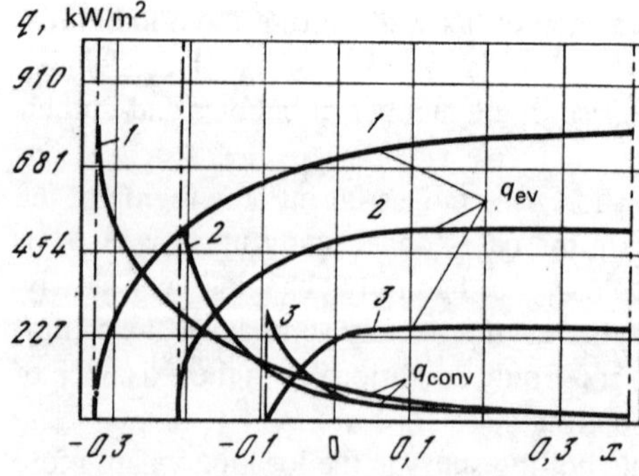

Figure 5.19 Plot of q_{ev} and q_{conv} vs. vapor quality; p = 13.7 MPa, ρw = 2000 kg/m$^2\cdot$sec. 1, q = 1136 kW/mkg/m^2; 2, 568 kW/m^2.

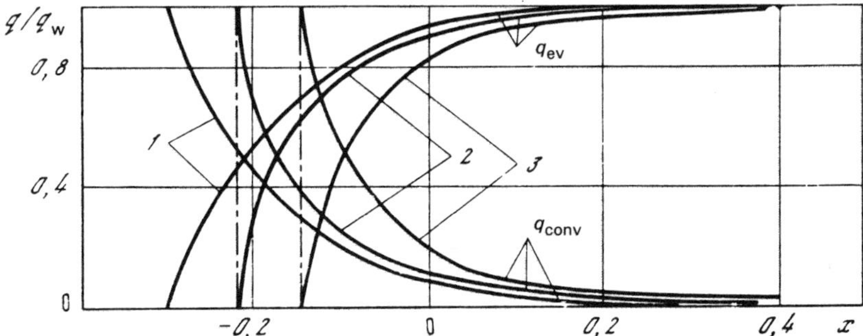

Figure 5.20 Effect of pressure on the convective and evaporative components of heat removal; $a = 580$ kW/m^2, $\rho w = 1000$ kg/m$^2 \cdot$ sec. 1, $p = 16.7$ MPa; 2, 13.7; 3, 9.8 MPa.

that upon reduction in q_w the length of the subcooled boiling region will decrease primarily due to the fact that vaporization starts closer to the point where $\bar{x} = 0$. An analogous situation is also observed with reduction in pressure (Fig. 5.20). With constant values of ρw and q_w, vaporization starts earlier at $p = 16.7$ MPa than at $p = 13.7$ MPa, and later at $p = 9.8$ MPa.

Over a wide range of relative enthalpies, the value of q_{ev} is significantly higher than that of q_{conv}. The value of q_{ev} increases significantly with increasing $\bar{x}$ and has already become the principal component in the subcooled boiling region and is responsible for the high value of the heat transfer coefficient there. The convective component of the heat flux is controlled by the subcooling of the liquid in the flow core and its superheat in the wall layer. Over most of the boiling region, it has low values and decreases with increasing flow quality.

Heat q_{ev} is the sum of the heat fluxes transmitted from the wall by the condensing vapor q_{cond} and vapor bubbles q_v

$$q_{ev} = q_{cond} + q_v \tag{5.11}$$

Using the above results, we can write

$$q_{cond} = q_w \left[\frac{1}{1 + (n_c - 1)(q_w/\alpha - \Delta T_{sub}) c_p^{sup}/r + n_c \bar{x}_{sub}} - \frac{\Delta x_{act}}{\Delta \bar{x}} \right] \tag{5.12}$$

$$q_v = q_w \frac{\Delta x_{act}}{\Delta \bar{x}} \tag{5.13}$$

Analysis of these components (Fig. 5.21) shows that, over a wide range of quality, most of the heat transfer in subcooled boiling occurs by condensing vapor. Only a small fraction of the heat is transmitted as latent heat in the vapor bubbles in this region; only in the region of nucleate boiling is the bulk of the heat transmitted by vapor bubble latent heat.

The convective component of the heat flux can also be subdivded into two components: heat transmitted to the flow core with superheated liquid

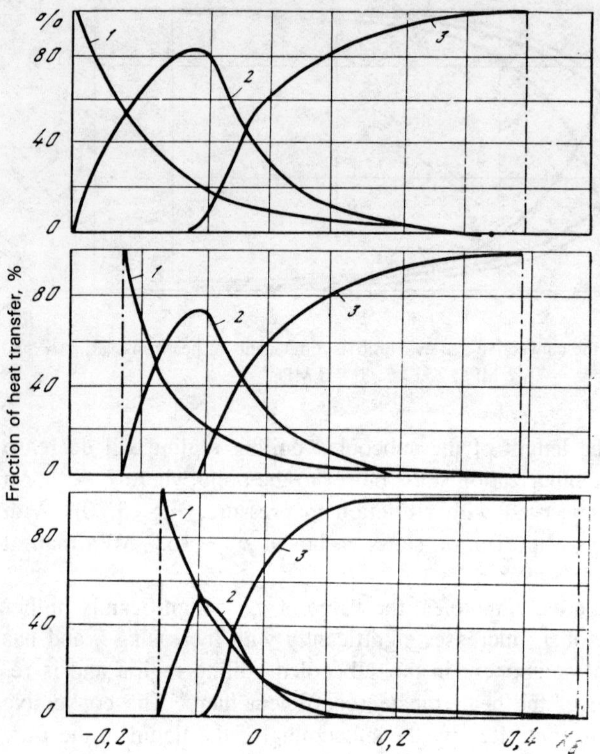

Figure 5.21 Effect of pressure on heat transfer from the wall layer to the flow core; $q = 580$ kW/m²; $\rho w = 1000$ kg/m²·sec. 1, $q_{sub} + q_{sup}$; 2, q_{conv}; 3, q_v.

$$q_{sup} = q_w \frac{(n_c - 1)(q_w/\alpha - \Delta T_{sub}) c_p^{sup}/r}{1 + (n_c - 1)(q_w/\alpha - \Delta T_{sub}) c_p^{sup}/r + n_c \Delta x_{sub}} \tag{5.14}$$

and the heat expended for superheating the liquid flowing from the core to the boiling temperature

$$q_{sub} = q_w \frac{n_c \Delta x_{sub}}{1 + (n_c - 1)(q_w/\alpha - \Delta T_{sub}) c_p^{sup}/r + n_c \Delta x_{sub}} \tag{5.15}$$

Considering the above, the entire heat transmitted to the flow core can be represented as a sum of components

$$q_w = q_{sup} + q_{sup} + q_{cond} + q_v \tag{5.16}$$

By analogy with the heat flux, the heat transfer coefficient can be represented by the sum

$$\alpha = \alpha_{conv} + \alpha_{ev}, \tag{5.17}$$

where

$$\alpha_{conv} = \alpha \frac{(n_c - 1)(q_w/\alpha - \Delta T_{sub})c_p^{sup}/r + n_c \Delta x_{sub}}{1 + (n_c - 1)(q_w/\alpha - \Delta T_{sub})c_p^{sup}/r + n_c \Delta x_{sub}} \tag{5.18}$$

$$\alpha_{ev} = \alpha \frac{1}{1 + (n_c - 1)(q_w/\alpha - \Delta T_{sub})c_p^{sup}/r + n_c \Delta x_{sub}} \tag{5.19}$$

$$\alpha = q_w/(T_w - T_0). \tag{5.20}$$

Figure 5.22 shows the results of calculation of α_{conv} and α_{ev} for $p = 16.7$ MPa, $\rho w = 1000$ kg/m^2·sec, and $q = 568$ kW/m^2 over the vapor quality range from x_{isb} to x_{cr}. Data on heat transfer coefficients are taken from Tarasova et al. [5.23]. It is seen from Fig. 5.22 that the evaporative component of the heat transfer coefficient increases steeply with vapor quality and that, even in the subcooled boiling range, it becomes the principal component, controlling the (high) value of α. The convective component of the heat transfer coefficient is controlled by the magnitude of the subcooling and superheating of the liquid.

5.3.2 The Diffusion Model

We now consider in the model [5.24] of the heat transfer process in a steam-generating channel as illustrated in Fig. 5.23 the heat flux is constant along the channel and subcooled boiling is occurring. The steam-generating channel is supplied

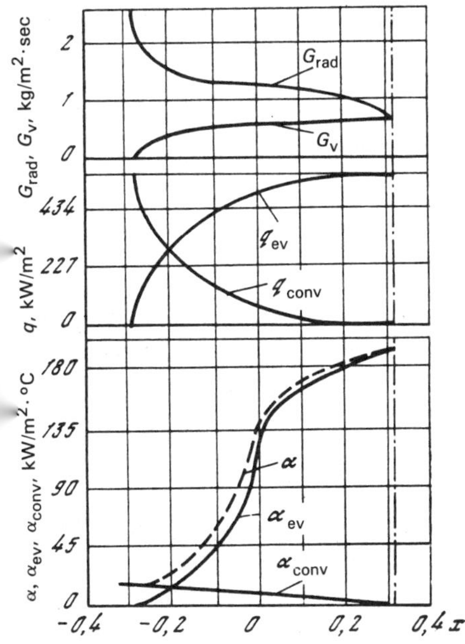

Figure 5.22 Heat and mass transfer parameters vs. vapor quality at $p = 16.7$ MPa, $q = 568$ kW/m^2, $\rho w = 1000$ kg/m^2·sec.

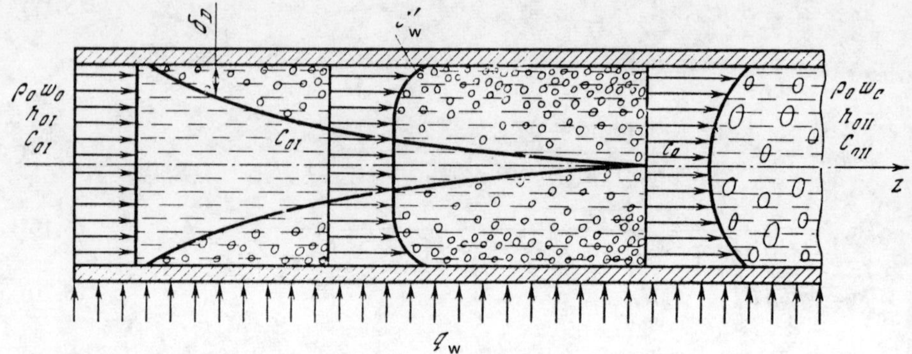

Figure 5.23 Physical model of the mass transfer process.

with liquid which has, for example, salt concentration of C_{0I}. As long as no vaporization occurs on the heated surface, the concentration of the salt over the cross section is constant, $C = C_{0I}$. In boiling, vapor bubbles separate from the heating surface, are carried away by the liquid flow and form a developing diffusion two-phase boundary layer. At some point along the channel, the boundary layers merge at the tube axis. Up to this point, the salt concentration changes only within the boundary layer in which the vapor bubbles are situated. The concentration in the flow core remains unchanged and equal to C_{0I}. The salt concentration in the liquid phase in the boundary layer is higher than on the flow axis due to the significantly lower solubility in the vapor. The maximum concentration in the liquid at the wall is C'_w, whereas that at the outer edge of the boundary layer is $C = C_{0I}$. This means that the vaporization is a source of concentration of impurities in the liquid phase at the wall. Under steady-state conditions the concentration of salt at the wall in the liquid phase will be controlled by diffusion processes in the boundary layer.

Following the merging of the diffusion boundary layers, the concentration in the liquid in the flow core will increase due to the presence of vapor in it

$$\bar{C}_0 = C_{0I}/(1 - \bar{x}_0) \tag{5.21}$$

where x_0 is the actual mean vapor quality of the flow.

The amount of salt forming at the wall due to vapor generating can be calculated on the assumption that the solubility in the vapor can be neglected and is given by the formula

$$j_{salt} = C'_w j_v \tag{5.22}$$

with increasing distance along the channel, the rate of vapor generation j_v rises, which results in a further rise in C'_w.

We introduce the nondimensional impurity flux St_{Dfl} (the diffusion Stanton number)

$$\mathrm{St}_{D\mathrm{fl}} = C'_w j_v / \overline{\rho_0 w_0} \, (C'_w - \bar{C}'_0) \tag{5.23}$$

and the nondimensional heat flux in the liquid phase (the thermal Stanton number)

$$\mathrm{St}_{\mathrm{fl}} = (q_w - j_v r) / \overline{\rho_0 w_0} \, (h'_w - \bar{h}'_0). \tag{5.24}$$

For turbulent flow we assume to a first approximation an analogy between heat and mass transfer. We assume that the elements (eddies) of liquid moving in the fluid are responsible for heat transfer

$$\mathrm{St}_{D\mathrm{fl}} = \mathrm{St}_{\mathrm{fl}} \tag{5.25}$$

The enthalpy h_w of the liquid at the wall is defined as the sum of the saturation enthalpy and the heat of superheating. The enthalpy of the liquid in the flow core shall be defined in terms of the actual mean value of vapor quality in the given section. The starting set of equations can be written in the form:

$$C_{\mathrm{sat}} j_v / \rho_0 w_0 \, (C'_{\mathrm{sat}} - \bar{C}'_0) = (q_w - j_v r) / \overline{\rho_0 w_0} \, (h'_w - \bar{h}'_0) \tag{5.26}$$

$$\bar{C}'_0 = C^*_{0\mathrm{I}} / (1 - \tilde{x}_0) \tag{5.27}$$

$$h'_w = h' + c'_p \Delta T_w \tag{5.28}$$

$$\Delta T_w = q_w / \alpha \tag{5.29}$$

$$\alpha = \alpha(p, q_w, \overline{\rho_0 w_0}, \tilde{x}_0, \ldots) \tag{5.30}$$

$$\bar{h}'_0 = (\bar{h}_0 - h' \tilde{x}_0) / (1 - \tilde{x}_0) \tag{5.31}$$

It was assumed that the salt precipitates at the wall when the concentration in the liquid phase at the wall attains the saturation concentration, $C'_w = C'_{\mathrm{sat}}$. Concentration $C_{0\mathrm{I}}$ then attains the limiting value $C^*_{0\mathrm{I}}$. Solving this set of equations for $\tilde{C}^*_{0\mathrm{I}}$, we have

$$\tilde{C}^*_{0\mathrm{I}} = (1 - \tilde{x}_0)(1 - \tilde{Z}) \tag{5.32}$$

where

$$\tilde{C}^*_{0\mathrm{I}} = C^*_{0\mathrm{I}} / C'_{\mathrm{sat}} \qquad \tilde{Z} = [(h'_w - \bar{h}'_0)/r]\,[\tilde{j}/(1 - \tilde{j})] \qquad \tilde{j} = i_v / j_{v.\mathrm{cr}}$$

$j_{v.\mathrm{cr}}$ is the flux of vapor from the wall in the dryout location. We assume that $j_{v.\mathrm{cr}} = q_w/r$, since at the dryout location $\Delta h'_w \gtrsim r$.

It follows from Eq. (5.32) that, to determine the limiting concentration $\tilde{C}^*_{0\mathrm{I}}$, one must have information on j_v. At present such information is virtually non-existent. Some experimental data for j_v are available for natural convection conditions at low pressure. There are no such data for forced convection conditions. It is hence impossible to use Eq. (5.32) for calculating $\tilde{C}^*_{0\mathrm{I}}$.

On the other hand, the available data for $\tilde{C}^*_{0\mathrm{I}}$ can be used to determine the amount of heat utilized in generating vapor at the wall in forced convection of a steam-water mixture. Solving these equations for j_v, we have

$$j_v = \frac{q_w \{1 - [C^*_{0I}/C'_{sat}(1 - \bar{x}_0)]\}}{(h'_w - \bar{h}'_0) + r \{1 - [C^*_{0I}/C'_{sat}(1 - \bar{x}_0)]\}} \tag{5.33}$$

Data on j_v allow one to correlate data and develop a method for calculating the limiting initial impurity concentration $\tilde{C}^*_{0I}$. For this we shall use certain results of the asymptotic theory of the turbulent boundary layer. It was shown by Kutateladze and Leont'yev [5.25] that, for such a boundary layer, the ratio of the Stanton number for heat and mass transfer for the boundary layer to that for fully developed flow is given by:

$$\Psi = (St/St_0)_{Re_T}^{**} = \left[\int_0^1 (\tilde{\rho}\tilde{q}_0/\tilde{q} \cdot \partial\omega/\partial\xi_T \cdot \partial\vartheta/\partial\xi_T)^{1/2} d\xi_T \right]^2 \tag{5.34}$$

To a first approximation, flow in a tube with vapor generation on the heating surface may be represented as the flow of a homogeneous fluid in a tube with permeable walls through which the liquid is removed. The rate of liquid removal is assumed to be equal to j_v. The maximum value, $j_{v.cr}$ will, in the assumed model, define the critical conditions with maximum coefficient of heat transfer to the liquid.

For the case of suction the limiting law of heat transfer is

$$\sqrt{\Psi} = \int_0^1 d\vartheta/\sqrt{1 - b\vartheta} \tag{5.35}$$

where $b = j_v/\rho_0 w_0 St$ is the permeability parameter.
Integrating Eq. (5.35)

$$\Psi = 4/[\sqrt{1 - b} + 1]^2 \tag{5.36}$$

or

$$b = 4(\sqrt{\Psi} - 1)/\Psi \tag{5.37}$$

we shall find the ratio of the current value of the permeability parameter to the critical value of this parameter

$$b/b_{cr} = (j_v/\rho_0 w_0 St)/(j_{v.cr}/\rho_0 w_0 St_{cr}) = j_v/j_{v.cr} St_{cr}/St \tag{5.38}$$

Since

$$b = 4(1 - \sqrt{St/St_0})/(St/St_0) \tag{5.39}$$

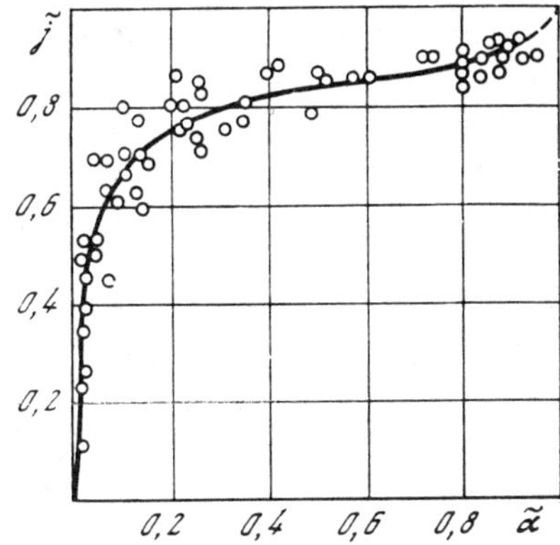

Figure 5.24 Plot of $\tilde{j} = f(\tilde{\alpha})$ [5.24].

$$b_{cr} = 4(1 - \sqrt{St_{cr}/St_0})/(St_{cr}/St_0) \tag{5.40}$$

Replacing St by α, we obtain

$$j_v/j_{v.cr} = f((1 - \sqrt{\alpha/\alpha_0})(1 - \sqrt{\alpha_{cr}/\alpha_0})) \tag{5.41}$$

Naturally, the form of the functional relationship in this case is unknown.

Equation (5.41) makes it possible to correlate available experimental data on C_{0I}^* and to develop an engineering method for calculating mass transfer in steam-generating channels.

5.4 METHOD FOR CALCULATING MASS TRANSFER IN STEAM–GENERATING CHANNELS

Equation (5.41) contains the quantity α_0, which has not been previously defined. This parameter is the coefficient of heat transfer under standard conditions. If it is assumed that this coefficient is equal to the coefficient of heat transfer at the start of subcooled boiling, $\alpha_0 = \alpha_{isb}$, then, if experimental data on coefficients of heat transfer and limiting values of the [impurity] concentrations are available, the above formula can be validated.

Figure 5.24 shows a plot of $\tilde{j} = f(\tilde{\alpha})$, where $\tilde{\alpha} = (\sqrt{\alpha} - \sqrt{\alpha_{isb}})/(\sqrt{\alpha_{cr}} - \sqrt{\alpha_{isb}})$. This plot was obtained by analyzing data over a wide range of variables: $p = 0.12$–16.7 MPa, $q_w = 291$–870 kW/m^2, $\overline{\rho_0 w_0} = 1000$–3500 kg/m$^2 \cdot$ sec. The vapor quality ranged from that at inception of subcooled boiling x_{isb} to the critical (dryout) quality x_{cr}. The values of α at high pressures were taken from data of Tarasova et al. [5.23] with the pressure correction $(p/16)^{0.43}$ and with correction for changes in the location of inception of subcooled boiling. The flow temperature was calculated from the equation

$$T_0 = T_w - q_w/\alpha \tag{5.42}$$

The value of T_w was determined according to recommendations of the first of the present authors with his coworkers [5.26]. For $p = 0.12$ MPa the values of α and the flow temperature were determined according to data of Romanovskiy [5.5]. The maximum values α_{cr} for this pressure were determined taking account of data obtained by Sterman [5.27]. The values of $\tilde{j}_v$ were determined from Eq. (5.33) from the measured C_{0l}^* values.

The experimental data are rather satisfactorily correlated, over the full range of variables covered, by a single curve. Figure 5.24 presents data for $\tilde{j}$ only up to 0.9. This is because at higher $\tilde{j}$ even an insignificant change in $\tilde{j}$ (within the limits of experimental accuracy) results in a significant change in $\tilde{Z}$ [see Eq. (5.32)]. The establishing of a relationship between heat and mass transfer is of great scientific and practical significance. Study of mass transfer in steam-generating channels started only recently, whereas heat transfer has been investigated rather fully. Consequently, using this expression, it is possible, on the basis of data on heat transfer (i.e., knowing α) to determine mass transfer conditions, or conversely, knowing the nondimensional vapor flux, it is possible to calculate α. This allows a significant reduction in the number of experiments and reduces them basically to individual experimental checks.

The relationship illustrated in Fig. 5.24 makes it possible to suggest the following method for calculating the limiting concentration below which impurities are not precipitated on the wall. The values of α_{isb}, α and α_{cr} are determined from specified values of p, $\rho_0 \overline{w_0}$, q_w and $\bar{x}_0$. The value of α_{isb} is calculated from the expression

$$\alpha_{isb} = 0{,}023\ \frac{\lambda'}{D}\ \mathrm{Re}^{0{,}8}\ \mathrm{Pr}^{0{,}43} \tag{5.43}$$

from parameters at the threshold of the inception of subcooled boiling. The cross section where this inception occurs is determined from published data, for example, from the data of Tarasova [5.28].

The values of α are determined from data of [5.29]

$$\alpha = \alpha^* \sqrt{1 + 7 \cdot 10^{-9}(w_{mix} r \rho' \, 3600/q_w)^{1{,}5} (0{,}7\alpha_{p.b}/\alpha^*)^2} \tag{5.44}$$

where

$$\alpha^* = \sqrt{\alpha_{isb}^2 + (0{,}7\alpha_{p.b})^2} \tag{5.45}$$

$$\alpha_{p.b} = 3q_w^{0{,}7} (p^{0{,}14} + 1{,}83 \cdot 10^{-4} p^2) \tag{5.46}$$

The value of α_{cr} is also calculated from Eq. (5.44). The mixture velocity w_{mix} is determined at $\bar{x}_0 = \bar{x}_{cr}$. The latter is calculated in accordance with [5.29].

Knowing α_{isb}, α and α_{cr}, and consequently $\tilde{\alpha}$, we can determine $\tilde{j}$ from Fig. 5.24. Substituting the value of $j_v/j_{v.cr}$ in Eq. (5.33), we find the value of $\tilde{C}_{0l}^*$; h' is found from recommendations by Miropol'skiy [5.30].

A special series of experiments was performed to check the above method. These were performed over a range of variables for which no mass transfer experiments had been performed. The conditions selected were: (1) p = 1.86 MPa, q_w = 350 kW/m^2, $\overline{\rho_0 w_0}$ = 1620 kg/m$^2 \cdot$sec; (2) p = 6.9 MPa, q_w = 250 kW/m^2, and $\overline{\rho_0 w_0}$ = 1000 kg/ m$^2 \cdot$sec.

Figure 5.25 shows the results of these experiments in coordinates $\bar{C}_{0l}^*/C_{sat}$ = $f(\bar{x})$. Heat transfer data for the analytic calculations were taken from the paper by Dzarasov [5.31], whereas data on the critical vapor quality was taken from reference [5.29]. As seen from Fig. 5.25, the experimental points fit rather satisfactorily the analytic curves. The agreement for p = 6.9 MPa is somewhat poorer which is possible due to the fact that $\bar{x}_{cr}$ according to references [5.29] and [5.31] differ somewhat from one another.

Accordingly, the suggested method has been checked out over a rather wide range of variables. A shortcoming of the method is the difficulty in determining α_{cr}. This occurs because Eq. (5.44) does not reflect with sufficient accuracy the manner in which α varies as a function of the vapor quality. Under conditions for which reliable experimental data, including those on α_{cr}, are available it is best to use experimental results.

Equation (5.32) can in principle also be used for determining the threshold salt concentration at the tube inlet (corresponding to inception of deposits) for the case when the heat flux distribution along the tube is arbitrary. However, when these calculations are performed, one must have information on the effect of the heat flux distribution on the values of the local heat transfer coefficient. Unfortunately, such information is presently not available. Given the insensitivity of the properties of the turbulent boundary layer to changes in boundary conditions, it can be assumed that the heat transfer coefficient is controlled solely by the local variables at the given tube location. Equation (5.32) can then be used for determining the threshold inlet concentration for an arbitrary lengthwise distribution of the heat flux.

5.5 FEATURES OF MASS TRANSFER
IN THE TRANS– AND POST–DRYOUT REGION

As previously noted, post-dryout heat transfer may occur in nuclear power plant heatexchangers. If this happens at low vapor qualities and high heat fluxes (i.e., the critical phenomenon is "boiling crisis of the first kind" such as is found in transient conditions in the core of water-cooled nuclear reactors), then the rise in the metal temperature clearly exceeds that permissible (even for a short time) and then the problem of corrosion does not arise at all. In cases when the temperature rise following dryout does not exceed the short- and long-term limits established from the point of view of strength of the metal, the equipment may operate continuously under post-dryout conditions. In this case the problem of corrosion in the trans- and post-dryout regions is extremely important (as was previously noted in Chapter 1).

Under conditions when the reliability and longevity of equipment is controlled by corrosion, one must consider the possible degree of concentration of corrosive impu-

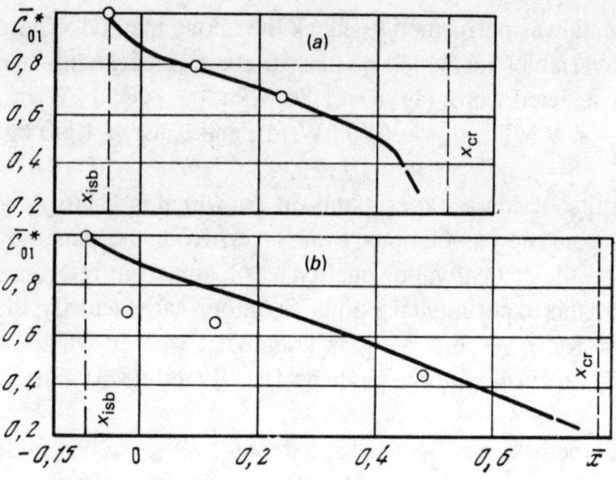

Figure 5.25 Comparison of analytic and experimental results on an impermeable wall. (a) p = 1.86 MPa, q_w = 350 kW/m²; ρw = 1620 kg/m²·sec. (b) p = 6.9 MPa, q_w = 250 kW/m², ρw = 1000 kg/m²·sec; the solid lines represent analytic results.

rities under post-dryout conditions [5.32–5.39]. It is important to note here that the most dangerous compounds are those having a high solubility with positive temperature coefficient of solubility, in particular NaCl and NaOH, since the operation of thermomechanical equipment at nuclear power plants may, in principle, be accompanied by formation of solutions of excessively high concentration, which induce active corrosion even of very stable materials. In particular, at the pressures and temperatures characteristic of the steam generators of nuclear power plants with water-cooled reactors (p = 6–7 MPa, T_w to 310–325 °C), NaCl may form solutions with concentrations of 50%, whereas under conditions prevailing in the reheaters of such nuclear power plants (p = 0.5–1.2 MPa, T to 260 °C) solutions with concentrations to 30% may be found in contact with the heating surfaces.

Corrosion in NaOH solutions is even more rapid. Under conditions typical of nuclear power plants with water-cooled reactors, NaOH concentrations to ~40% are possible in the steam generator and to ~60% in the reheater (~1 MPa). The concentrations in the superheaters of reactors with liquid-metal cooling, where the pressures are of the order of 15 MPa and wall temperatures are to 520–530 °C, may theoretically be as high as ~90%. However, under real conditions the rise in concentration in the liquid phase is limited by vaporization of the NaCl and NaOH, controlled by the starting concentration of these compounds in the wet-steam flow and conditions of mass transfer between the core of the two-phase flow and the wall layer of the liquid. As shown in this chapter, under conditions of normally developing nucleate boiling, mass transfer occurs at a very high rate and the degree of concentration does not exceed 5-10%, which is entirely safe from the point of view of corrosion, since the initial mean concentrations in two-phase flows at nuclear power plants are in the worse cases not more than several tens of ppm. However, under film dryout conditions, when, as shown previously in Chapter 3, the removal of water from the film by entrainment at the wave crests by entrainment due to bubbling and by evaporation, the film tends to zero flow and the degree of concentration may be exceedingly high.

Such conditions may arise, as previously noted, in the region of "crisis of the second kind" and the post-dryout region associated with it (i.e., in once-through steam generators and superheaters). Dryout conditions for the once-through steam generators used in the Pressurized Water Reactors produced by the Babcock and Wilcox Company are typified by low mass velocities from 200 to 300 kg/m^2, moderate pressures from 5 to 7 MPa and heat fluxes of the order of 100 kW/m^2. Note that the heat flux drops steeply after virtually complete evaporation of the water, since at low-pressure, α decreases manyfold, and the temperature of the heating fluid rises only slowly in the direction of flow of the steam being heated. Reheaters in nuclear power plants with water-cooled reactors operate typically at low pressures (0.5–1 MPa), low initial percentage moisture of the two-phase flow ($<1\%$) and also very low heat flux density. In the usual two-step reheating arrangement the heating-fluid temperature in the first stage is 220–230 °C, whereas the saturation temperature on the steam side may range from ~ 150 to 180 °C, depending on the pressure. For the geometries usually employed in reheaters (transverse steam flow over tube bundles in which condensing steam flows), data on the heat transfer coefficient in the initial region, where heat transfer is enhanced by droplets precipitating on the heated surface, are scarce. However, it can be assumed that the mean wall temperature in the initial zone is only just below the condensation temperature of the heating steam. Unfortunately (as far as is known to us) no experiments have been performed on this region, particularly not on mass transfer. Thus, one must use experimental data obtained under different conditions, in particular, at constant heat flux density (electrical heating) along straight circular tubes, and, as a rule, at much higher pressures. Before using such experimental data its applicability to the conditions prevailing at nuclear power plants must be assessed.

Mass transfer in this region has been subjected to the most thorough analytic and experimental study by Mann [5.32]. Let us consider this study in somewhat more detail.

We consider the physical model of the situation illustrated in Fig. 5.26. As the two-phase flow moves along the steam-generating channel, the liquid film on the channel wall gradually thins out. At some location ($z = 0$) dryout occurs and the wall temperature starts rising. However, the liquid film does not vanish immediately. This is because the concentration of impurities in the film increases as evaporation pro-

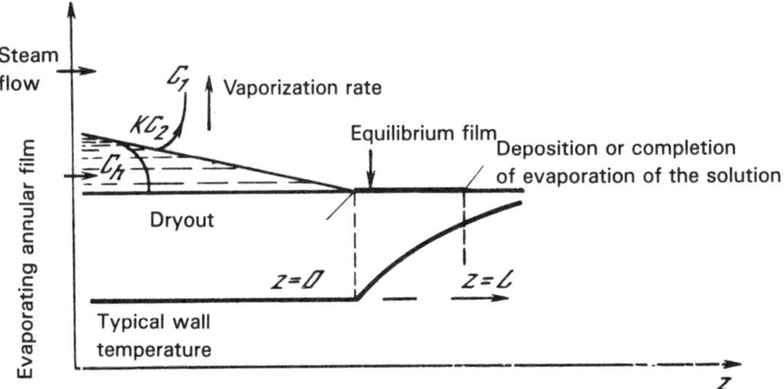

Figure 5.26 Model of mass transfer for solutions of impurities such as NaCl or NaOH [5.32].

ceeds and, consequently, the boiling temperature of the solution rises. The rise of concentration in the film is controlled by the distribution coefficient of impurities between the phases and their diffusion to the flow core. The length of the concentrated film $z = L$ is determined by the properties of the given impurity. In the case of NaCl it is limited by the inception of precipitation of the salt on the wall as a solid phase. For steam generators at $p > 0.1$ MPa the existance of a concentrated solution of NaOH in the form of a film on a wall is limited only by the length of the steam-generating channel.

The above model assumes that the concentration in the film is a function only of the wall temperature in the post-dryout region, whereas, actually, the length of the film depends on the impurity concentration and flow variables which control the solubility of impurities in the vapor. Mann investigated the case when $x_{cr} \cong 1$ and T_w is a linear function of the length.

The rate of transport of the dissolved substance from the film to the vapor was defined as [5.32]:

$$dS_1/dz = k\,(KC_2 - C_1)\,\rho_1 \pi D_{int} \tag{5.47}$$

where k is the mass transfer coefficient, and K is the distribution coefficient.

The flowrate of the impurity in the vapor is defined by the expression

$$S_1 = M_1 C_1 \tag{5.48}$$

where M_1 is the mass flowrate of the steam; for constant steam flowrate

$$dS_1/dz = dC_1/dz \cdot M_1 \tag{5.49}$$

Combining Eqs. (5.47) and (5.49) and substituting for the velocity of the steam

$$w = \frac{4 M_1}{\rho_1 \pi D_{int}^2}$$

we write Eq. (5.47) as

$$\frac{w D_{int}}{4k}\,\frac{dC_1}{dz} + C_1 = KC_2 \tag{5.50}$$

This equation can be solved if product KC_2 is known as a function of z. For a linear variation of temperature and for small temperature differences, C_2 was defined as

$$C_2 = C_{zL}\,\frac{z}{z_L} \tag{5.51}$$

where z_L pertains to the start of deposition of salt on the wall or to completion of evaporation of the film.

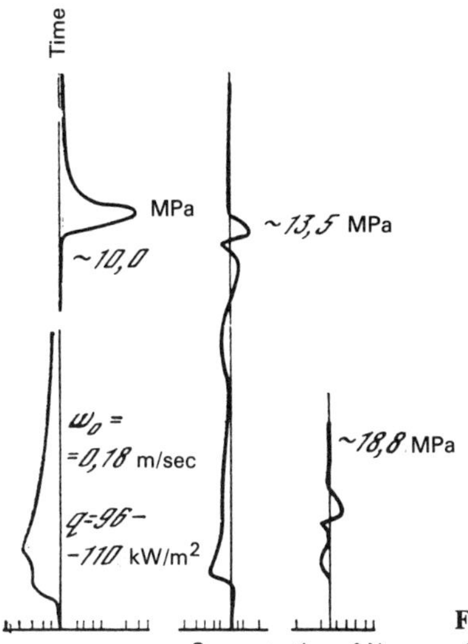

Figure 5.27 Variation in the Na concentration in the discharged steam as a function of time [5.32].

If K is regarded as constant, then the modified Eq. (5.47)

$$\frac{wD_{int}}{4k}\frac{dC_1}{dz} + C_1 = KC_{2L}\frac{z}{z_L} \tag{5.52}$$

has the solution

$$\frac{4kz_L}{wD_{int}}\frac{C_1}{KC_{2L}} = \frac{4kz}{wD_{int}} - 1 + \exp\left(\frac{4kz}{wD_{int}}\right) \tag{5.53}$$

with the boundary conditions $z = 0$ and $C_1 = 0$.

Finally, the concentration C_{1L} was obtained by substituting $z = L$ into Eq. (5.53)

$$C_{1L}/KC_{2L} = 1/A \, [A - 1 + \exp(A)] \tag{5.54}$$

where

$$A = 4kL/(wD_{int})$$

To check the above considerations Mann performed experiments with a small model of a steam generator [5.32]. The principal results of these experiments are plotted in Figs. 5.27 and 5.28. As seen from Fig. 5.28, the model describes satisfactorily the results at high pressures. At low pressures the ratio of analytic and experi-

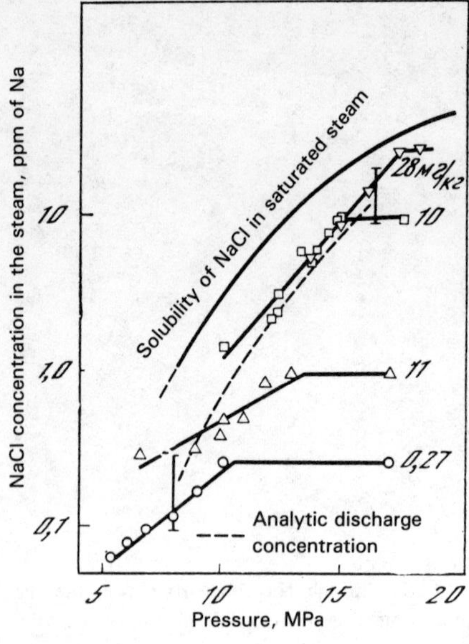

Figure 5.28 Equilibrium concentration of Na in the discharged steam as a function of pressure, for different feedwater NaCl concentrations [5.32].

mental values may be as high as 4-10. One can judge the order of magnitude of concentration in the film and of the concentration factor* from data in Table 5.1.

The above model can be used for obtaining an estimate of the upper limit of the concentration factor and for establishing validated standards for the feedwater of once-through steam generators with a post-dryout zone.

It is important to note that both the analytic and experimental values of the concentration factor were significantly lower than those obtained on the assumption of complete equilibrium—deposition started at concentrations in feedwater lower than the solubility of NaCl in the steam. The difference, based on experimental data, was only 30% at 17 MPa, but increased to 12-fold at 10 MPa. The predicted values were perceptibly closer to the equilibrium values than were the experimental ones (the

*The concentration factor is the ratio of impurity concentrations in the film to that in the feedwater.

Table 5.1

Concentration in feedwater, ppm	Maximum pressure at which deposition was observed, MPa	Solubility of NaCl in the liquid phase	Concentration factor, n	$1/n$
0.27	10.2	15	$5.5 \cdot 10^5$	$1.8 \cdot 10^{-6}$
1.0	12.5	16	$1.6 \cdot 10^5$	$6.3 \cdot 10^{-6}$
10.0	15.2	17	$1.7 \cdot 10^4$	$5.9 \cdot 10^{-5}$
28.0	17.5	18	$6.4 \cdot 10^3$	$1.6 \cdot 10^{-4}$

difference did not exceed fivefold) and also increased (although at a less pace) with reduction in pressure. Note that, under the particular experimental conditions, the kinetic factor results (particularly at low pressures) in a perceptible increase in the degree (of concentration as compared with simple calculations made on the assumption of complete equilibrium. The implication is that, when the pressure decreases further to 6–7 MPa (which is a level typical of steam generators in water-moderated, water-cooled power reactors), the role of the kinetic factor will not increase. Thus, in selecting standards for the feedwater of once-through steam generators of such reactors one must assume an approximately tenfold increase in the maximum concentration in the film as compared with the value obtained on assumption of thermodynamic equilibrium.

Evidently, such a conclusion must be regarded as preliminary, given the difference between conditions prevailing in experiments and in real steam generators, although the difference in the principal parameters in the paper by Mann [5.32] was smaller than in studies by others.

In the above study the impurity concentration in the film was calculated from the mass balance for the steam core of the flow. In the earlier published paper by Bakker and Hawtin [5.33] the problem was solved in an analogous formulation, but the conservation equations for mass were written for the liquid film. Dropping intermediate mathematical manipulations, we present the final results of [5.33]:

$$m_2/m_0 = \left(1 - \frac{\varepsilon_1}{\varepsilon_2} L\right)^{(\varepsilon_3/\varepsilon_1 - 1)} \tag{5.55}$$

where m is the molal fraction of the impurity, G_0 is the inlet liquid flowrate in the film, k_x is the mass transfer coefficient, and K_D' is the molal distribution coefficient

$$\varepsilon_1 = q/r, \quad \varepsilon_2 = G_0/\pi D_{int}, \quad \varepsilon_3 = k_x \left[1 - k_x/(k_x + \varepsilon_1 k_D)\right]$$

Using Eq. (5.55) Bakker and Hawtin calculated the changes in salt concentration in the liquid film along the superheater, and then results are listed in Table 5.2. The calculations were performed for the following conditions: pressure = 16.5 MPa, salt concentration in water = -1 ppm, molal distribution coefficient $k_D' = 9 \cdot 10^{-4}$, tube diameter $D_{in} = 1.18 \cdot 10^{-2}$ m, length of tube in the low heat flux zone (where the flux was $7.89 \cdot 10^3$ W/m^2) = 0.5 m, heat flux in the film vaporization zone = $3.79 \cdot 10^5$ W/m^2, flowrate of liquid at the start of the film $G_0 = 3.5 \cdot 10^5$ kg·mole/sec.

It should be noted in examining the analytic results that evaporation of the bulk of the liquid occurs within an extremely narrow zone. The last 10% of the film evaporates over a length of only 3 mm, whereas on the last two segments with an overall length of 0.6 mm (!) the concentration increased almost 40-fold. Under these circumstances, the diffusion of salt into the core of the steam flow is insignificant and the calculated final concentration exceeds the equilibrium value more than 300-fold. This result, which is in great contrast with experimental data of Mann [5.32], where the difference for 16.5 MPa was only 30%, can be partially attributed to the high value of q assumed in calculations for the film vaporization zone. Actually for the conditions applying in PWR steam generators, where the heating-fluid temperature rather than the heat flux is constant, the value of q is low (about 100 kW/m^2 toward the end of the nucleate boiling zone) and drops approximately tenfold after complete evaporation of moisture on the wall. Hence, the last remnants of the film will be evaporated much more slowly. In addition, the location where evaporation has terminated cannot be regarded as rigorously fixed in time (the boundary between the film and drywall region will fluctuate due to fluctuation in upstream conditions) or at a given tube perimeter. Generally, the data given in [5.33] cannot be used for conditions prevailing in nuclear power plants.

Cohen [5.34] considers two models—the Mann model [5.32] and the "rivulet" model. In the first he refined the approach to cover the case of critical vapor quality differing from unity. Calculations performed by Cohen for a steam generator with liquid-metal heating fluid and using the modified Mann model showed that the concentration factor is of the same order of magnitude as in the original Mann model [5.32].

According to the "rivulet" model, if we make allowance for transport of salt by volatility, then for the aforementioned conditions the concentration factor was calculated as $n = 6.55 \times 10^6$. For the mass transfer model used by Mann [5.32] the value $n = 4.2 \times 10^5$ was estimated and for dropwise transfer conditions the estimated value of n was 5.24×10^4. The above estimates of n were compared with experimental data on the corrosion of steel under post-dryout conditions. The experiments showed that n was equal to 5×10^4 or even less.

Pritchard et al. [5.35] performed an experimental study of relationships governing the formation of deposits and corrosion. They also present certain data on the concentration factor. They used radioactive solutions of NaCl and NaOH. Some of the data obtained is this study are listed in Table 5.3.

It is seen from Table 5.3 that the count in the post-dryout region is perceptibly higher than in the region preceding dryout. Here near the dryout zone, and directly

Table 5.2

Tube length, m	Mean mass transfer coefficient	Flowrate of liquid in the film, kg/sec	Salt concentration, ppm	Tube length, m	Mean mass transfer coefficient	Flowrate of liquid in the film, kg/sec	Salt concentration, ppm
0		$6.30 \cdot 10^{-1}$	1	9.522		$4.71 \cdot 10^{-5}$	13.38
	0.612				1.58		
0.5		$4.43 \cdot 10^{-4}$	1.423	0.523		$2.91 \cdot 10^{-5}$	21.65
	0.75				2.02		
0.51		$2.62 \cdot 10^{-4}$	2.395	0.524		$1.11 \cdot 10^{-5}$	56.71
	1.01				3.22		
0.52		$8.31 \cdot 10^{-5}$	7.585	0.5245		$2.09 \cdot 10^{-6}$	299.2
	1.25				5.98		
0.521		$6.51 \cdot 10^{-5}$	9.68	0.5246		0	2151
	1.38						

Table 5.3

	Count rate in location of			
Experimental conditions	Thermocouple No. 35	Thermocouple No. 37	Thermocouple No. 35	Thermocouple No. 37
	NaOH		NaCl	
Dryout between thermocouples Nos. 36 and 37 (50–70 mm past No. 35)	168 ± 19	642 ± 33	136 ± 17	458 ± 28
Dryout between thermocouples Nos. 42 and 43 (540 mm past No. 35)	85 ± 13	128 ± 16	132 ± 16	216 ± 19
NaOH ratio	2.06 ± 0.54	5.13 ± 0.90		
NaCl ratio			1.06 ± 0.26	2.15 ± 0.32

upstream of it, the count was higher than at significant distances from the dryout location. However, these data cannot be regarded as numerical values of rise in concentration in the wall layer, since the count was obtained not only from possible deposits but also from the concentrated solution in the wall layer and from droplets of the solution in the flow.*

As a whole, in spite of a number of contradictions in treating mass transfer in the trans- and post-dryout region, there is no doubt that severe impurity concentration occurs in this region, particularly at low pressures, where solution in the steam is made difficult by the very low solubility in steam as compared with water. It also appears probable that the concentration in the film can exceed values corresponding to equilibrium between the liquid and the mainstream steam. This is particularly so at high heat fluxes in the evaporation zone when the removal of liquid in the form of droplets leaving the film is virtually absent. In any case, at moderate pressures (as in steam generators) and low pressures (as in moisture separator/reheaters), one must take into consideration the appearance of highly-concentrated solutions even when the initial impurity content in the flow is low. To obtain more reliable data on the degree of concentration, it is necessary to perform experiments under operating conditions close to those in the relevant plant. Particular attention has to be paid to the temperature conditions in the dryout zone and in the post-dryout region immediately beyond it. This requires using heating methods which come close to those occurring in the actual power plant equipment.

*An indirect indication of the formation of deposits could be the existence of a high count sometime after metering of the radioactive salt to the flow ceased.

INVESTIGATION OF MASS TRANSFER IN CAPILLARY–POROUS STRUCTURES

A great deal of attention has been paid lately to the study of mass transfer in steam-generating systems. As previously mentioned, conditions of mass transfer in channels with an impermeable wall have been investigated in considerable detail. Reliable data are available which allow the calculation of the concentration of coolant impurities at the wall, which determines the rate of corrosion and scale formation for a relatively wide range of parameters. However, these have been only limited studies of mass transfer in steam-generating channels with permeable layer on their walls. This is not withstanding the fact that, as noted in a large number of investigations, the problem of concentration of impurities in boiling in capillary-porous (wick) structures is one of the principal problems. This problem controls to a significant measure the reliability of the operation of steam generators and boiling-type reactors.

At present there exists only a very limited volume of experimental data [6.1–6.10] on mass transfer in steam-generating channels with porous scale on the wall. It follows from studies performed by American workers [6.1–6.5] that the degree of concentration under these conditions is three to four orders of magnitude higher than in boiling on an impermeable surface. Soviet investigators [6.7–6.10] also noted a rise in the degree of concentration if iron-oxide deposits in a thin layer; however, this increase was less than one order of magnitude. The significant difference in the results is attributed to the complexity of the process under study and to differences in experimental technique.

It is known [6.11–6.15] that, recently, certain nuclear power plants in Europe,

USA and Japan have suffered massive damage to steam generatior tubes. The bulk of the damage occurred on parts of tubes situated in the vicinity of the lower tube sheet where the region of greatest temperature differences occur, i.e., in the zone of maximum possible concentration of aggressive impurities in the reactor coolant and greatest deposition of products of corrosion.

As a result of the extensive damage to the steam generators, a large number of 300–500 MW units had to be taken off line. Repairing steam generators contaminated with radioactive corrosion products is a lengthy process; the units were out of operation for up to 6 months, and sometimes even longer. Thus, damage to steam-generator tubes results in very high loss in power production. The most common cause of damage to steam-generator tubes is inter-crystalline corrosion. Inconel-600 alloy, in which the tubes in many steam generators are fabricated, is subject to corrosion cracking at alkali concentrations from 10 to 40%, i.e., corrosion of the metal occurs due to a significant concentration (by several orders of magnitude) of water impurities. In a properly designed circulation loop and with normal operating conditions such degrees of concentration are impossible. However, if there is inadequate circulation, or if dryout occurs, then "dry patches" will appear. Water droplets that impinge on the slightly superheated surfaces of the dry patches ($\Delta T = 30\text{--}40\,^{\circ}\mathrm{C}$), may be evaporated to a concentration of approximately 20–30%, resulting in intense corrosion.

A similar physical situation may come about also in very small clearances between the tube and the tubesheet, and also in the thick layer of porous deposits collecting on the tubesheet. In this case liquid is supplied only by the capillary head and mass transfer is made difficult. The mass transfer limitation will become particularly significant when the capillary passages become very long (hundreds of equivalent diameters). In fact the existence of such capillary passages is confirmed by the latest data, where it is noted that the thickness of the layer of deposits on the tubesheet is as high as 10–20 cm. The corrosion of tubes was quite significant in these cases [6.15].

The above confirms the importance of the study of mass transfer in the course of steam generation. The possibility of a significant deterioration of mass transfer is not confined to operating conditions of the steam generators mentioned above. The problem under study is a generic one, for which reason the questions discussed here are timely also in designing other types of equipment.

The increasingly large ratings of nuclear and conventional power plants puts ever increasing requirements to their reliability and continuity of operation. To prevent possible breakdown in the operation of steam-generating facilities, it is necessary to perform a comprehensive cycle of studies of mass transfer in boiling in capillary-porous structures. For this it is necessary to perform experiments as close as possible to those prevailing in operating power plants. In addition it is useful to perform simulation experiments under conditions modelling the principal features of the capillary-porous layers found in practice. The first type studies make it possible to obtain specific recommendations regarding the design and physico-chemical operating conditions of specific power generating units. The second group of studies aid in gaining deeper insight into the process, in developing a model, in obtaining a mathematic description and thus, to frame general recommendations on the physico-chemical conditions of operation of steam-generating surfaces.

6.1 EXPERIMENTAL STUDIES OF MASS TRANSFER IN STEAM–GENERATING CHANNELS WITH POROUS IRON–OXIDE SCALE

Very little has been published on mass transfer in steam-generating channels with porous scales. No systematic study of this problem has been undertaken in the USSR or in the West. One of the rare studies of the problem was the experimental investigation of Picone et al. [6.1]. The experiments were performed with a closed circulation loop; most of the tests being carried out at $p = 12.3$ MPa, $\rho w = 2280$ kg/m$^2 \cdot$sec, heat fluxes up to $q = 1250$ kW/m^2 and with quality $x = 0$. An oxide scale was deposited on the test rod by boiling on it (in a separate test loop) water which contained iron hydroxide as an impurity.

Radioactive Na^{22} was used as an indicator of the degree of impurity concentration (concentration ratio) at the wall. This isotope has a sufficiently large half-life and high gamma radiation energy. In addition, the use of sodium in the loop does not interfere with the required water regime of the facility, since there was an intention to use NaOH together with boric acid for controlling the chemical properties of the reactor coolant circuit being simulated. Such measurements using an external counter made it possible to determine the variation in the amount of Na^{22} near the wall in the course of the experiment. The measurements were made over a small tube segment, whose length was determined by the collimator dimensions. It was impossible to determine whether the Na is present in the form of a concentrated solution in the pores of the scale or in the form of the solid phase on the wall or within the iron oxide scale layer.

The degree of concentration n was calculated as follows. The activity D_{flow} of the flow having a concentration C_0 in control volume V was determined from the expression

$$D_{flow} = k\,(VC_0)_{flow} \tag{6.1}$$

the overall activity taking account of concentration in the wall layer is given by

$$D_\Sigma = \kappa\,(VC_0 + FC_{surf})_{dep}, \tag{6.2}$$

where K is a constant, F is the surface area of the deposits and C_{surf} is the surface concentration, equal to

$$C_{surf} = \left(\frac{D_\Sigma}{D_{flow}} - 1\right)\frac{V}{F}\,C_0 \tag{6.3}$$

The volumetric concentration C_w of Na^{22} in the iron oxide deposits was defined as:

$$C_w = C_{surf}/V_{int}\rho \tag{6.4}$$

where $V_{int} = \epsilon\delta_{dep}/2$. It was assumed in calculating the internal volume V_{int} (pore volume) that salt concentration occurs within one half of the thickness of the deposits layer ($\delta_{dep}/2$), and that the porosity is $\epsilon = 0.70$.

Experiments were performed on test sections without iron oxide scales for the systems $NaOH-H_3BO_3-H_2O$; $NaOH-Na_3PO_4-H_2O$ and in sections with iron oxide scales in systems $NaOH-H_3BO_3-H_2O$ and $NaOH-H_2O$. The thickness of the scale ranged from ~ 2.5 to $165\ \mu m$.

In the first series of experiments with non-scaled tubes (the $NaOH-H_3BO_3-H_2O$ system) no Na^{22} deposits were detected over the entire range of conditions under study which included both subcooled and saturated nucleate boiling. The same theme applied in the second series of experiments (the $NaOH-Na_3PO_4-H_2O$ system) at low PO_4 concentrations. Only when the PO_4 concentration was raised to 530 ppm, did Na^{22} start precipitating on the walls. Picone et al. [6.1] note here that the deposits were not only in the form of Na_3PO_4, but also in the form of other compounds, which did not readily dissolve in the water used for washing the loop.

In the first experiment with scaled tubes the thickness of the iron oxide scale on heater 1 (Fig. 6.1) was $12.7\ \mu m$ and $q = 665\ kW/m^2$ and $p = 12.5\ MPa$. The count rate of the Na^{22} counter increased with time for $T_{in} = 238\,°C$ and remained constant at $T_{in} = 121\,°C$. In experiments with experimental tube 2 the scale thickness was about $28\ \mu m$ and the experiments were performed at two heat fluxes. It is seen from Fig. 6.1 that the activity of Na^{22} in the deposits increased with the time of the experiment at a rate proportional to the heat flux. A particularly rapid growth was observed during the first one half to two hours of operation. Then the rate of accumulation of the Na^{22} decreased significantly. Figure 6.2 shows the results for washing off the deposits from test section No. 2 after termination of heating. A rather prolonged time (~ 500 min) was needed for the activity in the region with deposits to drop to the background activity level.

In runs 3 and 4, the thickness of deposits was respectively $\sim 5\ \mu m$ and $6.25\ \mu m$; no concentration of Na^{22} was observed in these runs. In run 5 the thickness of deposit was about 12.5 vm. Very little scale formation was observed during the first part of

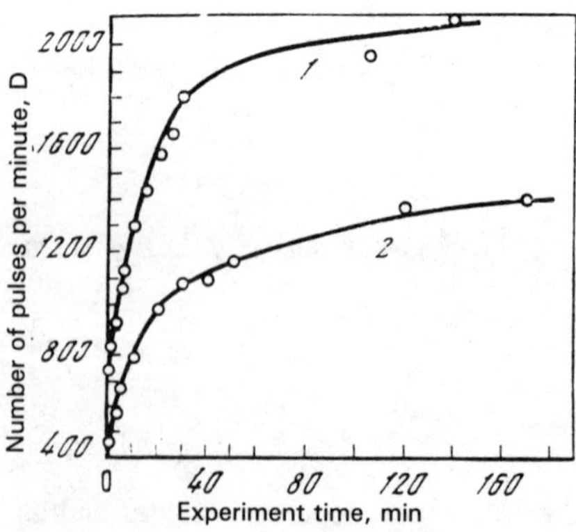

Figure 6.1 Number of count per minute vs. time at different heat flux densities [6.1]. 1, $p = 12.2$ MPa, $q = 974\ kW/m^2$; $T_0 = 295\,°C$; 2, $p = 12.2$ MPa, $q = 660\ kW/m^2$, $T_0 = 293\,°C$.

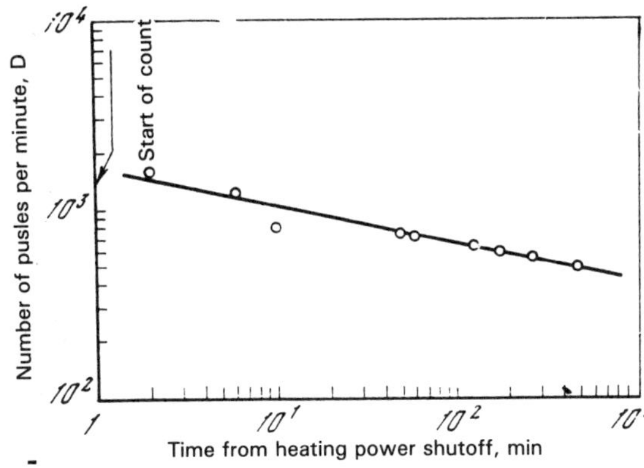

Figure 6.2 Number of counts per minute vs. the time from heating power shutoff [6.1].

the experiment. When the pressure was reduced from 12.4 to 4.4 MPa, the precipitation of Na^{22} ceased entirely. These runs were performed at two values of T_{in}. In the first case, rather insignificant deposits of Na^{22} were observed, whereas in the second there were no deposits at all.

The effect of test conditions was investigated in greatest detail in run 6. The mean thickness of iron oxide scale was 22.8 μm. When the heat flux density was changed, with the wall temperature remaining constant, the count rate rose. Increasing the wall temperature also increased the count rate. Comparison of results for test sections with different scale thickness shows that the count rate increases with δ_{dep}.

Figures 6.3 and 6.4 show the results from two parts of run 6 of the Picone et al. experiments. In the first part the wall temperature T_w was kept constant after first increasing the heat flux; in the second part (Fig. 6.5) both temperature and heat flux varied. At the beginning of this part of the test, the heat flux was reduced to a level of 6–7 kW/m^2 and subsequently increased. In the first part the pulse rate increased with time, including the period of q = const. In the second part the reduction in q was accompanied by a drop in the count rate, whereas at constant q, the count remained unchanged. The authors calculated the concentration ratio and found that $n = 8 \times 10^3$.

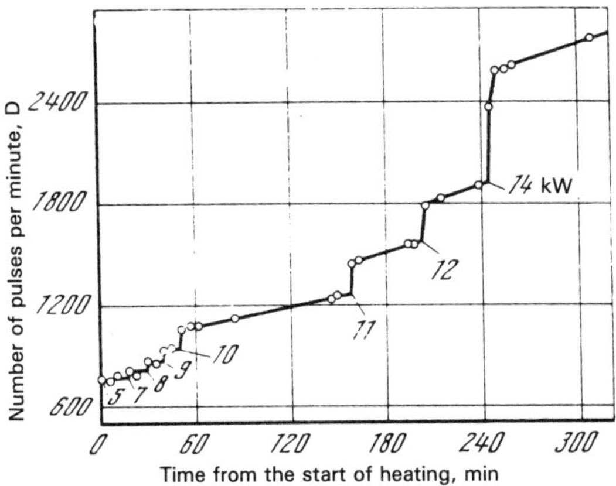

Figure 6.3 Effect of heat flux density on the Na^{22} concentration at constant heater wall temperature [6.1]; p = 12.4 MPa, T_0 = 287 °C, T_w = 330 °C.

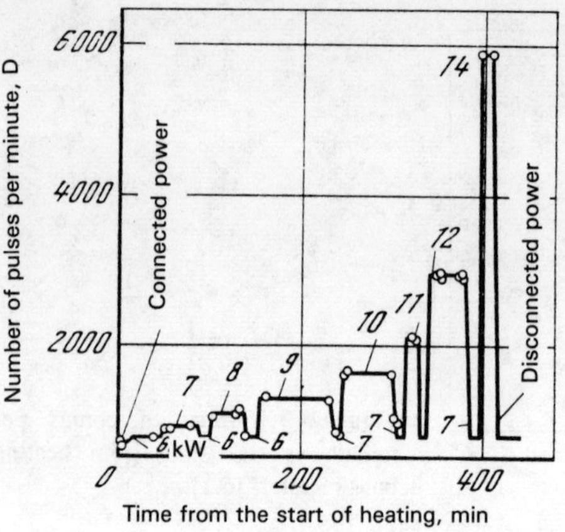

Figure 6.4 Concentrating of Na^{22} under nuclear boiling conditions at different heater wall temperatures [6.1]. Run 6, flow temperature 275 °C wall temperature 290 °C; run 7, 276 and 296; run 8, 277 and 301; run 9, 278 and 304; run 10, 279 and 311; run 11, 280 and 314; run 12, 280.5 and 320; run 14, 282 and 330 °C.

In the seventh run the deposit thickness was about 7.56 μm; this run was performed to determine the effect of removing the boric acid from the loop. The count rate at all the measurement points fell with removal of the anion. After reinjection of boric acid the count rate rose, but did not reach the previous level. As the boric acid was removed, the pH value of the coolant naturally rose. It is claimed in [6.1] that the change of pH has an effect on the degree of concentration. To confirm this claim, Picone et al., performed an eighth run in which the iron oxide scale was produced in a separate loop at high pH. The scale thickness was 15.3 μm. They could not detect any concentrating of Na^{22}. According to Picone et al., this run confirmed the hypoth-

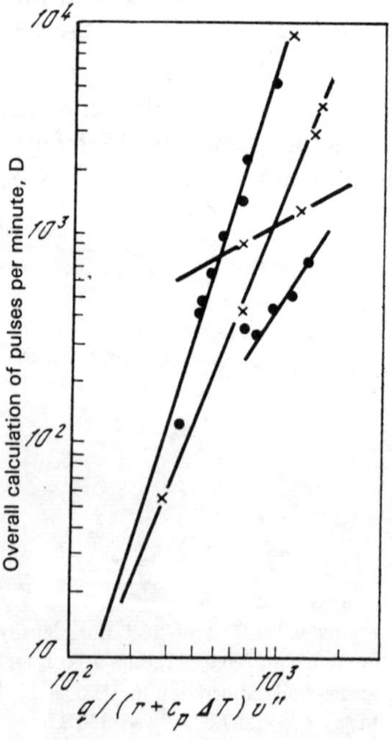

Figure 6.5 Total counting rate vs. $q/(r + C_p \, \Delta T)v''$.

esis that pH had an effect on the morphology of the scale structure and consequently on the concentration of impurities within the scale. However, they could not determine what actually happened to the scale structure at high pH.

The aforementioned experiments suggested the conclusion that the principal mechanism responsible for the increase in count rate is concentration of Na^{22} in the pores of the scale. The regime of boiling is of great importance. Other conclusions by Picone et al. [6.1] which are worth mentioning are: (1) a small difference in the scale thickness may bring about significant differences in the degree of concentration; (2) the minimum thickness that has a significant effect on the concentration ratio is of the order of 7.5–16 μm; (3) the concentration ratio increases with the rate of vaporization and with reduction in the specific volume of the steam (Fig. 6.5).

Analysis of the graphical data of Picone et al. [6.1] shows that the Na^{22} count rate increases continuously with time. The time behavior $D_\Sigma = D(t)$ of the count frequency is characteristic of conditions for scale deposition. At the same time the concentration of the indicator in solution near the wall would establish itself very rapidly. This means that the overall activity of radiation from the wall layer is determined not only by the rate of radiation from dissolved Na^{22} but also by radiation from Na^{22} deposited as a solid on the wall surface and in the pores of the iron-oxide scale. This situation is confirmed by the Picone et al.'s seventh experimental run. After boric acid was removed from the loop, the Na^{22} count started dropping and actually fell to the background values. the NaOH solution remained in the loop and one should have expected that it would concentrate in the iron oxide scale pores. However, the count rate did not rise. This implies that the increased Na^{22} count rate in the runs with boric acid was not likely to be due to solute concentration but, rather in the pores of the iron-oxide scale. It can at least be claimed that, in the experiments without boric acid, the degree of concentration did not exceed magnitudes at which saturation and precipitation of sparingly soluble sodium compounds occurred in solid form.

Another specific feature of these experiments consisted in the use of boric acid at high concentrations (more than gram per liter). As a result, only a moderate (several-fold) increase in concentration in the wall layer is needed to cause the salt content near the wall to attain its critical value C_{cr}. The mass transfer rate then decreases significantly and the degree of concentration becomes rather high. This conclusion is confirmed, in part, by experiments, also performed by Picone et al. [6.1], with sodium phosphates. At $p = 12.3$ MPa, $\rho w = 2280$ kg/m$^2 \cdot sec$, $q = 628$ kW/m^2 and $x \simeq 0$) deposits of Na_3PO_4 were observed at $C_{Na_3PO_4} = 916$ ppm. Assuming the solubility of Na_3PO_4 for these conditions to be $C_s' = 19{,}800$ ppm, we obtain a degree of concentration of $n = 21.6$. For approximately the same conditions ($p = 13.7$ MPa, $q = 580$ kW/m^2, $\rho w = 2000$ kg/m$^2 \cdot sec$, and $x \simeq 0$) data obtained at the Moscow Energetics Institute indicated that $n = 1.75$ (see Chapter 5). The latter experiments were performed at $C_w < C_{cr}$. In other experiments performed at $C_w > C_{cr}$ the Moscow Energetics Institute studies indicated a value of $n \simeq 18$, i.e., very close to that observed by Picone et al.

A significant shortcoming of the study by Picone et al., is the fact that the wall concentration C_w was calculated at a location half way through iron-oxide scale. Actually the salt would concentrate not only deep within the iron-oxide scale, but also

in its outer layers. As a result, the concentration ratio has been significantly overestimated. Another possibly significant error in their estimates of n could be due to the
fact that they did not measure the porosity, but simply assumed an arbitrary value,
$\epsilon = 0.7$.

A rather thorough study, which has significantly affected the development of
physical understanding, was that performed by Rice, Goldstein et al [6.2–6.5] on
corrosion of carbon steel steam-generating channels coated by a layer of porous iron
and copper deposits. The experiments were performed in a closed circulation loop
employing two test sections. The tubes were heated nonuniformly over their length
and perimeter, so as to simulate the characteristic profile of heat fluxes in the water-
wall surfaces of steam-generator furnace chambers. Iron and copper oxides were
added by periodic injection over several days of a suspension containing 50 grams
portions of each substance.

A large program of studies was implemented: the experiments were performed
with loop feedwater having various treatments (ammonium, phosphates and alkali),
and on surfaces coated with iron and copper oxides. In addition, for the surface
coated tubes, experiments were performed with metered injection into the loop of
fresh and sea water, and also of magnesium chloride, calcium sulfate and calcium
chloride.

It was established as a result that in the case of a clean surface, the corrosion was
rather moderate and did not exceed 0.025 mm/year. Experiments with iron and copper oxide scales showed that the corrosion rate is so high that the tubes may fail quite
rapidly (for example, in the experiments when alkali was added to the water, the tube
failed within 12 days). The rate of corrosion was seen to be proportional to the
hydrogen content of the water.

It is seen from Fig. 6.6 that the thickness of deposit increases with time, which
raises the hydrogen content at the wall. This means that the corrosion rate rises with
the thickness of deposit. The rise in wall temperature due to deposits was in general
rather small (see Fig. 6.6a). Thus, in experiments when ammonia was added to the
loop water (Fig. 6.6b) the changes in wall temperature ΔT did not exceed 3.9 °C. In
the case of alkali addition ΔT was approximately an order of magnitude higher (see
Fig. 6.6c). This is because the deposit thickness increases in this case and, in addition, the deposits contain other compounds, such as SiO_2, Ca, Al, PO_4, etc. The
presence of silicon in the scale raised the density and strength of its structure and
decreased its effective thermal conductivity.

The presence of these impurities in the scale indicates that the iron and coppe
oxide deposits have a porous structure, and the presence of the highly soluble SiO
and PO_4 points additionally to high degree of concentration in the liquid layer in
direct contact with the metal surface. Rice, Goldstein et al. [6.2–6.5] note that th
concentration of the solution on the metal-scale interface is a function of the temperature difference across the scale. In their opinion, the concentration on the meta
surface is a function of four factors: heat flux, thickness and thermal conductivity c
the scale, and concentration of the dissolved impurities.

The following additional conclusions which are important from the point of vie
of the physics of the phenomenon under study: the chemical composition of deposit

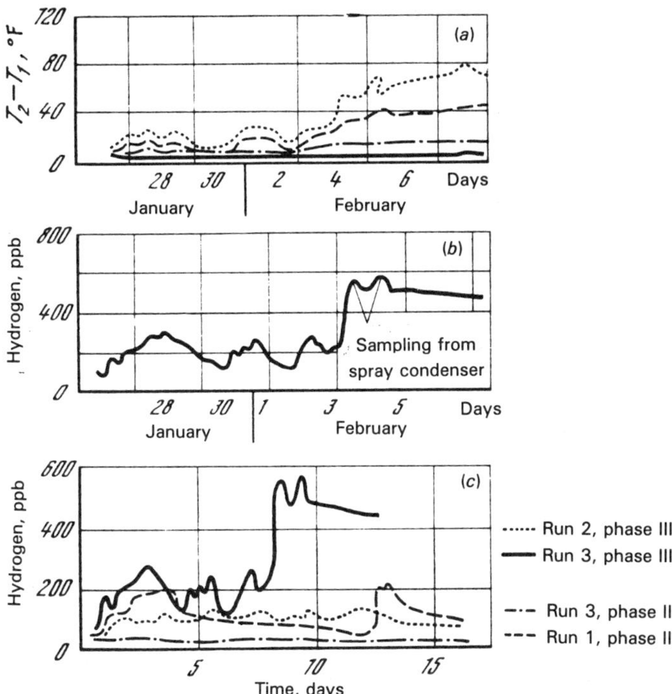

Figure 6.6 Variation in tube temperature and the hydrogen content of water (*a*) and comparison of experimental data on the corrosion rate for different feedwater treatment procedures (*b*, *c*).

on heated surfaces differs significantly from those on surfaces which are not heated; the addition of aggressive impurities (for example, addition of sea water to the loop water) significantly accelerates the corrosion; finally, addition of sodium phosphate to the loop water is an effective means for reducing the rate of corrosion of the heating surfaces.

Rice, Goldstein et al. do not present quantitative data on the degree of concentration. The values of n can be judged only from indirect data (on the rate of corrosion, on the hydrogen content of loop water) or from approximate calculations (by virtue of the complex ion composition of water), which are listed in Table 6.1.

Table 6.1

Run	Degree of concentration	Calculations based on compound	Run	Degree of concentration	Calculations based on compound
III-1	~420	SiO_2	IV-1	~1185	SiO_2
III-2	~608	Na_3PO_4	IV-8	524–876	Na_2SO_4
III-3B	~1500	Na_3PO_4			

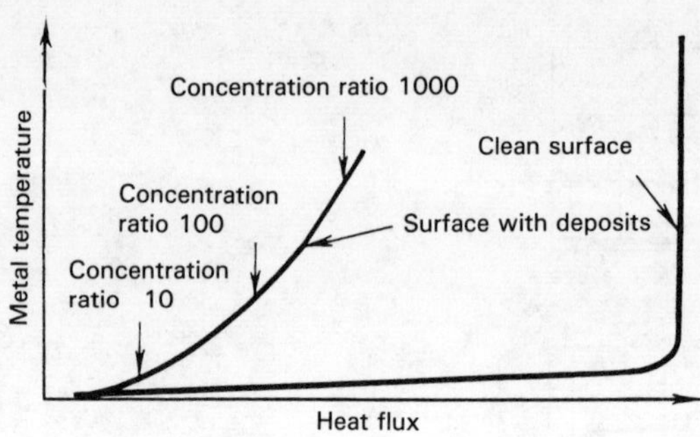

Figure 6.7 Qualitative dependence of the metal temperature of a clean surface and a surface with scales as a function of the heat flux [6.6].

The above experiments [6.2–6.5] were performed under the following conditions: p = 18.1 MPa, ρw = 750 kg/m$^2 \cdot sec$, q = 476 kW/m^2 and x = 0.3 (the information presented is for loop A for unit No. 20). The thickness of the scale was several millimeters. For approximately the same conditions (p = 16.7 MPa, ρw = 1000 kg/m$^2 \cdot sec$, q = 580 kW/m^2 and x = 0.3), data obtained at the Moscow Energetics Institute and at the Institute of High Temperatures of the USSR Academy of Sciences (see below), indicate that the concentration ratio on a wall coated by iron oxide scale is only about 7.0. It should be noted that the thickness of the iron oxide scale in the latter experiments was, at the most, 300 μm.

Castle and Mann [6.6] consider three causes of concentration of impurities in the wall layer, as a result of which the rate of corrosion of steam-generating channels may increase: (a) breakup of the annular liquid film due to hydrodynamic factors; (b) presence of cracks, bends, projections and other factors, perturbing the boundary layer structure on the steam-generating surface; (c) the existence on the steam-generating surface of porous deposits of various impurities, for example, products of corrosion of the structural materials.

Figure 6.7 shows the variation in temperature as a function of heat flux for a clean surface, and for a surface with scale. These deposits not only affect the temperature conditions under which the heating surface operates, but also induce a significant concentration of the impurities at the wall. In the opinion of Castle and Mann, the concentrations of impurities at the heating surface can exceed by several orders of magnitude their concentrations in the flow core.

According to Castle and Mann the corrosion rate is controlled by the solubility of iron in the loop water. The higher this solubility, the higher the corrosion rate. In their opinion the corrosion mechanism is based on the transport of soluble products of corrosion through the porous scale with subsequent crystallization in the surface crystalline layer. In this context, they consider the feasibility of treating the loop water with lithium hydroxide. They showed experimentally that the solubility of iron in LiOH is significantly lower than, for example, in NaOH. The above authors also note that scale formed at high heat flux regions may propagate to the regions of low q, thus widening the region in which intensive concentration of impurities at the metal surface may occur within the channel.

In the USSR, studies in this area were performed at the Moscow Energetics Institute and at the Institute of High Temperatures of the USSR Academy of Sciences

[6.7–6.10]. These experiments were performed with two circulation loops. The degree of concentration was determined by the salt method (see Chapter 5). Calcium sulfate was used as the indicator. The experiments were performed in the following manner. Under given operating conditions iron oxide deposits were accumulated on the surface of the experimental tube over a specified number of hours. The magnitude of deposits was monitored on the basis of the wall temperature gain and changes in the hydraulic drag. Then the indicator was added to the loop and the threshold of the start of salt deposition was monitored from changes in the indicator concentration.

The experiments were performed with both subcooled and quality boiling over a wide range of variables: p = 9.8–16.7 MPa, q = 580–1160 kW/m^2 ρw = 1000–3500 kg/m$^2 \cdot$ sec. The additives were iron oxides in the form of magnetite (Fe_3O_4) and hematite (Fe_2O_3). The pH of the working fluid was 8.5–9.2 or 5.0. The bulk of iron oxide consisted of particles several microns in size. The concentration of iron oxide in the flow ranged from several tenths of a ppm to several ppm. The experiments were performed at scale thicknesses which have corresponding temperature gains (ΔT) of 5 to 30 °C. In the majority of runs ΔT was around 5–10 °C and was independent of the heat flux.

There is virtually no data available on the effect of the scale thickness on the threshold at which deposition of truly dissolved impurities starts. For this reason we performed a series of experiments aimed at determining the threshold of deposition of calcium sulfate on a heating surface coated by an iron oxide layer; this layer was accumulated over 2, 4, 7 and 30 hr, respectively. The experiments were performed at p = 13.7 MPa, q = 580 kW/m^2, ρw = 1000 kg/m$^2 \cdot sec$ *and* x_{out} between -0.15 and -0.17. The heating was uniform along the tube. The results of these experiments are plotted in Fig. 6.8 as the concentration of calcium sulfate in the feedwater and the concentration ratio as a function of vapor the equilibrium vapor quality. The results shown in this figure were referred to point obtained previously on a clean surface (see Chapter 5). It is seen from these data that the threshold for deposition of calcium sulfate starts to decrease abruptly as the iron oxide scale starts accumulating, and then the decrease slows down. After approximately seven hours of scale accumulation an

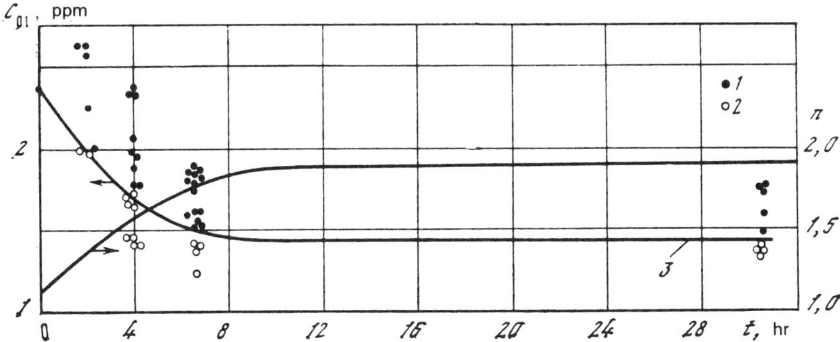

Figure 6.8 Variation in the threshold concentration C_{0I} for deposit initiation and of the concentration ratio n on heated surfaces with a porous layer of iron oxide deposits as a function of the time of accumulation of this layer. 1, Deposits present; 2, deposits absent; 3, threshold for deposition of $CaSO_4$ on a clean surface.

asymtotic level is attained which is maintained (within the limits of accuracy of determining the deposit threshold) for the rest of the 30-hr test. The behavior of n as a function of vapor quality is analogous.

The above experiments were performed in the subcooled boiling region with qualities (x_{out}) at the exit from the test section between -0.15 and -0.17. To clarify the effect of the time of accumulation of the iron oxide scale on the threshold of the start of calcium sulfate deposition in quality nucleate boiling prolonged experiments were performed at $x = 0.3$ during which iron oxide accumulation continued for 24 hr. The concentration for the onset of deposition obtained in these experiments coincided with the threshold curve obtained when the scale was accumulated for 7 to 10 hr.

To clarify the effect of the form of iron oxides on mass transfer, magnetite (Fe_3O_4) was used in addition to the hematite (Fe_2O_3). Experiments with magnetite were performed with a range of operating conditions and durations of iron-oxide scale deposition $(p, q, \rho w, t)$. The results of these experiments are plotted in Fig. 6.9. This figure also shows the results of a study of the effect of the following variations of the experimental technique: (1) adding the indicator salt to the loop simultaneously with deposition of the iron oxide scale; (2) addition of the indicator to the loop after the scale has already accumulated. The various conditions are indicated by appropriate symbols.

Analysis of experimental data leads to the conclusion that, over the range of variables under study, the form of iron oxide (magnetite or hematite) of which the scale is formed does not affect the threshold concentration for the start of deposition of calcium sulfate and on the behavior of mass transfer in a steam-generating tube. Neither the experimental technique (simultaneous or subsequent addition of $CaSO_4$) affect the final outcome of the experiments. For the main series of experiments described below, the runs were performed in stages: first the required iron oxide

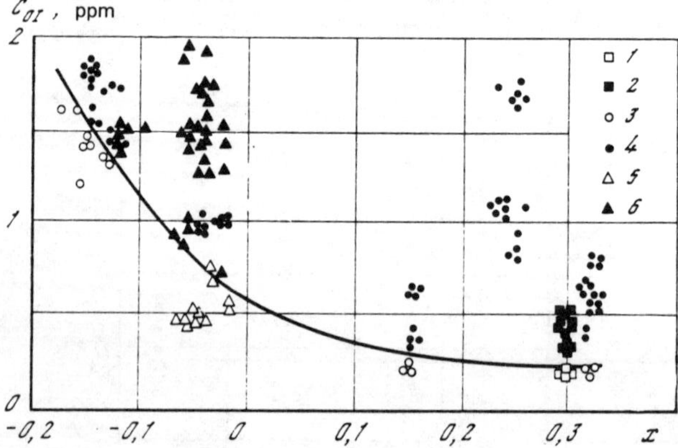

Figure 6.9 Threshold concentration for the deposition of $CaSO_4$ as a function of thermodynamic quality; $p = 13.7$ MPa, $q = 580$ kW/m^2, $\rho w = 1000$ kg/m$^2 \cdot$sec. Fe$_3$O$_4$ deposits: 1–4, addition of CaSO$_4$ to the loop after the iron oxide scale has formed; Fe$_2$O$_3$ deposits; 3–6, addition of CaSO$_4$ to the loop in the course of accumulation of the iron oxide scale.

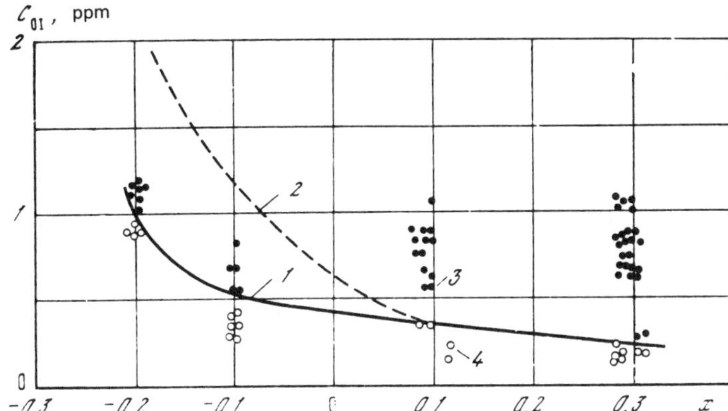

Figure 6.10 Threshold concentration for the start of $CaSO_4$ deposition as a function of the relative enthalpy; p = 13.7 MPa, q = 580 kW/m^2, ρw = 1000 kg/m$^2\cdot$sec. 1, pH = 5.0; 2, 9.2; 3, scale present; 4, scale absent.

scale was accumulated for 7 to 10 hrs, and then the electrolyzer was switched off and $CaSO_4$ was added; thereupon the threshold concentration for the start of calcium sulfate deposition was determined.

The results of the study to determine the effect of pH on C_{0I} are plotted in Fig. 6.10. Decreasing the loop water pH from 9.2 to 5.0 decreased C_{0I} in the subcooled boiling region. No effect of pH was observed for quality nucleate boiling. This means that the curve of C_{0I} = $C(x)$ for pH = 5.0 (see curve 1) is shallower than the analogous curve for pH = 9.2 (see curve 2). Note that the pH value of 5.0 in the feedwater was maintained only in the course of accumulation of the iron oxide scale. After the electrolyzers were switched off, which resulted in cessation of supply iron oxide suspension to the loop, the feedwater pH rose to 9.2. Hence the shift in the threshold value for calcium sulfate deposition in this case is a function only of the structure of iron-oxide scale obtained at pH = 5.0.

It follows from analysis of the above data that, in the presence of iron oxide deposits, the threshold concentration for $CaSO_4$ deposition decreases. The magnitude of this shift, and also the nature of variation in the threshold concentration curve is a function of the experimental conditions. At locations near the start of subcooled boiling and at the onset of dryout, the magnitude of the shift in the majority of cases is smaller than for nucleate boiling. With increasing quality, deposits occur at lower calcium sulfate concentrations in the feedwater. The slope of this curve is a function of the experimental conditions. At high heat fluxes the threshold curve for nucleate boiling is very shallow, almost horizontal. The value of the threshold concentration for calcium sulfate deposition decreases with rising pressure and with reduction in mass velocity.

Let us consider the effect of p, q and ρw on the concentration ratio. The saturation concentration in calculating n and n_c was taken as being equal to the solubility at saturation temperature; this introduces little error considering the insignificant difference between T_w and T_{sat} for the range of variables under study. The mass transfer rate on the surfaces with porous scale is significantly smaller than with a clean surface. In quality nucleate boiling the concentration ratio for water impurities in the wall layer increases 4- to 5-fold and attains, in certain cases, values of 8–10 at a sufficient distance from the dryout location, i.e., in the quality region of practical importance in steam-generating facilities.

Near the start of subcooled boiling, the deviation of the concentration ratio in a steam-generating tube with deposits from the corresponding value for a clean heating surface is insignificant. The difference increases with increasing quality. The concentration ratio changes rapidly in the vicinity of the inception of subcooled boiling (more rapidly than in the case of clean surfaces) and soon attains virtually constant values, characteristic of quality nucleate boiling. The subsequent change in n is insignificant.

The experimental results for q = 580 kW/m^2, ρw = 1000 kg/m$^2 \cdot$ sec and p of 9.8 and 13.7 MPa are plotted in Fig. 6.11. It is seen that, all other conditions remaining equal, the rate of mass transfer decreases with rising pressure. This apparently occurs because raising the pressure at the same quality reduces the linear flow velocity. This gives use to a reduction in the intensity of turbulent fluctuations, and consequently, to a decrease in the supply of liquid from the core to the wall layer. The capillary head, which ensures circulation of coolant in the scale pores, also decreases due to decrease in surface tension.

The effect of heat flux on mass transfer can be analyzed on the basis of Fig. 6.12. The effect of q on n and n_c is complex. At the start of the subcooled boiling region the concentration ratio increases steeply at high values of q and exceeds the value at small q. However, this growth then virtually ceases and the values of n at low q become higher. Apparently, the change in heat flux could affect the structure of the scale, which results in a change in the degree of indicator concentration. In addition, increasing the heat flux could also amplify the circulation of coolant through the porous scale, which reduces the degree of concentration of the indicator at the heating wall [6.9].

Analysis of the curves plotted in Fig. 6.13 shows that the rate of mass transfer increases with the flow velocity, i.e., the degree of impurity concentration in the wall layer decreases. The higher the flow velocity, the higher the intensity of turbulent fluctuations and the velocity of liquid droplets directed to the wall layer. This increases the rate of exchange of liquid between the flow core and the wall layer and aids in reducing the additive concentration in the wall layer.

Analysis of the effect of water pH on n (see Fig. 6.10) showed that reducing the pH of feedwater at which the scale accumulated on the wall from 9.2 to 5.0 was significantly detrimental to mass transfer in surface boiling. The degree of concentration at the start of the subcooled boiling region rose approximately twofold. The effect of pH decreases with increase in quality, and there was no effect of pH during scale deposition on the mass transfer results obtained for quality nucleate boiling. These results are a qualitative confirmation of the results of Picone et al. [6.1], obtained for pH = 8.5 and 10. At pH = 8.5 the Na^{22} concentration rose, whereas at pH = 10 this did not happen in spite of the fact that the scale thickness was greater.

The above experiments were performed with heating flux constant along the

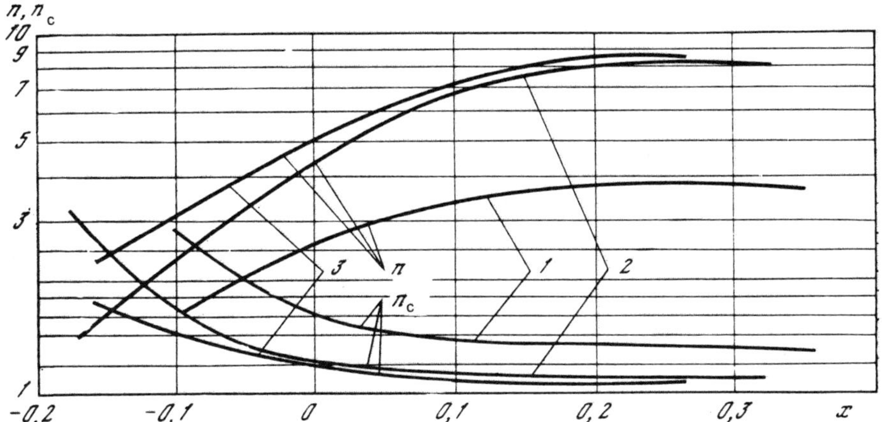

Figure 6.11 Effect of pressure on mass transfer behavior in the presence of iron-oxide scale; $q = 580$ kW/m^2, $\rho w = 1000$ kg/m$^2 \cdot$sec. 1, $p = 9.8$ MPa; 2, 13.7; 3, 16.7.

channel. Under plant conditions this flux is variable. For example, in nuclear reactor channels, the heat flux distribution is approximately cosinusoidal. It is advisable to consider the effect on mass transfer of the heat flux distribution along the channel.

Figure 6.14 shows experimental results for the case of a cosinusoidal variation in heat flux along the channel. The test section was ohmically heated and the cosinusoidal distribution was approximated by using six-segment heating. The minimum heat flux density (segments I and VI) was maintained at 330 kW/m^2, the maximum (segments III and IV)—at 582 kW/m^2. The heat flux density in segments II and V was 437 kW/m^2 [6.10].

The deposition of iron oxides and of the CaSO$_4$ which served as the indicator was monitored from the gain of wall temperature and from chemical analyses. Due to the insignificance of calcium sulfate deposits in segments I and II, the concentration of the sulfate at the start of segments III and IV was virtually equal to the feedwater

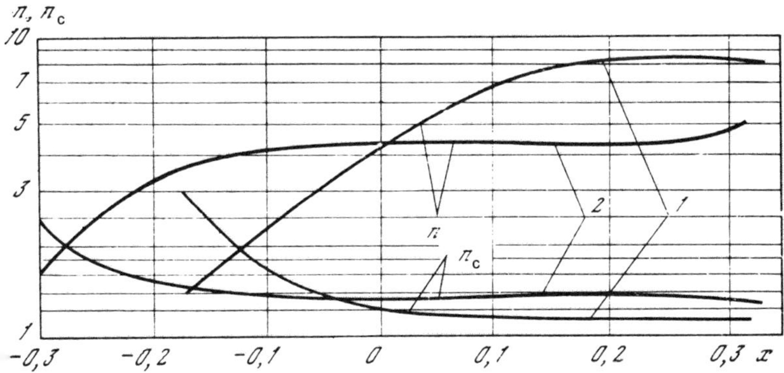

Figure 6.12 Effect of heat flux on mass transfer behavior in the presence of iron-oxide scale, $p = 13.7$ MPa, $\rho w = 1000$ kg/m$^2 \cdot$sec. 1, $q = 580$ kW/m^2; 2, 872 kW/m^2.

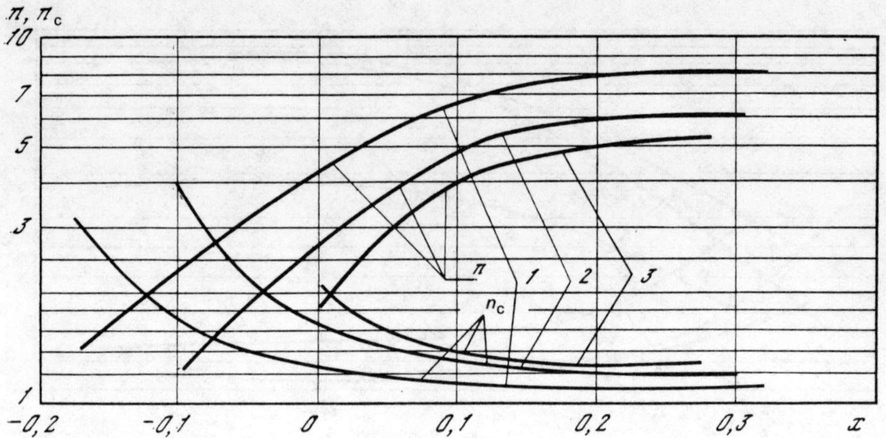

Figure 6.13 Effect of mass velocity on features of mass transfer in a steam-generating tube in the presence of iron-oxide scale; $p = 13.7$ MPa, $q = 580$ kW/m^2. 1, $\rho w = 1000$ kg/m$^2 \cdot$ sec; 2, 2000; 3, 3500.

concentration C_{0I}. In experiments with high vapor qualities the rise in wall temperature due to iron oxide deposits was $\Delta T = 3-5\,°C$. The thickness of these deposits was approximately the same as in the majority of experiments with uniform heat flux distribution. In experiments with low vapor quality the change in wall temperature ΔT_{Fe} due to iron-oxide deposition was $11-12\,°C$ in the "hot" part of the channel (segments III and IV). This experiment also had another purpose—to check the effect of iron oxide deposits on the threshold subcooled boiling quality x_{isb}.

The gain ΔT_{Ca} in wall temperature due to calcium sulfate deposition was $2-12\,°C$. The time for sulfate accumulation ranged from 3.5 to 8.5 hr. The experiments were performed at low sulfate concentrations (see Fig. 6.14, the open symbols correspond to absence of $CaSO_4$ deposition) and at high concentrations (in the same figure the blackened symbols correspond to $CaSO_4$ deposits).

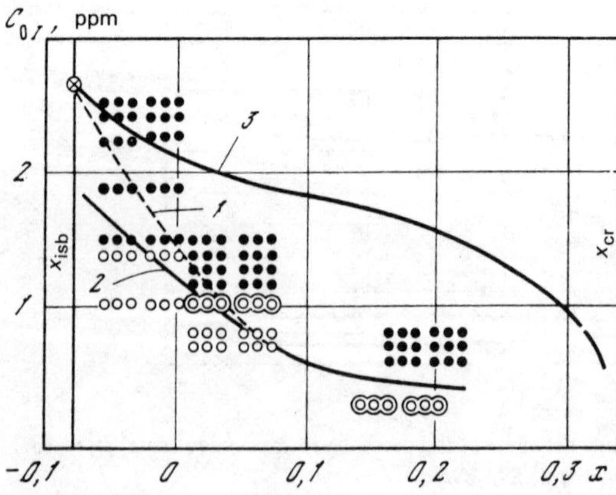

Figure 6.14 Plot of $C_{0I} = C(x)$ in a cosinusoidally heated channel at $p = 13.7$ MPa, $\rho w = 3500$ kg/m$^2 \cdot$ sec, $q = 580$ kW/m^2.

Figure 6.14 shows data for the "hot" part of the test section (q_{hot} = 580 kW/m^2), the threshold curve 1 for analogous conditions from experiments with uniform cooling and threshold curve 3 from experiments with a clean wall. Curve 2 drawn from experiments with cosinusoidal heat flux distribution at $x < 0$ lies to the left of curve 1. At $x > 0$ these curves virtually merge. Comparison of data at $x < 0$ can only be qualitative. In the non-uniform heat flux experiments, the thickness of the iron oxide scale was significantly greater than under similar conditions with uniform heating. As a result of the threshold concentration at the inception of subcooled boiling shifted to the left (to the region of lower qualities). In experiments with uniform heating at $x \cong x_{\text{isb}}$ the effect of deposits on x_{isb} has not been specifically investigated. The threshold curve, extrapolated to point x_{isb} corresponding to inception of subcooled boiling on a clean wall, is somewhat arbitrary. For exact calculations, allowance must be made for the shift in the location of boiling inception, which may be significant in the case of high ΔT_{Fe}.

The agreement between results for different heat flux density distributions along the channel ($x > 0$) points to the independence of parameter C_{01}^* on the heat flux distribution. Naturally, this conclusion, as in the case of a wall without a porous deposit layer (see Chapter 5) pertains only to the part of the channel in which the heat flux rises (segments I and IV). In the present case the physical situation for ensuring independence is more favorable. This is because boiling occurs in isolated elementary cells of the porous scale. There is no flow along the surface of the channel (within the scale layer) and the effect of flow history in this case should not manifest itself.

Figure 6.15 shows all the results of experiments with iron-oxide scale n in coordinates $n = n(X)$, where

$$X = \frac{x - x_{\text{isb}}}{x_{\text{cr}} - x_{\text{isb}}}$$

The value of X ranges from 0 to 1. When $x \to x_{\text{isb}}$, $X \to 0$, and when $x \to x_{\text{cr}}$ then $X \to 1$. This parameter is remarkable in the fact that it makes it possible to

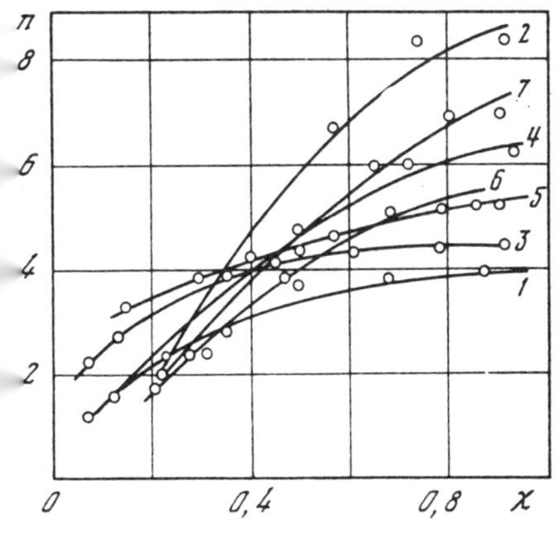

Figure 6.15 The degree of concentration as a function of χ. 1, p = 9.8 MPa, ρw = 1000 kg/m$^2 \cdot$sec, q = 580 kW/m^2; 2, 13.7 MPa, 1000 kg/m$^2 \cdot$sec, 580 kW/m^2; 3, 13.7 MPa, 1000 kg/m$^2 \cdot$sec, 870 kW/m^2; 4, 13.7 MPa, 2000 kg/m$^2 \cdot$sec, 580 kW/m^2; 5, 13.7 MPa, 2000 kg/m$^2 \cdot$sec, 1160 kW/m^2; 6, 13.7 MPa, 3500 kg/m$^2 \cdot$sec, 580 kW/m^2; 7, 16.7 MPa, 1000 kg/m$^2 \cdot$sec, 580 kW/m^2.

present experimental results obtained under significantly different conditions in convenient coordinates.

The value of n (see Fig. 6.15) changes significantly in the vicinity of $x = x_{isb}$, and then changes little. If one uses estimates of the effective thermal conductivity of deposits [6.16, 6.17], then in the experiments under study the thickness of the iron-oxide deposits δ_{dep} did not exceed 300 μm. Assuming $D_{s.c}$, the diameter of the "steam channels" found in the porous deposit, as the characteristic dimension, we can calculate the nondimensional thickness of the deposit $\bar{\delta} = \delta_{dep}/D_{s.c}$. According to data of Macbeth et al. [6.16], $D_{s.c} = 5$ μm, whereas according to those of Rassokhin et al., it is 10 μm. Consequently, $\bar{\delta}$ ranges between 30 and 60. The calculations show that, for those tests, $n < 10$. From this follows an important practical conclusion. At $p = 9.8$–16.8 MPa, $\rho w = 1000$–3500 kg/m$^2 \cdot sec$, $q = 1160$ kW/m^2 and $x_{isb} < x < x_{cr}$ and iron oxide scale thicknesses $\tilde{\delta} < 60$ the concentration ratio does not exceed an order of magnitude. The data of Picone et al. [6.1] obtained for conditions included in the above range are very much on the high side.

6.2 INVESTIGATION OF MASS TRANSFER
IN ARTIFICIAL CAPILLARY–POROUS STRUCTURES

Analysis of published data thus indicates that contradictions exist in results for concentration of impurities in boiling in channels with thin porous deposits of products of corrosion on walls. According to data of Picone et al. [6.1], the concentration ratio in a layer of iron oxide deposits (where $\tilde{\delta}$ amounts to several units) may be four orders of magnitude. On the other hand, according to the first of the present authors with his coworkers [6.10], under approximately the same conditions the concentration ratio does not exceed a single order of magnitude ($n < 10$). Picone et al. note that the minimum thickness which has a significant effect on the degree of concentration is $\delta_{dep} = 7.5$–10 μm. Let us again take diameter $D_{s.c}$ within the deposit layer as the scale for calculating the nondimensional thickness. Then the minimum nondimensional thickness at which the deposits have a significant effect on the concentration ratio is $\tilde{\delta}_{min} = 1$–2. It is advisable to check this using a capillary-porous body of different structure, whose relative thickness is of approximately the same order of magnitude. It is desirable that this structure would not change in the course of the experiment, i.e., that it retains a constant geometry.

In the experiments described here we used a filter mesh of stainless steel wire with effective cell diameter $D_{ef} = 0.2$ mm and nondimensional thickness $\tilde{\delta} = 2$ as such a structure. The experiments were performed under forced convection conditions, using the equipment described in [6.10]. The flow vapor quality ranged from that for inception of subcooled boiling x_{isb} to $x = 0.65$. The coolant moved upward. The entire length of the test section was subdivided into two parts of 1000 mm length each, the lower part with a plain surface and the upper part with a mesh covered surface.

The mass transfer conditions in the steam-generating channel were monitored by the salt method. The deposition of the indicator salt was monitored using the gain in

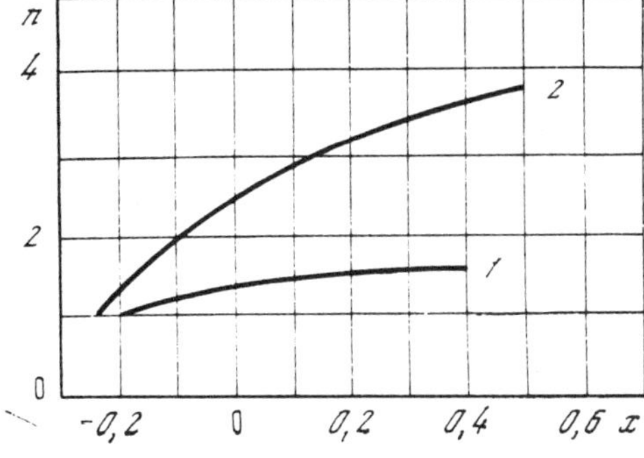

Figure 6.16 Concentration ratio n vs. vapor quality in boiling in a capillary-porous structure of fixed geometry. 1, $p = 6.9$ MPa, $q = 250$ kW/m^2, $\rho w = 1000$ kg/m$^2 \cdot$sec; 2, $p = 1.86$ MPa, $q = 350$ kW/m^2, $\rho w = 1620$ kg/m$^2 \cdot$sec.

wall temperature and by chemical analyses. The experiments were performed under two sets of conditions:

	p, MPa	q, kW/m^2	ρw, kg/m$^2 \cdot$ sec
Mode 1	1.86	350	1620
Mode 2	6.9	250	1000

Figure 6.16 shows results of mass transfer experiments for the segment with the mesh, as well as data obtained with the plain wall. It is seen from the figure that the maximum values of the concentration ratio n do not exceed 4 for the first set of conditions. For the second set of conditions concentration ratio is even lower. Note that the effect of porosity is higher at $p = 1.86$ MPa than at 6.9 MPa. In addition, the region of nucleate boiling in the tube with the mesh covering widened significantly both due to shift of the point of inception of subcooled boiling to a lower quality region, and to movement of the critical (dryout) point to higher quality. For the first set of conditions, the inception of boiling shifted to lower quality by an amount $\Delta x = 0.17$ and for second set by $\Delta x = 0.12$.

The study reported here showed that the value of n in a capillary-porous structure with fixed geometry and $\tilde{\delta} = 2$ does not exceed an order of magnitude. Given the experimental technique used, the results of Picone et al. [6.1] are seen to give a significant overestimate.

The physical situation for boiling in thick capillary-porous structures ($\tilde{\delta} \simeq 10^3$) is significantly different. Such conditions were investigated under pool boiling conditions. The equipment used consisted of a cylindrical vessel having a volume of about 2 liters, made of stainless-steel and textolite [6.18]. The vessel was capable of working at atmospheric pressure. The test section consisted of a copper rod, equipped with two rows of thermocouples and was soldered into the center of the vessel. A nichrome heater was installed into the hollow part. The upper part of the test section had a cylindrical bore. The diameter of the working part of the test section was 25 mm. Spherical copper powder with diameter $D_{c.p} \simeq 60$ or $D_{c.p} \simeq 300$ μm was poured into the bore. The powder was pressed to the rod surface by a stainless-steel mesh.

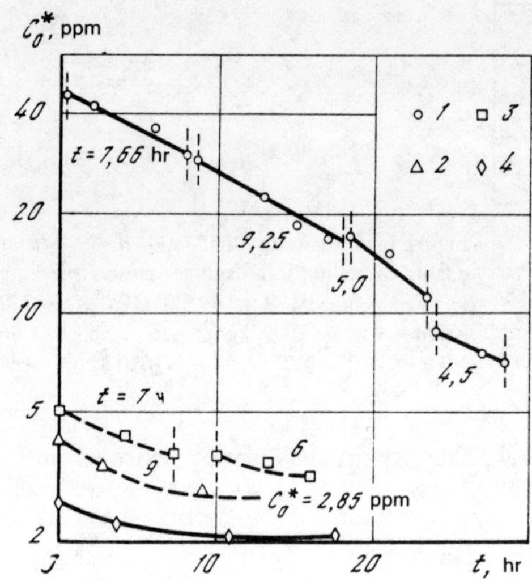

Figure 6.17 Variation in indicator salt concentration C_0^* as a function of time. 1, 2, $D_{c.p}$ = 300 μm, δ_b = 6450 μm; 3, 4, $D_{c.p}$ = 63–80 vm, δ_b = 6400 μm; 1, q_w = 11 W/cm²; 2, 15.7; 3, 11; 4, 12.6 W/cm² [sic].

The experiments were performed as follows. A $CaSO_4$ solution was poured into the vessel and the heat flux was preset at the specified level. The $CaSO_4$ concentration in the vessel dropped with time. When the concentration was significantly higher than its equilibrium value C_0^* the lowering of the concentration may require an extended time period. In this case the experiments were performed in several stages (Fig. 6.17).

At each successive stage the starting concentration was equal to the final concentration of the preceding stage. After attaining concentration C_0^*, which is the equilibrium concentration for the given conditions, the concentration did not change with time. This concentration was used as the basis for calculating the degree of concentration, with the solubility calculated on the basis of the wall temperature. The crystal form which is most stable to 100–150°C is the calcium sulfate semihydrate $CaSO_4 \cdot 0.5H_2O$.

The results of these experiments are plotted in the coordinates $n = n(q)$ in Fig. 6.18. The heat flux effect is very significant for a copper powder particle diameter of $D_{c.p}$ = 300 μm. Whereas for q = 11.8 W/cm² we have n = 117, then for q = 15.7 W/cm², n = 320. In the second case ($D_{c.p}$ = 63–80 μm) the change in heat flux increased the value of n only 1.19-fold. At a given heat flux density (q = 11.8 W/

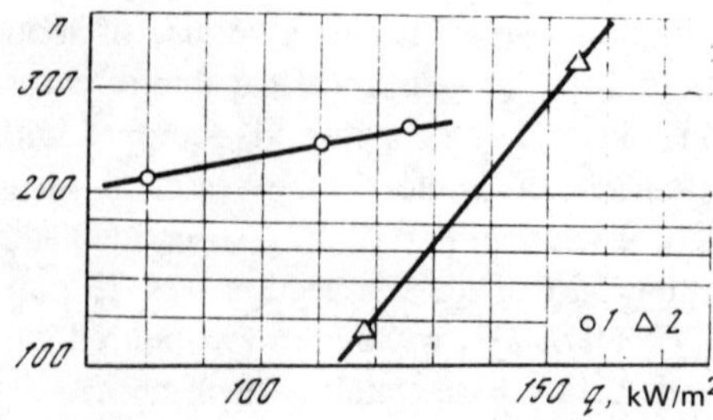

Figure 6.18 Effect of heat flux density on the degree of concentration under natural convection conditions. 1, $D_{c.p}$ = 300 μm, δ_b = 6400 μm; 2, $D_{c.p}$ = 63–80 μm, δ_b = 6400 μm.

cm^2) the value of n in a copper-powder bed with $D_{c.p} \simeq 60$ μm is 1.78-fold higher than for $D_{c.p} = 300$ μm.

To determine the nondimensional thickness of the copper powder beds the equivalent hydraulic diameter d_{hyd} was used as a reference dimension. The appropriate recommendations for bulk materials are used. Thus, for hexagonal packing (porosity $\sim 30.0\%$) the hydraulic diameters of channels for particles 60 and 300 μm in diameter are respectively $D_{hyd}^{60} = 8.6$ μm and $D_{hyd}^{300} = 42$ μm, whereas for cubic packing (porosity 48.0%) $D_{hyd}^{60} = 19.2$ μm and $D_{hyd}^{300} = 96$ μm. The relative thickness of the bed of spheres having a diameter of 300 μm $\tilde{z}^{300} = \delta_{dep}/D_{hyd} = 67-153$, and that for a bed of spheres 60 μm in diameter $\tilde{z}^{60} = 336-750$.

The increase observed in the concentration ratio is of great practical significance. Whereas in a particle bed with $\tilde{z}^{300} = 153$ the degree of concentration was as high as $n = 320$, in the scale formed by iron oxides, for example, on the tubesheet of a Westinghouse steam generator with $\tilde{z} \simeq 10^4$ the degree of concentration should be significantly higher. Thus, the results of the copper powder had experiments confirm the feasibility of accidents as a result of tube corrosion in locations with significant deposits. This means that the operating personnel and designers must be charged with determining the possible locations of accumulation of products of corrosion and to make operational and design arrangements to prevent dangerous growth of the scale.

6.3. THEORY OF BOILING MASS TRANSFER IN CAPILLARY-POROUS STRUCTURES

Two alternative physical models have been postulated for analyzing the concentration of impurities in the course of boiling in capillary-porous structures. In the model due to Macbeth [6.19], developed initially for heat transfer calculation and subsequently modified by Cohen [6.20-6.22] and also by Collier and Kennedy [6.23] for the problem of mass transfer, use is made of physical concepts previously developed in investigating the processes occurring in heat pipes. The salient feature of this model is its postulation of the presence of a vapor chimney without a liquid film. The models developed in the High-Temperatures Institute of the USSR Academy of Sciences consider two cases: namely, cocurrent and countercurrent flow of vapor and liquid film in the vapor chimney respectively. Let us consider the features specific to each model.

The principal ideas about the fundamental processes governing the circulation of coolant in the pores of deposits in steam-generating channels were formulated first by Macbeth [6.19]. The physical model (Fig. 6.19) reduces in main outline to the following. We consider a unit cell of the capillary porous body, containing a vapor chimney of relatively large diameter and an ensemble of smaller diameter liquid channels communicating with the chimney. The liquid vaporizes in a purely molecular manner from the surfaces of menisci in liquid channels. The greatest part of the vaporization and also the supply of liquid occurs through passages adjoining the chimney near the heating surface, where the temperature difference is greater. The

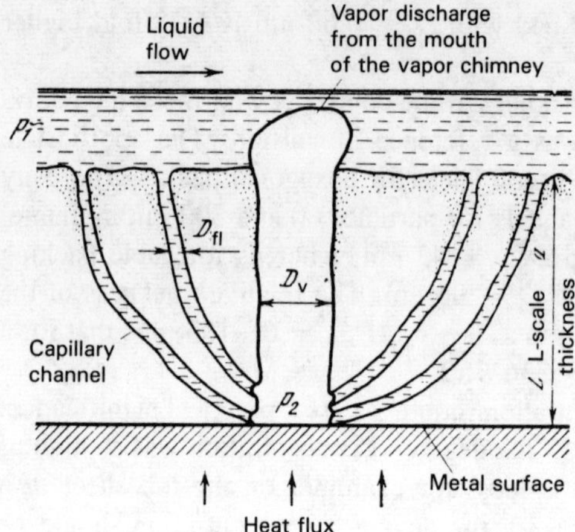

Figure 6.19 Model of boiling in a wick.

vapor moves along the dry chimney and enters the main stream. It is postulated that all the heat supplied to the heating surface is utilized for vaporization.

Circulation within the loop is provided by the capillary pressure. The capillary force has to overcome the pressure difference, the drag of the liquid channel and provide the velocity head

$$\frac{4\sigma}{D_{fl}} = (p_2 - p_1) + 32L\left(\frac{\mu w}{D^2}\right)_{fl} + (\rho w^2)_{fl} \tag{6.5}$$

The model assumes laminar flow and the drag is calculated from the Poiseuille formula. The equation of motion for the chimney is written by analogy

$$p_2 - p_1 = \frac{4\sigma}{D_v} + 32L\left(\frac{\mu w}{D^2}\right)_v + (\rho w^2)_v \tag{6.6}$$

Substituting $\Delta p = p_2 - p_1$ into the preceding equation, we have

$$4\sigma\left(\frac{1}{D_{fl}} - \frac{1}{D_v}\right) = 32L\left[\left(\frac{\mu w}{D^2}\right)_{fl} + \left(\frac{\mu w}{D^2}\right)_v\right] + (\rho w^2)_{fl} + (\rho w^2)_v \tag{6.7}$$

Expressing the velocities of the liquid, vapor and the effective porosity in terms of geometric parameters

$$w_{fl} = 4G/(\pi D_{fl}^2 N (1 - y) \rho_{fl}) \tag{6.8}$$

$$w_v = 4G/\pi D_v^2 N y \rho_v \tag{6.9}$$

$$\varepsilon = \frac{\pi D_v^2}{4} N y + \frac{\pi D_{fl}^2}{4} N (1 - y) \tag{6.10}$$

where ϵ is the effective porosity of the deposits, N is the total number of pores, y is the fraction of the vapor channels, G is the flowrate of liquid through the deposits, and writing

$$A = yD_v^2 + (1-y)D_{fl}^2 ; \qquad B = \mu_{fl}/(1-y)\rho_{fl}D_{fl}^4 ; \qquad C_c = \mu_v/y\rho_v D_v^2$$

we can obtain finally

$$4\sigma \left(\frac{1}{D_{fl}} - \frac{1}{D_v} \right) = \frac{32LG}{\varepsilon} A(B + C_c) + \frac{G^2}{\varepsilon^2} A^2 \left[\frac{B}{\mu_{fl}(1-y)} + \frac{C_\kappa}{\mu_v y} \right]$$

$$(6.11)$$

Macbeth's study is based on Schindler's model for heat pipes. The principal feature of the Macbeth model is the incorporation of velocity heads for both phases. Otherwise the model is analogous to that of Schindler.

Cohen's refined physical model [6.22] is shown in Fig. 6.20. The energy equation is written as:

$$\lambda_{ef} S \frac{\partial^2 T}{\partial z^2} = \alpha P [p(T,C) - p_0] + \rho' c_p' S \frac{\partial}{\partial z}(w'T) \qquad (6.12)$$

Assuming $p - p_0 = \beta\vartheta$ and $\alpha = \kappa_1 r$, where λ_{ef} is the effective thermal conductivity, S is the net cross section for the liquid in the elementary cell, z is the longitudinal coordinate, $\vartheta = T - T_0$ is the temperature difference, α is the heat transfer coefficient, w' is the velocity of the liquid, and neglecting the heat used for heating the liquid, Cohen obtains the expression:

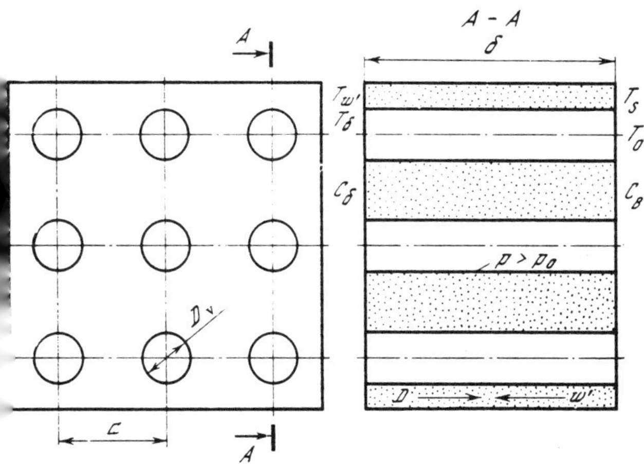

Figure 6.20 Model of the concentration process [6.22].

$$\frac{\partial^2 \vartheta}{\partial z^2} - a^2 \vartheta = 0, \tag{6.13}$$

where $a^2 = (\kappa_1 r P \beta)/(\lambda_{fl} S)$ whereas κ_1 is the mass transfer coefficient.

The boundary conditions are $z = 0$, $\vartheta = 0$; $z = \delta$ and $\vartheta = \vartheta_\delta$. The solution of Eq. (6.13) is

$$\vartheta = A' \exp(az) + B' \exp(-az) \tag{6.14}$$

Since

$$\frac{q}{\varepsilon} = \lambda_{ef} \left| \frac{\partial \vartheta}{\partial z} \right|_{z=\delta} \tag{6.15}$$

where ϵ is the porosity, the final expression for ϑ is:

$$\vartheta = \frac{q}{\varepsilon \lambda_{ef} \, a} \frac{\mathrm{sh}\, az}{\mathrm{ch}\, a\delta} \tag{6.16}$$

The above equation leads to a particularly interesting conclusion. If the ratio of conductive heat fluxes at the wall and on the surface of the deposits is written in the form:

$$\left(\frac{\partial \vartheta}{\partial z} \right)_0 \Big/ \left(\frac{\partial \vartheta}{\partial z} \right)_\delta$$

then

$$\frac{q_{\delta=0}}{q_w} = \frac{1}{\mathrm{ch}\, a\delta} \tag{6.17}$$

It follows that, for conditions frequently encountered in practice, the bulk of the heat is transmitted by the phase transition. A significantly smaller part of heat (from several tenths a percent to several percent) is transmitted by means of thermal conductivity of the skeleton of the capillary-porous structure.

Cohen's diffusion equation employs only the convective and molecular terms

$$D \frac{\partial^2 C}{\partial z^2} - \frac{\partial}{\partial z}(w'C) = 0 \tag{6.18}$$

The boundary conditions are $z = 0$; $C = C_0$; $z = 0$ and $D\partial c/\partial z = w_0'C$.

The expression for w' was obtained from the material balance

$$S\rho'w' = P\varkappa_1\beta \int_z^\delta \vartheta \, dz \tag{6.19}$$

or

$$w' = \frac{P}{S} \frac{q\varkappa_1\beta}{\varepsilon\lambda_{ef}\rho'a^2} \left[1 - \frac{\mathrm{ch}\,az}{\mathrm{ch}\,a\delta}\right] \tag{6.20}$$

Using the last relationship, the concentration ratio can be written as:

$$n = \exp \frac{q}{D\varepsilon\rho'} \left[z - \frac{1}{a} \frac{\mathrm{sh}\,az}{\mathrm{ch}\,a\delta}\right] \tag{6.21}$$

and for $z = \delta$

$$n_\delta = \exp \frac{q}{D\varepsilon\rho'} \left[\delta - \frac{1}{a} \mathrm{th}\,a\delta\right] \tag{6.22}$$

Cohen compares the results calculated from his model with experimental data. In particular, the heat transfer data used by him for this purpose were those obtained by Macbeth et al. [6.16] and Rassokhin et al. [6.17]. The comparison was performed on the basis of values of a correction coefficient λ_{ef}. Cohen notes that the calculated values differ by a factor of 1.8 from those in [6.17] and by a factor of 3.8 from those in [6.16].

The mass transfer data of Picone et al. [6.1] had to be recalculated by Cohen before they could be compared with the model. This was done by determining the mean concentration ratio

$$\bar{C}_w/C_0 = \frac{1}{\Theta_\delta}(\exp \Theta_\delta - 1)\left(\frac{\gamma - 0,1}{\gamma}\right). \tag{6.23}$$

where

$$\Theta_\delta = \frac{q}{D\varepsilon\rho'r}\left[1 - \frac{1}{\mathrm{sh}\,a\delta}\right]$$

The concentration ratio (see Sec. 6.1) within the deposits was determined by Picone et al. on the basis of Na^{22} radiation. The quantity recorded was the total radiation, which made such an averaging necessary. The results of calculations are listed in Table 6.2.

It is seen from the table that the ratio $\bar{C}_w/C_0$ ranged from 10^2 to $3 \cdot 10^3$. The actual degree of concentration at the wall is obviously even higher. In particular, he presents as an example a mean concentration ratio $\bar{C}_w/C_0$ of 295, whereas at the wall $C_w/C_0 =$

Table 6.2

T_{sat}	T_0	$D_\Sigma - D_v$, Counts/min	Q, kcal/m^2·sec	W, kg/m^2·sec	$\bar{C}_w/C_0$
282.22	275.56	125	95.10	0.26	98
288.33	276.67	250	110.92	0.31	127
293.89	277.22	550	126.73	0.37	284
298.33	277.78	650	142.62	0.42	348
306.11	278.89	1025	158.66	0.47	545
308.89	279.44	1425	174.32	0.55	765
315.56	280	2250	189.76	0.62	1220
326.67	281.67	5250	221.83	0.79	2950

Note. $\delta_{dep} = 2.54 \cdot 10^{-3}$ cm, $\epsilon = 0.7$; background level, 650 counts/min.

2724. It must be noted here that the relative thickness of the deposit layer was only $\tilde{\delta} = \delta_{dep}/D_{v.c} = 5.1$.

The above estimates of the concentration ratio are very much on the high side. This, as was noted above, may be a result of the fact that the study by Picone et al. suffers from significant errors. The overall radiation in the deposits was due to the presence of Na^{22} both in the solution and in solid form, and the contribution of the solid phase to the total radiation was not determined in [6.1].

It follows from the model of the process that concentration of impurities at the wall is controlled by the boiling point depression for the compounds dissolved in the water, their passing into the steam in accordance with the distribution factor and also by inverse diffusion of impurities through the liquid channels. It is known that the rate of these processes is low and it can be claimed *a priori* that the degree of concentration should be several orders of magnitude when the thickness of the layer of iron-oxide deposits is less than 100 μm, which in fact was also proven by Cohen. Under these conditions there intense corrosion should occur of the heating surfaces when they are covered by even relatively thin porous deposits. However, in practice, intense corrosion occurs only in regions covered by very thick deposits or poorly washed by liquid (slots in rolled joints or between tubes and fastening elements).

The Cohen model has a number of significant shortcomings. In particular, it does not include the possibility of removal of concentrated solution in the form of liquid masses associated with the detaching vapor bubble and dilution of the solution at the wall due to influx of liquid through the vapor channel during the "waiting" period between bubbles. Both these processes, which produce a kind of "scavenging" will significantly reduce the degree of concentration of impurities contained in the loop water.

A salient feature of models developed in the Institute for High Temperatures of the USSR Academy of Sciences is they take account of the presence of liquid films in the vapor channels. It was necessary to check whether such liquid films exist under pre-dryout boiling conditions. For this purpose, an experiment was performed aimed at detecting microfilms in a porous structure through which superheated steam was bubbled into an aqueous solution of $CaSO_4$ which was at saturated temperature $T = T_s$ [6.26]. Jena-glass filters were used as the porous medium. The presence or absence of a film was determined from electrical contact between the steam-side filter surface and the liquid-side (solution) surface. This was established by means of a probe, moved to different points on steam-side surface. The maximum pore (vapor channel) diameter was 50 μm. The quantities measured were the vapor flowrate, temperatures of the vapor and liquid, and surface temperatures on both sides of the filter. The time was also determined for the destruction of electrical contact through the filter, i.e., rupture of the film or reduction in its limiting length to a size smaller than the thickness of the filter (3.5 mm).

The results of one series of experiments are shown in Fig. 6.21 in the form of a plot of concentration C_0^* of the solution (curve 1) at which the film rupture vs. $Q^* = G_v r/\pi R_\Phi^2$. The appearance of a film due to gravity forces is apparently impossible, since the pressure difference exceed significantly 7×10^3 Pa. It was established experimentally that, under certain conditions, the liquid film actually exists and is stable. Figure 6.21 also shows (curve 2) the concentration ratio $n = C_0^*/C_{sat}'$, attained in the experiments, as a function of Q' the surface heat flux equivalent of the superheated steam flow. In these experiments with a relative structure thickness of $\delta \cong 70$ and $q/q_{cr} \cong 0.1$ it was found that $n = 20$, whereas at $q/q_{cr} \cong 0.2$, $n = 200$, i.e., n was found to rise significantly as a function of Q^*.

The first of the present authors with his coworkers [6.24] consider the following model (Fig. 6.22) of boiling with a liquid film in the vapor channel (chimney) in the porous layer. The bubble initiates at the wall, grows in the vapor channel, moves out to the surface of the capillary-porous body, attains its departure diameter and moves into the liquid. A liquid film is constantly present in the vapor chimney, supplying liquid to the heating surface. The driving force ensuring flow of the liquid towards the wall is the surface tension gradient, which arises as a result of change in the surfactant concentration in the liquid along the film in the channel due to its vaporization. The rate of change of surface tension with concentration $\partial\sigma/\partial C$. Assuming that

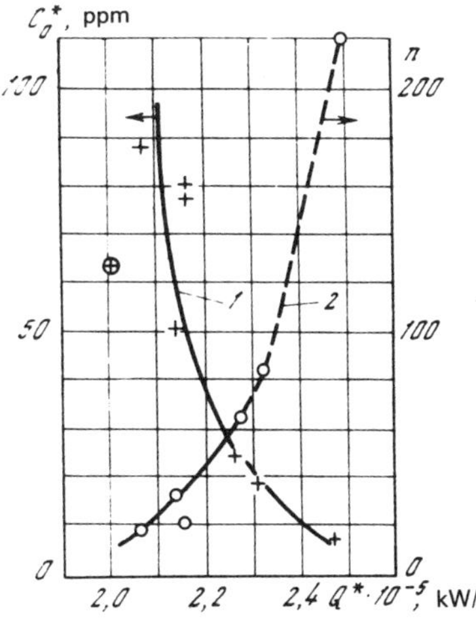

Figure 6.21 Effect of heat flux Q^* on C_0^* and n [6.26]. 1, C_0^*; 2, n.

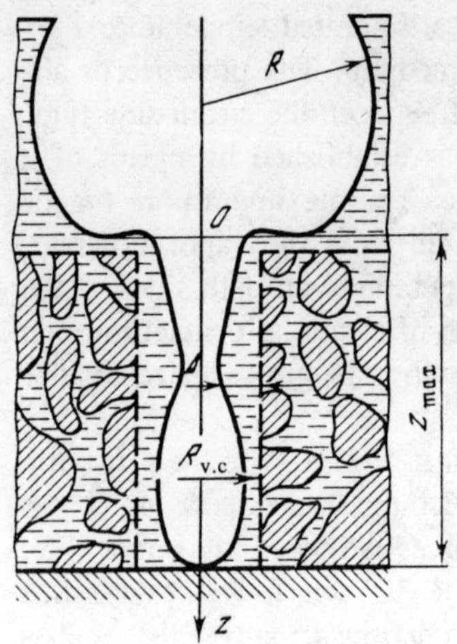

Figure 6.22 Model of the mass transfer process for boiling in a capillary-porous body [6.24].

near the heating surface the channel is constantly filled with vapor, and that the distribution of liquid film and of concentration along the channel is steady, the equation of motion for the film is written as:

$$\frac{\partial \tilde{p}}{\partial \tilde{z}} = \frac{1}{1-\eta}\frac{\partial}{\partial \eta}\left[(1-\eta)\frac{\partial \overline{w}}{\partial \eta}\right] \tag{6.24}$$

with the boundary conditions $\eta = 0$, $\tilde{w} = 0$, $\eta = \tilde{\delta}$ and $\partial \tilde{w}/\partial \eta = \partial \tilde{\sigma}/\partial \tilde{z} - (4\tilde{q}\beta/(1 - \tilde{\delta})^3$.

In Eq. (6.24) and henceforth we use the notation:

$$\tilde{w} = \frac{wR_{v.c}}{D}\;; \qquad \tilde{p} = \frac{pR_{v.c}^2}{\mu'D}\;; \qquad \tilde{\sigma} = \frac{\sigma R_{v.c}}{\mu'D}\;; \qquad \tilde{z} = \frac{z}{R_{v.c}}\;;$$

$$\eta = 1 - \frac{R}{R_{v.c}}\;; \qquad \tilde{q} = \frac{qR_{v.c}}{rD\rho'}\;; \qquad \tilde{\delta} = \frac{\delta}{R_{v.c}}\;; \qquad \omega = \frac{\rho''}{\rho'}\;;$$

$$\tilde{C} = \frac{CR_{v.c}\kappa}{\nu'\rho'D}\;; \qquad \tilde{g} = \frac{g}{\rho'DR_{v.c}}\;; \qquad \beta = \frac{\nu''}{\nu'}\;; \qquad \kappa = \left(\frac{\partial \tilde{\sigma}}{\partial \tilde{C}}\right)_{eq}$$

The boundary condition at $\eta = \delta$ was obtained by taking the shear stresses at the surface of the liquid film equal to zero. Since the vapor flow is laminar and

$$p' = p'' - \frac{\sigma}{1-\delta} \tag{6.25}$$

then we can write the following expression for the liquid flowrate in the film:

$$\tilde{g} = 2\pi \int_0^{\tilde{\delta}} \overline{w}\,(1 - \eta)\,d\eta \tag{6.26}$$

To within third-order differences $\tilde{\delta}^3$ the solution is

$$\frac{\tilde{g}}{\pi} = \left(\frac{\partial \tilde{\mathfrak{z}}}{\partial \tilde{z}} - 4\tilde{q}\beta\right)\tilde{\delta}^2 - \left(12\tilde{q}\beta + \frac{2}{3}\frac{\partial \tilde{\mathfrak{z}}}{\partial \tilde{z}}\right) \tag{6.27}$$

In the case at hand

$$\tilde{g} = \pi\tilde{q} \tag{6.28}$$

Assuming equality of the convective $(+\tilde{z})$ and diffusive $(-\tilde{z})$ mass fluxes, and also the equality of the ratios

$$\frac{\partial \tilde{C}}{\partial \tilde{z}} = \frac{\partial \tilde{\mathfrak{z}}}{\partial \tilde{z}} \tag{6.29}$$

we obtain the following distribution of $\tilde{\delta}$ and $\tilde{C}$:

$$\tilde{q} = \left(\frac{\partial \tilde{C}}{\partial \tilde{z}} - 4\tilde{q}\beta\right)\tilde{\delta}^2 - \left(12\tilde{q}\beta + \frac{2}{3}\frac{\partial \tilde{C}}{\partial z}\right)\delta^3 \tag{6.30}$$

$$q = 2\tilde{\delta}\left(1 - \frac{\tilde{\delta}}{2}\right)\frac{1}{\tilde{C}}\frac{\partial \tilde{C}}{\partial \tilde{z}} \tag{6.31}$$

Note that if molecular diffusion only is assumed to be the mechanism ensuring removal of vapor from the wall to the flow, then it can be assumed *a priori* that the degree of concentration at the walls may be of several orders of magnitude.

Solving these equations for two extreme cases of film flow—films with small liquid flowrate (long films) and with high flowrate (short) films, we obtained the following expressions.

For the first type of films at low $\tilde{C} < \beta$:

$$\tilde{C} = \frac{24}{11}\beta\left[\left(1 + \frac{11}{24}\frac{\tau_0}{\beta}\right)\exp\left(\frac{11}{6}\tilde{q}\tilde{z} - 1\right)\right] \tag{6.32}$$

$$\tilde{\delta} = \frac{75}{242}\left[1 - \left(1 + \frac{11}{24}\frac{\tau_0}{\beta}\right)^{-1}\exp\left(-\frac{11}{6}\tilde{q}\tilde{z}\right)\right] \tag{6.33}$$

The film length is limited to

$$\tilde{z}_{\max} < 1.7/\tilde{q}$$

and the maximum concentration

$$\tilde{C}_{\max} < 17\beta$$

Estimation of the concentration ratio n from the above expressions yields the following results. For $T_0 = 100\,°C$, $\kappa \simeq 10^{-7}$ kg·m^2/mole, $D \simeq 2\cdot10^{-5}$ m^2/hr, $C_0 \simeq 10^{-1}$ mole/m^3, $R_{v.c} \simeq 10^{-2}$ mm, $Q = 10^5$ W/m^2 (i.e., $q \simeq 10^6$ W/m^2) the impurity concentration in the film increases 10^3-fold over a distance of $\tilde{z} = 17$ and at $q \simeq 10^7$ W/m^2 it changes by a similar amount over $\tilde{z} = 1.7$.

For films of the second type at $\tilde{C} \gg 8\sqrt{\beta}$ the solution is

$$\tilde{C} = 4\sqrt{\beta}\left\{ \mathrm{tg}\left[\sqrt{\beta}\ \tilde{q}\,(\tilde{z} - \tilde{z}_0) + 1\right] + \frac{1}{\mathrm{tg}\left[\sqrt{\beta}\ \tilde{q}\,(\tilde{z} - \tilde{z}_0) + 1\right]} \right\} \tag{6.34}$$

$$\delta = \frac{2}{\sqrt{\beta}}\ \frac{\cos^2\left[\sqrt{\beta}\ \tilde{q}\,(\tilde{z} - \tilde{z}_0) + 1\right]}{\mathrm{tg}^2\left[\sqrt{\beta}\ \tilde{q}\,(\tilde{z} - \tilde{z}_0) + 1\right] - 1}\left\{ \mathrm{tg}\left[\sqrt{\beta}\ \tilde{q}\,(\tilde{z} - \tilde{z}_0) + 1\right] + \right.$$
$$\left. + \frac{1}{\mathrm{tg}\left[\sqrt{\beta}\ \tilde{q}\,(\tilde{z} - \tilde{z}_0) + 1\right]} \right\}$$

According to Eq. (6.35) at

$$\tilde{z} - \tilde{z}_0 \cong 0{,}6/\tilde{q}\sqrt{\beta}$$

the concentration in the film grows in an unlimited way, i.e., the film length is limited by $\tilde{z}_{\max} < 0.6\,\tilde{q}\sqrt{\beta}$.

Under actual plant conditions two types of film can apparently exist. Transition from one type of film to another occurs at $\tilde{C} \geq 8\sqrt{\beta}$. At this position, the film has its maximum thickness $\tilde{\delta}_{\max}$.

It is clear from the above that the model is primarily appropriate at high impurity contents of the water, when the diffusion head is sufficiently high. The model has the advantage that, over a certain range of concentrations, it ensures supply of liquid to the wall irrespective of the existence or lack of liquid channels. The model is also relevant under conditions when vapor is generated in single chimneys.

Let us now consider the second model developed in the High Temperatures Institute of the USSR Academy of Sciences for boiling in porous structures with a liquid film in the vapor chimneys [6.25]. We shall illustrate the model with an example of a unit cell, formed on the surface by iron-oxide deposits (Fig. 6.23). The cell includes a steam chimney and an ensemble of liquid channels. Initiation of the vapor bubble is possible on a vaporization site on the chimney wall at the point of contact of a particle with the surface of the wall, or at the contact points of particles in layers closest to the heating surface. The bubble is initiated and moves in the chimneys. The length of the chimneys and liquid channels is not equal to thickness δ_{dep} of the scale and differs from it by the tortuosity factor ϵ. The vapor is generated periodically. The vapor phase is nucleated due to the superheat of the liquid, whereas further growth of the bubble occurs due to evaporation into it of liquid in the "wedge" at its base and the film on the chimney wall.

A liquid film with δ_{fm} is present between the vapor bubble and the porous struc-

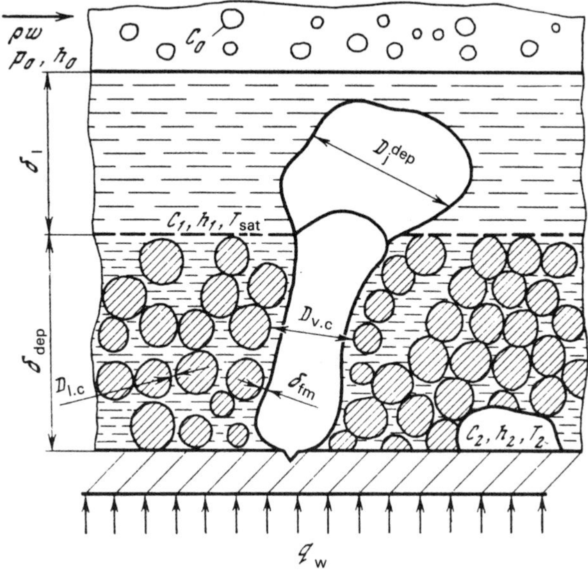

Figure 6.23 Physical model of boiling mass transfer in capillary-porous bodies [6.25].

ture. The outer edge of the film is acted upon by shear stresses exerted by the vapor. For this reason, if the effect of $\partial\sigma/\partial C$ is neglected, the liquid in the film will move cocurrently with the vapor. Liquid is supplied to the wall by capillary forces by way of liquid channels situated near the wall.

As the bubble grows, it moves into the wall layer of the flow δ_1 and is acted upon by an ensemble of all those forces arising in the course of forced convection boiling on a surface without a porous layer. As the bubble diameter grows, the capillary pressure tends to collapse the bubble cavity within the chimney. The bubble breaks off after attaining a certain size.

After the bubble departure the chimney is filled with liquid. There then starts the bubble "waiting" period t_1. During this time, the liquid is superheated to a certain temperature, at which it flashes. The duration t_2 of flashing of the superheated liquid is very short. Bubble development then takes place over a relatively long period t_3. The last period t_4—time of film closure—is also very short.

The above considerations make it possible to formulate a closed set of equations. However, to be able to perform calculations one must know such parameters as the bubble generation frequency, times t_1 through t_4, the departure diameter, etc., which for these conditions are virtually unknown. To be able to obtain numerical results, we shall simplify the problem as follows. We shall postulate that the principal period controlling the concentration of impurities is the bubble growth period t_3. In addition, we will neglect the molecular diffusion of mass in the liquid film. These assumptions may give an overestimate of the degree of concentration. The errors introduced by these assumptions are to some extent mitigated by the assumption of low hydraulic drag in the liquid capillaries.

Given the above we can write that the concentration ratio is

$$n_0 = C_2/C_1 = (G_v + G_{fm})/G_{fm} \qquad (6.36)$$

where G_v is the flowrate of vapor through the chimney and G_{fm} is the flowrate of liquid through the film. The flowrate of vapor through the chimney can be written as:

$$G_v = q/r\varepsilon a N \qquad (6.37)$$

where ϵ is the relative area of the chimney, N is the number of chimneys per unit surface, and $a = t_3/\Sigma_{i-1}^4 \, t_1$ is the relative duration of period t_3.

Equation (6.37) is written on the assumption that the entire heat flux is removed from the wall by the vapor (due to latent heat of evaporation). As shown by calculations performed by Cohen [6.22], the vapor transfers $\sim 99\%$ of the heat supplied to the wall (this pertains to the case when the circulation ratio of liquid through the porous layer tends to unit).

The flowrate of liquid through the film is given by the equation

$$G_{fm} = 0{,}25\pi D_{v.c}^2 \rho' \overline{w}_{fm} (1 - \varphi) \qquad (6.38)$$

where ϵ is the relative area of the chimney, N is the number of chimneys per unit surface, and $a = t_3/\Sigma_{i-1}^4 \, t_1$ is the relative duration of period t_3.

$$\overline{w}_{fm} = \frac{\tau_v \delta_{fm}}{2\mu'} \qquad (6.39)$$

τ_v is the shear stress, and μ' is the dynamic viscosity of the liquid.

To complete the set of equations we write the following relationships: relation between the shear stress and friction pressure losses

$$\tau_v = \frac{\Delta p_{fr} \, D_{v.c} \, \sqrt{\overline{\varphi}}}{\delta_{dep}} \qquad (6.40)$$

expression for Δp_{fr}

$$\Delta p_{fr} = \xi \frac{\rho'' \overline{w}_v^2}{2} \frac{\delta_{dep}}{d_{v.c} \, \sqrt{\overline{\varphi}}} \qquad (6.41)$$

friction factor

$$\xi = \frac{0{,}316 v''}{\overline{w}_v D_{v.c} \sqrt{\overline{\varphi}}} \qquad (6.42)$$

mean vapor velocity

$$\overline{u'}_{\text{v}} = \frac{\prime G_{\text{v}} v''}{\pi D_{\text{v.c}}^2 \, \varphi}$$ (6.43)

liquid film thickness

$$\delta_{\text{fm}} = 0{,}5 D_{\text{v.c}} \left(1 - \sqrt{\varphi}\right)$$ (6.44)

To be able to calculate n_0 one must know the friction pressure loss Δp_{fr}.

Let us assume that the entry of the film into the vapor chimney occurs due to the effect of the cylindrical meniscus. The departure diameter of the bubble can then be determined from the formula

$$D_{\text{v}}^{\text{dep}} = 2 D_{\text{v.c}} \sqrt{\varphi}$$ (6.45)

and the mean frictional pressure drop for period t_3 can be estimated from the equation

$$p P_{\text{fr}} = \frac{\sigma}{D_{\text{v.c}} \sqrt{\varphi}}$$ (6.46)

Solving the above equations for n_0, we obtain

$$n_0 = 1 + \frac{6\prime \varphi^{3/2}}{(1 - \varphi)(1 - \sqrt{\varphi})} \frac{v'}{v''}$$ (6.47)

where

$$\varphi = \left(\frac{q \delta_{\text{dep}} v''}{\pi D_{\text{v.c}}^3 \, r g a N \sigma} \right)^{2/3}$$ (6.48)

This model is now checked using experimental data in Fig. 6.15. If we use estimates of the effective thermal conductivity of the deposits [6.16, 6.17] and the gain in wall temperature ΔT due to iron oxide deposits typical for such experiments, then in the experiments [6.7–6.9] being considered we may estimate that δ_{dep} the thickness of the iron oxide layer, does not exceed 300 μm (it was assumed that $\lambda_{\text{ef}} = 35$ W/m$\cdot$°C, $\Delta T = 5$°C). Assuming that the dimensions of the elementary cell are the same as in [6.16], $D_{\text{v.c}} = 5$ μm, the spacing of the vapor channels $S = 14$ μm and $N = 5 \cdot 10^9$ per m^2, it is possible to calculate the degree of concentration n. In Fig. 6.24 the calculated values are compared with the experimental values n_{exp}. Here, in accordance with the results of Nesis [6.27] it was assumed that $a = 0.5$, i.e., the time of bubble growth is equal to one half of the bubble generation period.

Note that the n_{exp} measured in the experiments included concentration in the capillary-porous structure itself, as well as in the liquid film in the wall (thickness δ_l). On the other hand, the model considers concentration only in the capillary-porous body itself, n_0. To match these values, it is necessary to incorporate a concentration ratio n_l in the liquid layer δ_l. Then the predicted value of concentration ratio is

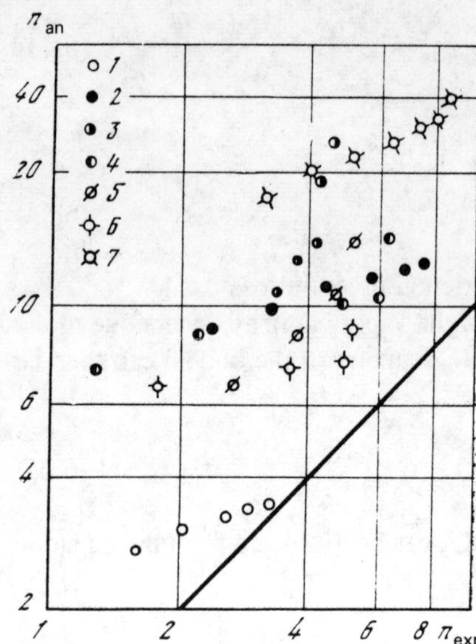

Figure 6.24 Comparison of analytic and experimental values of the degree of concentration [6.25]. 1–7, Represent points under conditions described in Fig. 6.15.

$$n_{an} = n_0 n_1. \tag{6.49}$$

The salt concentration at the outer surface of deposits depends on the rate of mass interchange with the flow core and of the vapor generation rate, irrespective of where it occurs, be it within the scale or on its outer surface. For this reason n_1 can be determined from experimental data on a wall without deposits (see Chapter 5). However, it must be remembered that Eq. (6.49) is somewhat arbitrary, since the generation of vapor on a wall without deposits and on a porous surface differ in the following ways:

1. In the wall covered by a porous structure the number of nucleation site depends little on q and x, i.e., the structure has a sufficiently constant number of channels for removal of heat near critical conditions. On a plain surface, the number of nucleation sites is a function of q and x.
2. The frequency of nucleation on a plain surface is higher than on a surface covered by a porous structure.

It is seen from Fig. 6.24 that the predicted values (n_{an}) in fact lie above the experimental values (n_{exp}), as indeed follows from the model. At the same time it should be noted that the agreement improves with increasing x. Under certain conditions, n_{an} and n_{exp} are virtually identical. The greatest difference occurs at high pressures and heat fluxes. This is natural, since processes occurring during periods t_1, t_2 and t_4 then become significant. This difference increases also at x_{isb}. It is assumed in the model that the entire heat flux is used for phase transition whereas it is clear that a significant part of the heat can be transferred by the liquid phase in the surface boiling region.

The set of equations of this model does not contain any additional constants. It uses only published data. It is possible that this is in part responsible for the large difference in n_{an} and n_{exp}, since under the computational conditions they could be somewhat different. In particular, the results could have been affected by the structure of deposits. In calculations it was assumed to be the same as in [6.16], and the question of stability of structures with changes in q, p, ρw, etc., has not been explored. It is advisable to perform experiments in structures with fixed geometry. On the whole it can be said that the agreement between n_{an} and n_{exp} is best under conditions which are similar to those assumed in the model. However, in order to be able to develop a rigorous engineering technique, it is necessary to further investigate the process of concentration. Here particular attention should be given to the nature of phenomena occurring in the course of boiling in capillary-porous structures during time periods t_1, t_2 and t_4.

FORMATION AND MOTION OF MOISTURE IN TURBINE FLOWPASSAGES

Nuclear power plant turbines which are supplied with wet steam operate under conditions which are different than those applying in conventional power plants. The main differences are due to the low initial steam temperature and pressure; expansion of steam plants with ANR-type and channel-type high-capacity reactors usually starts from saturation or from a steam superheat of 30–40 °C. Thus, all the high-pressure turbine stages operate on wet steam; a separator or a separating superheater is placed after the high-pressure part of the turbine. Also, the turbine operation is to a significant extent coupled to the reactor performance.

This means that the construction and operation of turbines at nuclear power plants requires the solution of a number of complicated problems of gas dynamics of two-phase flows. These problems are: formation of moisture at sub- and transonic velocities, formation of liquid films and large droplets, motion of moisture in turbine flowpassages, interaction between moisture and rotor blades, effect of the liquid phase on the principal parameters of turbine flowpassages and effect of impurity concentrations in the liquid phase on the corrosion of metal, particularly in the Wilson zone. The solution of these problems will make it possible to optimize the flowpassages of wet-steam turbines by improving their efficiency and reliability. In this and the following chapter we consider only the most important features of the flow of wet steam in nuclear power plant turbines and methods of moisture removal from the latter. Investigations and calculation of such turbines have been most comprehensively treated by Filippov et al. [7.1, 7.3] and by Troyanovskiy [7.2].

7.1 PRECIPITATION OF MOISTURE IN TURBINE STAGES

The location where the liquid phase precipitates in turbine flowpassages and the forms in which it occurs affect significantly the efficiency and reliability of turbine operation. Information is required on the location of condensation, the size of the droplets, droplet growth and on the formation of liquid films. Using this information, it is possible to more exactly calculate the effect of moisture on the reduction in turbine-stage efficiency, to estimate the erosion resistance of the blades and to more reliably select the design of moisture removal devices.

For nuclear plants, the expansion of wet steam in a turbine starts with the first high pressure stage at a moisture percentage between 0.1 and 0.7%. In the first low-pressure stage (in designs with an external moisture separator but without a super-heater) the percentage moisture is about 1%. When the steam is superheated the transition to wet steam occurs at the second or third stage (counting from the end of the flowpassages of the low-pressure part) if the turbine is equipped with a moisture separator reheater (which is the case for the overwhelming number of nuclear power plants).

In general the following forms of condensation may occur in turbine flowpassages: (1) spontaneous formation of moisture in the steam as a result of the flow attaining maximum subcooling; (2) formation of moisture in the near wake of the exit edges of stationary and rotating blading; (3) formation of moisture in the tip vortex or separation region of turbine cascades; (4) condensation of subcooled steam on surfaces of the turbine flowpassages; (5) condensation of steam in regions with elevated macroscale flow turbulence; (6) partial condensation on solid and liquid particles.

Spontaneous formation of moisture upon entering the Wilson zone (the region on the *i-s* diagram where the subcooling ΔT of the steam is the greatest) has been extensively investigated for laminar flow expansions. Spontaneous formation of moisture in the stationary-blading flowpassages can occur only at trans- and supersonic steam velocities. The location of the condensation shock, its strength and the size of the resulting droplets depend to a large degree on the rate of expansion of the steam $\dot{a} = -c/pdc/dz$, which for the case of superheated or fully subcooled steam, can be expressed in terms of other parameters, for example, $\dot{a} = \kappa M^2/(M^2 - 1)c/F$, where κ is the isentropic exponent, F is the cross sectional area, M is the Mach number, c is the steam velocity and z is the length of the nozzle.

Figure 7.1 shows correlations of the maximum subcooling $\Delta T_{max} = f(\dot{a})$ and size of droplets r_3 forming in the course of spontaneous condensation of steam for different initial pressures $p_{0\,in}$.

The rate of formation of condensate nuclei is usually calculated from the Zel'dovich-Frenkel' equation:

$$J = \left(\frac{P}{\kappa^* T}\right)^2 \frac{\alpha_{cond}}{\rho_2} \sqrt{\frac{2\sigma m_0}{\pi}} \exp\left[-\frac{\beta \Delta W_*}{\kappa^* T}\right] \tag{7.1}$$

where κ^* is the Boltzmann constant ($1.38 \cdot 10^{-23}$ $J/°C$), α_{cond} is the condensation coefficient, ρ_2 is the density of the liquid (kg/m^3), σ is the surface tension (N/m), m_0 is the molecular mass (kg), $\Delta W_* = 4\pi r_* \sigma/3$ is the work of formation of condensation nuclei, r_* is the critical droplet radius (m) and β is a correction factor.

According to experimental data of the last of the present authors with his cowor-

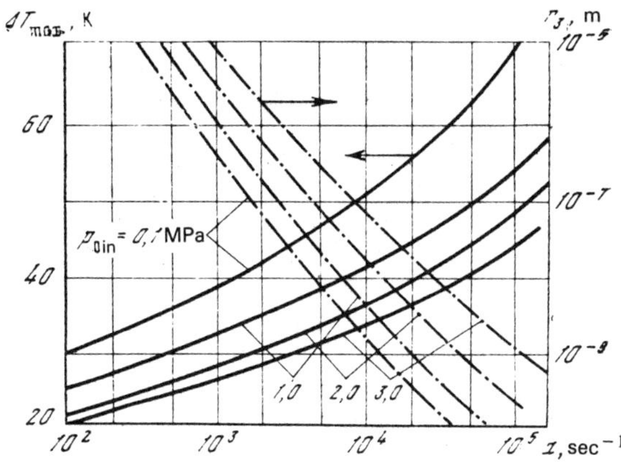

Figure 7.1 Maximum subcooling ΔT_{max} and radius of condensate droplets r_3 vs. rate of expansion of steam $\dot{a}$.

kers [7.4], the value of β rises with the initial pressure (β = 2.87, 4.12 and 4.87 at p_0 = 2.5, 18 and 32 kg/m^2, respectively).

The number of droplets forming in the course of spontaneous condensation is given by the expression

$$n = \int_{z_0}^{z} \frac{I\,(\xi)}{w\,(\xi)}\,dz$$

where ξ is the coordinate of the point where the droplet forms, and z is the current coordinate.

It is characteristic of turbine cascades that the supercooling of the steam does not exceed 45 °K and accordingly (see Fig. 7.1) the size of droplets arising at nucleation sites does not exceed $1 \cdot 10^{-4}$ mm.

The condensation shock in the case of condensation at transonic velocities (M < 1.1–1.3) may be unstable. Unsteadiness of the condensation shock is observed in single de Laval nozzles and in the convergent divergent passages of turbines cascades. Figure 7.2 shows experimental data on the variation in the relative static pressure p/p_{01}, density ρ/ρ_{01} and velocity of condensation and compression shocks [7.1].

The reason for the appearance of unsteady phenomena is the release of heat in the course of moisture formation. When heat is released at a very high rate (in the downstream-divergent-part of the nozzle) the static pressure and steam temperature rises. The net effect may be a reduction of the steam velocity. If the static pressure rise is sufficient, then the flow velocity may drop to M = 1, and a compression shock may arise in the region with high rate of heat release. The compression shock moves upstream to the region of the minimum cross section and then into the subsonic part of the nozzle. The motion of the compression shock reduces the supercooling of the steam flow in the zone where maximum condensation rate had originally occurred and condensation occurs first further downstream at a higher value of M. However, as the compression shock continues to move upstream, its effect on condensation

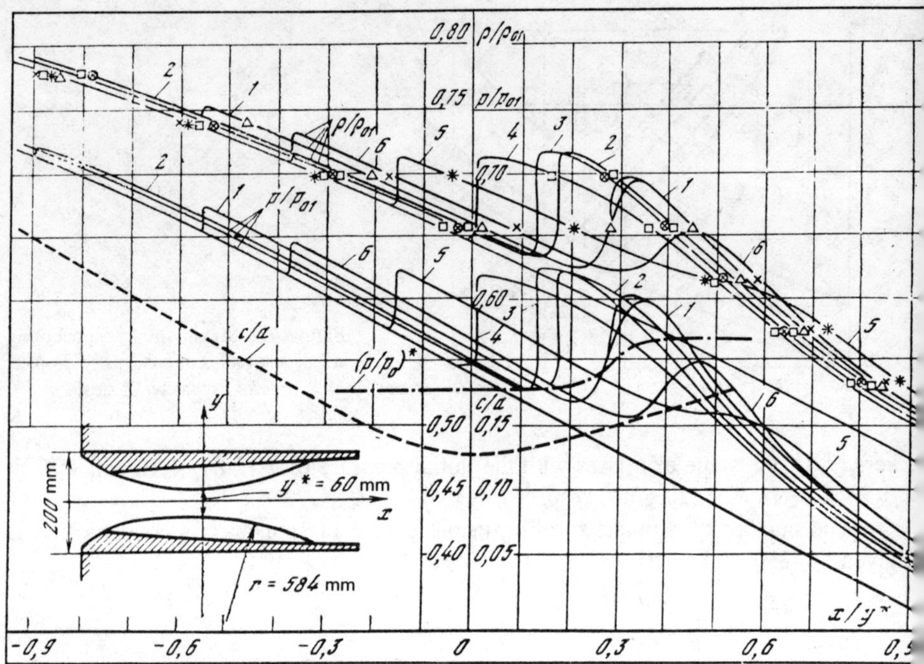

Figure 7.2 Schematic of nozzle, distribution of the relative static pressure p/p_{01}, density ρ/ρ_{01} and also the relative velocity of condensation and compression shocks c/a (the numerals designate mode numbers).

vanishes. The condensation zone then returns to its previous location, which induces the appearance of a new compression shock and the cycle is repeated. Thus, there exists a periodic fluctuation in the mode of flow of supercooled steam in nozzles. Filippov et al. [7.1] present a complete description of this phenomenon for Laval nozzles and also describe the results of optical studies. These studies showed that unsteady condensation phenomena occur at frequencies between 500 and 2000 Hz and induce changes in the flowrate of steam through the cascade and a reduction of the pressure downstream of it by 7–8%. In turbine cascades operating at high steam pressures (to 7 MPa), unsteady condensation phenomena may occur also at transonic velocities ($M > 0.8$).

Analytic studies have been carried out by Filippov et al. [7.1] of flows of steam with $p_0 = 0.6$ MPa and $T_0 = T_{in}$ for the C-9012A stationary-blading profile with subsonic flow; these studies showed that the steam expands along the entire nozzle with complete supercooling. Unsteady condensation arises first in the diverging exit section of the subsonic part. Condensation at this location near the throat induces significant changes in pressures and M, which vary with time. These changes also affect the quantity of moisture that forms. Calculations show that unsteady condensation of steam in such nozzle blading changes the percentage moisture downstream of the cascade by 1.5%, causes flowrate fluctuations $\Delta G/G_0$ of 1.3% and changes Δp by 0.2 MPa; the frequency of these fluctuations is $d = 1.31$ kHz.

Significant unsteady, variation in the flowrate and pressure of steam past the nozzle blading may result in failure of these elements.

Spontaneous formation of moisture in nozzle flows has of now been rather satisfactorily explored, both analytically and experimentally. However, steam flow condensation shocks may occur only in certain specific locations with the turbine flowpath. Spontaneous appearance of moisture and unsteady phenomena may occur at various locations during the condensation of steam in nuclear power plant turbines. At full load, the most likely location is in the nozzle blading of the first high-pressure stage. At partial load, it is probable only in certain parts of the low-pressure blading, in which the steam may expand from superheat to the point of intersection with the Wilson zone. This because the entire high-pressure blading operates at subsonic velocities, with moisture present in the steam from the very start and with developing turbulence. On the other hand, the low-pressure blading, which operates on wet steam has, in addition to elevated turbulence level, a developed three-dimensional flow and significant vertical variation in parameters, which significantly interferes with spontaneous formation of moisture. In individual cases, spontaneous condensation of steam may not occur uniformly over the moving blades, but only on the backwards facing side and in the vicinity of the oblique exit; this may be so even in the case when the expansion of the inlet steam occurs without intersection of the Wilson zone. The dimensions of the droplets may then range from 10^{-4} to 3×10^{-4} mm, i.e., it may be greater than for conditions of spontaneous condensation in nozzles where $r_3 \simeq 0.3 \times 10^{-4}$ mm.

The most probable form of condensate formation in turbines is the formation of moisture droplets in edge wakes and in the oblique exits from turbine cascades. Studies of moisture size distribution past the C-9012A two-dimensional nozzle cascade ($\bar{t} = t/b = 0.75$, $l/b = 0.7$, $\Delta = 1.7$ mm, M = 0.93) have been carried out using an optical (laser) method for measuring particle sizes. At the start of the expansion, the steam was superheated ($\Delta T_0 > 7\,°\mathrm{K}$) at the exit from the cascade; the calculated equilibrium exit moisture concentration was $y = 2.2\%$. Moisture forms past the exit edges and precipitates at the highest rate past the blades on the back side (Fig. 7.3). The largest droplets ($d_3 \geq 10^{-4}$ mm) precipitate in the edge wake, in the zone with developed vortex motion. Studies by Filippov [7.1] show that the vortex street past the exit edge of nozzle blading (Fig. 7.3) is a controlling moisture precipitation source in a turbine stage operating at M < 1.0. In addition, the vortex street exerts a significant effect on the entire course of steam flow in the clearance and in the rotating blading of the stage, since the accelerating vortices break up and the resulting moisture penetrates into the main stream. Experimental studies of the size distribution of moisture in a turbine [7.5] showed that, past the rotating cascade (the experiments were performed with an inverted turbine model), the moisture is distributed uniformly over the pitch, and the dimensions of particles were much higher than in spontaneous condensation ($r_3 \leq 1.3 \times 10^{-4}$–$1.7 \times 10^{-4}$ mm).

This means that the flow of steam in a turbine stage at subsonic velocities is characteristically accompanied by intense precipitation of moisture in vortices. This process subsequently results in virtually equilibrium flow of steam in turbine flowpassages, since the number of droplets arising in the vortices becomes sufficient for

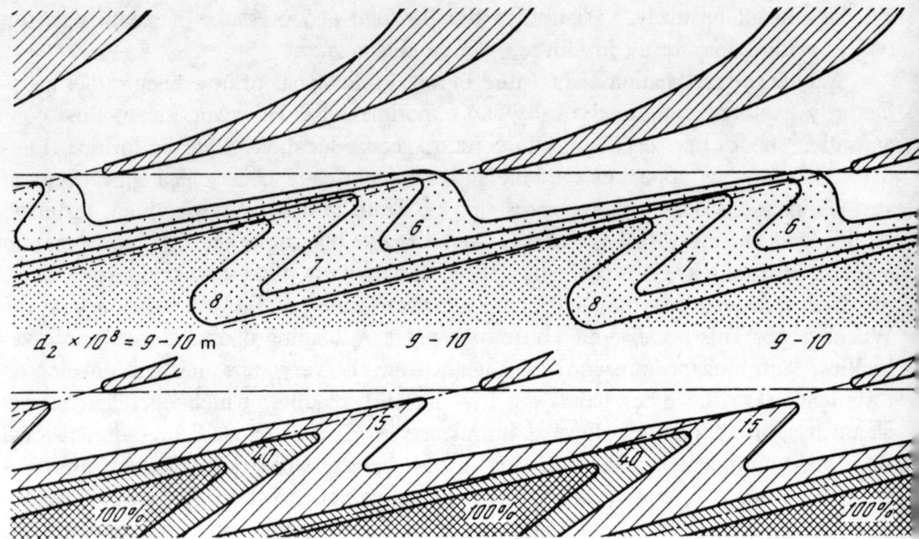

Figure 7.3 Droplet size and relative moisture content distribution fields in steam flow over a nozzle cascade [7.1].

condensation of steam on their surfaces. Naturally, the droplets, following formation, grow by condensation of steam on them as they move through the turbine flowpassages. The most extensively used form of description of droplet growth is the Hertz-Knudsen equation [7.4]:

$$\frac{dr}{dt} = \frac{\alpha p}{\sqrt{2\pi RT}} \left(1 - \sqrt{\frac{T}{T_{sat}}} \right) \tag{7.2}$$

where α is the condensation coefficient and R is the gas constant.

Analysis shows that, upon traversing the entire length of the turbine flowpassage, the droplets' size increases insignificantly (approximately twofold) by condensation of steam on their surfaces. In addition to the formation of moisture in vortices formed at the exit edge of a turbine stage, moisture forms also in the secondary flow zone in the region of flow separation. These processes of moisture formation are most characteristic for off-design turbine operation conditions (at partial load), when flow over the turbine cascades is accompanied by developed separation of flow at the roots of the blades and at the inlet edges of these blades.

In contrast to the results from experimental and analytic studies of spontaneous condensation of steam in Laval nozzles, moisture formation in real flows of steam in turbine flowpassages may occur partially at an earlier stage as a result of elevated flow turbulence. This induces local precipitation of moisture and a responding down-stream shift of the start of spontaneous condensation of steam. It can be assumed that under off-design conditions and, taking account of the elevated turbulence level, the supercooling of the steam does not attain the maximum value and spontaneous moisture formation does not occur.

7.2 PARAMETERS OF TWO–PHASE FLOW IN A TURBINE STAGE

Above, we considered the formation of moisture with particle diameters less than 1 μm and, naturally, such moisture follows the steam flow. The efficiency of turbines flowpassages and the occurrence of erosion failure of rotating blading, depend to a significant degree on the percentage content of large-size moisture droplets. Large-size moisture droplets form from liquid which precipitates on the solid surfaces of the flowpassages forming liquid films. These large droplets move at velocities which differ significantly in magnitude and direction from those of the steam. They form in turbine flowpassages basically as a result of breakup of liquid films which flow along the stationary and rotary blading.

As of now, the least explored problems are the formation of large-size droplets from smaller ones, the precipitation of small droplets on the surfaces of stationary and rotating blading and the formation of thin liquid films.

Large-size droplet formation in turbine flowpassages occurs in the following stages (as shown in Sec. 7.1). Fine droplets form in the flow of supercooled steam where they are induced to precipitate on the surface of the nozzle blading and the cylinder casing by turbulent fluctuations in the boundary layer regions. These droplets combine into liquid films which subsequently acquire a wavy surface, generate additional turbulence in the wall zone and thus speed up the precipitation of liquid droplets. The moisture thus gradually accumulates in the film.

Subsequently there are possibly two cases of breakup of films and formation of large droplets as a result: (1) shedding of liquid films from the exit edges of rotating blades and shrouds; (2) breakoff of large droplets from the surface of waves on the films. In addition to this, films and large droplets can form on the surface of rotating blading as a result of condensation of steam.

Deych et al. [7.8] observed that for the case when moisture precipitation occurred within a single stage, there was a vertically nonuniform distribution of moisture flow and drop size in the region downstream of the rotating blades. This occurs because steam condenses on the rotating blading and forms a continuous film or separate rivulets, which are injected into the steam flow. In spite of the small thicknesses of the liquid films (rivulets), the droplets which are shed from the exit edges of rotating blading have diameters exceeding by several-fold the thickness of the films from which they formed; this is because the velocities of moisture along the rotating blading in this case are low and the liquid accumulates gradually on the exit edges.

The liquid particles are torn off from the exit edges of the blades when the surface tension force holding the droplets to the blade surfaces falls below the breaking off force due to centrifugal actions.

When moisture is present in the steam supplied to the stage inlet (i.e., at $y_0 > 0$), both fine and large size drops are present, the velocity and direction of the latter

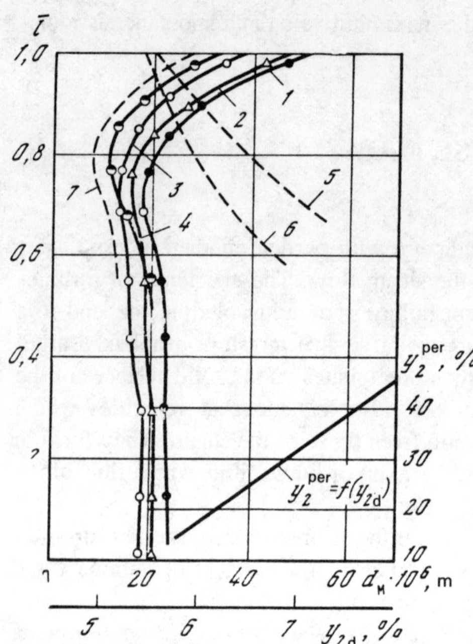

Figure 7.4 Variation in mean droplet diameter over the height of a blade at different turbine rotor ppm and y_0 = var. 1, 2, n = 4000 rpm, 1, y_0 = 6.1%; 2, y_0 = 9%; 3, n = 6000 rpm, y_0 = 6%; 4, n = 8000 rpm, y_0 = 9%; 5, 6, n = 7000 and 9000 rpm (experiments at the Khar'kov Turbine Plant [7.8]; 7, n = 10,000 rpm.

differing significantly from those of the steam. Naturally, the fraction of the moisture in the form of large droplets in a given flowpassage cross section is controlled by a number of factors, among which are the location where the condensation starts, the efficiency of turbine stages, the distribution of enthalpy dissipation, the values of Re and M, the rotor rpm, etc.

When the steam entering the turbine stage is wet, several types of moisture flux will exist past the rotating cascade: liquid shed from the inlet and exit edges of rotating blading; liquid particles forming upon impact on the blades or those shed from film surfaces; moisture that passed the flowpassages without coming into contact with the blades, etc. (see below). Naturally, the processes of formation of large moisture droplets in all these cases are also different and will depend on the geometry and operating conditions.

Figure 7.4 shows results of measurements of droplet size distribution under various operating conditions (ω = var' y_0 = var) over the radius ($\bar{l}$) of a rotor blade following the last (seventh) turbine stage. The droplet size increases toward the periphery of the blades (curves 1 through 4). Also, the bulk of the moisture is concentrated at the periphery of turbine stages. Thus, for example, when the percentage moisture in the steam leaving the fourth stage in experiments of the Khar'kov Turbine Plant (as calculated from steam charts) was y = 8%, the percentage moisture at the periphery (y_2^{per}) was more than 40% (Fig. 7.4).

However, the variation of mean drop size with radial position [d_M = $f(\bar{l})$] does not always follow the variation of y = $f(l)$. Thus, in experiments performed at the Khar'kov Turbine Plant with a multistage experimental turbine [7.8] under various operating conditions with the circumferential velocity-varying curves 5 through 7 in

Fig. 7.4), increases in droplet-size at the periphery were not always observed. This can be attributed to two factors: (1) vertical variation in the shape of the rotating airfoil, i.e., the nature of motion of liquid on the surface of the blades and (2) increased concentration of moisture at the inlet to the stage under study (at $\bar{l} > 0.6$–0.8) due to the significant increase in the turbine flowpassage area.

The mean diameter d_M of the drops is significantly affected by the rotor rpm (curves 2, 4 and 5 through 7 in Fig. 7.4). With increasing rotor rpm (i.e., circumferential velocity u of the rotor blades) the modal size of droplets falls at all the moisture contents (see curves of $d_M = f(y)$ in Fig. 7.4). The increase in rotor rpm increases the normal component of the velocity at which the drops impact on the exit parts of rotor blades. Consequently, more drops are broken up, the intensity of irrigation of the rotor blade surfaces decreases and, finally, the rate of shedding of moisture from the inlet edges of the rotor blades increases. The influence of the last factor on the change in the variation in droplet size distribution can be confirmed by results of experiments on a rotating disk with water supplied to its center. As in experiments with a turbine stage, the value of d_M increases with increasing moisture flowrate Q (shaded curves in Fig. 7.5). It is interesting to note, however, that with increasing circumferential velocity u the droplets shed from the edge of the disk (the thickness of the edge was 0.5 mm) are increasingly smaller. The satisfactory agreement of experiments results (Fig. 7.5) for a disk and a multistage turbine confirms the fact that the shedding of moisture from the exit edges of rotating blades is the controlling process in establishing the dimensions of moisture droplets in the flow of steam.

Figure 7.6 shows the results of investigation of the effect of percentage moisture y downstream of a turbine stage on d_M. As y increases the location where condensation starts moves upstream, i.e., moisture starts forming earlier. Naturally, the earlier start of concentration results in the formation of larger moisture droplets toward the end of the flowpassage. Over a wide range of operating conditions, d_M increases almost linearly with y. Figure 7.6 shows the results of several measurements; the curves show the behavior of $d_M = f(y_{2d})$ at the periphery of various stages and also at positions downstream of stages 4, 6, and 7, respectively. The measurements were made at constant rpm but at different pressures and circumferential velocities. Curves 4 through 6 in Fig. 7.6 represent the variation of d_M with u downstream of the seventh (last) stage at $y_{2d} = $ const.

The above studies show that the drop size spectrum changes significantly with

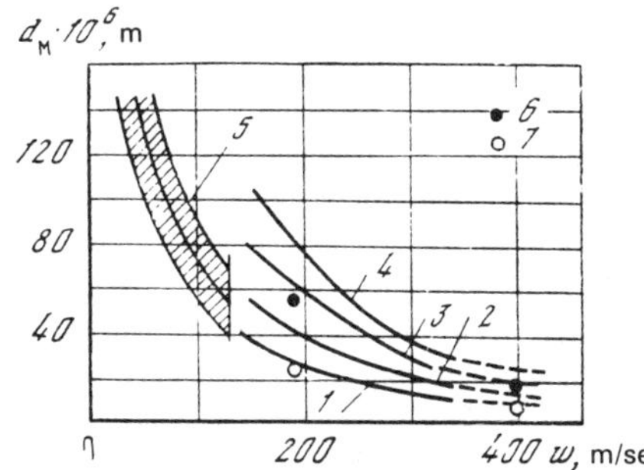

Figure 7.5 Variation in mean drop diameter as a function of circumferential velocity. 1, $y = 4$; 2, 6%; 3, 8%; 4, 10%; 5, disk spraying $Q = 0.7 \cdot 10^{-6}$ m^3/sec; 6, 7, experiments at the Khar'kov Turbine Plant [7.8].

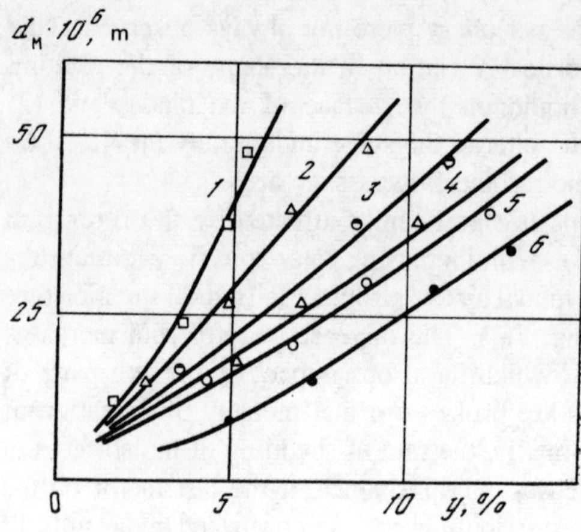

Figure 7.6 Variation in mean diameter of droplets downstream of a turbine stage as a function of the percentage moisture in the steam.

turbine operating conditions. An accumulation of a large volume of experimental data showed that the form of the droplet size distribution downstream of the rotating blading is practically independent of turbine geometry and operating conditions. The drop size distribution is satisfactorily approximated by normal Gaussian distribution function:

$$m_i/m_M = \exp\left[-\pi\left(\frac{d_i}{d_M} - 1\right)^2\right]$$

where m_i is the mass of droplets with diameter d_i, whereas m_M is the mass of droplets with modal diameter d_M.

Figure 7.7 shows the effect of turbine operating conditions and droplet size on the angle of droplet discharge from the rotor cascade. Note that in the plane of measurements (the probe could not detect moisture moving in radial direction) and over the entire range of probe placement (from 0 to 180°) the droplet size was large. However, the largest particles move in a direction close to tangential ($\Delta = 0.7$–1) and exit the cascade at an angle which differs most from the exit angle of the steam.

Consequently, at low u the controlling droplet flux past the rotating blading is the moisture shed from the exit edges of the rotating blading (curve 1 in Fig. 7.7). With

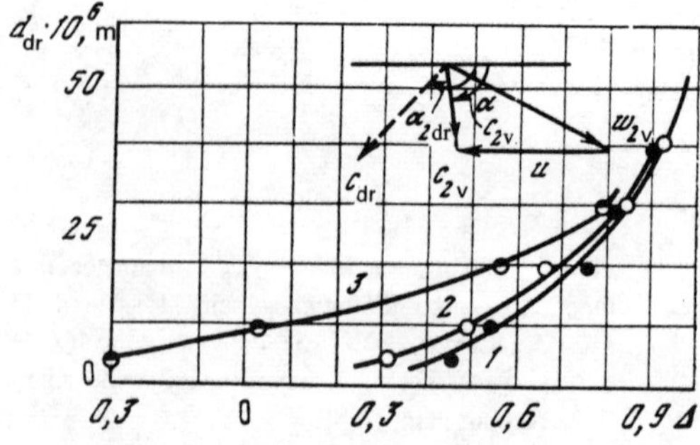

Figure 7.7 Effect of droplet diameter on deviation Δ of the motion of moisture vs. the flow direction α_{21}. 1, $u = 160$ m/sec; 2, 240 m/sec; 3, 330 m/sec.

increasing rpm, the extent of droplet breakup on impact increases, and the nature of moisture motion over the rotating blading changes; for example, the range of angles of exit of droplets from the rotating blading widens (curve 3). Here, one already observes fluxes of moisture particles which leave the stationary to rotating blading clearance at an angle of $\alpha_{21} < \alpha_{23}$ ($\Delta < 0$).

Data on the percentage content of large-size droplets is sparse. Unfortunately, in the studies described above, only indirect estimates of this quantity were obtained from data on the moisture-removal efficiency. It is hence of interest to analyze other direct measurements of this quantity in turbine flowpassages. These measurements were performed by measuring the size of impressions made by the drops on an oxide film; even though allowance for the exposure time was made, such have a significant error. Nevertheless, the results obtained in measurements by Deych et al. [7.6] and Kiryukhin et al. [7.7] allow analysis of the effect of operating conditions on the content of large-size droplets and give a practical indication of the actual processes of moisture separation in turbine stages.

The fraction λ of large-size droplets is given by $\lambda = G_{ldr}/yG_m$, where G_{ld} is the flowrate of large-size droplet moisture, y is the total percentage moisture and G_m is the flowrate of wet steam per unit time at the location under study. λ in general depends on location, on the value of y, on the geometry of the cascade, etc. Investigations show that the distribution of λ over the flowpassage is close to the corresponding distribution of total percentage moisture. In all cases the percentage of large-size moisture droplets increases with the overall percentage moisture. At low moisture content (as calculated for equilibrium from the steam tables) ($y_{2d} < 3\%$), the moisture contained in large droplets is only a few percent of the total moisture, since in this case, moisture forms only within the stage under study. However, under off design conditions (rotor rpm $\omega < 0.5\omega_{nom}$ at $y_{2d} = 3$–4%) the amount of moisture contained in the large droplets increases significantly, since, due to the low efficiency, several stages will be operating on wet steam.

In general the following classification of droplet sizes can be made for the two–phase flow past the nozzle blading at relatively low pressures ($p_0 < 2\text{kg/cm}^2$).

1. Small particles, with $d_M < 1$ μm, which pass through the flowpassage without coming into contact with the blading surfaces. These particles can either be present in the flow entering into the cascade, or form within the cascade as a result of condensation. The motion of these droplets is close to that of the steam, both in velocity and direction.

2. Medium-size particles, with 1 μm $< d_M < 10$ μm, which were present in the flow upstream of the cascade, but which for various reasons did not collide with the blades (the cascade pitch is large, the angle of entry of the droplets is close to the angle of exit of flow from the cascade, etc.). Such droplets move with slip, increasing the mean exit angle of the flow.

3. The largest particles, which form in the edge wake.

4. Particles forming by breakup of large droplets within the flowpassage due to loss of their stability (without contact with the flowpassage walls).

5. Particles forming as a result of reflection into the flow of a part of the mass of large droplets impacting on the surfaces of the aerofoil (or film).
6. Droplets which form as a result of breakoff and breakup of a film flowing over the aerofoil surface. Of greatest interest for this study are droplets forming as a result of shedding of the film from the convex part of the aerofoil. These particles, in addition to their large size, have an exit angle α_{12} at the channel exit greater than that of the steam phase. This makes their effect, in reducing the efficiency and causing erosion of the stages, similar to that of particles moving in the edge wake.

Figure 7.8 shows a typical pattern of breakup of a liquid film shed from the exit edge of a nozzle blade. Such shedding is characterized by the fact that, at low percentage moisture in the steam entering the turbine stage, the film does not break up simultaneously over the entire height of the blade. Rather, large droplets are ejected locally at various frequencies.

Figure 7.9 shows the structure of moisture downstream of the nozzle blading and the principal features of the motion of moisture droplets along the nozzle blading pitch. The mass mean diameter of droplets at the inlet, upstream of the cascade, was greater than 55 μm. The initial percentage moisture was $y_0 = 2.5\%$. It is seen from Fig. 7.9 that multipeak distributions occur at many points. This is primarily due to the presence of diverse sources for the droplets. These are zones where film separa-

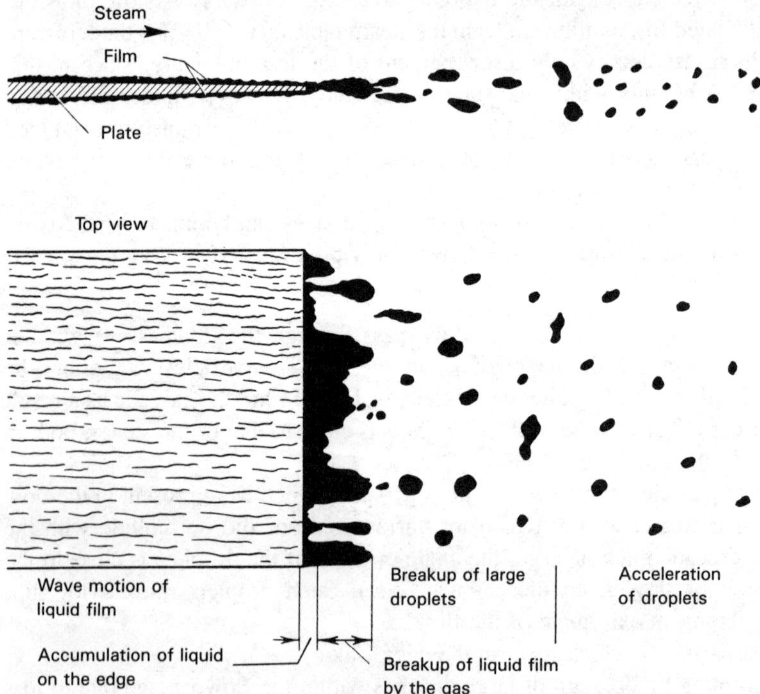

Figure 7.8 Breakup of liquid films shed from the exit edge of nozzle blading.

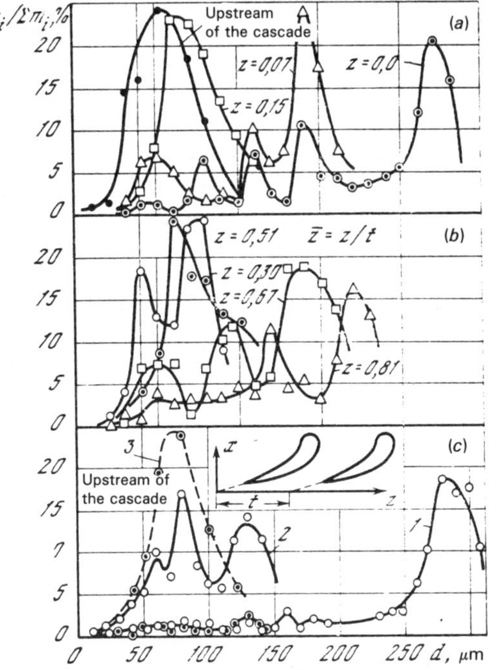

Figure 7.9 Droplet size distribution downstream of nozzle blading. (*a*) and (*b*) Plot of $m_i/\Sigma m_i$ vs. droplet diameter along the pitch of nozzle blading (C-9012A air-foil cross section) at $\bar{x} = x/t = 0.1$, $y_0 = 2.5\%$, $\epsilon = 0.72$, Re $= 7 \cdot 10^5$, $\bar{t} = 0.75$, $b = 72.5$ mm; (*c*) effect of edge separation of moisture on the droplet size distribution; 1, without film suction; 2, with film suction ($m_i/\Sigma m_2$); 3, with film suction ($m_i/\Sigma m_{1[\text{sic}]}$); $y_0 = 4.5\%$, $\epsilon = 0.65$, $\bar{t} = 0.75$, $x = 5$mm.

tion, breakup and reflection of incident droplets, etc. occurs. Naturally, in each case, particles appear which have a characteristic size, the magnitude of which depends on the specific conditions at the point of droplet formation (velocity of steam and droplets, film thickness, dimensions of incident droplets and their angle of incidence). Droplets with different dimensions and velocities are not swirled in the same manner in the boundary layer, the directions and magnitudes of the lift forces acting on them are not the same. As a result, the particles move along different paths and, in any given volume within the channel downstream of the cascade, there may exist at any time droplets which formed under different conditions.

7.3 MOTION OF MOISTURE PARTICLES IN TURBINE–STAGE PASSAGES

The principal features and separation capacity of a turbine stage depend significantly on the patterns of motion of various liquid particles both in the nozzle and rotor blading.

Analysis of forces acting on a single droplet shows that, in analyzing the paths of moisture motion in a turbine stage, only the aerodynamic forces exerted by the steam flow on the droplet need be considered.

The equation of motion of a single droplet is:

$$m_3 \frac{d\mathbf{w}_3}{d\tau} = C_D S_M \rho_1 \frac{w_{rel}}{2} \mathbf{w}_{rel} \tag{7.3}$$

where d_3 is the droplet diameter, $\mathbf{w}_{rel} = \mathbf{w}_1 - \mathbf{w}_3$; $S_M = \pi d_3^2/4$, and C_D is the drag coefficient of the droplet; if it is assumed that the droplet is spherically shaped, then $m_3 = \pi \rho_3 d_3^2/6$ (the values of C_D are given in Table 2).

In considering the two-dimensional problem, we can write the equations of motion for the droplet in nondimensional form, taking as the reference dimension the chord b of the turbine blade airfoil, and as the reference velocity either the velocity of steam upstream of the nozzle blading c_0 or the velocity $\mathbf{w} = \mathbf{c}_1 - \mathbf{u}$ upstream of the rotating blading:

$$\frac{d\mathbf{c}_{3x}}{d\bar{\tau}} = \frac{C_D(\mathrm{Re})}{\tau_m}(\mathbf{c}_{1x} - \mathbf{c}_{3x})$$

$$\tag{7.4}$$

$$\frac{d\mathbf{c}_{3y}}{d\bar{\tau}} = \frac{C_D(\mathrm{Re})}{\tau_m}(\mathbf{c}_{1y} - \mathbf{c}_{3y})$$

where d_3 is the droplet diameter, $\mathbf{c}_{3x} = dx/d\tau$; $\mathbf{c}_{3y} = dy/d\tau$; $\mathbf{c}_{3x} = c_{3x}/c_0$; $\mathbf{c}_{3y} = c_{3y}/c_0$; $\mathbf{c}_{1x} = c_{1x}/c_0$; $\mathbf{c}_{1y} = c_{1y}/c_0$; $\mathbf{x} = x/b$; $\mathbf{y} = y/b$; $\tau = \tau/(b/c_0)$; $\mathrm{Re} = (\rho_1 d_3 c_0)/\mu_1 (\mathbf{c}_{1x} - \mathbf{c}_{3x})^2 + (\mathbf{c}1y) - \mathbf{c}_{3y})^2]^{1/2}$; and $\tau_m = 4/3(d_3^2 c_0 \rho_3)/(\mu_1 b)$ is the conditional relaxation time of the motion.

It is seen that the part of droplet motion within the cascades is a function of the nondimensional quantities Re and τ_m. To be able to determine the radial motion of droplets with the cascade it is necessary to consider the droplet motion in cylindrical coordinates r, z, φ.

By analogy with Eq. (7.4) we obtain the following equations of motion for the three-dimensional problem:

$$\frac{dc_{3r}}{d\tau} = \omega^2 r + 2\omega c_{3\varphi} + \frac{c_{3\varphi}^2}{r} + \frac{3}{4} c_D \frac{\rho_1}{\rho_2} \frac{c_{rel1}}{d_3}(c_{1r} - c_{3r})$$

$$\frac{dc_{3z}}{d\tau} = \frac{3}{4} c_D \frac{\rho_1}{\rho_3} \frac{c_{rel2}}{d_3}(c_{1r} - c_{3r}) \tag{7.5}$$

$$\frac{dc_{2\varphi}}{d\tau} = -2\omega c_{3r} - \frac{c_{3\varphi} c_{3r}}{r} + \frac{3}{4} \frac{\rho_1}{\rho_3} \frac{c_{rel\,\varphi}}{d_{dr}}(c_{1\varphi} - c_{3\varphi})$$

where $\mathbf{c}_{3r}$, $\mathbf{c}_{3\varphi}$, $\mathbf{c}_{3z}$, $\mathbf{c}_{1r}$, $\mathbf{c}_{1\varphi}$, $\mathbf{c}_{1r}$ are the nondimensional projections of the droplet and steam velocities (referred to velocity c_0 or ω_1; $\mathbf{r}, \mathbf{z}, \boldsymbol{\varphi}$ are nondimensional coordinates; $\tau = \tau/(x_0/c_0$ is the nondimensional time; ω is the angular velocity; x_0 is the characteristic dimension, and τ_m is the nondimensional time of relaxation of motion.

Equations (7.5) for nozzle and stationary guide vanes can be simplified to the form

$$\frac{dc_{3r}}{d\tau} = \frac{c_{3\varphi}^2}{\bar{r}} + \frac{c_D(\mathrm{Re})}{\tau_m}(c_{1r} - c_{3r})$$

$$\frac{dc_{3\varphi}}{d\tau} = \frac{c_{3\varphi}c_{3r}}{\bar{r}} + \frac{c_D(\mathrm{Re})}{\tau_m} \qquad \frac{dc_{3z}}{d\tau} = \frac{c_D(\mathrm{Re})}{\tau_m}(c_{1r} - c_{3r}) \qquad (7.6)$$

Equations (7.3)–(7.6) make it possible, once the steam velocity field and the initial conditions are known, to determine the paths of moisture motion in different parts of turbine flowpassages.

In predicting moisture motion from Eqs. (7.3), the velocity of gas in any point of the channel may be estimated by assuring plane-parallel incompressible flow and employing the generalized Cauchy formula [7.9]:

$$w(z) = w_\infty + \frac{1}{2t_i} \int_L w(\xi)\, \mathrm{cth}\,\frac{\pi}{t}(z - \xi)\, d\xi \qquad (7.7)$$

Expressing the complex function in Eq. (7.7) in parametric form, the expression for velocity may be expanded into its real and imaginary parts and expressions for the velocity components obtained in nondimensional form for any point in the channel

$$w_x = 1 - \frac{1}{2}\int_0^{2\pi} \frac{1}{2}\, \frac{\sin 2\pi(y - \mathbf{x})\, w(\Theta)\, \Omega(\Theta)}{\mathrm{ctg}^2\,\pi(x - y) - \cos^2\pi(y - \mathbf{x})}$$

$$(7.8)$$

$$w_y = \frac{1}{2}\int_0^{2\pi} \frac{1}{2}\, \frac{\mathrm{sh}\,2\pi(x - y)\, w(\Theta)\, \Omega(\Theta)}{\mathrm{ch}^2\,\pi(x - y) - \cos^2\pi(y - \mathbf{x})}\, d\Theta - \Gamma$$

where $\mathbf{x}$ and $\mathbf{y}$ are the coordinates of the point at which the velocity is determined (x and y are referred to the cascade pitch), Γ is the circulation velocity, $w(\Theta)$ is the nondimensional velocity over the surface of the cascade, defined by the solution of integral equations using the method of Zhukovskiy et al. [7.9].

The error in determining the velocity over the airfoil does not exceed 3%, which is also the error in determining the velocity in any point in the flowpassage.

The differential equations [7.5] describing the motion of a droplet were solved by the method of characteristics. When the droplet moved from the guiding to the working passage the absolute motion was replaced by relative motion.

Experimental studies of the paths of solid particles in a model of a turbine stage [7.10] indicate the existence of the various (and often very complex) forms of motion of which liquid particles might be expected to follow in passages within the cascades and in the spaces between them. The paths of the particles depend on inlet conditions and on the location of rotor blades relative to the nozzles. The motion of the particles was investigated by means of high-speed cine photography (filming rate of 6000 frame/sec). These studies made it possible to not only follow the paths of particle motion, but also to determine the particle velocities at different parts of the turbine stage.

Figure 7.10*a* shows the principal dimensions of the cascades and the experimental paths of solid particle motion for the case where the particle entered the nozzle flow passage without impacting on the inlet edge of the stationary blading. Below, in Fig. 7.10*b*, is shown the case when the solid particles collide with the inlet edges of the stationary (nozzle) blading and then collide with the concave surface of nozzle blading and with rotor blading. In Fig. 7.10 the time between identically numbered points (particles) is 3.3×10^{-4} sec, and the initial velocity of the particles at entry to the nozzle blading is 45 m/sec. It is seen that the moving particles may be either accelerated or decelerated as a result of collision and their direction of motion may change. It follows from the experiments that solid particles may impact with the blading in the passages within the cascades and may undergo multiple collisions, for example, with the inlet edges or with the concave part of the rotor blades. Naturally, studies of the motion of solid particles in the flowpassages of model turbine stages is only an approximation of the real pattern of movement of moisture droplets in real turbines. However, these studies made it possible to determine the possible droplet motion patterns.

To be able to calculate the motion of droplets in turbine-stage passages one must know the real vertical distribution of moisture at the inlet, the angle of inlet to the stationary blading, the size and velocity of droplets and the relationships governing the reflection of the droplets upon collision with stationary and rotating blades.

The percentage moisture in the steam upstream of a turbine stage is plotted in Fig. 7.11 [7.3]. The paths of the moving moisture depend significantly on the mean droplet diameter d_M of the droplets upstream of the stage. As an illustration, Fig. 7.12 shows very characteristic curves of $dn/d\bar{S}$ at three cross sections for the circumferential velocities actually encountered in long blades.

Note that the largest droplets move in a plane close to normal relative to the turbine axis ($\alpha_2^{dr} \rightarrow 180°$) (see, for example, the data in Fig. 7.8). For example, according to Stiasny and Taič [7.11], droplets with d_M between 50 and 150 μm move past the last stage of a turbine ($N = 200$ MW) at angles α_3 ranging from 150 to 180°. At the same time, the angles of motion of fine droplets vary over a rather wide range.

Obviously, at very high steam pressures (in the high-pressure stages) the curves of $d_3 = f(\Delta)$ and $\alpha_2^{dr} = f(d_M)$ will have an entirely different shape. However, at present no data are available on the droplet size distribution and direction of motion over high pressure turbine stages. It can be assumed that, for large-size droplets, the shape of the curves will not change significantly with rising p_0 provided that it is assumed also that the minimum size of large droplets will change with rising steam pressure.

Calculations show that at low pressures (i.e., $p_0 \leq 0.5$ MPa, characteristic of the low-pressure stages) the diameter of droplets in the stationary blading does not exceed 20 μm, and that in the rotating blades is around 5 μm. For the high-pressure stages ($p_0 > 1.5$ MPa), on the other hand, the minimum diameter of large droplets may be as high as $d_M = 50$ μm for the nozzle blading and 30 μm for the rotor. Naturally, such a significant increase in d_M for large droplets in the high-pressure stages of the turbine

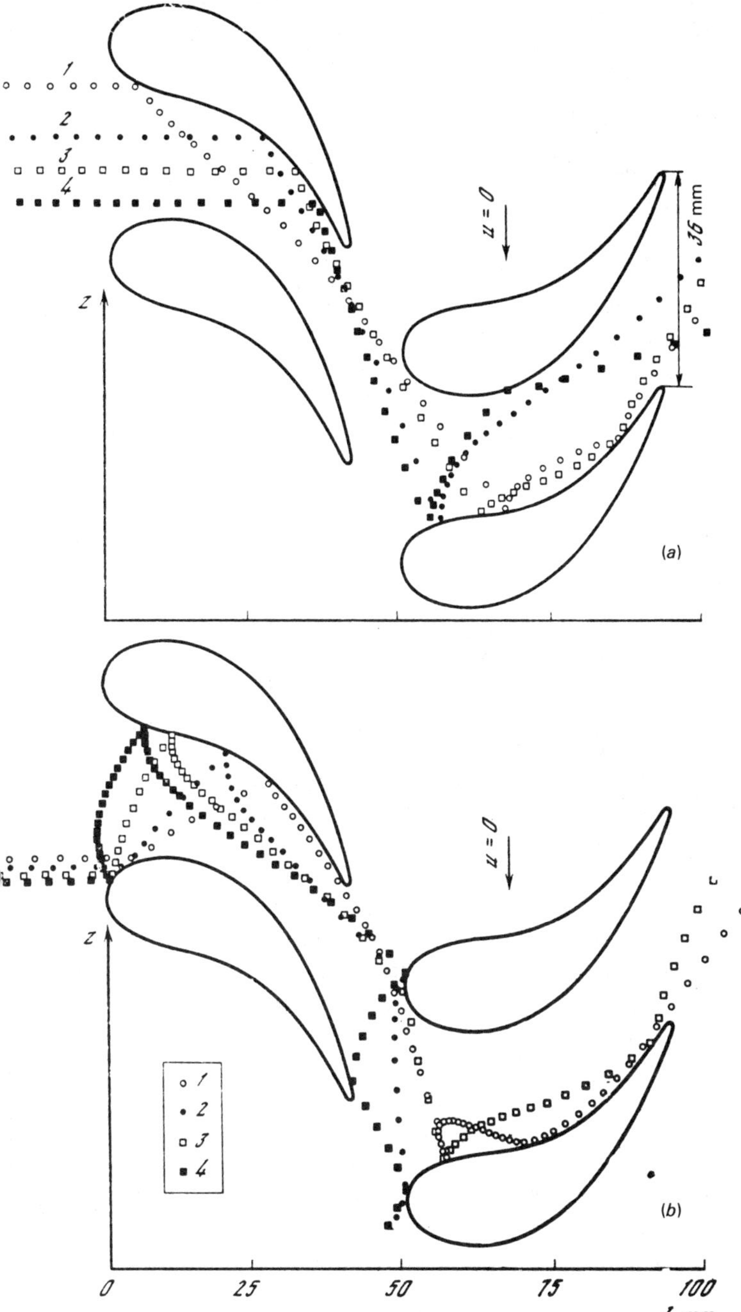

Figure 7.10 Paths of solid particles (1–4) in a turbine-stage model.

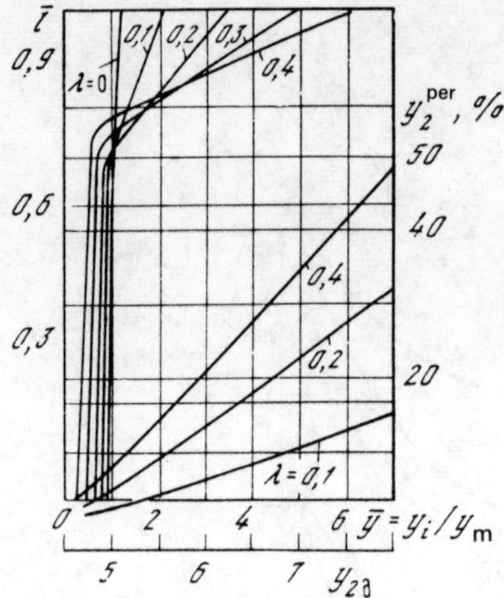

Figure 7.11 Distribution of moisture $y = \bar{y}_i / y_m$ downstream of a stage at height $\bar{l}$ along the blade and the percentage moisture y_2^{per} at the periphery of a stage vs. the mean percentage moisture y_{2d} calculated from steam tables for different values of λ, the fraction of the moisture in the form of large droplets.

will result in a significant reduction in the fraction of these droplets in the flow and will have a corresponding effect on the power-output and separation characteristics of the wet-steam flow.

Let us consider the theoretical effectiveness of moisture precipitation on the surface of a nozzle blading for moisture distributions actually encountered at the inlet to such blading. The initial conditions for $y_0 = 9$ and 3% were taken from Figs. 7.6 through 7.12 with allowance for the droplet size distribution function. Figure 7.12 shows the results of calculation of the variation in precipitation of moisture along the back and the concave parts of a nozzle blade for these cases. It was assumed here that the moisture which comes into contact with the blade surface is not returned to the flow, i.e., remains on the blade. It is seen that precipitation on the back is observed only in the vicinity of the inlet edge. On the concave part of the blade moisture is deposited over the entire length with virtually constant rate of deposition ($d\eta/\partial s =$ const). Four times as much moisture precipitates per unit surface of the concave part than on the convex part (at $y = 8\%$). With reduction in the percentage moisture in the steam, the difference in the effectiveness of precipitation on the convex and concave sides of the blade increases even more.

The results of calculations shown in Fig. 7.12 were obtained at steam pressures entering the nozzle blading typical of the last low-pressure stages. With increasing density of the steam ($p_0 > 0.5$ MPa) the content of large-droplet moisture decreases; for example, at $p_0 = 3$ MPa the fraction of such moisture, as compared with calculations presented above, decreases sixfold. The above circumstance not only reduces the rate of moisture precipitation on turbine blade surfaces, but increases the difference between the amounts of moisture precipitating on the back and front sides of the blade.

Precipitation of moisture was also calculated for three cross sections over the height of a last stage blade (at $y_0 = 8\%$ entering the nozzle blading). The results of these calculations are shown in Fig. 7.13. As follows from these data, at the root section III (Fig. 7.12) the greatest precipitation occurs at the inlet edge on the back side, and also on the exit edge of the front part of the blade. Precipitation in the middle section II of the rotor blade is qualitatively analogous to the precipitation on a stationary nozzle blade. At the same time, peripheral section I in the assumed model has the lowest precipitation rate, and moisture precipitates in a rather narrow zone on the back of the blade on the inlet-edge side. Note that, allowing for all the real processes inherent to motion of liquid droplets in cascades passages (for example, coalescence, breakup, boundary-layer motion, etc.), "dry" zones on blade surfaces, as predicted in these calculations, will be absent.

The real process of moisture deposition on the surface of rotating blading will differ significantly from that considered above, since, at all locations, there are present fluxes of droplets which are reflected after collision with the blades.

At the present time, there are very few data on reflection of moisture particles upon collision with solid surfaces. One of the rare investigations of this problem was that performed at the Moscow Energetics Institute with both moving and nonmoving surfaces. These studies show that the impacting drops break up and are reflected over an angular range of 15°. For moisture particles impacting on rotating blading at normal incidence and with velocities in excess of 30 m/sec, the residence time on the surface is very short and the moisture is not entrained by the centrifugal force field. Thus, the reflected moisture has virtually no radial velocity component [7.1]. Hence

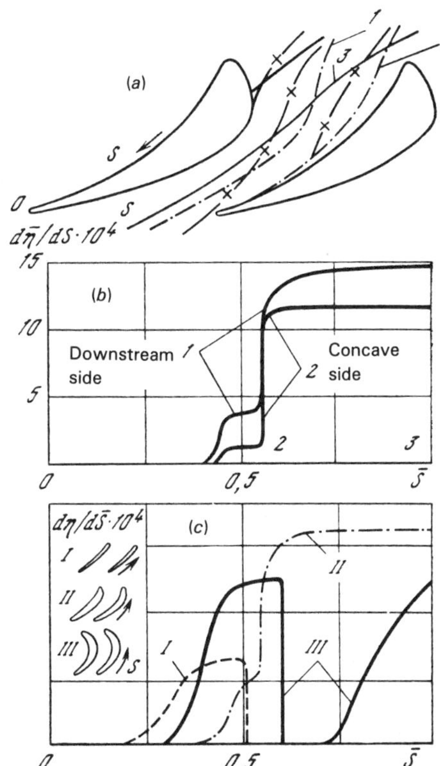

Figure 7.12 Paths of motion and deposition of moisture in turbine cascades. (a) Calculated paths of characteristic moisture fluxes in nozzle blading: 1, d_3 = 75 μm; 2, 40 μm; 3, 10 μm. (b) Distribution of the deposition of moisture over the surface of a nozzle blading cascade: 1, y_0 = 8~; 2, y_0 = 3%, $\dot{M}_{lt}$ = 0.8%. (c) Distribution of moisture precipitation over the peripheral (I), center (II) and root (III) cross sections of a rotor blade.

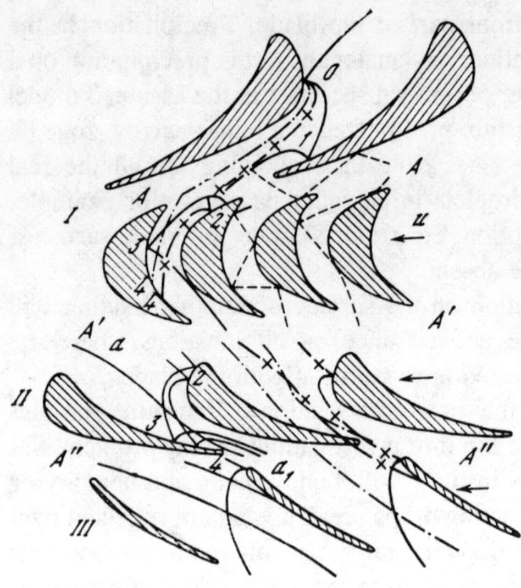

Figure 7.13 Analytically calculated paths of moisture drops reflected from blade surfaces in a turbine stage. 1, d_3 = 75 μm; 2, 30 μm; 3, 5 μm; 4, path of solid particles; I, velocity blading; II, reaction blading; III, peripheral section of long rotor blades [sic].

no fundamental difference was observed in the experiments between the reflection of drops from moving and nonmoving surfaces respectively. Note that we are considering here cases of collision when the effect of the steam boundary layer on the droplet is negligible.

Studies of the reflection of moisture show that the velocity of reflected particles at normal impact velocity of 30 m/sec ranges between 0.2 and 0.8 of the incident velocity [7.1]. The paths of particles in a turbine stage were calculated (Fig. 7.13) assuming that the velocity of reflected particles $c_{dr.ref}$ to be 0.8 of the incident velocity $c_{dr.in}$.

A computer program was used to determine the total droplet path through the nozzle blading, across the clearance between the nozzle and rotor blading and in the rotor cascade. Up to section A-A (see Fig. 7.13), the calculations were performed for absolute motion and, further downstream, the relative motion was estimated.

Let us consider the path of a droplet, taking account of its possible breakup and reflection at the surface. A particle 100 μm in diameter is incident at point 0 on the surface of a nozzle blade (Fig. 7.13). As a result of impact, it breaks up and is reflected into the flow. The ensuing droplet paths are significantly affected by droplet size. Calculation of the paths of three reflected particles of different diameter shows that large particles move across the channel after being reflected and reach the concave surface of the adjoining blade at point 1, whereas small particles are carried out of the nozzle-blading passage by the flow. Droplets which reach the blade surface at point 1 are also reflected into the flow. Numerous calculations show that, in a channel formed by TC-1A blades, it is highly improbable that the moisture reflected from the concave side of the blade will cross the passage and reach the back.

Large droplets that cross the clearance are incident on the surface of inlet edges on the back side of the rotor blades and are reflected. It is of interest to consider the case when the reflected moisture is capable of crossing the axial clearance and reaching the exit edges of the stationary cascade. According to calculations, such motion of moisture is possible; however, the probability of incidence of moisture on the exit edge of nozzle blades is low at high steam densities. Thus, for example, when the steam pressure in the clearance is 0.5 MPa, droplets with diameter smaller than 75 μm are reflected into the upstream stream by not more than 4 mm. Hence the known cases of failure of surfaces of exit edges of stationary blading in high-pressure parts of wet-steam turbines cannot be attributed to the effect of moisture reflected from the inlet edges of the rotating blades. Possibly this failure is a result of reflection of large (d_M = 200 μm) solid particles, which is also confirmed by results obtained by Filippov and Povarov [7.3].

A moisture particle (d_M = 75 μm, Fig. 7.13) which is incident at point 2 of the inlet edge of the rotor blade moves into the flowpassage of the rotor cascade and is incident onto the adjoining blade at point 3. The normal velocity of droplet impact at point 3 is high and reflection of moisture is possible here also. In Fig. 7.13 reflection of the droplet at points 1 through 3 is shown for this case breakup into smaller particles does not occur. In fact, in each collision zone there will appear a cluster of increasingly smaller droplets, as shown in starting point O.

However, moisture needs not always be reflected from the concave side of rotor blades. Depending on the angle and normal velocity of impact, the droplets may not be reflected from the blade, and may separate out on the blade surface (for example, at point 4, Fig. 7.13). The region of rotor blading where collision occurs without reflection of liquid particles should have the maximum effectiveness of moisture separation [7.3].

Droplets approaching rotor blading at an incidence angle of $\alpha_1 \leq 20°$ will also have significantly smaller angles of reflection and accordingly different paths of motion.

For comparison, Fig. 7.13 shows the paths of motion of solid spherical particles (path 4). It is seen that the solid particles perform multiple collisions with the rotor blading and with the exit edges of stationary blading. These analytic results are in satisfactory agreement with experimental data (Fig. 7.10). It should, however, be remembered that the experiments were performed with nonmoving rotor blading, whereas the calculations were made for $\omega > 0$.

For the same conditions at the exit of the nozzle blading as in the preceding cases, similar calculations were performed for reaction-type rotor blading and for channels characteristic of the peripheral cross sections of "long" blades. The typical paths obtained as a result are plotted in Fig. 7.13 (I and II). It can be concluded from analysis of these data that the transition from velocity to reaction blading is accompanied by a gradual reduction in the fraction of the droplets of a given diameter which are incident on the inlet edge. With reaction blading droplets of significant diameter may, after reflection, reach the opposite side of the passage. In addition, in reaction blading, and within the range of inlet angles and initial location actually encountered, large droplets may pass through the passage without impinging on the surface (path a-a, Fig. 7.13).

The above data on the motion of drops in turbine passages conform to the known facts on erosion wear on the inlet edge. In fact, a particle moving from the initial entry location and colliding with the inlet edge at point 2 (Fig. 7.13), has a maximum velocity component equal to the relative velocity of the particle. A small shift in the initial position reduces the angle of impingement of the particle on the surface, which increases the impulse received by the surface. Note that, due to the small curvature of the peripheral-section airfoil, a significant proportion of the large droplets pass through the channel without contact with the surface.

Naturally, the motion, collision and precipitation moisture depends on the operating conditions and on the geometry of the turbine stages. The analysis and method of calculation presented above make it possible, if the initial conditions (dimensions, velocity and direction of motion of droplets) are known, to describe with sufficient accuracy the principal relationships governing the motion of moisture in turbine-stage passages and to outline ways for improving turbine stages in order to raise their efficiency, reliability and separation capacity.

7.4 MOTION OF LIQUID FILMS IN A CENTRIFUGAL FORCE FIELD

A significant portion of the moisture formed in nuclear power plant turbines flows over the nozzle and rotating blading in the form of films. The physical processes of motion of liquid films over the nonmoving surfaces are analogous to those discussed in Chapter 2.

The separation of moisture, the drop size and the reduction in turbine efficiency are significantly influenced by the flow of liquid films on the surface of the rotor blading due to centrifugal and aerodynamic forces. Analysis of the forces acting on the liquid on surfaces of rotor blades shows that the centrifugal force is predominant [7.13].

Experimental studies of liquid films moving on the surface of a rotating disk were performed at the Moscow Energetics Institute [7.13] and will now be described. A film form on the disk as a result of spreading of liquid from the center to the periphery by inertia forces. The film motion was investigated using an electrical system for measuring local liquid film parameters (film thickness and phase velocity of waves), developed at the above Institute [7.12].

Observations in stroboscopic light, synchronized with rotation of the disk and also using cine and still photography point to the existence of a clearly developed wave structure of the phase interface. The waves propagate in a radial direction close to a phase velocity exceeding 8- to 10-fold the bulk radial velocity of the liquid. For fully developed wave motion, the phase velocity is virtually independent of the radius, whereas the height of the wavecrests is 2–3 times the thickness of the film in the troughs.

The following five typical zones were observed in the spreading of liquid over the disk (see Fig. 7.14): (I) a small zone at the center of the disk with a smooth or only slightly undulated film surface; (II) a zone of developed wave flow of the liquid film; (III) an annular zone in which the film begins to break up; (IV) a zone of jet

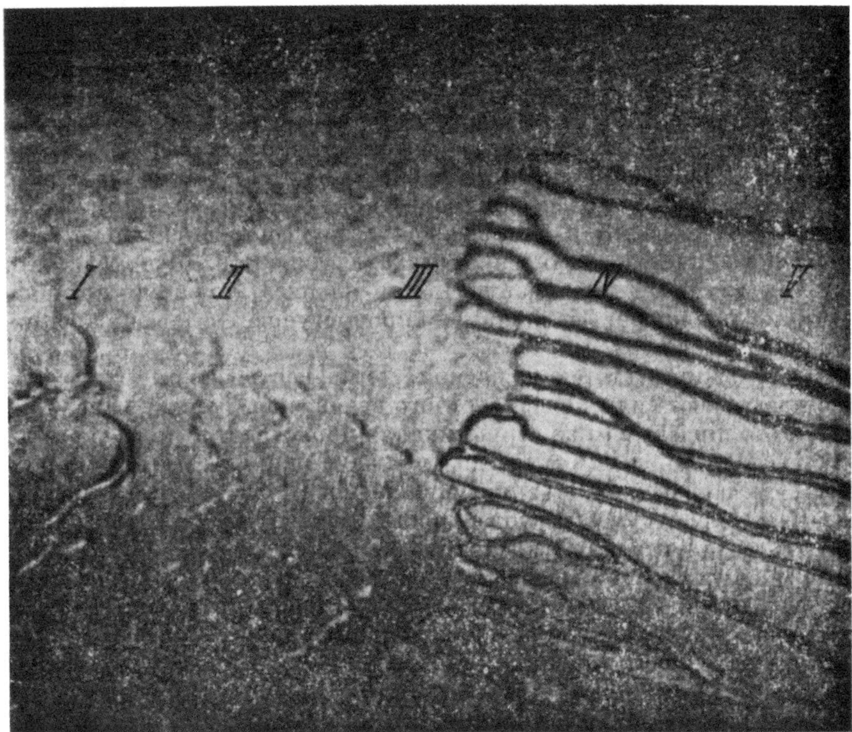

Figure 7.14 Typical zones of liquid flow over the surface of a rotating disk at $\omega = 300$ rps and $Q = 4 \cdot 10^{-6}$ m^3/sec.

flow of liquid over the disk surface; (V) a zone of breakup of jets into individual elements.

Experimental studies show that breaks in continuity in the film flow are possible both in the flow of thick films (low disk rpm) and of thin films (high rpm).

At low rpm, when the thickness of the liquid film exceeds significantly the height of microprojections on the disk surface, there forms on the disk surface a layer of liquid, whose thickness for a given rpm is determined by the state of surface and by physical properties of the liquid. The excess liquid is shed intermittently from the edge of the disk surface due to surface tension. In this case liquid at the disk surface accumulates in roll form, remaining in this state until the centrifugal force does not exceed the surface tension force. The thickness of the roll is controlled by the capillary pressure at the edge of the disk, which is a function of the surface curvature and the dynamic contact angle Θ. Liquid breaks off from the unstable liquid ring in the form of droplets at points where the waves leave the disk. Each droplet entrains a liquid jet, i.e., the liquid flows off the disk in the form of periodically-breaking-off jets. The presence of these jets at the edge of the disk may result, upon reduction in flowrate, in the rolling up of the film into filaments in the outer region of the disk and in the formation of dry patches. The formation of filaments is initiated at the front of large waves, the troughs to both sides of which are unstable to the formation of dry patches.

An annular liquid roll, from which liquid filaments flow out, forms also on the resultant continuity boundary (see Fig. 7.14). These filaments have a wavy structure and, as they continue moving, may break up into individual liquid elements. In developed wave motion this boundary may turn out to be unstable and move in different radial directions, depending on whether the wave element approach the roll is a crest or a trough. Such an unsteady motion of the three-phase front in the presence of a temperature difference may induce the formation of microcracks in the disk material.

The breakup of the film depends significantly on the state of the disk surface, and on the flowrate and physical properties of the liquid (Fig. 7.15). A reduction in viscosity improves the stability of film flow; a similar result is produced by an increase in the flowrate. Surface roughness improves wetting and in the case of films whose thickness significantly exceeds the height of the microprojections, results in increasing flow stability.

At high rpm, when the centrifugal forces exceed significantly the surface tension forces, the film is shed from the disk continuously, and its breakup into jets and droplets occurs outside the disk, which eliminates the occurrence of dry patches on its edge. However, at high rpm the film thicknesses may become commensurable with the size of microprojections on the disk surface. In this case additional eddies, induced by the large resistance to flow may appear at the microroughness elements, since the film now flows over individual roughness elements, which may serve as initiators for film breakdown.

The functional relationship between the lower stability boundary for film flow, the structure and kinematic parameters of the film and the state of the disk surface, appears to be complicated. Evidently, for films having a thickness commensurable with the height of microprojections, this relationship will control the maximum permissible height of microprojections Δ_{max} for continuous film flow. It has the general form:

$$\Delta_{max} = f\,(\mathrm{Re}_2,\ \omega,\ \sigma,\ \Theta)$$

It must be remembered that the rupture of a liquid film is a process which requires a finite time for its development. In each point of the disk, when regarded as a potential rupture zone, the stability threshold will be the state when the kinetic energy of a large wave is able to compensate for the loss of energy in stretching the film and thus prevent the subsequent rupture. If the time between the passage, of two waves is smaller than the film rupture time, then waves whose amplitude only slightly exceeds δ_{max} will prevent the rupture of the film.

Figure 7.16 shows the radial variation in film thickness at different water flowrates Q and at variable disk rpm. The experimental data on the film thickness are compared with the most widely used relationship

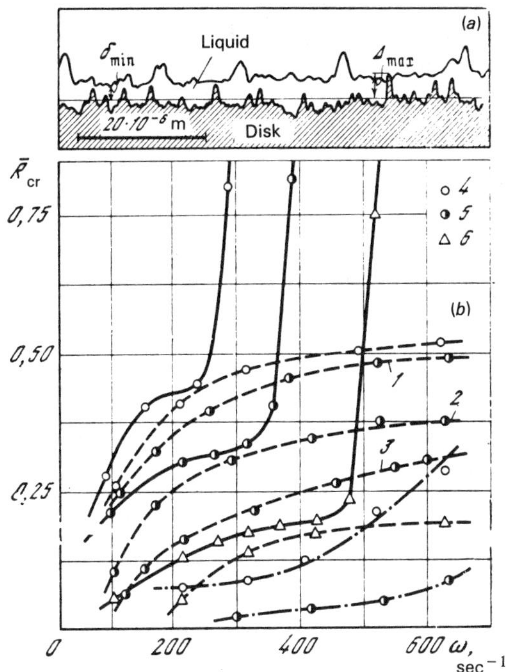

Figure 7.15 Schematic of the motion of a liquid film over a rough surface (*a*) and the effect of the state of the surface and viscosity of the liquid on the stability limit of film flow on rotating disks (*b*) [7.5]. 1, $v \cdot 10^6 = 0.48$; m/sec; 2, 0.81; 3, 1.31; the solid lines are for $\Delta_{max} \leq 20$ μm [sic], the dashed lines—$\Delta_{max} \leq 1 \mu$m, the dash-dotted line corresponds to a nonwetted surface; 4, $Q \cdot 10^{-6} = 1.54$ m^3/sec; 5, 10.04; 6, 0.47 m^3/sec.

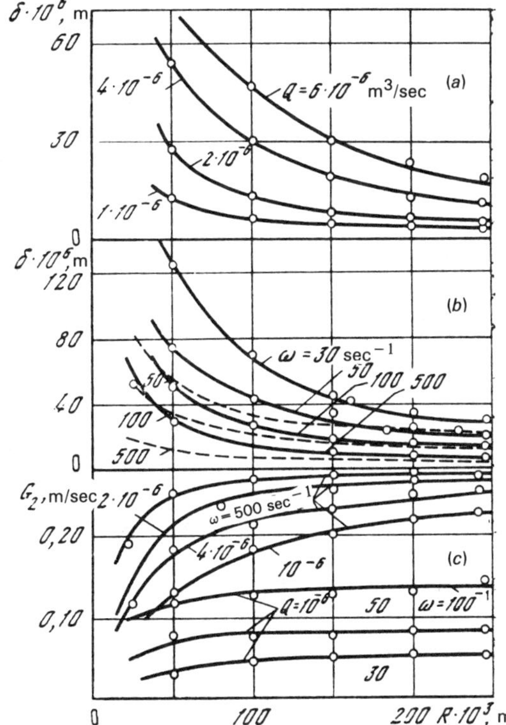

Figure 7.16 Variation in the bulk film thickness over the disk radius. (*a*) At variable liquid flowrates $\omega = 500$ rps; (*b*) at [variable] disk rps [and] $Q = 2 \cdot 10^{-6}$ m^3/sec; (*c*) effect of liquid flowrate Q and disk rps ω on the bulk radial velocity of the liquid film; the solid curves represent the experimental and the dashed curves the analytical results.

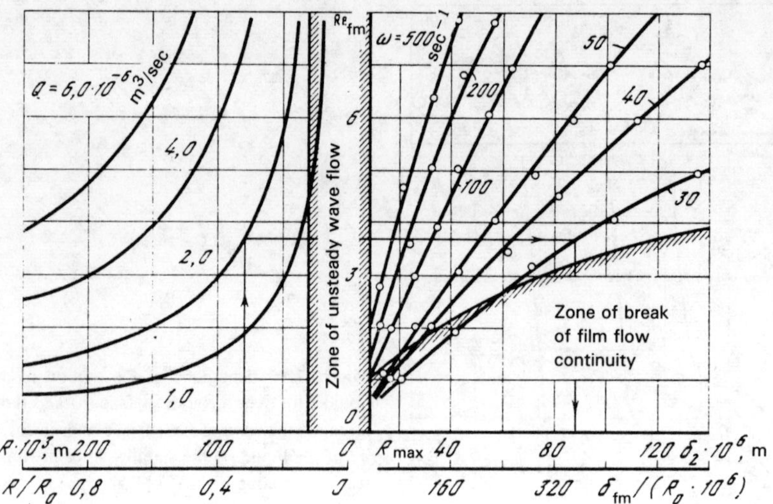

Figure 7.17 Nomogram for determining the film thickness on a rotating disk ($\nu = 0.48 \cdot 10^{-6}$ m²/sec) class 8 surface finish).

$$\delta = \left(\frac{15Q\mu_2}{\pi\rho\omega^2 r^2} \right)^{0,33}$$

where Q is the liquid flowrate, μ the viscosity, ρ the density of the liquid, and r is the radius at which the film thickness is determined. These comparisons exhibit significant differences in thickness; the actual values may differ by two orders of magnitude from those predicted. The radial variation in film thickness decreases with reduction in the flowrate.

Determination of film thickness at a given film radius makes it possible to also determine, for a given liquid flowrate, the film average velocity. The film velocity latter did not change significantly in the radial direction at low δ, and its values do not exceed 1–2 m/sec. The film velocity is significantly affected by ω and Q.

The waves on the film surface move at velocities higher than the bulk velocity (c_2). The wave velocity exceeds c_2 by a factor of 8 to 10 and increases with increasing ω. The wave velocity varies little with radial position.

Figure 7.17 shows a nomogram for determining the thickness of a liquid film over the radius of a disk at variable ω and Q. At $R = 0.15$ (left quarter) there exists a zone with unsteady liquid flow. Under these conditions it is difficult to determine the film thickness. It is also impossible to determine the film thickness (δ_2 when the value is less than 8 μm since thinner films are then of the same order of size as the roughness of the disk surface. In spite of the generalized form of the nomogram, it applies only to a limited range of liquid flow in a centrifugal force field, since film thickness δ_2 is also a function of the physical properties of the liquid and the surface roughness.

The functional relationship describing the effect of structural and kinematic parameters of the film flow and the state of the disk surface, on the lower boundary of stability of liquid films appears to be very complex. The rupture of liquid films in the centrifugal force field exhibits a clear hysteresis. Waves on the film surface affect not only the stability of motion, but also the energy transfer with the surroundings. This is analogous to the case of a film-covered nonmoving wall. The waves may significantly exceed the height of disk roughness elements and increase the mean tangential force applied to the interface of the two-phase boundary layer on the disk, then aiding the transmission of momentum to the gas. The wave structure at the phase interface distorts the velocity profile in the gas and increases the hydraulic resistance of the disk. Thus, waves on the film surface serve as moving roughness elements. Evidently, the wave structure of the film surface also increases the fluctuations of instantaneous-velocity components and the turbulence level of the gas.

It is most probable that the liquid film on the surface of rotor blades of wet-steam turbines is unstable, and the moisture may move in the form of jets (see Fig. 7.13) or of individual liquid elements. Analysis of forces acting on an element of liquid moving over the surface of rotating rotor blades shows that in the LP section of turbines, the centrifugal forces exceed by almost 100-fold the aerodynamic forces. For rotating blades in the HP section, for example, at steam pressures of 6 MPa, the ratio between the centrifugal and aerodynamic forces approaches five. Neglecting the aerodynamic effect on the liquid film, we shall consider the motion of a liquid element subjected to centrifugal and Coriolis forces and to the force of interaction between the liquid and the solid surface.

It must also be remembered, in considering the flow of liquid in a centrifugal force field, that the latter changes along the radius; i.e., in this case there occurs an unstable flow with a constant variation in the film thickness and, thus, with a variable Coriolis force. The nature of flow of liquid jets and individual liquid elements is significantly affected by the boundary between the wetted and dry solid regions on the rotating surface.

Studies of the flow of liquid films in a centrifugal force field make it possible to calculate the motion of moisture over the surface of rotor blades with cylindrical as well as variable shape of the airfoil over the height of the blades [7.3]. Detailed calculations of the motion of moisture over the surfaces of rotor blades make it possible to analyze the most typical paths of moisture motion and determine the fraction of the moisture shifted to the periphery.

7.5 EFFECT OF THE LIQUID PHASE ON THE PRINCIPAL PARAMETER AND DESIGN OF WET–STEAM TURBINE STAGES

Let us consider the effect of moisture on the parameters of turbine stages using the method of integral relationships. These methods are simple, yield an acceptable accuracy and are recommended for design purposes.

These methods employ equilibrium i-s diagrams or steam tables. The detailed flow structure and the nonequilibrium of expansion are incorporated by using appro-

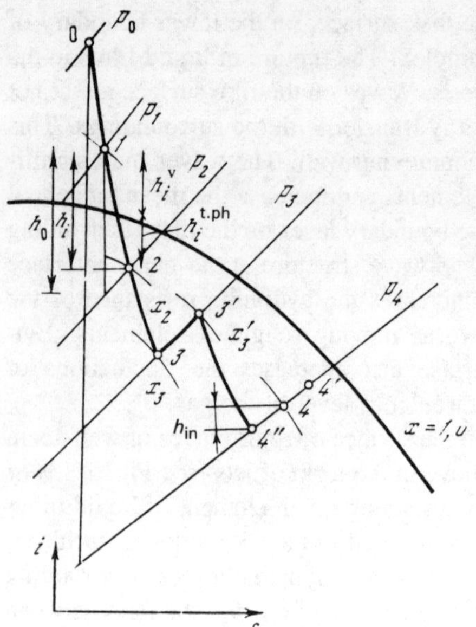

Figure 7.18 Expansion of [wet] steam in turbine stages under different conditions.

priate coefficients. The final states are assumed to be equilibrium. Detailed design of stages is performed by standard methods (as in the single-phase region) and the effect of moisture is incorporated by integral corrections for the flowrate coefficient, degree of reaction and the turbine efficiency η_{0i}. The amount of moisture in the stages is determined from the actual (equilibrium) expansion of steam on the i-s diagram with correction Ψ for moisture separation from the turbine (the separation factor Ψ is equal to G^s/G^m—the ratio of moisture removed in the separator to the total flowrate of moisture upstream of the stage).

The expansion of wet steam in individual stages of a multistage turbine differs significantly depending on whether the expansion intersects the saturation curve, or occurs in the two-phase region. We shall consider the expansion of steam on an i-s diagram for four stages (Fig. 7.18). The expansion in the first stage occurs in the single-phase region.

The expansion in the second stage intersects the saturation curve. There is no initial (primary) moisture. The enthalpy dissipated in the single-phase region shall be denoted by h^v, and that dissipated in the two-phase region, by $h^{t.ph}$.

The state of the wet steam at the start of expansion in the third stage, represented in Fig. 7.18, coincides with the state of the wet steam past the second stage in point 2 (Fig. 7.18). The final percentage moisture in the steam $y_2 = 1 - x_2$ downstream of the second stage serves as the initial value for the third stage. There is virtually no separation of moisture past the second stage, since the moisture is present in the form of small droplets, irrespective of whether it formed in the condensation shock or on the surfaces of the rotor blades. This means that calculations for the third stage must be based on initial moisture content of $y_2 = 1 - x_2$ and a gain in this quantity $\Delta y = y_3 - y_2$ in the course of expansion ($y_3 = 1 - x_3$).

The expansion in the fourth stage starts in point $3'$, which lies at lower percentage moisture than point 3. A part of moisture contained in large droplets downstream of the third stage can be removed in a separator. The amount of moisture just removed is $\Delta y_{\text{sep3}} = \Psi_3 y_2$, where y_2 is the initial percentage moisture upstream of the third stage, and Ψ_3 is the separation factor in the third stage. If the exit velocity losses are high, which is usually the case for the last stages of condensing turbines, then the percentage moisture used in the calculations must be based on the actual expansion without allowance for exit velocity losses (point $4''$ rather than 4 in Fig. 7.18). With allowance for removal of moisture in the separator the expansion upstream of the following stage occurs at point $4'$. The amount of moisture separated is $\Delta y_{\text{sep4}} = \Psi_4 y_3$.

The distribution of moisture over the height of long blades is clearly nonuniform and is a function of a number of parameters: location and process of moisture formation, percentage content of large-droplet moisture (moisture whose paths differ significantly from the streamlines of the steam phase and which precipitates on the surface of stationary and rotating blading), and geometry of the stage proper. It is at present still impossible to take all these factors into account. In conjunction with this, one may use in the first approximation, in designing the final stages of turbines, the generalized distribution of moisture over the blade height downstream of a stage. This applies with a large fan ratio for different values of percentage content of large droplet moisture λ (Fig. 7.11). This relationship was obtained on the basis of the previously cited experimental studies of the distribution of moisture downstream of the last stage of full-size and model turbines [7.7, 7.8].

Usually the bulk of moisture in last turbine stages is concentrated at the periphery. Hence for convenience in calculating the upper cross sections of stages, Fig. 7.11 shows the variation in percentage moisture at the periphery of blades y_2^{per} as a function of the equilibrium percentage moisture y_{2d}, downstream of the stage under study for different values of the fraction of large-droplet moisture in location $0.9\bar{l}_r$.

The most difficult aspect in determining the distribution of moisture over the various blade cross sections is estimating the fraction of large-droplet moisture. To a first approximation, on the basis of experimental studies of turbine stages, rotor and nozzle blading, and also of detailed calculations of multistage wet-steam turbines, it is possible to suggest a method for determining the fraction of large-droplet moisture $\lambda' = f(z)$ as a function of the location in the turbine flowpassage where the moisture originates (Fig. 7.19). If the moisture appeared in the fourth stage, counting from the end of the turbine ($z - 3$), then, according to Fig. 7.19, the value of λ' past the last stage is 0.28. This curve was constructed for a final pressure downstream at the last stage of 0.005 MPa. The value of δ' will decrease with rising pressure in accordance with the values of the correction factor κ_p (Fig. 7.19b) and then $\lambda = \lambda' \kappa_p$.

The droplet size distribution in each stage is of great importance for detailed calculations. This distribution over the height of a turbine stage depends to a large extent on the location and form of moisture formation in the turbine flowpassage. Thus, for example, if the moisture is formed in a given stage as a result of a condensation shock, the size of droplets at different pressures and expansion gradient will lie between the limits of $0.05~\mu\text{m} < d_3 < 0.2~\mu\text{m}$. If the moisture forms in vortex

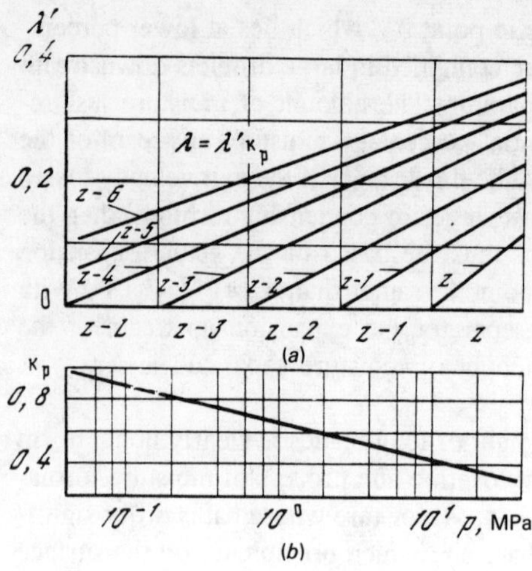

Figure 7.19 Fraction λ' of large-droplet moisture as a function of the point of moisture formation (a) and correction κ_p for the effect of pressure on λ (b).

streets, in regions of separation and secondary flows, then the modal diameter d_M of such droplets may be somewhat greater and as much as 0.05 to 0.5 μm. In this case larger droplets, which form from moisture condensed on the surface of the rotor blades, will also be present downstream of the turbine, but their percentage content is small.

There are no direct measurements of the droplet size distribution of moisture forming within a given stage, for which reason the variation in d_M over the height of a blade, which is affected by many parameters, can be represented approximately basically only as a function of the location of moisture inception in the turbine flowpassages (Fig. 7.20). These data are approximate and integral, since a change in the steam parameters modifies the operating conditions of the stages, and this may exert a significant effect on the distribution and size of droplets over the height. Figure 7.20 shows the results of experimental studies of the size distribution of moisture past a

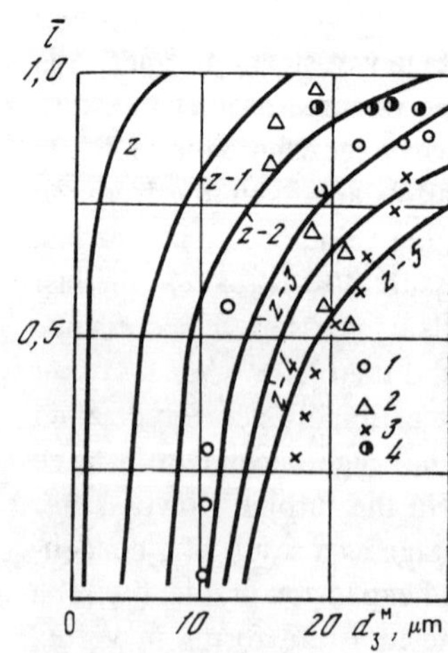

Figure 7.20 Variation in the mean droplet size past a turbine stage for different location of inception of condensation. 1, Experiments of the Polzunov Boiler and Turbine Institute; 2, of the Khar'kov Turbine; 3, Moscow Energetic Institute; 4, EEC.

turbine stage with a high fan ratio, for the case when moisture started forming in different preceding stages.

It is at present possible, with sufficient accuracy, to make allowance for the effect of moisture in determining the inlet and exit angles, kinetic energy losses, discharge coefficients, efficiency, degree of reactivity and critical velocity, and also to suggest a separation arrangement for the last turbine stages.

However, the lack of experimental data for a wide class of various profiles, obtained for real (usually different) phase velocity ratios both with respect to magnitude and direction, and distributions of droplet size, do not allow one to recommend reliable methods for calculating cascade parameters.

Loss factor. Two forms of kinetic energy losses are possible in multiphase flows: due to nonequilibrium of the process (ξ_{noneq}) and due to irreversible losses.

The loss factor for two-phase flows can be defined as:

$$\zeta = \frac{H_{0d} - H}{H_{0d}} = \zeta_v - (1 - \zeta_{noneq})\zeta_v \tag{7.9}$$

where $H = H_{0d} - \Delta H_0 - \Sigma \Delta H_i$ is the actual enthalpy drop; H_{0d} is the enthalpy available for equilibrium expansion; H_0 is the available enthalpy for nonequilibrium expansion; ΔH_0 is the reduction in this enthalpy due to nonequilibrium, and $\Sigma \Delta H_i$ are the irreversible losses of flow energy (in interphase friction and friction in the boundary layer as a result of phase transition and heat transfer at finite temperature difference, losses in condensation and compression shocks, etc.).

In general analytic calculation of individual loss components involves great difficulties. Since all that we need here is to define the expressions for velocities c_1 and w_1, we shall use only approximate equations, based on experimental studies of turbine cascades.

The overall coefficient of kinetic energy loss in stationary and rotor blading is given by the expression

$$\zeta^{wet} = \zeta^{sup} + \Delta\zeta$$

where ζ^{sup} is the loss factor for a cascade with superheated steam, whereas $\Delta\zeta$ is the loss increment due to the presence of moisture.

The quantity $\Delta\zeta$ is determined from Fig. 7.21 as a function of the initial steam superheat ΔT (or initial percentage moisture y_0) and the fraction of large-droplet moisture λ. The values of coefficients δ^{sup} were determined on the basis of generalized characteristics [7.1].

Figure 7.21 shows values of $\Delta\zeta$ [sic] for $y_0 = 12\%$. For y_0 between 10 and 40% and $\lambda = 0.3$ the calculations should be performed from the expression $\Delta\zeta^{wet} = \zeta^{sup} + \Delta\zeta$.

Inlet and exit angles of two-phase flows. Acceleration of wet steam in cascade flowpassages results in the appearance of differences in the magnitudes and directions of phase velocities. The larger the droplets, the smaller the curvature of their streamlines and the greater their exit angle from the cascade. An increase in the exit angle of the liquid phase is also induced by shedding of the film from the convex surface of the airfoil and the motion of the removed droplets at an angle greater than the direction of motion of the steam. Under these conditions the exit angle for the steam also increases. This is due in the first place to deflections of the steam streamlines in the skewed cross section of the cascade due to a similar deflection of the moisture droplets and a rise in the kinetic energy losses of the steam.

In general the exit angles of the steam and water depend on many factors: steam-flow parameters, droplet size distribution, phase velocities, type of cascades, etc. In approximate calculations, as in determining the loss factor, one considers only the effect of the initial percentage moisture (y_0) upstream of the cascade and the fraction of large-droplet moisture (λ), on increase in the exit angles of the steam $\Delta\alpha_{11}$ ($\Delta\beta_{21}$) and $\Delta\alpha_{12}$ ($\Delta\beta_{22}$) are given by the expressions

$$\alpha_{11}\,(\beta_{21}) = \alpha_1\,(\beta_2) + \Delta\alpha_{11}\,(\Delta\beta_{21})$$

$$\alpha_{12}\,(\beta_{22}) = \alpha_1\,(\beta_2) + \Delta\alpha_{12}\,(\Delta\beta_{22})$$

where $\alpha_1(\beta_1)$ are exit angles of single-phase flow, to be equal to the effective angles $\alpha_{1ef}(\beta_{2ef})$.

Figure 7.22a shows experimental data on the change in the steam exit angle for the TC-9012A nozzle airfoil with $t = 0.75$ at $b/l = 0.5$ and different M.

The exit angle for wet steam is given by the expression

$$\alpha_1^{\text{wet.s}} = \alpha_1 + \kappa_p\Delta\alpha_1'$$

where α_1 is the exit angle for single-phase flow, $\Delta\alpha_1'$ is the deviation in direction for wet steam, obtained from Fig. 7.22b, and κ_p is the pressure correction (Fig. 7.22c).

The presence of moisture past the nozzle blading increases the "unswirling" of the steam flow in axial direction, for which reason the inlet parts of the rotor blading

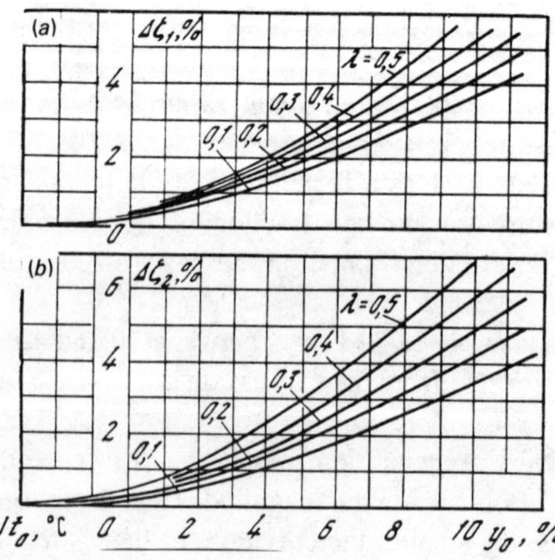

Figure 7.21 Rise in steam energy losses vs. the percentage moisture y_0 for nozzle (a) and rotor (b) cascades.

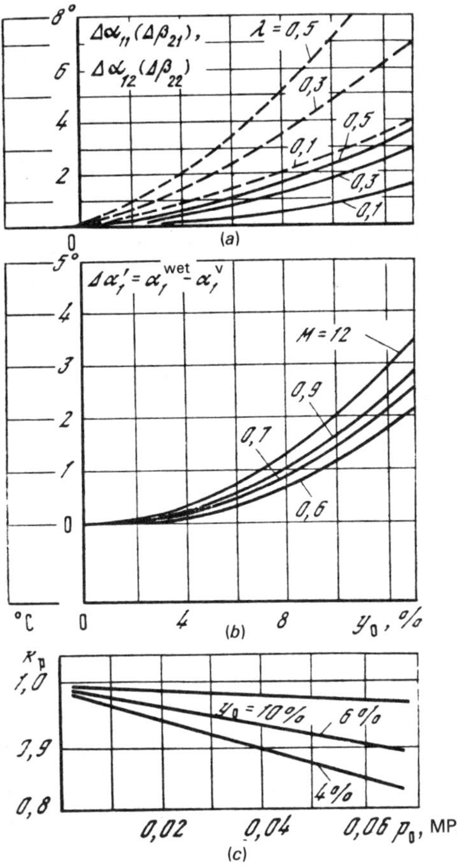

Figure 7.22 Variation in $\alpha_1(\beta_1)$ as a function of the fraction of large-droplet moisture (*a*), initial percentage moisture y_0 (*b*) and pressure ahead of a stage (*c*). The solid curves are a plot of $\Delta\alpha_{12}(\Delta\beta_{22})$, the dashed curves—$\Delta\alpha_{11}(\Delta\beta_{21})$.

must be configured with correction for changes in α_1 and β_1. This is particularly important for peripheral sections of the rotor blading, since the significant moisture content in this zone has a significant effect on the direction of motion of the mainstream steam. If the calculations are not correct for changes in angles α_1 and β_1 for peripheral sections, then the energy lost on impact on the rotor blades increases significantly. Consequently the erosion wear of the surfaces will increase due to the rise in the impact component of drop velocity.

The optimum inlet angle for a wet-steam rotor cascade β_1^{wet}, obtained by correcting β_1, can be found analytically for a single-phase fluid. However, difficulties in determining a number of initial parameters, and primarily the phase velocity slip $\nu = c_1^{\text{wet}}/c_1^{\text{sup}}$ make it at present impossible to recommend reliable expressions for determining $\Delta\alpha_1$. For this reason for $y_{0i} = 25\text{--}50\%$ and $\lambda = 0.3$ the inlet angle of the peripheral cross sections of rotor cascades should be determined from the expression:
$$\beta_1^{\text{wet}} = \beta_1^{\text{sup}} + (3 - 5)\,[^\circ].$$

The flowrate factor. The two-phase flowrate through a passage of arbitrary shape

$$\dot{m} = \rho w F = \rho F\,(x_1 w_1 + x_2 w_2) = F\,(\rho_1 \varphi_1 w_1 + \rho_2 \varphi_2 w_2) =$$

$$= \frac{F w_1\,[x_1 + (1 - x_1)\,v]}{v_1\,[x_1 + (1 - x_1)\,\bar{v}]} \tag{7.10}$$

where x_1 is the actual vapor quality in the minimum cross section F; v is the actual specific volume of the steam, $\bar{v} = v_2/v_1$ is the ratio of specific volumes of the liquid and steam phases, and $\nu = w_2/w_1$ is the velocity slip.

The two-phase flowrate can be expressed in terms of the flow quality $x_{\mathrm{flwo}} = \dot{m}_1/\dot{m}$

$$\dot{m} = \frac{F w_1 \nu}{v_1[x_{\mathrm{flow}}\nu + (1 - x_{\mathrm{flow}})v]} = \frac{F w_{1\mathrm{an}}\nu\,\sqrt{1 - \xi}}{v_{1\mathrm{an}}\chi[x_{\mathrm{flow}}\nu + (1 - x_{\mathrm{flow}})\bar{v}]}$$

where $w_{1\mathrm{an}}$ and $v_{1\mathrm{an}}$ are the analytically obtained velocity and specific volumes of the steam in the minimum cross section; $\chi\;v_1/v_{1\mathrm{an}}$ is the ratio of the actual to the calculated volume of steam (degree of nonequilibrium) and ξ is the overall loss factor.

This means that the two-phase flowrate can be calculated if ν, x_{flow}, χ and ξ are known. However, their determination is rather difficult. Hence the actual two-phase flowrate for each cross section in the stationary and rotor blading is calculated from experimental values of the flowrate factor. The calculated flowrate in this case is determined from data of the equilibrium i-s diagram:

$$\dot{m}_t = F_1 w_{1\mathrm{an}}\rho_{1\mathrm{an}} = F_2 w_{2\mathrm{an}}\rho_{2\mathrm{an}}$$

In Eq. (7.10) the analytic velocities $w_{1\mathrm{an}}$ $(w_{2\mathrm{an}})$ and densities $\rho_{1\mathrm{an}}$ $(\rho_{2\mathrm{an}})$ refer to the minimum cross sections $F_1(F_2)$. However, calculation of the actual flowrate requires that corrections be made for the difference in static pressures in the minimum cross section and past the cascade, and also for the nonuniformity of the velocity fields. Then $G = \mu\bar{g}G_{1\mathrm{an}}$ and the flowrate factor for wet steam is

$$\mu^{\mathrm{wet}} = \mu^{\mathrm{sup}} + \kappa^{\varepsilon}\Delta T \Delta\mu \tag{7.11}$$

where μ^{sup}, the flowrate factor for superheated steam, which is a function of the relative cascade height $\bar{l} = l/b$ and the convergence ratio of the channel, etc. (it can be determined using data given by Abramov et al. [7.14]); $\Delta\mu$ is the change in μ due to wetness of steam for different values of y_0 and channel convergence ratios $\sin\alpha_0(\beta_1)/\sin\alpha_1(\beta_2)$ (can be determined from Fig. 7.23a).

The correction κ^{ε} for the effect of pressure ratio on the flowrate factor is determined from Fig. 7.23b, whereas the coefficient ΔT reflecting the degree of nonequilibrium of the expansion process as a function of the mean droplet size distribution in the section under study is obtained from Fig. 7.23c.

The rise in μ^{wet} as the start of the process approaches the saturation curve is due to the fact that the actual specific volume is decreased by supercooling as compared with the equilibrium value $v_{1\mathrm{an}}$ whose predicted fall $c_{1\mathrm{an}}(w_{2\mathrm{an}})$ is significantly greater than the reduction in the real velocity c_1/w_2. Upon transition through the saturation curve the actual steam velocity c_1 remains virtually unchanged, whereas $c_{1\mathrm{an}}$ decreases

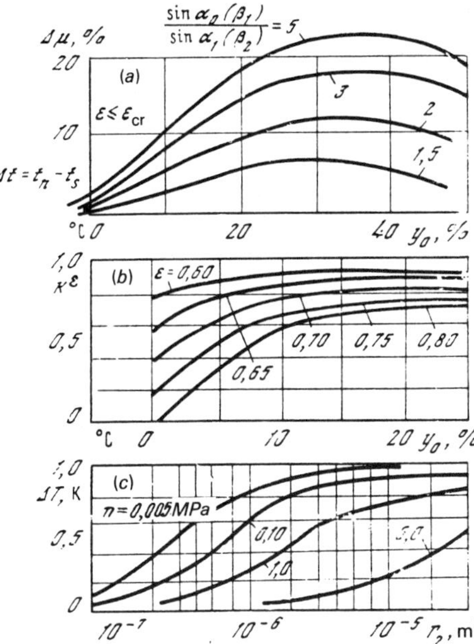

Figure 7.23 Increment $\Delta\mu$ in the discharge factor with rise in percentage moisture y_0 at $\alpha \leq \epsilon_{cr}$ (a) and correction factors κ^ϵ (b) and ΔT (c).

significantly due to the reduction in the dissipated enthalpy. The latter results in a further rise in μ^{wet}. However, an increase in the percentage moisture involves an energy loss with consequent drop in c_1. Hence the rate of rise in μ^{wet} first decreases, and then the values of μ^{wet} fall.

Speed of sound. The significant nonuniformity in the distribution of moisture over the height of those turbine stages with high fan ratio, makes it necessary to correct for changes in the speed of sound in wet as compared with superheated steam. Analytic and experimental studies described by Deych and Filippov [7.15] show that the propagation of sound in wet steam is a process much more complex than in superheated steam. The speed of sound in wet steam is a function of the steam and liquid temperatures, of pressure, of relaxation time, of interphase processes transfer, of steam and liquid velocities. Figure 7.24 shows the results of calculations performed at the Moscow Energetics Institute of the relative speed of sound ($\bar{a} = a^{wet}/a^{sup}$, where a^{wet} and a^{sup} are the speed of sound in wet and superheated steam) as a function of percentage moisture y downstream of the cascade and nondimensional

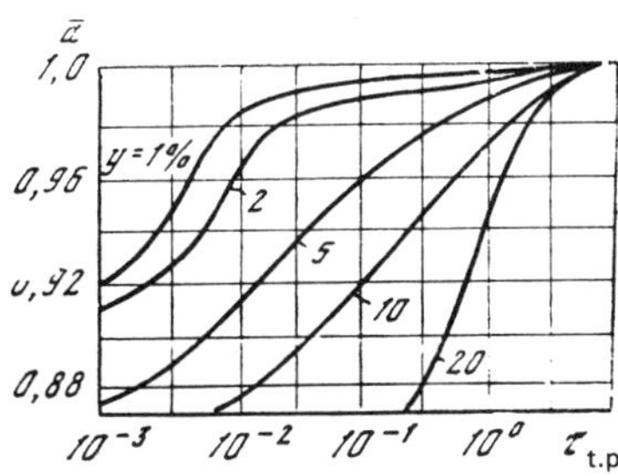

Figure 7.24 Variation in the relative speed of sound in wet steam $\bar{a}$ as a function of τ_t.

relaxation time of interphase momentum transfer $\tau_{\text{t.ph}} = \frac{4}{3}(d_3^2 c_0 \rho_3)/(\mu_1 b)$. These results are for the steam temperature range between 35 and 70 °C; ρ_2 is the density of the liquid phase, d_3 is the bulk diameter of the droplets, c_0 is the mean velocity of steam in the passage, μ_1 is the dynamic viscosity of the steam, and b is the blade chord.

Thus, if the parameters at the cascade exit are known, the change in the speed of sound in wet steam as compared with that for superheated steam can be determined from Fig. 7.24.

The reaction of a stage. In calculating the individual cross sections of the last stages of wet-steam turbines allowance must be made for changes in the degree of reaction of the turbine stage due to the presence of moisture. Experimental studies performed at the Moscow Energetics Institute on various turbine stages typical of individual sections of long twisted blades, show that the degree of reaction in all the cross sections increases with increasing moisture content. The degree of reaction is affected most by moisture in the root cross sections (Fig. 7.25) [7.3].

The increase $\Delta\rho$ in the degree of reaction can be calculated using Fig. 7.25 which shows values of $\Delta\rho$ for different cross sections as a function of the percentage moisture upstream of the stage in the section under study. This figure also gives the correction κ^ϵ for the effect of the pressure ratio per stage $\epsilon = p_2/p_0$. Then

$$\rho^{\text{wet}} = \rho^{\text{sup}} + \kappa^\epsilon \, \Delta\rho \tag{7.12}$$

Efficiency. It was found in a large number of experiments that the efficiency of turbine stages in wet-steam turbines is affected primarily by the droplet size distribution, the fraction of large-droplet moisture, values of M and Re, velocity ratio u/c_{ph}, etc. [7.1–7.3]. The reduction in the efficiency of individual sections within a stage can be determined by the following formula, which was verified repeatedly for description of various stages with average fan ratio:

$$\Delta\eta_{0i}^{\text{wet}} = \eta_{0i}^{\text{sup}} - \eta_{0i}^{\text{wet}} = 2u/c_0 \left(\kappa_1 y_0 - \kappa_\zeta \Delta y \, \frac{h_0^{\text{wet}}}{h_0} \right) \tag{7.13}$$

where h_0^{wet} is the fraction of enthalpy dissipated in wet steam, y_0 is the percentage moisture entering the section under study, Δy is the increase in percentage moisture within the stage, κ_2 is a constant (equal to 0.35), and κ_1 is a coefficient which depends on fraction λ of large-droplet moisture and the degree of reaction.

For the case of a last stage operating at $y_0 > 0$ and $\lambda > 0.35$ the efficiency reduction of individual cross sections can be estimated in accordance with experimental data of Filippov and Povarov [7.3] (Fig. 7.26).

Selection of the optimum available enthalpy difference. Losses due to wetness of steam reduce the stage efficiency and the optimum velocity ratio u/c_0. These differences increase with rising moisture content and may become significant, which makes it necessary to correct for the effect of moisture on the maximum available enthalpy difference of a stage.

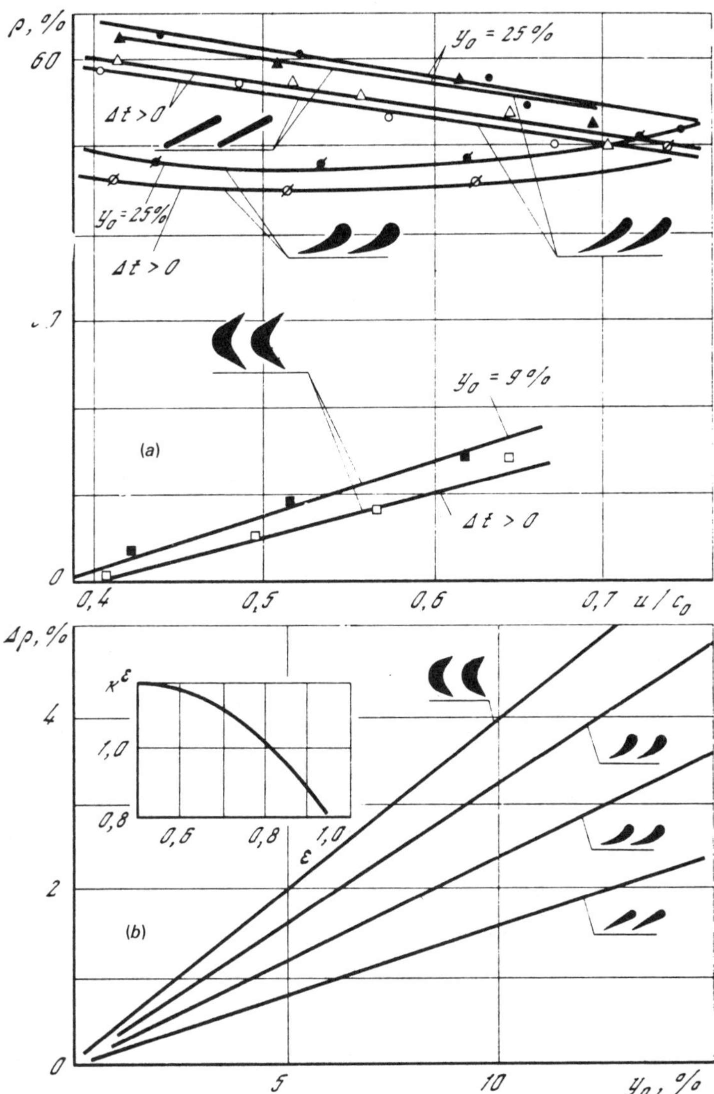

Figure 7.25 Variation in degree of reaction ρ as a function of the relative velocity u/c_0 for different cross sections of turbine stages (*a*); increase $\Delta\rho$ with rise in the initial percentage moisture y_0 (*b*).

We shall write the change in stage efficiency caused by small deviations from optimum u/c_0 in the form of a parabolic relationship, using Eq. (7.13) for h_0^{wet}/h_0:

$$\Delta\eta_{0i}^{u/c_0} = (\eta_{0i})_{\text{max}}^{\text{sup}} - \eta_{0i} = \kappa_3\left[\frac{u/c_0}{(u/c_0)_{\text{opt}}^{\text{sup}}} - 1\right]^2 \tag{7.13a}$$

where, according to investigations of various stages with $d/l = 2.5\text{–}5.0$ $\kappa_3 = 0.77$;

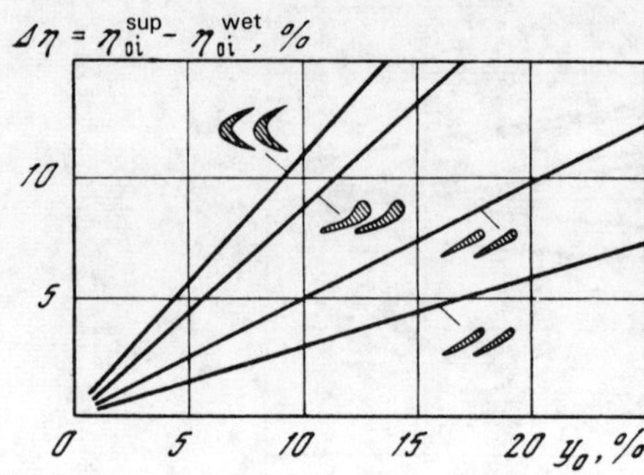

Figure 7.26 Reduction Δ in efficiency of individual cross sections of a turbine stage with rising initial percentage moisture y_0.

$(u/c_0)_{\text{opt}}^{\text{sup}}$ is the optimum value of u/c_0 at the maximum values of efficiency $(\eta_{0i})_{\text{max}}^{\text{sup}}$ of a turbine stage operating on superheated steam.

Equations (7.13) and (7.13a) can be used for determining the optimum value of u/c_0 for operation with wet steam.

Specification of a larger enthalpy dissipation for the last stage of a wet-steam turbine will, in addition, yield the following positive effects: increasing the enthalpy dissipation will increase the steam velocity and consequently reduce the size of water droplets which, in its turn, will improve the stage efficiency and decrease the erosion wear of the blades; when the termination of expansion is specified, an increase in h_0 will shift the start of condensation to the last stages, resulting in a reduction in λ; selection of a larger enthalpy dissipation will ensure reliable operation of the stage at partial overloads.

Moisture separation. Any of the final stages operating under conditions when moisture is present at the inlet to the nozzle blading ($\lambda > 0.2$) should be designed with in-channel (suction through hollow nozzle blades) and peripheral moisture separation devices. The actual arrangement for in-channel separation depends on the experience and technology developed at the plant. Such designs should attempt to provide for the maximum number of individual moisture-collecting slots on the nozzle blade surface [7.3]. In-channel separation should not be used over the entire height of the nozzle blade, but only over its peripheral part (up to $0.5l_n$). The effectiveness of in-channel and peripheral separation methods can be determined using the techniques described in the book by Filippov et al. [7.1].

The use of in-channel separation makes it possible to improve the efficiency of the corresponding cross section of the stage in which moisture removal is provided. The efficiency of individual sections with separation arrangements $(\eta_{0i}^{\text{wet}})_{\text{sep}}$ will be higher than η_{0i}^{wet} without separation, by the amount $\Delta\eta_{0i}^{\text{sep}}$, i.e.,

$$(\eta_{0i}^{\text{wet}})_{\text{sep}} = \eta_{0i}^{\text{wet}} + \Delta\eta^{\text{sep}} \tag{7.14}$$

On the basis of experimental data, the increase in stage efficiency as a result of removal of moisture through hollow nozzle blades can be determined from the expression $\Delta\eta_{0i}^{\text{sep}} = 0.02\Psi y_0$, where Ψ is the effectiveness of in-channel moisture separation, % [7.1]; y_0 is the percentage moisture ahead of the stage in the section under study, %.

For the case when the steam upstream of the last stage is superheated, and wet steam forms only in the course of expansion within the stage, the use of in-channel separation is not justified, and then it is better to provide for peripheral separation of moisture above the rotor blades. The use of peripheral separation of moisture in the final stages improves in all cases the operating efficiency of the exhaust piping. In conjunction with this, it is best to remove moisture not only above rotor blades, but also downstream of the last stage at the inlet to the exhaust piping.

Turbine stages with large fan ratio in a wet steam turbine are calculated from the equilibrium *i-s* diagram. The sequence of selection and calculation of principal geometric dimensions does not differ from that used for superheated steam. The turbine stage is calculated to a first approximation in several sections without correction for the moisture content, but then the available enthalpy dissipation selected is larger in accordance with Eqs. (7.14).

If in-channel separation is used, then the path of expansion of the steam on the *i-s* diagram is corrected for moisture removal. This calculation is an approximation. Further calculations are performed allowing for the effect of moisture on the performance of the sections under study. For this, one determines the percentage moisture, droplet size distribution and fraction of large-droplet moisture ahead of the stage in each section under study. A correction is then made for the change in degree of reaction and stage geometry, for which one determines the speed of sound $\bar{a}^{wet}$ in each section. the losses ξ, the inlet and exit angles, and also the discharge factor. The above corrections may require redesign of the entire stage.

7.6 BEHAVIOR OF IMPURITIES IN TURBINE FLOWPASSAGES

It was found that in the course of operation of turbines at nuclear power plants with water-cooled reactors, where the expansion starts from the saturation curve, there exist two most prevalent problems which reduce the reliability of turbine flowpassages and which are associated with the formation of moisture and flow of wet steam at high pressures ($p_0 > 1.0$ MPa).

The operation of the flowpassages in the high-pressure stages operating on wet steam is, as opposed to turbines operating at fossil-fuel plants, accompanied by erosion by wet steam and erosion-corrosion failure of orifices, casings, packings and rotors [7.1, 7.2]. This phenomenon is observed virtually in all turbines and is as of now controlled basically using the highest quality materials, i.e., stainless steels.

A great deal of attention is currently attracted by the large number of turbine blade, diaphragm and rotor failures associated with expansion of steam in the Wilson zone.

The problem of appearance of microcracks in metals and the failure of stressed components (rotor blades and diaphragms) in the moisture formation region is as yet insufficiently explored.

On the one hand, such failures of stationary and rotating blading can be attributed to the intermittent forces exerted by the steam flow as a result of instability of spontaneous moisture formation [7.1, 7.2]. On the other hand, unsteady phenomena associated with moisture formation in the flowpassages of real turbines in nuclear power plants, occur only in nozzle-blading steam distribution systems and at partial loads in nozzles of the first stages in the high-pressure parts, for example, in the first K-220-44 turbines. The high turbulence of the flow, its periodic unsteadiness and the three-dimensionality of flow virtually prevent the inception of spontaneous condensation in the stages of the low-pressure zone of the turbines.

The most probable explanation put forward at present is that the failure of rotor blades and sometimes also of disks and rotors are the result of formation of corrosion microcracks on the surface of highly-stressed materials due to the effect of highly-concentrated solutions of corrosive impurities, precipitating from the steam in the course of its expansion. There is no concensus at present on this conclusion because no comparison was made of the water regime and composition of salts for those turbines for which corrosion failure was observed and those where this did not happen. Numerous studies of the salt content and water regime were performed primarily for fossil-fuel fired power units [7.16], and at present there are very little data on the water regimes and concentrations of individual impurities along the flowpassages of nuclear power plant turbines.

However, a number of major accidents due to failure of rotor blades in the low-pressure region of turbines have been reported, these failures occurring when the steam expands through the Wilson zone. Examples are those at the Hinckley Point A (England) and the Rancho Seco (US) nuclear power plants (the latter having Babcock and Wilcox once-through steam generators) [7.17, 7.18]. Failure is attributed to the formation of microcracks due to impurities in the steam. It was established in analyzing these and other accidents that contact had occurred between turbine materials and in the turbine flow a working fluid containing NaCl and NaOH and, sometimes, extremely high concentrations of acids. At present the behavior of impurities in turbines has not yet been sufficiently explored to allow quantitative estimation of the effect of given variables on the extremely high concentration level needed to obtain a corrosively acting liquid phase. Clearly, such concentration can occur at nuclear power plants despite the use of current advanced water-treatment technology.

Until recently, complete demineralization of the entire condensate was regarded by a number of companies as necessary only for single-loop units, it is currently used in the majority of new plants even for two-loop units. Ever increasing use is made of precoat filters, which allow a significant reduction in the possibility of sodium "hop-over" through the condensate demineralization unit. In spite of all this it was found that nuclear power plant feedwater and steam contain more than 50 various compounds—oxides, silicates, sulfates, carbonates, chlorides, etc., true, with extremely low concentrations.

These compounds cannot perceptibly affect the hydrodynamics and thermodynamics of the wet and saturated steam, for example, the formation of moisture in supercooled steam; however, in a number of cases, their presence in turbine flowpassages may be dangerous from the point of view of reliability of blades and disks.

The corrosive capacity of the liquid depends to a large extent on the values of the pH, and also on the redox potential, at certain values of which there starts appearing pitting and particularly corrosive stress cracking.

As is known, the water regime in two-loop nuclear power plant turbines is maintained alkaline by using volatile alkalies, usually ammonia and hydrazine. These alkalies, which hide out primarily in the steam phase, do not present any danger with respect to electrochemical corrosion of turbines. Boiling-reactor units use a neutral water regime, without water additives. Under these conditions the content of corrosive impurities in the steam supplied to the turbines in extremely low and very high concentrations of these impurities in the liquid phase are needed in order to induce corrosion damage of flowpassage components. The complexity of physico-chemical hydrodynamic and mass transfer processes, particularly given the high rate of variation in the steam parameters in the course of expansion in the turbine, make it difficult to perform an even qualitative analysis of the concentration mechanism of impurities in the liquid phase formed in the course of steam flow through the turbine. Under certain conditions, as shown in Chapter 1, corrosive impurities may exist in the form of traces of the liquid or solid phase already at the inlets to the low-pressure part of the turbine.

In analyzing the impurity-induced corrosion failure of stressed elements of turbine flowpassages one should not consider the overall impurity concentration, but the probability of local precipitation of individual aggressive substances on the metal surfaces.

Corrosion failure of metals naturally requires a long time period, for which reason the precipitation of impurities on the metal surface is a necessary, but not sufficient condition for the appearance of microcorrosion cracks. For this to occur it is also necessary that these impurities form highly concentrated solutions at the metal surfaces and that the latter be free of films which would dilute these solutions.

In the experience of the present authors, thin water films or local liquid spots may exist on the surface of rotor blades, in spite of the fact that the latter is subjected to centrifugal forces; the thickness of these liquid formations may be commensurable with microprojections of the rough metal surface, i.e., for turbine blades $\delta_{fm} \leq 15$ μm. In this case the forces of adhesion of the liquid to the surface are virtually equal to the centrifugal and aerodynamic forces acting on the liquid film. Hence, in spite of the action of the centrifugal and aerodynamic forces, the water, and the more so viscous concentrated solutions of impurities precipitating from the steam onto the surfaces of rotor blades may remain stationary and may not be entrained by the steam flow. This may provide for prolonged interaction of chemically active substances with rotating, highly stressed, components of the turbine, which are frequently subjected to alternating stresses.

Naturally, the appearance of moisture proper at the inlet of the cascade results in constant precipitation of water and formation of water films of various thickness. These water films will dilute the concentrated films of chemically active substances. Then the liquid films will be propelled by the steam flow in the nozzle blading or by aerodynamic and centrifugal flow in the rotor blading and disks (see Sec. 7.4). From this time on, the probability of failure of metal surfaces in turbine flowpassage due to impurities becomes negligible, since their concentration drops steeply and the time of interaction between them and the metal also decreases.

For this reason the most dangerous locations for corrosive attack by impurities in nuclear power plant turbines are in the zone of superheated steam immediately upstream of its interaction of the saturation curve, or immediately past it.

The steam flowing in these stages is supercooled at times by as much as 40 °K. If the start of expansion of steam in the nozzle blading lies above the saturation curve on the i-s diagram, then liquid water films cannot form on the blade surfaces, since the steam in all points of the nozzle blading surface will be superheated relative to its stagnation parameters. However, in such dry supercooled steam there may exist liquid impurity particles, whose saturation temperature is higher than that of pure water at a given pressure. Figure 7.27 shows as an illustration above the curve of $x = 1.0$ on the i-s diagram a zone of inception of formation and stable existence of liquid NaCl particles. The concentrated solution of NaCl is one of the most active substances inducing pitting and corrosion cracking of stressed components.

The narrow shaded region above $x = 1.0$ in Fig. 7.27 reflects the zone on the i-s diagram which is most dangerous for corrosion failure in turbine flowpassages.

If the expansion of steam in a nuclear power plant turbine starts from superheated steam (points $0'$ in Fig. 7.27) (for example, in the case of nuclear power plants with once-through steam generators) then the most dangerous locations are the first high-pressure stage, the first part of the moisture separator reheater and one of the low-pressure stages.

Above the shaded region, NaCl exists in the form of dry salt or of a true solution, whereas below $x = 1.0$ all the salts will actively dissolve in water, forming low-

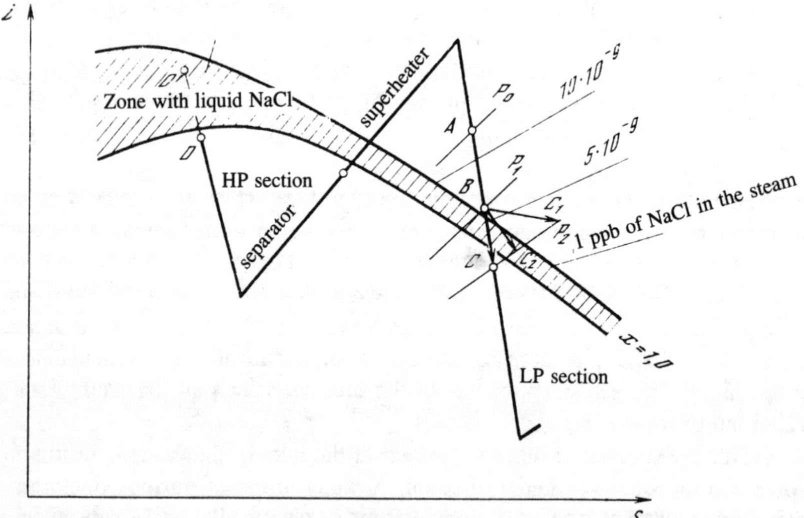

Figure 7.27 Possible steam expansion modes in one of the low-pressure stages upon crossing of the saturation curve.

concentrated compounds, which are not dangerous from the point of view of corrosion failure.

Let us consider possible steam expansion modes in a low-pressure stage. Note that expansion in such a stage can always occur along points *ABC* shown in Fig. 7.27 (where *AB* is expansion from p_0 to p_1 in the nozzle blading, whereas *BC* is the same for the rotor blades), if not under design conditions, then at partial load.

If expansion in the rotor blading starts in point *B* ($\Delta t > 0$) and terminates in the region of wet steam ($y > 0$) (point *C*), then the expansion will be accompanied by concentration of salts in the supercooled steam, formation of moisture and condensation of steam on the surfaces of rotor blades (see Sec. 7.1). Hence the formation of thin nonmoving films of aggressive solutions is possible only at low percentage moisture [sic] in the steam.

In addition to expansion of steam along line *BC*, expansion is possible also at lower efficiencies (see curves *OC* and BC_2). For example, a part of the steam that passes the radial clearance above the shrouds may expand to p_2 at constant i (see line BC_1). At the same time the steam loses energy in the secondary flows at the root and the periphery, which means that expansion can occur along curve BC_2. This case is the most probable for the formation and existence of concentrated liquid solutions of NaCl, which may precipitate onto the rotor blading surfaces at the root and on the periphery, if the stage is shrouded. Zones of the labyrinth packing between the diaphragm and the shaft, where the probability of precipitation of the salt solution is high, may also be dangerous. This figure also shows the solubility of NaCl in the steam. When the turbine operates under varying conditions, the expansion of the steam is shifted to the right due to reduction in efficiency of the flowpassage and the solubility of NaCl in the steam may range between 10 and 1 ppb. This means that in one of the low-pressure stages there exist conditions favorable to the precipitation and adhesion of concentrated liquid impurities on the surfaces of rotor blading. The most dangerous and probable region for salt deposition is the root of rotor blades, and, if it is remembered that the maximum local supercooling of steam occurs on the back of the blade near the oblique edge then microcracks most probably form close to the exit edge on the reverse side of the rotor blade.

The overall concentration of impurities in the steam is extremely low, for which reason diffusion precipitation of droplets may occur only very slowly. In the presence of solid particles, their small size excludes their inertial precipitation on the solid surfaces of disks, nozzle and rotor blading. The most probable precipitation of impurity particles on the metal surface will occur due to turbulent transfer of particles to the wall in boundary layers. All these processes will cause transfer of impurities to the metal surface at an extremely low rate. However, with time the local accumulation of salts in the film become perceptible.

However, if the film that thus formed will not be removed as a result of the elevated turbulence of the steam flow, corrosion will occur most actively in the region of secondary flows, i.e., at the root and periphery of the nozzle and rotor blading.

It must be remembered that when turbines operate under varying loads the precipitation of salts is accelerated on the one hand by increasing flow turbulence and on the other due to the increased probability of washing off of deposits of easily soluble impurities in moisture separator reheaters and of subsequent inertial collision of moistened salt particles and the elements of turbine flowpassages.

In addition to the above mechanism of precipitation of liquid and solid impurities on the surface of metals, there exists the probability of settling of liquid water forming by condensation of the steam. As is known, the size of the droplets forming in the flow does not exceed $d_{dr} < 0.4$ μm. These droplets may precipitate on the surface of a turbine stage only as a result of turbulence-inertial and diffusive processes. However, the water particles arising in the turbine stage actively absorb the salts dissolved in the water, forming chemically aggressive solutions. As a result of the absorption of impurities dissolved in the steam by the water, fine liquid particles are transformed into highly concentrated acids and alkalies, which corrode the metal. Note that the condensation of steam and formation of fine droplets occurs in a narrow zone of the turbine flowpassage, with the condensation zone shifting over the height of blades; this zone may traverse simultaneously (depending on the diameter) through one or two stages and, as was noted above, its location may shift upstream under partial load.

The behavior of NaCl in a turbine as described above depends relatively little on the design and parameters of the nuclear power plant turbine, since the zone in which concentrated NaCl solutions exists is limited on the i-s diagram by a narrow band, whose width depends little on the pressure at which the expansion curve intersects the interface between water and steam.

On the other hand, the behavior of NaOH may be significantly affected by the thermal parameters of the nuclear power plant. Since it is impossible to perform quantitative calculations, particularly those making allowance for the kinetics of the rapidly occurring expansion of steam in a turbine, it is sufficient to consider several typical designs of nuclear power plant turbines.

Figure 7.28 shows two versions of the expansion of steam in turbines on an i-s diagram for modern nuclear power plants with water-cooled reactors: (1) the two-loop unit No.1 5 of the Novovoronezhskaya Nuclear Power Plant employing a PWR-type reactor (a VVER-1000 unit); the parameters of the unit (initial pressure about 6 MPa and two-stage reheating to 250 °K at 1.2 MPa) are typical also of the majority of highly-rated Western units; (2) unit No. 1 of the Leningrad Nuclear Power Plant with an RBMK-1000 boiling-water reactor and two high-rpm turbines with reheating at a pressure of $p_{ch} = 0.35$ MPa.

The i-s diagram also shows the thermodynamic equilibrium concentrations of aqueous NaOH and approximate values of NaOH concentrations in steam, in equilibrium with the aqueous concentrations.

It is seen from the i-s diagram of the VVER-1000 unit that, at this elevated reheater pressures in the moisture separator reheater and the NaOH concentrations corresponding to standards or continuous operation, even on the highly unlikely assumption that the entire Na is presented in the steam in the form of NaOH, the concentration of the latter does not exceed its solubility in the steam.

It can hence be expected that the steam leaving the reheater will be a true solution. Since the first expansion stage in the low-pressure part of the turbine terminates

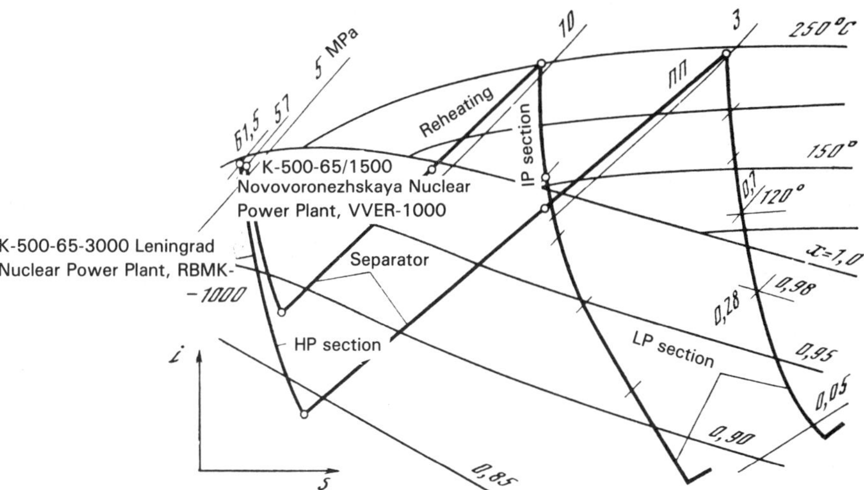

Figure 7.28 Expansion of steam in nuclear power plant turbines.

near the saturation curve for pure water, it is very difficult to believe that droplets of concentrated NaOH solution succeeded in forming from a real solution with the exceedingly low concentration (of the order of 10 ppb) prior to intersection of this curve. The formation of NaOH droplets in the second and third stages is then virtually harmless.

Evidently, such a conclusion can be only approximate, since it considers only the specific conditions for the flow of only the mainstream steam, and not of the flows mentioned above, for example, in labyrinth packings and in the tip regions of the cascades.

The heat conditions for the RBMK-1000 unit employing an 1500 rpm Khar'kov Turbine Plant unit, and the more so for unit No. 1 of the Leningrad Nuclear Power Plant, will be different due to the lower pressures. On the other hand, the higher percentage moisture at the separator inlet will result (for a given final percentage moisture) in a greater reduction in the content of all the impurities at the reheater inlet, due to removal of impurities in the separator. On the other hand, the content of all the impurities (excepting NaOH) after the reheater will be lower due to reduction of their actual solubility in the lower-pressure steam. However, in addition to these conditions, which are more favorable for operation of the LP section of the turbine, there occurs, firstly maximum concentration of NaOH in droplets forming in the reheater, which results in the possibility of corrosion of tubes of the moisture separator reheater, and secondly, the probability of retention of a part of NaOH which leaves the reheater in the form of highly-concentrated droplets and their suction together with the steam into the low-pressure part increases significantly. Such droplets are by themselves corrosive and, in addition, serve as nuclei for condensation of solutions of impurities in the steam. In addition, for example in the No. 1 unit of the Leningrad nuclear power plant, intersection of the pure water—steam saturation curve does not occur in the second, but rather in the fourth stage, which also increases the danger of corrosion of the LP section.

Obviously, all these comparisons are rather tentative. In addition, the probability of corrosion damage may depend significantly not only on the overall concentration of Na in the steam, but also on the form in which it is present. For example, the condensers at the Leningrad Nuclear Power Plant are cooled by sea water. The bulk of Na will be in the form of NaCl. The danger of chloride corrosion is associated with lower pressures, which may outweigh the increase in danger due to the behavior of NaOH. Still, this very approximate analysis shows that parameters of the steam-turbine cycle at nuclear power plants should be selected also with consideration of the water regime and corrosion. Evidently, in order for such calculations to be quantitatively reliable, it is still necessary to perform a tremendous volume of scientific studies and to accumulate operational experience.

To complete the question of operation of saturated-steam nuclear power plant turbines, we wish to note that the questions of water regime and corrosion could take an entirely different from if one would use separating stages built integrally with the turbine, described in Chapter 8. In this case there would be no need for reheating and the entire expansion could be obtained with wet steam, which would entirely eliminate the problem of impurity concentration. This should be taken into account in estimating the prospects for introducing multistage separation without reheating.

Turbines operating at nuclear power plants with fast-neutron, liquid-metal cooled reactors operate at initial steam parameters which differ only insignificantly from typical fossil-fuel fired units. Thus, the only large unit of this type currently in operation in the USSR (BN-600 unit) operates at a pressure of 13 MPa and temperature of 505 °C at the turbine inlet; it uses three typical turbines rated at 200 MW each with a moderate reduction in initial temperature. This unit is equipped with a liquid-sodium heated reheater with the same final steam temperature (505 °C) as after the primary superheater.

Reheating occurs at 2.6 MPa. The entire curve of expansion up to the LP section occurs in the superheated steam region. Under these conditions it can be assumed that even if the Na content specifications for the steam are somewhat exceeded, the steam in the reheater will contain NaOH as a real solution. It is seen from Fig. 7.29, which represents the i-s diagram for the BN-600 nuclear power plant turbine, that under conditions of thermodynamic equilibrium the appearance of droplets of concentrated solution is possible only when the pressure drops to 0.5–0.7 MPa, i.e., at the inlet to the medium pressure part. Actually, at least for the mainstream steam, the zone of precipitation of the liquid phase will lag significantly and may, in spite of the perceptible supersaturation, attain a pressure of 0.50 MPa, at which the expansion curve intersects the pure water—steam interface.

On the whole, the conditions of operation of these turbines are virtually identical to those prevalent in large fossil-fuel fired units. This means that the specifications concerning the Na content of steam supplied to the turbine should be more demanding than for turbines using saturated steam (there is no washing of steam in the separator and the pressure of the steam at the intersection of the steam/water interface curve is low).

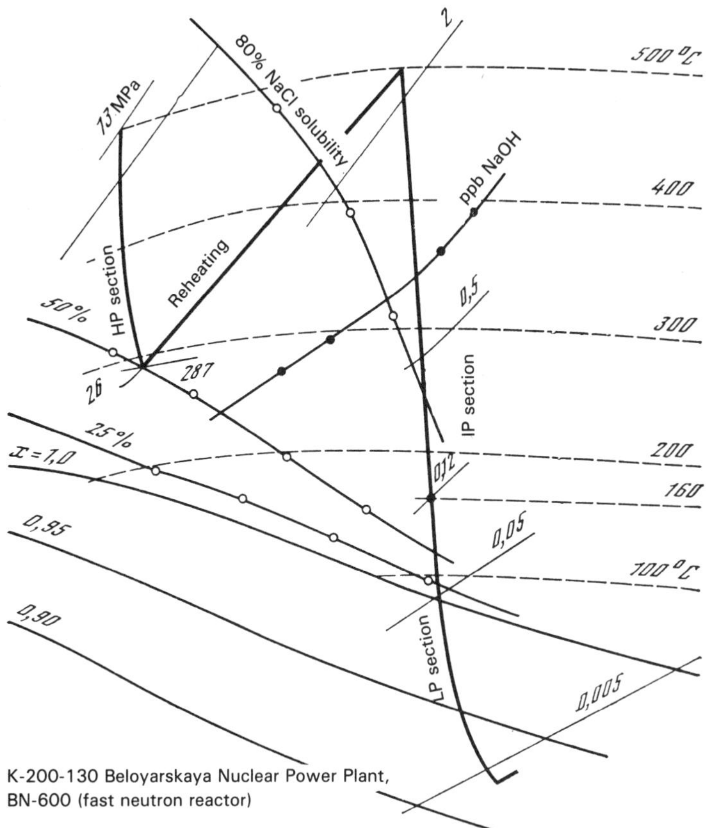

Figure 7.29 Expansion of steam in a turbine at the Beloyarskaya Nuclear Power Plant with a BN-600 [fast neutron] reactor and the NaOH solubility curves on the *i-s* diagram.

Another example of a large liquid-metal cooled unit is the Superphenix Nuclear Power Plant (France) rated at 1200 MW employing two 600 MW turbines.

This unit uses another steam-power cycle with higher initial pressure—18 MPa and a somewhat lower temperature—490 °C. However, the main difference is the absence of sodium reheating and expansion of steam in the turbine to ~ 1.0 MPa. At this pressure and moderate percentage moisture (3–4%) the steam passes the separator and is supplied to a reheater, heated by bled steam.

This means that the turbines at this nuclear power plant operate under conditions where the possibility of formation of concentrated solutions is almost identical to that in conventional units with saturated-steam turbines. It should, however, be noted that when the moisture content of the steam entering the separator is lower, the washing of the steam will be less effective, since the steam expansion curve intersects the inter-face curve on the *i-s* diagram at a slightly higher pressure than in turbines employed with VVER-1000 units. This may, on one hand, require somewhat stricter specifications concerning the NaOH concentration, but on the other, the depth of the super-heated steam zone itself is somewhat less and accordingly the kinetic lag may be more effective.

Of course, a more detailed analysis, which will become possible after obtaining more reliable thermodynamic data and study of the kinetics of moisture precipitation in turbine stages, will result in modification of turbine designs and distribution of enthalpy dissipations by stages.

It should be noted in conclusion that we have considered here several possible versions of the inception of metal failure in turbine in the course of formation and evaporation of condensate particles over a narrow range of percentage moisture values. At present there are still not sufficient data on these salt precipitation processes and on the appearance of microcracks in nuclear power plant turbines for comprehensive analysis of the accidents occurring in turbines and moisture separator reheaters. The latent results obtained by the Moscow Energetics Institute [7.18] and by a number of Western companies [7.19, 7.20] show that there is a need for further study of the formation of microcracks, corrosion, and vibration-corrosion cracking of components of turbines flowpassages. The solution of this problem requires, in addition to improvement of the water treatment technology, rigorous conformance to pH value standards, specifications on the maximum permissible content of the most aggressive impurities, and a new approach to the design of flowpassages of nuclear power plant turbines.

SEPARATION OF MOISTURE
AT NUCLEAR POWER PLANTS

The efficiency and reliability of turbines in which expansion of steam occurs from the saturation curve depend to a large extent on the effectiveness of moisture removal along the entire path of the wet steam. The principal moisture removal stages are: moisture removal in reactors and steam generators, separation of moisture upstream of the turbine, moisture removal in the flowpassages of the high- and low-pressure sections, removal in exhaust piping and crossovers between the HP- and LP-sections, removal of moisture in internal separators and in moisture separator reheaters (MSR).

Reliable and effective turbine operation requires that the steam entering the HP section has minimal percentage moisture (less than 0.2%), since an elevated percentage moisture entering the HP section not only increases the radiation danger in nuclear systems but also reduces the efficiency of turbines and has a significant effect on the operation of the first turbine stage and the entire process of formation and the dimensions of moisture particles in the flowpassages of the HP section. At the same time, separation of moisture in the course of steam generation should be such that no moisture at all would enter the turbine.

The operation of turbines at nuclear power plants is complicated by the fact that the effectiveness of moisture separation is not monitored, since at present no devices for measuring the percentage moisture are used at inlets to the HP sections. At the same time, a significant part of the moisture at the turbine inlet moves in the form of a film, and no provisions are made for its removal. The formation, motion and separation of moisture in turbine flowpassages are much better explored. Removal of moisture from the turbine flowpassages improves the efficiency and reliability of the turbine, and also reduces the erosion wear of blades in the LP section and erosion of diaphragms in the HP section.

8.1 MOISTURE REMOVAL IN CHEVRON ELEMENTS AND CENTRIFUGAL SEPARATORS

Chevron demistor and cyclone-type separation elements are extensively used at nuclear power plants. These elements are used in channel-type reactors, steam generators at nuclear power plants with water-cooled reactors, and also in separators and moisture separator reheaters of turbines. Lately there is a tendency toward increased use of chevron separating elements. As of now, about thirty various configurations of chevron-type separating elements are known (Fig. 8.1). The shapes of the separation elements are extremely varied. Note that Soviet-produced elements have a smooth wavy surface (type 2 in Fig. 8.1). A number of Western companies employs louvers with projections and corners on the plate surface. It is assumed that the steam flowing through passages in the chevron (types 17, 18 and 28) will be turbulized and this will improve the effectiveness of precipitation of fine droplets. Individual separation elements, while forming chevron-type channels, are themselves not in the shape of chevrons.

The following problems must be solved in selecting separation channels: (1) obtaining high moisture removal efficiency; (2) ensure minimum losses of steam energy in the passage (losses of energy by the steam passing the chevron); (3) ensure adaptability to repetitive manufacture of the elements. In spite of the extensive use of these elements at nuclear power plants, there is at present very little data on their effectiveness.

To analyze the effectiveness of deposition of droplets in a channel formed by chevrons as a function of operating conditions and geometry, we shall consider a simplified scheme of the motion of moisture in a channel with sign-variable curvature and constant cross sectional area [8.1]. We are assuming a model of a two-phase steady-state flow, consisting of an incompressible gas and a discrete system of spherical moisture particles of different sizes. The calculations do not include heat and mass transfer between phases, the interaction between individual moisture particles and the effect of the liquid on the steam phase. The ideal-liquid assumption here does not extend to the mechanism of flow over the droplet.

The motion of a droplet propelled by the steam flow is described by the equation

$$m \frac{d\mathbf{w}_2}{d\tau} = C_\mathrm{D} S \frac{\rho_\mathrm{v} \Delta w_1 \Delta \mathbf{w}_1}{2} \tag{8.1}$$

where m is the mass of the droplet, C_D is its drag coefficient, S is its median cross sectional area, ρ_v is the density of the steam, Δw is the velocity of the latter relative to the droplet, i.e., $\Delta \mathbf{w} = (\mathbf{w}_1 - \mathbf{w}_2)$.

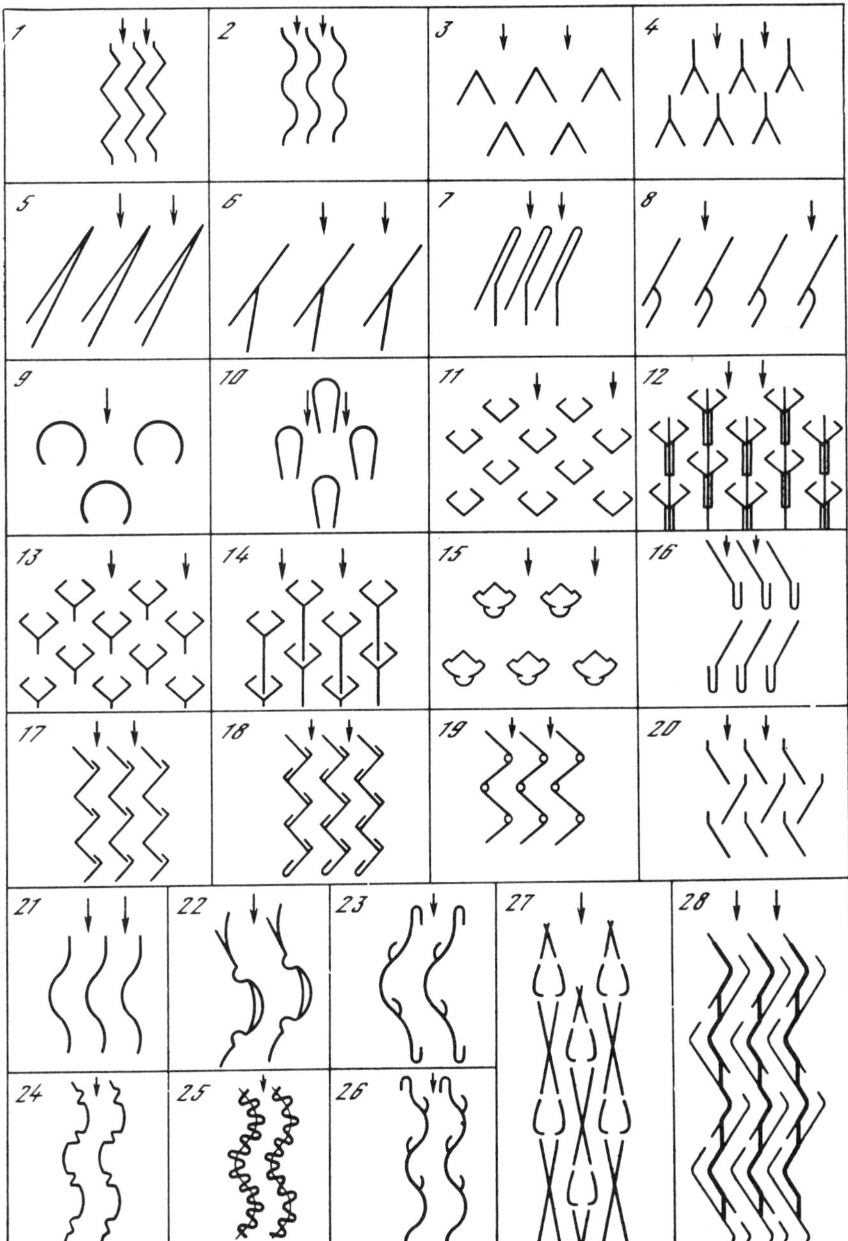

Figure 8.1 Chevron separating elements (the wet steam motion is downward).

The calculations are performed with a power law Reynolds number dependence of the drag $C_D = \tilde{f}(\mathrm{Re})$.

To examine the two-dimensional problem of motion of a droplet in a curved channel, we write Eq. (8.1) in polar coordinates

$$\frac{dw_{3x}}{d\tau} = \frac{w_{3\varphi}^2}{R} + \kappa w_1 w_R C_D(\mathrm{Re})\, \mathrm{Re}^{-n(\mathrm{Re})}$$

$$\frac{dw_{3\varphi}}{d\tau} = -\frac{w_{3x} w_{3\varphi}}{R} + \kappa w w_\varphi C_D(\mathrm{Re})\, \mathrm{Re}^{-n(\mathrm{Re})} \tag{8.2}$$

where w_{3x} and $w_{3\varphi}$ are the absolute radial and angular velocities of the droplet; Δw_{1R} and $\Delta w_{1\varphi}$ are the radial and angular velocities of the steam relative to the droplet; R is the absolute value of the current radius vector of the droplet, $\kappa = {}^3\!/_8(\rho_1/\rho_3)(1/r_3$; ρ_1, ρ_2 are the densities of the steam and the droplet, and r_3 is the droplet radius.

The droplet paths were obtained by solving Eqs. (8.2) on a computer, employing the Merson modification of the Runge-Kutta method. The droplets are regarded as having precipitated if their paths come into contact with the passage surface. The effectiveness of droplet precipitation on the passage walls was determined by analyzing the droplet paths.

The effectiveness of precipitation of moisture particles is defined by the function $\eta_\Phi = f(d_{\mathrm{dr}})$, each value of which defines the fraction of droplets of a given size which reach the channel walls. In general, for a known inlet particle size distribution function the effectiveness of droplet precipitation as a function of the initial moisture content is defined as

$$\Psi = \frac{\displaystyle\int_0^{r_{\max}} \tilde{F}(r)\, \eta_\Phi(r)\, dr}{\displaystyle\int_0^{r_{\max}} \tilde{F}(r)\, dr}$$

where $\tilde{F}(r)$ is the radial droplet distribution function. It is usually difficult to determine $\tilde{F}(r)$ analytically, for which reason one usually uses experimental data on the droplet size distribution function at the inlet to the chevron passage. It then becomes possible to obtain a qualitative and quantitative estimate of the rate of moisture precipitation on the walls of a channel for a given geometry, or to determine the spacing and shape of the element arrangement to give the required moisture precipitation rate.

Figure 8.2 shows the effect of geometry (spacing of chevrons and curvature of the passage) on the effectiveness of precipitation of droplets of different size on the passage walls (the passage is depicted schematically in the right part of the figure). Naturally, when the bends in the passage meet at a greater angle, the rate of precipitation will increase. Thus, for example, at the maximum joining angle of $\varphi = 90°$ almost all the droplets with $d_{\mathrm{dr}} = 10\ \mu\mathrm{m}$ will precipitate. Reduction of φ to $45°$ is not accompanied by a significant reduction in η_Φ (curves 1 through 3 in Fig. 8.2), which

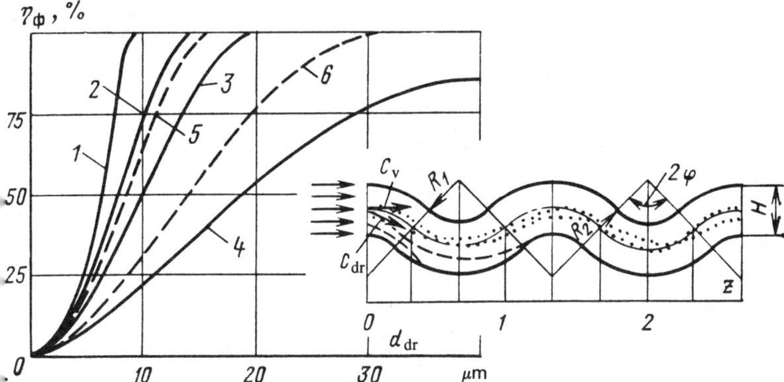

Figure 8.2 Effect of geometry and wet-steam parameters on the precipitation effectiveness as a function of droplet size; $v_m = 20$ m/sec, $p = 0.1$ MPa, $\bar{H} = H/R_1$. 1, $\varphi = 90°$, $\bar{H} = 0.45$; 2, $60°$ and 0.45; 3, $45°$ and 0.45; 4, $30°$ and 0.45; 5, $45°$ and 0.25; 6, $45°$ and 0.9.

leads to the conclusion that the optimum louvered passage is that with $\varphi \cong 45°$, since it has smaller power energy losses than a passage with $\varphi = 90°$. Further "straightening" out of the passages (curve 4) results in a significant drop in η_Φ and almost 50% of droplets with diameters of 20 μm pass through without coming into contact with the passage walls. The same figure (curves 3, 5 and 6) show the effect of channel width (relative spacing of the chevron arrangement) on the moisture precipitation effectiveness η_Φ. It is seen that at low relative spacings ($\bar{H} < 0.5$) the effect of channel width on η_Φ for droplets with $d_{dr} > 8$ μm is small. However, decreasing $\bar{H}$ to the point where $\tilde{H} < 0.5$ results in a reduction in η_Φ, and most of moisture with $d_{dr} < 15$ μm does not precipitate on the walls.

Above, we considered the case of precipitation of moisture in various chevron passages at constant ρ_v/ρ_{dr}, i.e., at constant steam density, velocity and ratio of phase velocities at the passage inlet.

Changes of the flow variables of the wet steam will affect the nature of droplet motion and the droplet precipitation on the walls. The density of the steam, more precisely, the ratio ρ_v/ρ_{dr} is an important parameter controlling the separation of moisture in such channels, since knowledge of the effect of this ratio on the separation of moisture in the channel will make it possible to estimate the separation capacity of chevron elements at different pressures and to extend the results of experimental studies performed at low pressures to the high-pressure region.

With increasing steam pressure its capacity to entrain the droplets increases and the effectiveness of droplet precipitation in the louvered passage decreases (curves 1 through 5, Fig. 8.2). The value of η_Φ is affected also by the velocity of the steam in the channel. Thus, for example, when velocity v_m is reduced from 20 to 1 m/sec (curve 5) the effectiveness of precipitation of droplets with $d_{dr} < 50$ μm and decreases more than fourfold. The precipitation of droplets is affected less by the phase slip $\nu = V_{dr}/V_v$ at the inlet to the passage. As follows from calculations, variation in ν between 0.1 and 1 does not change η_Φ by more than 5%. Differences in the directions of motion of the liquid and steam phases at the inlet to the louvered channel also does not result in a significant change in the effectiveness of droplet precipitation in the channel.

To be able to determine the length of the passage, it is important to know the value of η_Φ as a function of the number of turns in the passage (see Fig. 8.2). Calculations show that the bulk of the moisture precipitates at the start of the passage [at the first complete turn, see Fig. 8.2, curves 1 through 3 $\eta_\Phi = f(z)$].

Detailed experimental studies of chevron separation elements were performed at the Polzunov Boiler and Turbine Institute [8.2]. Figure 8.3 shows the design of a separation unit and the shapes of the passages under study. Air-water mixtures were used in the experiments. The critical velocity of the humid air c_{cr} was determined as a function of the percentage moisture y_0. The term critical velocity c_{cr} is defined here as the velocity of air at which removal of moisture from the channel surface starts.

To determine the value of c_{cr} at variable ρ_v/ρ_{dr} one usually uses the Kutateladze number as a function of the initial percentage moisture of the steam:

$$K = c_{cr} \sqrt{\rho_v} / \sqrt[4]{g^2 \sigma (\rho_{dr} - \rho_v)}$$

where c_{cr} is the velocity of the air, ρ_v and ρ_{dr} are the specific densities of the air and liquid, σ is the surface tension coefficient, and g is the acceleration of gravity.

In addition to the above parameters of chevron stacks, the effectiveness of moisture separation is significantly affected by the stacking angle α. Hence the maximum permissible steam velocities are established on the basis of the selected α. The air-water studies [8.2] show that inclined stacks permit a lower critical air velocity.

In fact, if we consider these characteristic stacking angles ($\alpha = 0$, 45 and 90°), we see that when $\alpha = 0$ the direction of steam flow is upward, whereas the liquid films move downward due to gravity, for this reason in this case entrainment of liquid particles from the wavy wall surface starts at lower air velocities. When the stacking angle is 45° the air in the stacks flows close to normal to its inlet, i.e., also here partial counterflow occurs between the liquid (film) and the air. However, it must be remembered in selecting the value of α that when it is reduced, the permissible velocity of the air (steam) decreases, the net cross sectional area of the stack increases, and the permissible thoroughput rises accordingly. According to Ryzhkov et al. [8.2], the permissible thoroughput of a stack placed at $\alpha = 45°$ is 30–40% percent higher than for the case of $\alpha = 90°$. At the same time the loss of energy by the air (steam) upon passing inclined ($\alpha < 90°C$) chevron stacks is somewhat higher than for vertical stacks.

Another important parameter of chevron channels, in addition to their separating capacity, is the drag of the separating elements. Figure 8.4 shows experimental data of Demidova and Sorokin [8.3] on the variation of drag for stacks I through IX as a function of the air (steam) velocity.

It is seen from the experiments that energy losses $\Delta\bar{p} = (p_0 - p_{01}/p_0)\,100\%$ (where p_0 and p_{01} are the stagnation pressures up and downstream of the louvered stack) may change several-fold as a function of the passage shape and air velocity. For this reason chevron elements must be optimized with allowance for their aerodynamic drag.

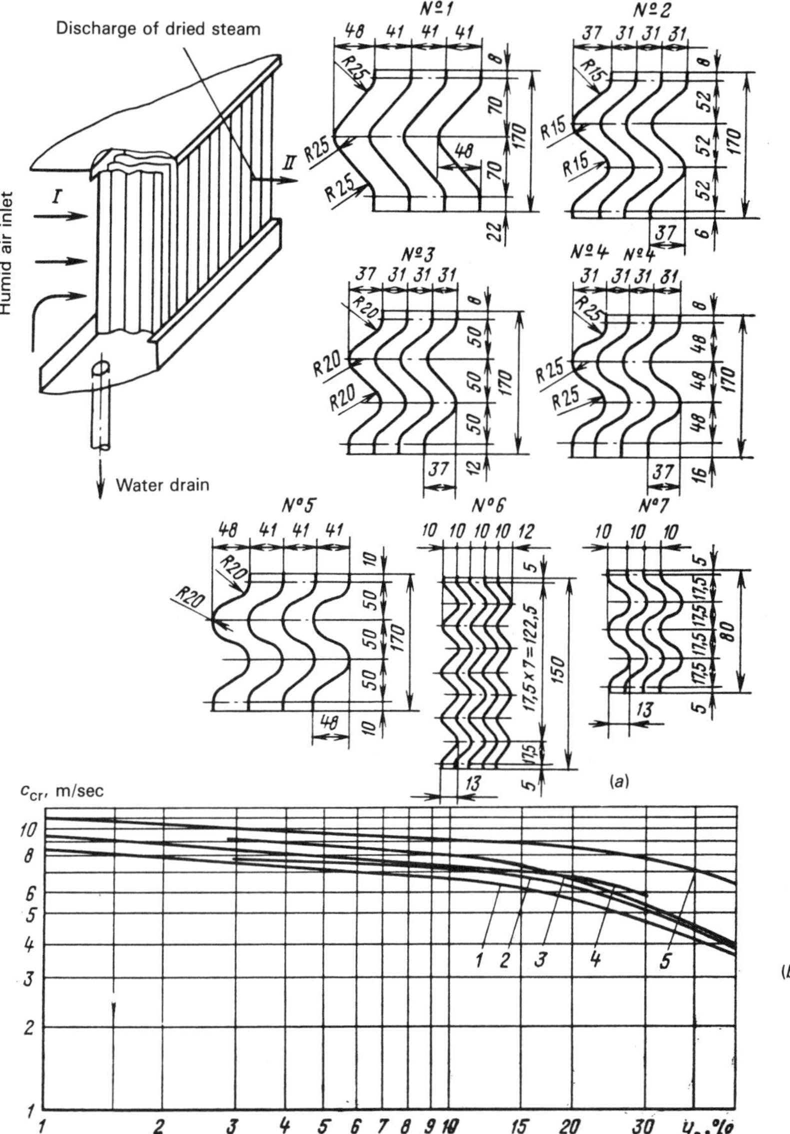

Figure 8.3 Separator unit, louvered elements (*a*) and values of critical velocities for the first five stacks (*b*) as a function of the initial humidity of the air (the numbers of the curves correspond to the number of stack).

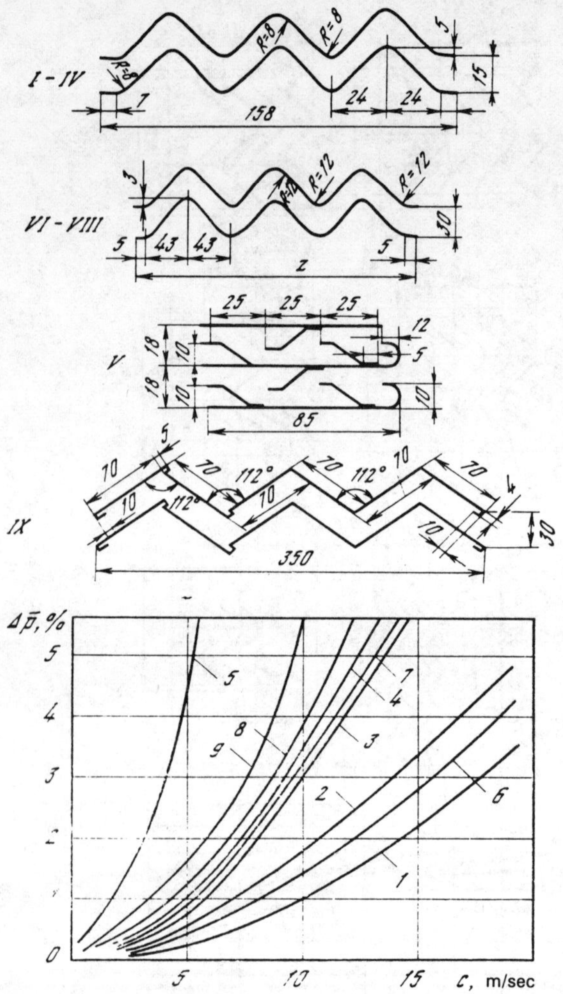

Figure 8.4 Aerodynamic drag $\bar{\Delta}p$ of louvered stacks vs. the velocity of air in them (the curve numbers correspond to those of the stacks).

The effectiveness of moisture extraction by chevron stacks depends significantly on the droplet size distribution function and the ratio of the densities of the steam and water. The existing experimental data were obtained primarily with moisture atomized by pressure nozzles, and the droplet size distribution in these experiments is different from those encountered in turbines. It is of interest, therefore, to investigate the effect of the size of droplets on the extraction of droplets from a chevron stack.

Figure 8.5 shows the results of studies performed at the Khar'kov Turbine Plant [8.6] on a chevron stack extracting moisture from wet steam at $p_0 = 0.15$ and 0.3 MPa and with varying droplet size at the stack inlet. The stack was 600 mm high and 200 mm wide. The experiments were performed with a stand employing multistage cooling of steam by injection of air into the steam using pressure nozzles. The critical steam velocity decreases with rising density of the steam (curves 1 through 3) and at $p_0 = 0.3$ MPa does not exceed 6 m/sec. The initial percentage moisture (which ranged from 5 to 30%) was not found to have a significant effect on the variation in moisture content past the separator. It was noted, as in experiments with the air-water system, that the removal of liquid through the louvered stack depends significantly on the droplet size. Thus, for example, variation in the modal droplet diameter d_{dr}^{M} from 40 to 300 μm (this was obtained by using different spray nozzles) results in departure from the curve of $y_1 = f(c_0, d)$ in the supercritical region (curves 1 through 3), whereas in experiments at $p_0 = 0.15$ MPa changing the droplet size from 35 to 350 μm changes the moisture removal over the entire range of steam velocities under study.

Centrifugal separators are extensively used at nuclear power plants as the first moisture removal stage. Moisture extraction in them occurs due to precipitation of moisture on the outer surfaces of cylinders brought about by centrifugal forces in the "swirled" two-phase flow. Centrifugal separators are more compact than louvered or mesh-type separators; however, they have a much poorer moisture separation efficiency and high losses of steam energy.

Investigations performed at the Moscow Energetics Institute of centrifugal wet-steam separators with the test stand shown in Fig. 8.6a [8.4], make it possible to demonstrate the overall nature of moisture extraction for three types of centrifugal separators. Separator I (Fig. 8.6b) consists of a swirling device (guide vane) and two cylinders. Separator II has, additionally, a 3 mm wide annular slot. The latter allows removal of a part of the moisture in the form of a film forming on the inner surface of the cylinder. Separator III had not one, but two 3 mm wide annular slots, with tapered second and third part of the inner housing. Studies with wet steam at $p_0 = 2$ MPa showed that with increasing load (inlet steam quality x_{in}) the percentage moisture first drops (Fig. 8.7a), which signifies transition from gravity to centrifugal separation; then depending on the velocity c_v^{wet} of wet steam it remains constant over a certain range of loads, and then increases steeply. The steep rise in y_1 in the steam downstream of the separator is due to the start of intensive entrainment of moisture from

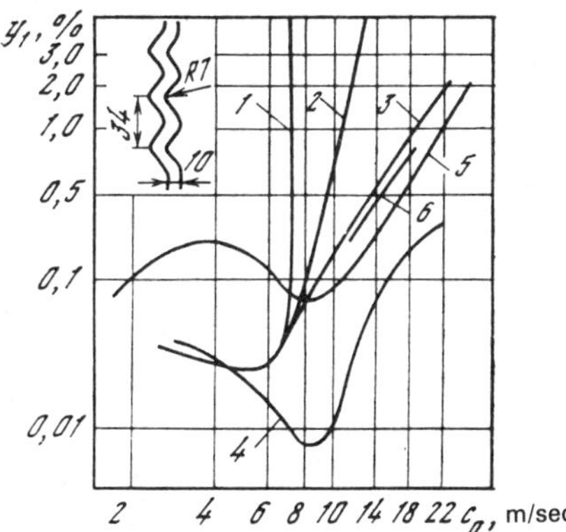

Figure 8.5 Variation in percentage moisture in steam y_1 past a louvered stack as a function of the steam velocity for different droplet sizes at the inlet. $p_0 = 0.3$ MPa: 1, $d_{dr}^{M} = 40$ μm; 2, 80 μm; 3, 300 μm; $p_0 = 0.15$ MPa: 4, 35 μm; 5, 70 μm; 6, 350 μm.

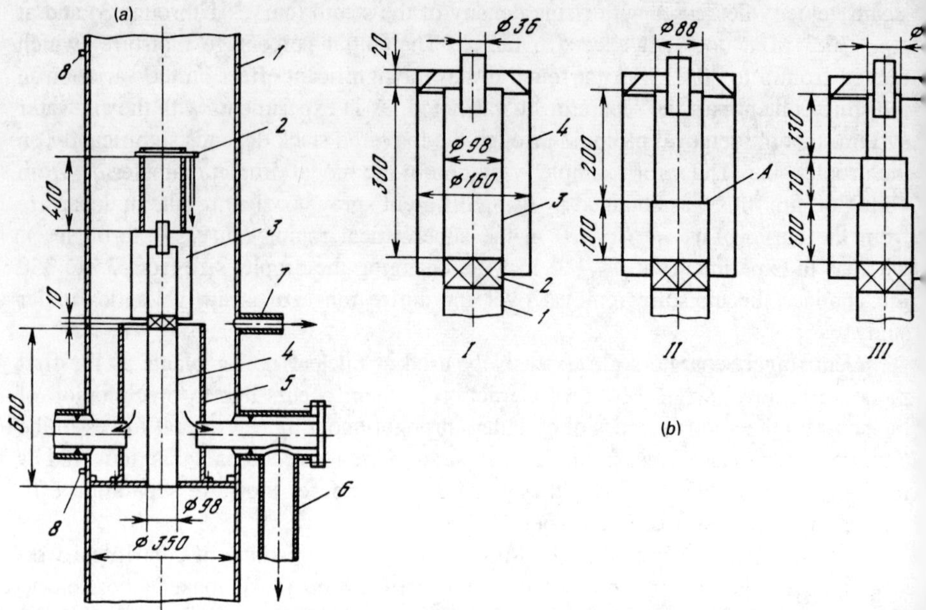

Figure 8.6 Schematic of test stand (*a*) and models of separators under study (*b*). (*a*): 1, column; 2, moisture-meter sampler; 3, separator; 4, standpipe; 5, inner casing; 6, downtake pipe; 7, drain line; 8, level measuring stub. (*b*) 1, Inlet stub; 2, swirl generator; 3, casing; 4, jacket. A, Annular slot.

the film surface. Studies of separators with annular slots (models II and III) showed that a part of the moisture is removed along the flow of wet steam to the periphery. This makes it also possible to reduce the entrainment of steam into the drain system. Note that, in spite of the extensive use of centrifugal separators, the processes of moisture extraction in them have not yet been sufficiently explored.

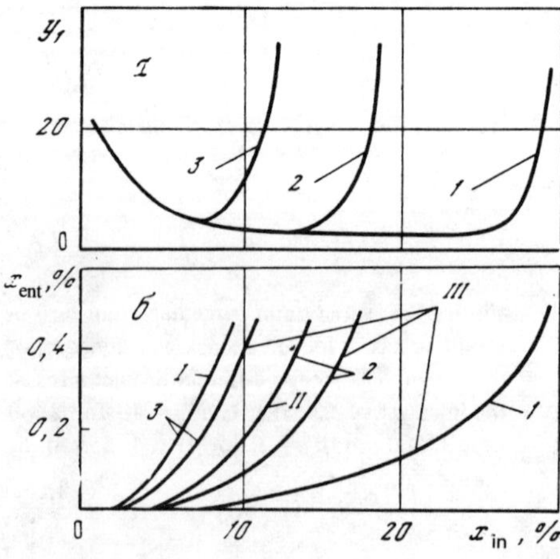

Figure 8.7 Variation in percentage moisture in steam y_1 past a centrifugal separator (*a*) and x_{ent}, fraction of steam entrainment (*b*) as a function of load x_{in}. I–II, Model numbers; 1, $c_v^{wet} = 1.5$; 2, 2; 3, 10.

Table 8.1

	Steam generator	
Quantity	Unit III and IV of a VVER-440 PWR reactor	Unit V of a VVER-1000 reactor
Unit electrical rating, MW	73	250
Specific steam duty, kg/sec·m^2 3.6	8.0	
Type of secondary separator	Inclined chevron without drain devices	Vertical chevron with drain devices
Minimum distance from tubes to separator, m	0.75	0.85
Maximum permissible level of water above the tubes, m	0.15	0.15
Ratio of the limiting velocity of steam entering the chevron stack to the mean nominal	1.2	1.5
Ratio of the maximum local heat flux density to the mean density	2.1	2.1

8.2 MOISTURE SEPARATION IN REACTORS AND STEAM GENERATORS

A separation system should be compact and exhibit a high effectiveness of moisture removal, since these qualities control to a significant degree the efficiency and reliability of turbines and the radiation safety of operation of nuclear power plant equipment.

Shell- and channel-type boiling water reactors employ gravity separation of moisture in the reactor housing and separation of moisture in the reactor housing by means of centrifugal separators in uptake and downtake wet-steam flows. The design and operating principles of separation systems depends on the type of reactor and of the assumed method of steam generation and separation.

In the USSR, nuclear power plants with PWR reactors make extensive use of horizontal-type steam generators with chevron separators (see Table 8.1) (for example, at the Novovoronezhskaya Nuclear Power Plant). USSR fabricated steam generators have two moisture separation stages. Separation of moisture at the first stage occurs due to gravity forces, whereas the second separation stage has chevron separation units (Fig. 8.8).

Studies of the steam generators at the Novovoronezhskaya Nuclear Power Plant showed that the location of the actual level (level swelling) and entrainment of moisture over the length (width) of the steam generator differ significantly for each characteristic zone of the vaporization surface and are controlled by the local steam velocity.

The moisture content of the steam in the space upstream of louvered separators in

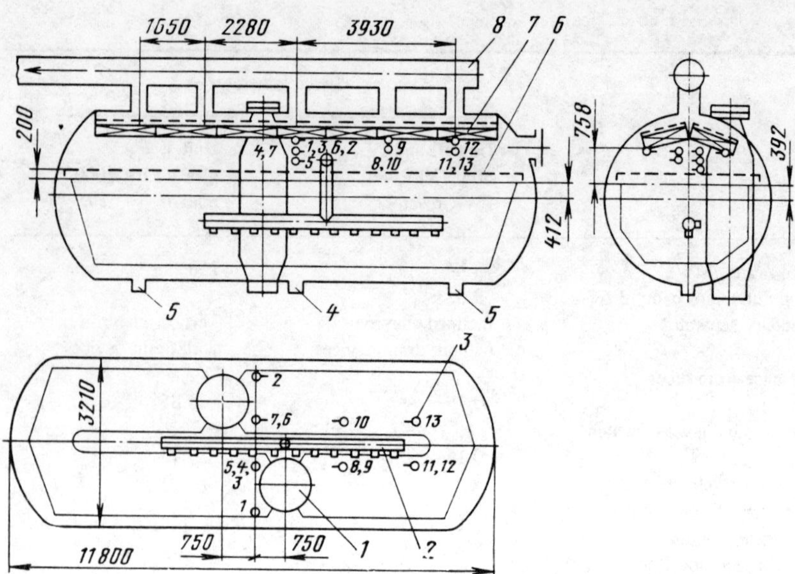

Figure 8.8 Steam generator at a nuclear power plant employing VVER-400 PWR reactor. 1, Reactor coolant supply manifold; 2, feedwater manifold; 3, samplers 1 through 13; 4, central blowdown pipe; 5, end blowdown pipes; 6, submerged perforated plate; 7, louvered separator; 8, steam manifold (fine-print numerals designate steam sampling points).

steam generators at nuclear power plants employing VVER-440 reactors is controlled by the actual location of the level (interface between the steam-water mixture and steam) [8.6]. Figure 8.9a shows the position of the level in steam generators with submerged preferential plate (curve I) and without it (curve II). It is seen that in the second case level H_c in the steam generator rises by almost 400 mm and the steam-water mixture approaches the inlet to the chevron separator. The most dangerous case is that of "flooding" of separator stacks and reduction of effectiveness of moisture removal in them. This may result in penetration of moisture into the steam line and its entry into the high-pressure part of the turbine.

Figure 8.9b shows the results of measurement of percentage moisture at the inlet to different separator stacks as a function of level H_c at the nominal load of the turbine. It is seen that the moisture content of steam at different points at the same distance from the top of the evaporator tube bundle differs several-fold.

The mean velocities of steam in the chevron stack are about 0.3 m/sec (in horizontal type steam generators for the VVER-1000 this velocity is 0.45 m/sec), which eliminates the possibility of critical flow of liquid films in the separator. However, studies performed by Kiryukhin et al. [8.7], show that the moisture content of steam in the steamline downstream of the separator (curve 1, Fig. 8.9b) can be higher than at various points upstream of the louver stacks (points 1 through 4). The most probable reason for intake of moisture into the steam line are clearances made necessary to manufacturing needs and the reduced effectiveness of individual separation elements.

Measurements of the moisture content of steam leaving the steam generator as a function of level H_c, for three steam generators, showed that at $H_c > 250$ mm the percentage moisture of steam in the steam line exceeds the permissible values $y_{perm} \leq 0.25\%$ (Fig. 8.10).

The use of a submerged perforated plate makes it possible to reduce the moisture content of steam leaving the steam generator over the entire range of loads and level

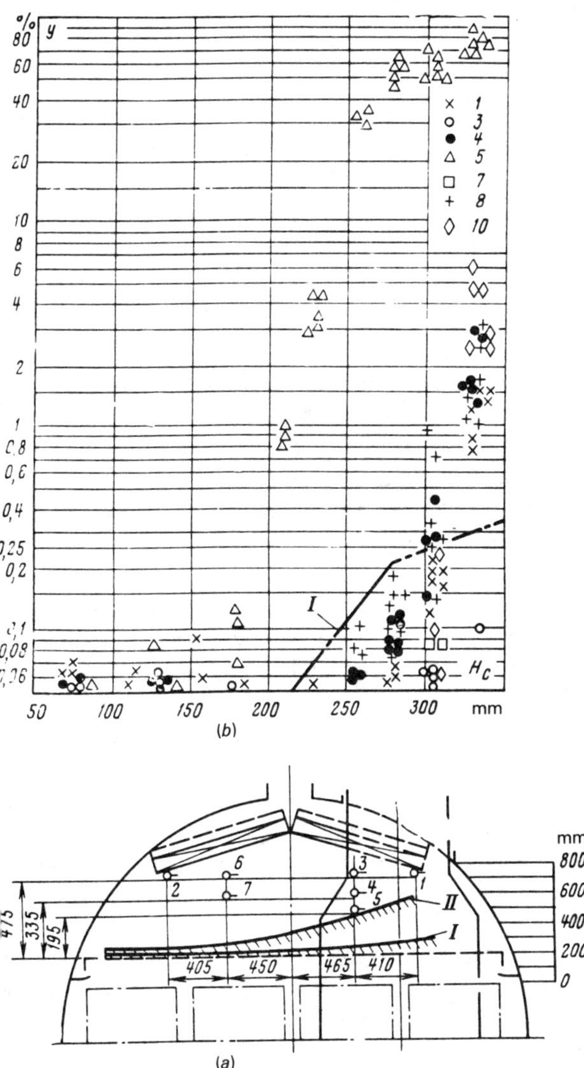

Figure 8.9 Effect of actual level in the steam generator on the percentage moisture of steam upstream of the separator and in the steam line. (a) Location of actual level H_c in a steam generator of a nuclear power plant with a VVER-440 reactor; I, with submerged perforated plate, II, without a submerged perforated plate. (b) Percentage moisture in steam in different points upstream of a louvered separator at nominal load (the numbers of points correspond to the steam sampling points, see Fig. 8.8).

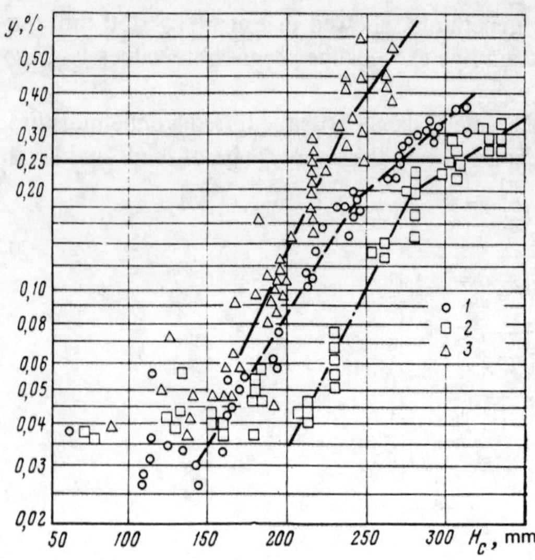

Figure 8.10 Percentage moisture in the steam leaving a steam generator vs. water level H_c at steam-generator pressure p_0 between 4.5 and 4.8 MPa. 1, Separation units with intentional moisture removal system; 2, steam generator with submerged perforated plate; 3, steam generator without submerged perforated plate.

H_c. Some improvement of the effectiveness of moisture removal was obtained also with a steam generator with drain channels in the separation units (curve I).

Calculations of separation parameters of a steam generator include determination of characteristics of the water space and calculation of the steam space. The presence of a submerged-type evaporator and a feedwater manifold in the water space complicates the hydrodynamic situation and results in significant nonuniformity of heat fluxes.

Analysis of separator elements in reactors and drum separators shows that chevron elements do not differ significantly from separation elements of moisture-separator reheaters (MSR) and their use is based basically on the same experimental results.

Vertical steam generators are extensively used in the West. These steam generators have certain advantages with respect to moisture removal, since they are not subject to rigorous limitations on the height of the separation volume. However, due to the reduced cross sectional area, the separator height is greater.

High steam duties for large vertical steam generators require at present the use of special devices for first-stage moisture separation such as baffles, diaphragms, small- and large-diameter cyclones.

The separation arrangement of large vertical steam generators can be examined for the case of the *Westinghouse* steam generator (Fig. 8.11). Separation in these generators occurs in two stages. The first stage contains axial cyclones, whereas the second has chevron stack units.

Usually cyclones are placed above steam-generating surfaces. They occupy a significant part of the steam space.

Most frequently, use is made of small-diameter cyclones, although the concept of "small" is relative, since the diameter of the cyclones may be as much as 0.3 m and their height up to 1.5 m. Cyclone separators are used for preliminary (primary)

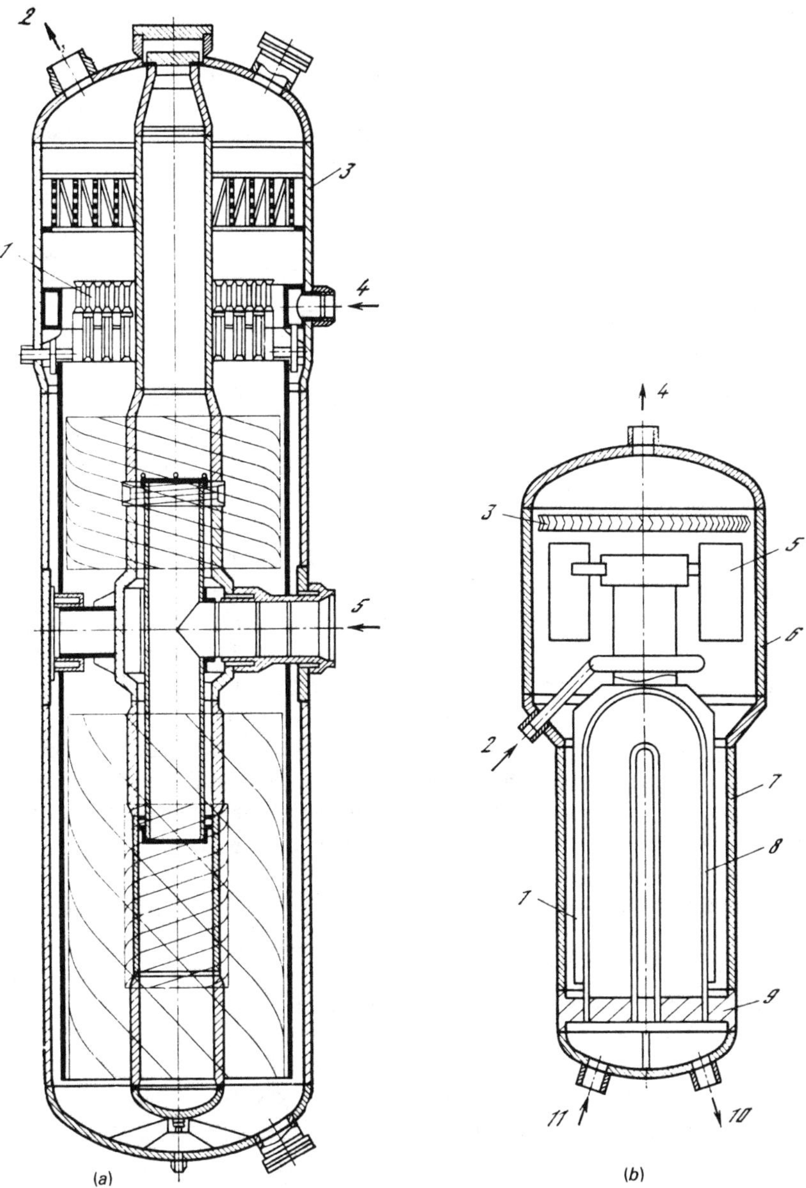

Figure 8.11 Vertical nuclear power plant steam generators with spiral- (*a*) and U-shaped (*b*) tubes. (*a*) 1, First-stage separator; 2, steam discharge; 3, second-stage separator; 4, feedwater inlet; 5, heating-water inlet. (*b*) 1, Inner casing; 2, feedwater inlet; 3, second-stage louvered separator; 4, steam discharge; 5, first-stage separators; 6, top of shell; 7, bottom of shell; 8, U-shaped tubes; 9, tube-sheets; 10, 11, exit and inlet of heating water, respectively.

Table 8.2

Steam generator	Pressure in MPa	Specific load on the shell, $kg/sec \cdot m^2$	Specific duty of one separator, kg/sec	Filling ratio
Obrigheim Nuclear Power Plant	5.5	28	92	0.3
Palisades Nuclear Power Plant	5.3	23	42	0.55
Westinghouse Co.	5.3	36	90	0.4

drying of steam and reduce the percentage moisture to $y < 30\%$. Such a value of y is rather high. To ensure a moisture content at the turbine inlet of not more than 0.25% rigorous specifications are put to the separating capacity of the second stage.

Table 8.2 lists some data on separators of vertical steam generators in nuclear power plants.

In nuclear power plants with channel-type boiling water reactors, phase separation and steam generation occur in external drum separators. The development of drum separators has been significantly affected by designs of drums of steam boilers for conventional power plants. However, there exists a number of nuclear-system specific factors affecting the selection of the engineering solution in fabricating drum separators: (1) significant radioactive background in the premises of the drum separator as well as within the latter; (2) the presence of free oxygen, products of hydrolysis in the reactor coolant; (3) the need to ensure backup water in the case of operation under accident and transient conditions; (4) high unit steam capacity; (5) relatively low reactor-coolant pressure (6–9 MPa); and (6) high content of reactor water in the drum separator.

The presence of a radioactive background inside of the latter makes it necessary to simplify in-shell devices, in order to reduce the inspection and repair time to a minimum. In-shell devices include receiving boxes, submerged and ceiling perforated plates and feedwater manifold with mixers (Fig. 8.12) [8.8].

The steam-water mixture flows into the water space of the drum separator through a system of inlet tubes, situation in two rows to each side of the horizontal drum. The design velocity of the mixture at the inlet is 8–9 m/sec. The mixture from inlet tubes flows to receiving boxes, whose overall area amounts in the plane to about 55% of the evaporation-surface area. The steam-water jets from the lower row of the steam-conducting tubes are directed onto the inclined wall of the box and the jets from the upper row of the steam-conducting tubes passes above the edge of weir, touching it. Water from the submerged perforated stack plate is drained through slots and placed at the wall of the drum separator. Reflecting baffles are placed below the weirs and prevent the passage of steam through the slots (bypassing the submerged plate). Water is drained through 12 tubes in the lower part of the drum. The steam separated in the steam space passes through the ceiling perforated plate and flows into the intake manifold through 14 steam outlet tubes.

The relationship between the separation, hydrodynamic and heat transfer processes is particularly significant in shell-type boiling-water reactors with natural circulation. One has to deal primarily with two problems: obtaining the dryest possible steam and reducing the amount of steam removed by the circulating water into the downcarrier system ("carryunder").

A shell-type boiling-water reactor is usually a single-loop system with circulation of reactor coolant through the core, where it is partially evaporated (Fig. 8.13). The two-phase mixture flows into primary separators, and then the water is returned to the downcarrier piping. After primary separation the wet steam is supplied to the secondary separation stage. Steam that leaves the free evaporation surface may entrain liquid droplets. At high steam velocities the inertia force may exceed the force of gravity, with the result that the separation will be ineffective (entrainment of moisture by the steam) and conversely, steam bubbles may be entrained by water into the downcarrier channel (entrainment of steam).

The length of the stabilized region is significantly affected by the water head and by the vapor quality.

Boiling water reactors with forced circulation are extensively used in the West. They employ a two-stage moisture separation system.

The first stage employs centrifugal separators (cyclone type), which reduce the percentage moisture to $y \leq 10\%$. It is assumed that the entrainment of steam by the circulation water does not exceed 0.3%. The first moisture separation stage is placed above the reactor core, on up to 1.9 m high vertical supports. A mixing chamber is situated in the region from the upper grid of the core to the inlet to the separator.

The second separation stage in these reactors consists of vertical chevron units. A perforated plate is usually placed at the inlet to the chevron separator in order to equalize the velocity field.

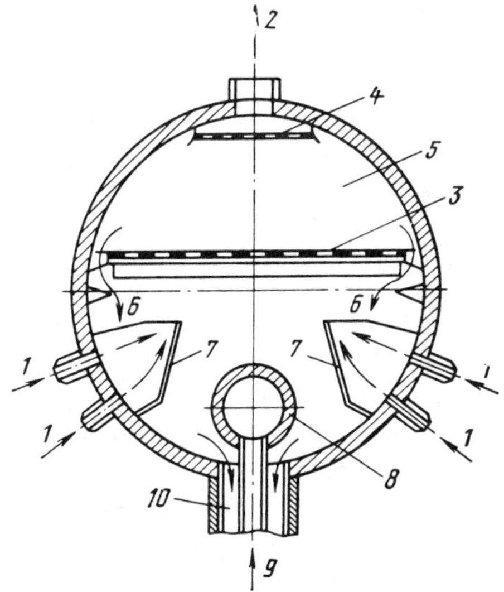

Figure 8.12 Schematic of drum separator in a high-power channel-type reactor. 1, Steam-water flox inlet; 2, steam discharge; 3, submerged perforated plate; 4, ceiling perforated plate; 5, steam space of precipitation moisture separation; 6, draining of water from perforated plate; 7, receiving box; 8, feedwater manifold; 9, feedwater inlet; 10, draining of separated moisture.

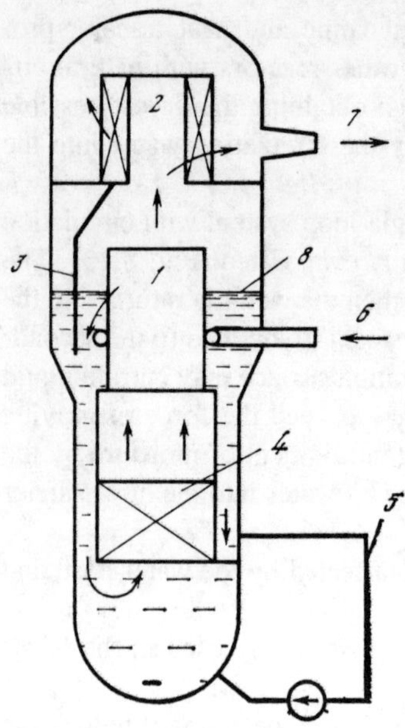

Figure 8.13 Moisture separation arrangement in a shell type boiling water reactor with induced circulation. 1, Primary separation of moisture; 2, second separation stage; 3, drain tube of the second separation stage; 4, core; 5, forced-circulation loop; 6, feedwater inlet; 7, steam discharge from reactor; 8, water evaporation surface.

8.3 SEPARATION OF MOISTURE WITHIN THE TURBINE FLOWPASSAGES

Currently, moisture removal through slots in hollow nozzle blades has become the accepted practice both in the USSR and the West. Figure 8.14 shows as an illustration sections through the low-pressure part of the largest nuclear power plant turbines. Turbines of the Khar'kov Turbine Plant, which are produced for nuclear power plants for 1500 rpm operation, have hollow blades on the last two diaphragms, whereas the KWU Company produces turbines with low-pressure parts having one diaphragm of hollow nozzle blades. According to various data, the effectiveness of this method is not the same. This is because the formation of moisture and liquid films is a complex process which depends on many factors. Studies show that the effect of operating conditions (Re and M, u/c_0, y_0 and λ_0, where the latter is a fraction of large-droplet moisture, and y_0 is the percentage moisture in the steam upstream of the stage under study) on the moisture removal effectiveness is much stronger than the shape and location of slots on the surface of nozzle blades [8.9].

Studies of static conditions (on a two-dimensional cascade) in experimental turbines and full-size equipment made it possible to determine the manner in which the individual parameters affect the moisture removal effectiveness [8.10]. Thus, it was found for example, that the moisture removal effectiveness ψ in nozzle blading decreases with rising u/c_0 (u) of the preceding stage (Fig. 8.15). The variation in ψ as a function of u/c_0 (u) is in full conformance with the change in the size of droplets upstream of the diaphragm under study as a function of u/c_0 of the preceding stage. The size of droplets and the direction of their motion at the inlet to the nozzle-blading flowpassage, control the amount of water which forms the film in the region of the slot on the blade surface. In connection with this, even at constant y_0 and enthalpy dissipation per stage, a change in the rotor rpm will affect the moisture extraction effectiveness.

The moisture extraction effectiveness decreases with reduction in y_0 upstream of the stage under study (see Fig. 8.15). This occurs because, in a multistage turbine, reduction in the moisture content of steam upstream of a stage occurs simultaneously with the shift of the start of moisture formation to this stage. This shift changes the droplet size distribution upstream of the diaphragm under study and leads to a change of moisture extraction effectiveness.

It is seen in comparing the effectiveness of moisture removal calculated on assumption of complete precipitation of moisture for known initial conditions and a specified steam velocity field) with experimental results that these data do not agree. This is because the actual formation of liquid films depends on the breakup of water droplets upon impact on the nozzle blading and entrainment of moisture from the wave surfaces of liquid films. Experiments with flat plates showed that on most of the

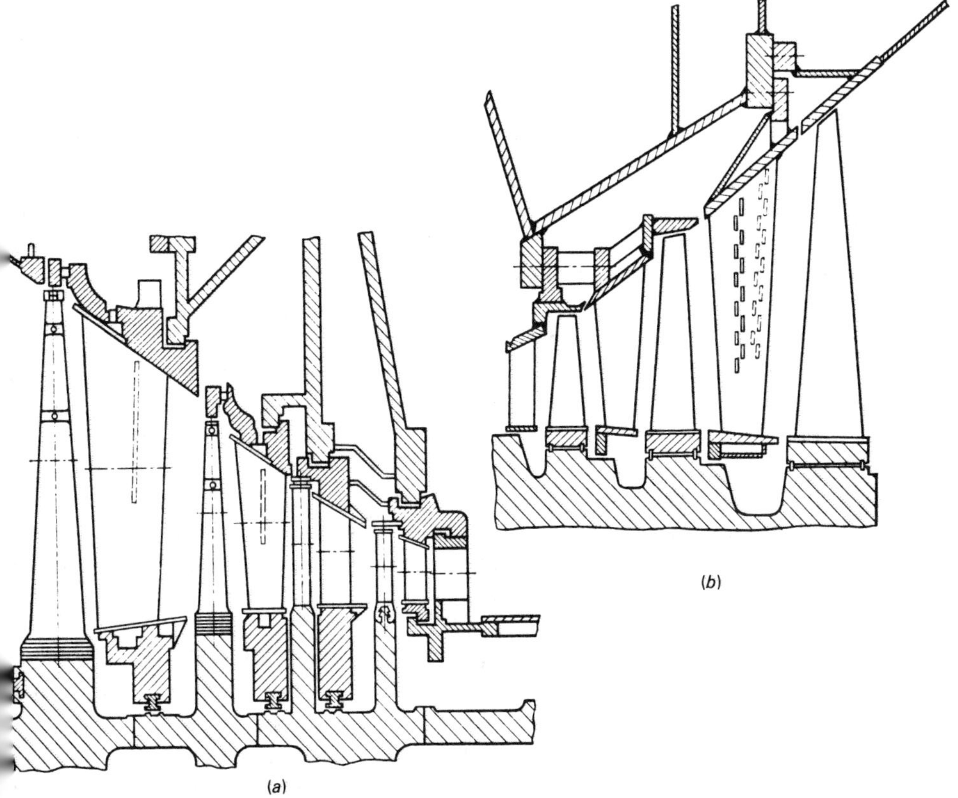

Figure 8.14 Flowpassages of low-pressure parts of nuclear power plant turbines. (a) Low-speed (1500 rpm) turbines produced at the Khar'kov Turbine Plant (height of last [rotor] blade l_r = 1450 mm). (b) KWU turbines (1500 rpm, l_r = 1365 mm).

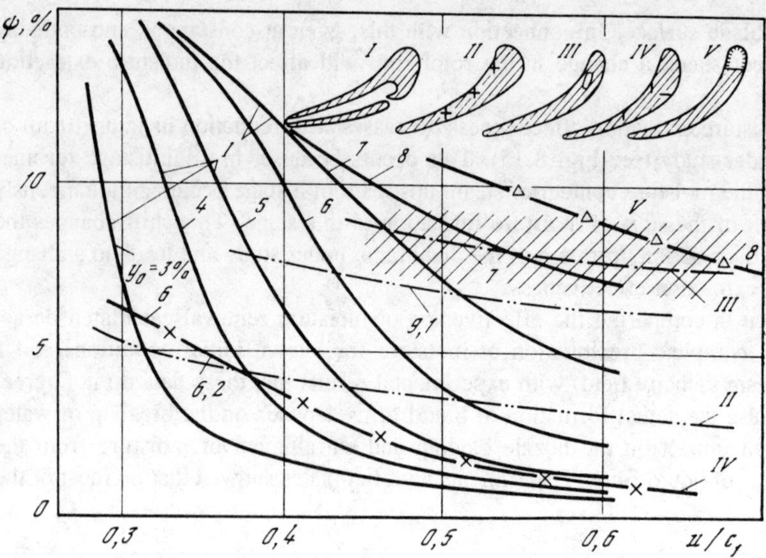

Figure 8.15 Effectiveness of in-channel moisture separation as a function of the turbine operating conditions. I, Last diaphragm of a seven-page turbine; II, second diaphragm of the one-and-a-half stage of an experimental turbine (artificially produced initial moisture); III, last diaphragm of turbine; IV, single turbine stage on moist air; V, second diaphragm of experimental turbine.

surface of nozzle blades liquid films attain the critical flowrate, i.e., the flowrate of water at the extraction point does not depend on the film formation history. In conjunction with this, removal of moisture, for example, in the region of the exit edge, does not correspond to the flowrate of water which precipitated on the blade surface. However, if we consider available experimental data on the effectiveness of moisture extraction in nozzle blading under similar conditions (Re, y_0, ρ_v/ρ_{fl}, u/c_0, M) we can see that the data of different investigators are in rather satisfactory agreement (shaded region in Fig. 8.15). For similar conditions of wet-steam flow in nozzle blading ($y_0 = 8\%$, Re, $u/c_0 = 0.5$) the moisture extraction effectiveness ψ depends little on the location of the slot on the blade surface. This conclusion has been rather well verified experimentally.

In addition to suction of moisture from the surface of nozzle blading, more extensive use should be made of moisture removal from the peripheral surface of the diaphragm, in the region of developed secondary flows and elevated moisture concentration.

It follows from Fig. 8.15 that slots on the surfaces of nozzle blading allow removal of 6–9% of moisture, which flows to the nozzle flowpassage. Naturally increasing the number of slots, using suction from the periphery of the diaphragm and other methods can increase further the effectiveness of moisture extraction.

Of definite interest for improving the overall effectiveness of film removal from the surface of blades is the use of suction of moisture through several slots. The lack of intereaction between slots due to the existence of critical flow in this liquid films on the blade (maintained by shedding of moisture), results in the fact that simultane ous suction of moisture through several slots, situated in different parts of the airfoil is quite effective. Hence, arrangement of moisture removal through several slots with identical static pressure at the point of moisture extraction, in spite of design difficulties, requires further attention.

It is seen from Fig. 8.16a that for a typical distribution of static pressure over the airfoil there exist several versions of design solutions for removing moisture by means of slots 1 through 4 and others, connected by a common chamber. This eliminates flow of moisture and steam from one slot to another. Figure 8.16b shows the location of slots, whereas Fig. 8.16c [sic] shows a photograph of a hollow blade of an experimental turbine.

At present a sufficient number of studies have been performed of in-channel separation under low-pressure conditions ($p < 0.1$ MPa). The use of in-channel moisture separation, while it complicates the fabrication technology of turbine diaphragms, improves the efficiency of the turbine and reduces the erosion of rotor blades. Lack of experimental data on the effectiveness of in-channel moisture separation makes it as yet impossible to fully recommend this method for moisture extraction in the low-pressure sections of nuclear power plant turbines. However, analysis of flow of liquid films under high pressure conditions shows that it can be expected that the use of in-channel moisture separation in the high-pressure parts of turbines operating on steam with y_0 in excess of 5–7% will be effective and may raise the efficiency of these stages by 1–2%. It must be remembered here that arrangements for in-channel separation in the high-pressure sections of turbines is simplified by the relatively low blade height, massive diaphragm edges and simplicity of selection of the necessary pressure difference per slot. Evidently, the steam water mixture removed through the hollow blades will be directed into regenerative taps in accordance

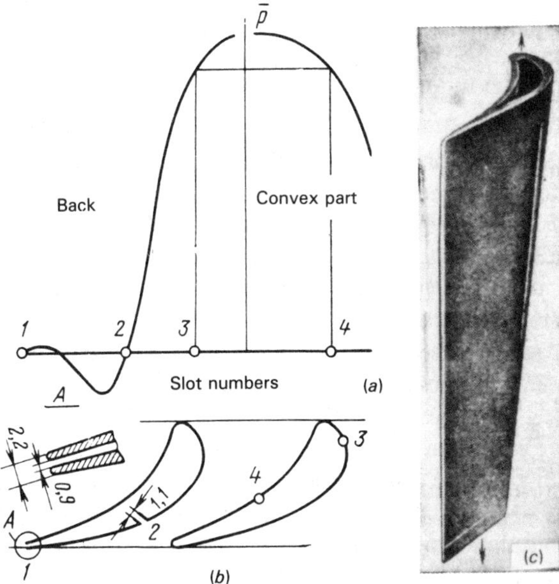

Figure 8.16 Removal of moisture into a hollow nozzle blade simultaneously through several slots. (a) Static pressure distribution over the airfoil outline. (b) Slot arrangement. (c) View of the blade.

with the overall thermal arrangement. In conjunction with this, in-channel moisture separation in the last stages of HP sections of wet-steam turbines is currently recommended for industrial experimentation.

In addition to direct removal of moisture by suction from turbine flowpassages through hollow blades, it is of interest to examine evaporation of liquid films on the nozzle blade surfaces by heating the latter by hotter steam. Studies performed at the Moscow Energetics Institute and the Leningrad Polytechnic Institute with two-dimensional spools. The liquid films were partially evaporated and the droplet size downstream of the exit edges of nozzle blading confirmed that this method holds promise. Lately the KWU company performed studies of the effect of heating the diaphragm of the last stage of the low-pressure section, on the erosion wear of inlet edges of rotor blades. It was noted that this method makes possible a significant reduction in the erosion of blades.

Supply of high-quality steam for heating nozzle blading in the LP section, on the one hand, reduces the cycle efficiency, whereas on the other hand, it improves the efficiency of turbine stages located downstream of the heated diaphragm. Calculations show that the heating of nozzle blading by higher quality steam does not improve the turbine efficiency, but reduces the erosion failure of rotor blades. Hence this method of moisture removal can in certain cases be recommended for the LP section.

In addition to removal of moisture in nozzle blading use is made of methods of moisture separation in the clearance between the nozzle and rotor cascades. This method of moisture extraction is used in turbines which employ separators built into an enlarged axial clearance. In fact, calculation of the motion of droplets toward the periphery as a result of swirling of the wet-steam flow, which exits at angle α_1, without correction for the actual pattern of steam flow in the enlarged axial clearance indicates a high effectiveness of moisture extraction (up to 70–80%). However, under real conditions of wet-steam flow past the nozzle blading the value of α_1 increases, for which reason the gradient of the radial shift of droplets to the periphery decreases with an increase of the axial clearance. For this reason a significant increase in $\bar{z} = z/l$ should not improve the effectiveness of moisture removal past the nozzle blading.

Detailed studies of moisture separation at different distances from the inlet edges of third-stage blades were performed at the Moscow Energetics Institute [8.11]. The studies were performed with a two-shaft experimental turbine, which allows studies with very large radial clearances (up to 1.65 of the height of nozzle blading). The moisture removal effectiveness obtained in these experiments was relatively low (Fig. 8.17); in all the four chambers the moisture removal effectiveness depends little on u/c_0, and the overall value of ψ does not exceed 15%. In the case when moisture in the first stages forms naturally, the effectiveness of moisture extraction (at moisture content past the second stage of $y \leq 3\%$) through all the chambers is even lower ($\psi \leq 5\%$). It is seen that the moisture extraction effectiveness past the nozzle blading does not exceed the values of separation coefficients $\bar{\psi}$ obtained previously [8.9] at moderate relative clearances $\bar{z} = z/l = 0.1$–0.3 and does not confirm the prediction of possible high efficiency of moisture separation in a turbine stage with enlarged $\bar{z}$. This happens because a large axial clearance is responsible for a significant increase

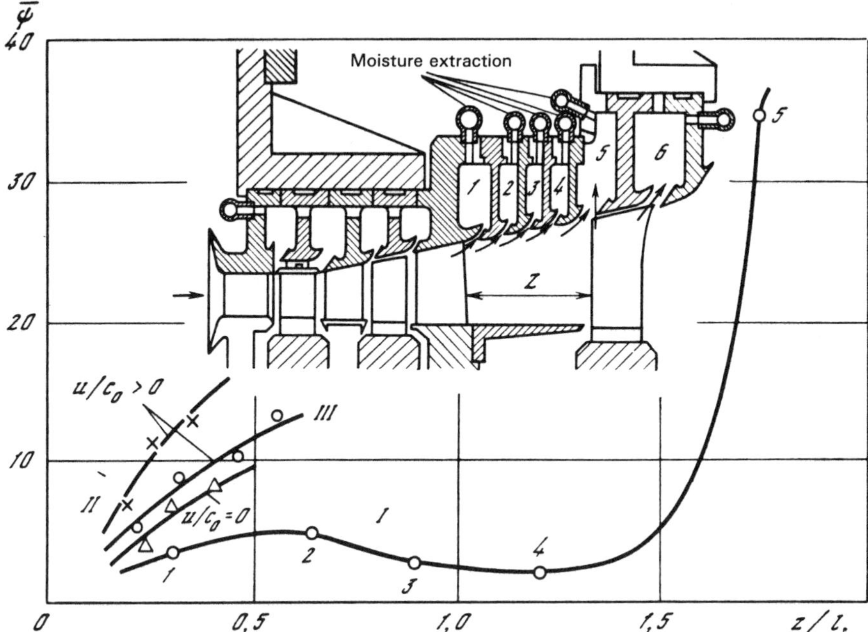

Figure 8.17 Moisture extraction in the clearance between the nozzle and rotor cascades at $\epsilon = 0.85$ and $y_0 = 6\%$. I, Extraction of moisture into five simultaneously operating chambers; II, separation in a single turbine stage of the Moscow Energetics Institute; III, [sic] separation in a single turbine stage with [humid] air as the working fluid.

in α_1 for steam, which reduces the separation of moisture towards the periphery. At the same time, an increase in the distance between the rotor blades and moisture extracting chambers reduces the effectiveness of removal of moisture shed or reflected from inlet edges of rotor blades. However, arrangement of moisture extraction in the region of this inlet edge of the rotor cascade makes it possible to obtain quite high values of separation coefficient ψ (fifth chamber in Fig. 8.17). Increasing $\overline{z}$ reduces the stage efficiency. For example, in experiments with the turbine whose flowpassage is shown in Fig. 8.17 the efficiency of the last stage was reduced by 8% due to increasing axial clearance.

In principle, a turbine rotor cascade can serve as a satisfactory moisture separator. However, extensive studies of moisture separation showed that the effectiveness of moisture removal from the space above ordinary rotor blades is small. This, in the first place, is due to unfavorable conditions for moisture separation, which usually exist in full-size turbine stages: high values of Re and ρ_v/ρ_{fl}, fine-particle moisture, etc. The main processes controlling moisture separation in the rotor cascade is: entry of moisture into the cascade, deposition of droplets on the rotor blades, collision and reflection of droplets, and finally, the motion of moisture over the surface of rotor blades. A large number of studies has been performed of these processes.

Experimental studies at the Moscow Energetics Institute of the flow of liquid films in the field of centrifugal forces show that the thicknesses of films on the surface of rotor blading δ_{fm} ranges between 7 and 12 μm. It was found in these studies

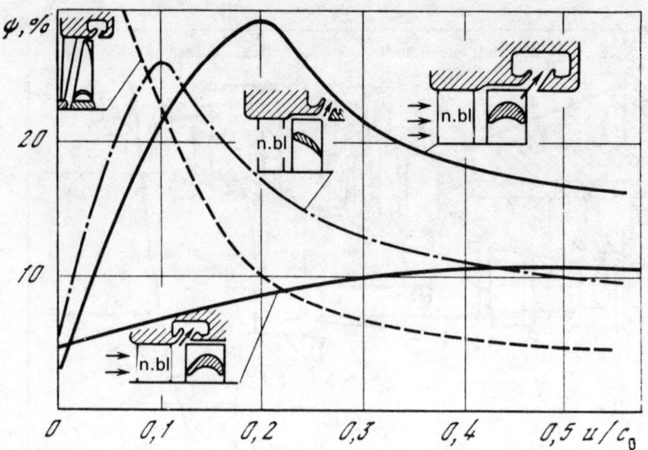

Figure 8.18 Typical curves of $\psi = f(u/c_0)$ for different types of turbine cascades ($y_0 = 6\%$, $p_0 < 0.1$ MPa).

that the liquid film on the surface of rotor blading is broken up by centrifugal forces into individual jets, which in its turn may break up into individual liquid elements. Breakup of liquid films on rotor blades due to surface roughness can occur up to thicknesses $\delta_{fm} \leq 40\ \mu$m. The separating capacity of rotor blades (shedding of moisture to their periphery) is to a large extent controlled by the mode of flow of liquid films in a centrifugal force field and the airfoil shape. The paths of moisture motion over the surface of varying-profile rotor blading can be calculated. However, the forces of friction between the liquid and the blade are insufficiently accurately determined in these calculations. Physical studies of the flow of liquid on the surface of a rotating disk show that, depending on the surface roughness, the film velocity does not exceed 5 m/sec even at high circumferential velocities. This reduces the contribution of Coriolis forces, deflecting the paths of moisture motion from the radial direction.

Analytic and experimental studies show that the separation capacity of various turbine cascades depends significantly on the shape of the blade airfoils.

Figure 8.18 shows characteristic curves of moisture removal effectiveness ψ as a function of velocities u/c_0 for different types of rotor cascades. It is known that the absolute values of separation coefficient ψ can be different depending on the operating variables (Re and M), ratio of phase densities ρ_v/ρ_{fl}, moisture size distribution at the inlet to the turbine stage, etc.

Studies show how long rotor blades with variable profile have a poor ability to move moisture to the periphery. In conjunction with this, turbine cascades with long blades (for example, the last low-pressure stages) have a low moisture extraction effectiveness over virtually the entire operating range.

The separation capacity of a reaction rotor blade with medium fan ratio is higher than that of long rotor blades with variable profile. However, the moisture extraction is most efficient in impulse-type rotor cascades. It is typical of such a cascade that at

$u/c_0 > 0.3$ the effectiveness of moisture removal in the region of the inlet and exit edges is higher than for other turbine stages.

The effectiveness of moisture extraction above rotor cascades can be improved by using more refined moisture extraction chambers, special shrouds and intentionally producing roughness patterns on blade surfaces or special moisture shedding grooves. Studies performed at the Moscow Energetics Institute confirmed the prospects of using special shrouds for increasing the moisture removal in high-pressure turbine stages and of producing directional roughness patterns or moisture shedding grooves (Fig. 8.19). Naturally, the use of special shrouds (with negative overlap at the exit edges) and of moisture-shedding grooves results in some reduction in the efficiency of turbine stages operating with superheated steam. The aerodynamic imperfection of such a cascade decreases with the appearance of moisture, and it is possible to improve the effectiveness of moisture extraction in it without reduction in efficiency.

However, it must be remembered in designing moisture removal systems, which reduce the efficiency in the high-HP sections of nuclear power plant turbines, that, due to the high density of the steam (high values of ρ_v/ρ_{fl}) the possibility of moisture extraction from flowpassages in the high-pressure section is limited [8.12]. In conjunction with this, in certain cases, a reduction in the efficiency of a cluster of stages due to design features which improve moisture extraction, may not compensate for the reduction in the detrimental effect of the removed moisture.

In addition to providing for moisture extraction in intermediate turbine stages, it appears advisable to use moisture extracting systems in the immediate vicinity of the last turbine stages, since removal of moisture at the inlet to the diffusor will improve its efficiency.

It appears quite promising to use turbine stages with improved separation capacity in wet-steam turbines.

Efficient moisture removal arrangements in turbine flowpassages using moisture separating stages significantly improves the efficiency and reliability of the turbine, and in certain cases makes it possible to entirely dispense with the use of the cumbersome and expensive external moisture separators [8.9, 8.13, 8.14].

In the USSR, comprehensive studies have been performed of various separating stages which are already in use in small turbines and ensure high reliability, efficiency and an allowable percentage moisture downstream of the last turbine stage, without external separators. An example of such turbines is that used for driving the feedwater pump at the No. 5 unit of the Novovoronezhskaya Nuclear Power Plant.

These separating stages can be subdivided into three types: a turbine stage with improved separating capacity, but with virtually the same efficiency as that of ordinary turbine stages; a turbine stage with very high moisture extraction efficiency, but with an efficiency lower than that of ordinary stages due to special devices which improve the moisture separation; a special separating stage whose rotor is placed on independent bearings and does not perform useful work, but exhibits maximum moisture extraction efficiency (to 80–95%), since it operates with low enthalpy dissipates and circumferential velocities [8.8, 8.9, 9.11].

Stages of the first type can be placed instead of ordinary stages in locations where

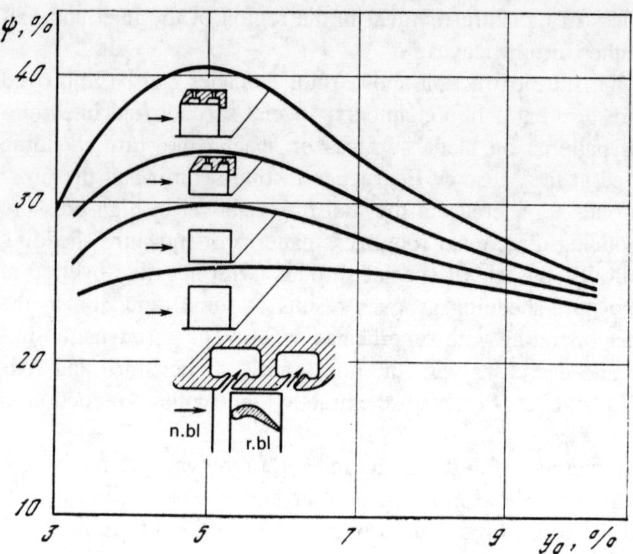

Figure 8.19 Separating capacity of various turbine separating stages at variable initial moisture content of steam ($u/c_0 = 0.5$, $\epsilon = 0.85$).

the percentage moisture in the steam is above 5–6%. Turbine stages of the third type can be placed as an intermediate stage in the turbine flowpassage or at its end (for example, the HP section of nuclear power plant turbines).

The main distinguishing features of separating stages of the second type are special profiling and machining of stationary and rotor blades, small relative pitch of the rotor cascade, larger than usual axial clearance, low enthalpy dissipations (per stage) and a developed system of moisture extracting devices [8.11]. Studies performed at the Moscow Energetics Institute of such a separating stage in a two-shaft experimental turbine [8.9] made it possible to establish an important fact—namely, the stability of the effectiveness of moisture separation by the rotor cascade to changes in u/c_0 (u) over a wide range (Fig. 8.20). It was found in the experiments that the effectiveness of moisture separation in the region of inlet (chamber A) and exit (chamber B) edges as a function of u/c_0 varies differently. With rising u/c_0 (at $u/c_0 >$ 0.3) separation of moisture above the inlet edges of rotor blades starts reducing, whereas that past the rotor increases. thus, the overall moisture extraction effectiveness remains virtually unchanged. Very high overall values of the separation factor were obtained in these experiments. Obviously, with changes in operating conditions (Re, y_0, λ, ρ_v/ρ_{fl}, etc.), and also with changes in the mode of moisture formation the values of ψ may drop. However, these studies show that in all cases, a separation stage has a significantly higher separation capacity than ordinary turbine stages.

In the third type of separating stages the rotor is placed on separate bearings and the circumferential velocities in the rotor blading do not exceed 20–30 m/sec. This ensures the most favorable manner of moisture precipitation on the rotor blade surfaces and prevents breakup, reflection and removal of a part of the moisture by the

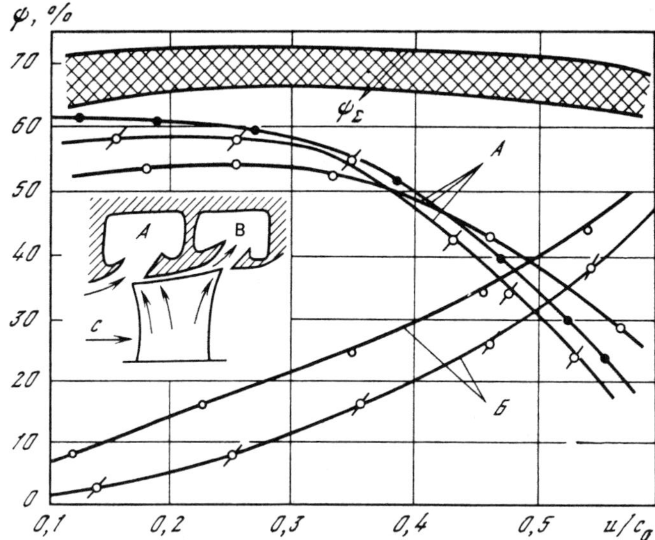

Figure 8.20 Moisture extraction effectiveness by a separating stage as a function of u/c_0.

steam flow after collision with the rotating blades [8.9]. At the same time the rotor is designed for maximum precipitation of moisture on the blades (for example, using narrow flowpassages with sign-alternating curvature; see Fig. 8.21).

Low circumferential velocities and small enthalpy dissipation (Re and M) make it possible to obtain high separation capacity in these cascades. Thus, for example, in experiments on the separating stage used at the Moscow Energetics Institute with an enthalpy dissipation of less than 8.4 kJ/kg and at circumferential velocities up to 100 m/sec, the moisture separation effectiveness was as high as about 90% over a wide range of operating conditions (re, u/c_0, y_0). The same stage, when operating with a

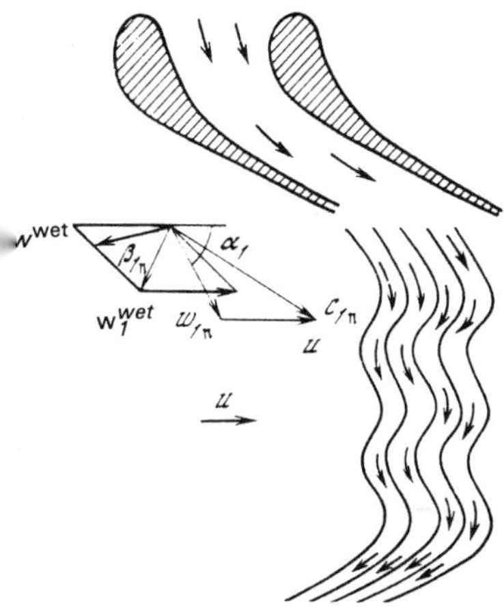

Figure 8.21 Schematic of a separating stage with sign-alternating curvature flowpassage.

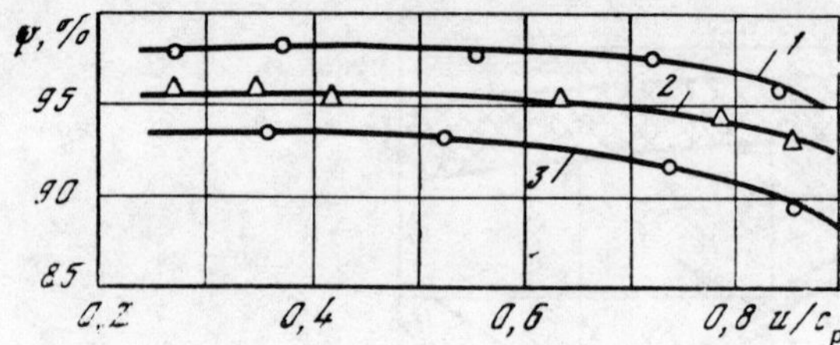

Figure 8.22 Separation capacity of a free running turbine cascade. 1, Rotor velocity 50 m/sec; 2, 75; 3, 100.

freely rotating rotor at circumferential velocities less than 50 m/sec provided a moisture removal effectiveness to 95% (Fig. 8.22).

The use of separating stages can radically change the moisture removal arrangements in nuclear power plant turbines, since these separating stages are simple, compact, light and cheap. In connection with this it is of interest to consider a wet-steam turbine without a reheater, but with two moisture separating stages.

Separating stages can also be placed upstream of the turbine (pre-turbine moisture separation) and at the end of the HP section flowpath, but the most likely location of these separating stages is in crossovers between the turbine sections. Naturally, the use of special separating stages should be combined with other, conventional moisture removal methods: removal of the liquid film through slots in steam lines and stationary blading, the use of extensive peripheral moisture extraction system, etc.

8.4 EXTERNAL SEPARATORS AND MOISTURE SEPARATOR REHEATERS

Currently, virtually all the nuclear power plant turbines operating on wet steam, whose expansion starts from the saturation curve, are equipped with external separators or moisture separator reheaters (MSR). These devices are used because of the low quality of steam supplied to turbines at nuclear power plants and accordingly because of the high moisture content of the steam that expands in the turbine.

Figure 8.23 shows several steam expansion cycles in turbines on an i-2 diagram. If nothing is done to remove moisture within the turbine, then the steam leaving the last low-pressure stage may contain up to 24% percent of water (Fig. 8.23, curve 1). It is for this reason that external MSRs are used at nuclear power plants both in the USSR and the West.

Separation of moisture in turbine flowpassages, as well as in external separators, not only improves the turbine reliability (reduces the erosion of blades, housings and diaphragms), but also improves their efficiency. The use of external moisture separation may reduce the percentage moisture in the steam leaving the separator to 0.2–1¢ and increases the turbine efficiency by as much as 1–4%. The effectiveness of using internal moisture separation or MSR depends significantly on the crossover pressure p_c (the pressure between the high- and low-pressure sections of the turbine), since p_c affects the efficiency of the flowpassages and the entire turbine. Increasing p_c increases the moisture losses in the low-pressure section and increases the efficiency of the high-pressure stages. The effect of p_c on the efficiency of a turbine with the medium-pressure section will be different. In this case the number of the low pressure stages does not depend on the selection of p_c.

Figure 8.24 presents analytic curves on the effect of p_c/p_0 on the efficiency of a 1000 MW turbine rate at 1500 rpm at $y_0 = 1\%$ and $p_0 = 6.4$ MPa [8.13].

Two separation stages can be used in the case of large turbines with $p_0 = 6$ MPa. In this case the turbine has an intermediate pressure [IP] section. Of greatest interest here is the turbine [8.9] which will employ special separator cascades (see Sec. 8.3). According to calculations, a turbine with two compact and high efficiency separation stages is more promising than an arrangement with a single separation stage and reheat.

Figure 8.25 shows possible arrangements of external separators in conjunction with nuclear power plant turbines.

In addition to these arrangements, provisions for external separation can be made even in the high-pressure section. Thus, for example, the Parsons company uses two external separation stages in its 800 MW turbine (Fig. 8.26). The double-pass HP section has two external moisture separators. Wet steam that passed several stages of the first pass is removed from the turbine into a separator from where it is directed into the second pass. The second moisture separation stage operates in the same

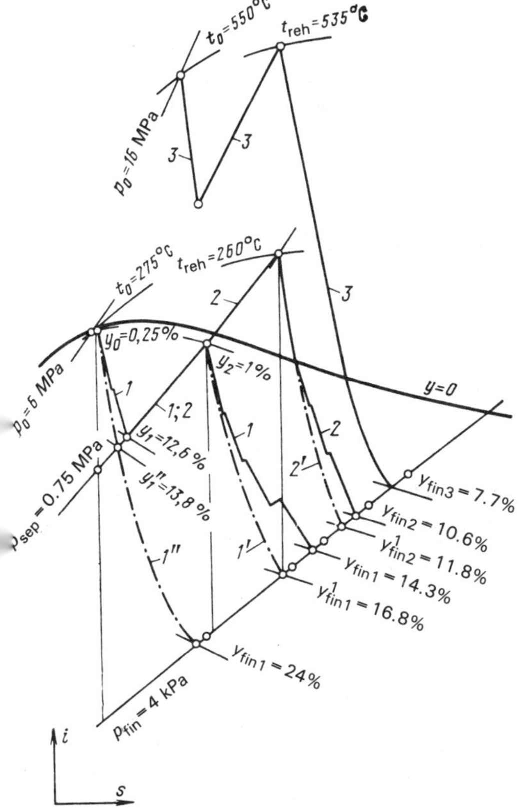

Figure 8.23 An i-s diagram of expansion of steam in a turbine. 1 and 1′, in a saturated-steam turbine, $p_0 = 6$ MPa, $p_{fin} = 4$ kPa, with external separator; 1″, same as above, without separator; 2 and 2′, same as above with external separator and reheat to $t_{reh} = 260\,°C$; 3, in higher-pressure, higher-temperature turbine; $p_0 = 16$ MPa, $t_0 = 550\,°C$, with reheating to $t_{reh} = 535\,°C$; the dash-dotted curves correspond to expansion using standard systems for moisture removal in the flow-passages, the solid curves correspond to expansion with a highly efficient moisture extraction system, including in-channel separation.

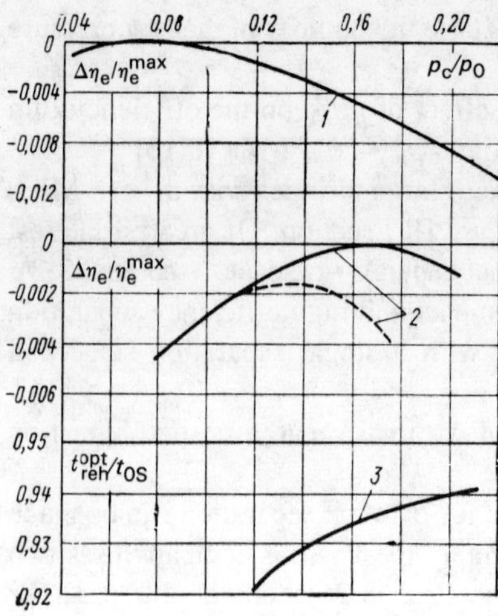

Figure 8.24 Effect of crossover pressure p_c on the performance of a turbine [8.13]. 1, Efficiency of a turbine with an external separator and reheater with a medium-pressure section ($t_{reh} = 250\,°C$); 2, same as above, without a medium-pressure section; 3, optimum reheat temperature t_{reh}^{opt} (calculated for a saturated-steam turbine) with $p_0 = 6.4$ MPa, $p_e = 1000$ MW, $n = 1500$ rpm.

manner, but in opposite direction. Unfortunately, experimental and operating information on this moisture separation system is not available.

At present there is no concensus concerning the selection of external separator arrangements. As an illustration Fig. 8.27 shows some designs of nuclear power plant turbine separators. It is seen that these separators employ various moisture separation methods (centrifugal forces in swirled wet-steam flow, precipitation of moisture on various surfaces of plates and chevron elements or meshes). For this reason external separators differ significantly by their design, dimensions and location in the turbine hall.

However, at present, nuclear power plants with water-cooled reactors use primary systems with moisture separator reheaters (MSR), since the availability of su-

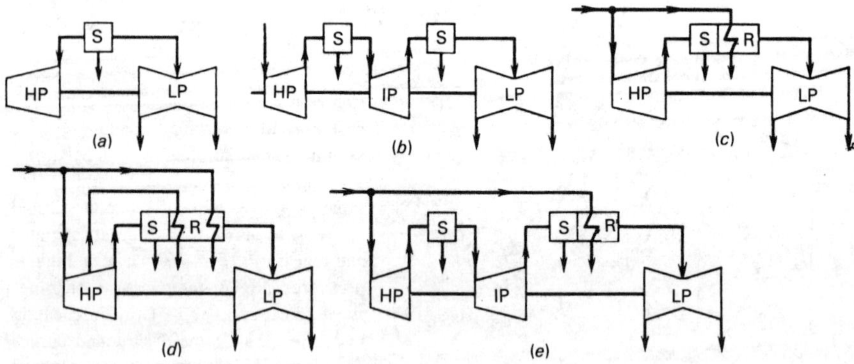

Figure 8.25 Possible arrangements of external separation and reheating of steam in turbines operating on saturated or slightly superheated steam. (*a*) Single separation stage. (*b*) Two separation stages. (*c*) Single separation stage and one-stage reheating by fresh steam. (*d*) Single separation stage and two-stage reheating by bled and fresh steam. (*e*) Two separation stages and single-stage reheating by fresh steam. *S*, designates a separator; *R*, designates a reheater.

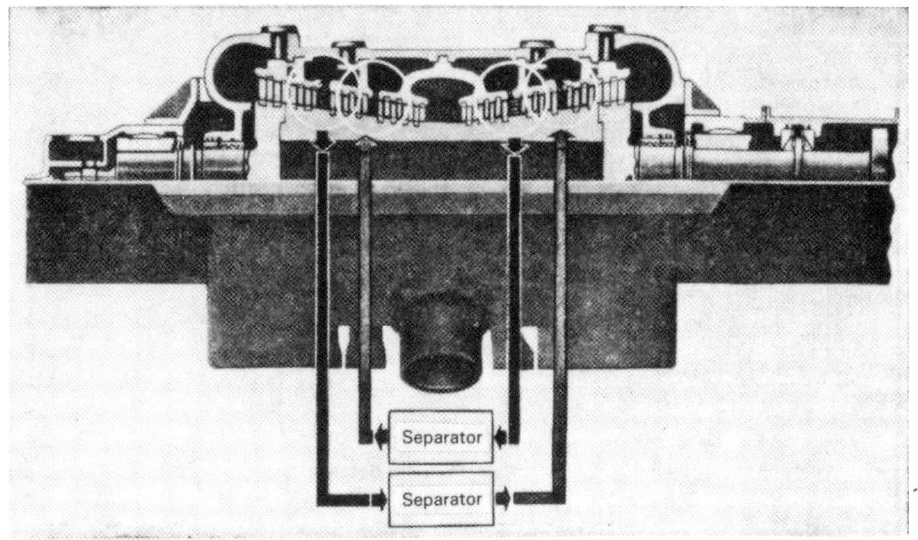

Figure 8.26 External separation arrangement used in the high-pressure section of a Parsons 800 MW nuclear power plant turbine.

perheated steam entering the LP section is easier to control and improves the reliability of the first low-pressure stages and their efficiency. However, reheating involves the danger of electrochemical corrosion, primarily due to the presence in the steam of NaOH, which enters it in the reheater in the form of droplets of highly-concentrated solution (see Sec. 7.5).

At present an MSR is very expensive (its cost may amount to one-third of that of the turbine), and at the same time the least reliable nuclear power plant component. In spite of the extensive use of MSR in nuclear power plants there is no agreement yet concerning the shape of the separating elements or the location of the MSR. Thus, for example, the KWU company employs both vertically and horizontally placed MSRs [8.9, 8.15].

The vertical MSR produced in the USSR, starting with MSR-220-1 have standard designs of the separator and reheater parts. Figure 8.28 shows a production run MSR for turbines of the Khar'kov Turbine Plant [8.13].

Steam leaving the HP section flows to the upper separator chamber and then into the separator. The latter consists of sixteen identical separating units, arranged radially between the MSR casing and the superheated steam discharge pipe. This design of the separator part is better than other arrangements used in the USSR and the West, since it ensures an effective and compact moisture separation system and can ensure minimum losses of steam energy [8.9].

The moisture is removed in the separator unit (Fig. 8.29) which consists of five stacks of louvered elements 1. Each such stack actually represents two separation stages. The first is changing the direction of steam flow ahead of louvered elements 2 by means of guide plates 3. Here the largest droplets are removed (the arrow shows

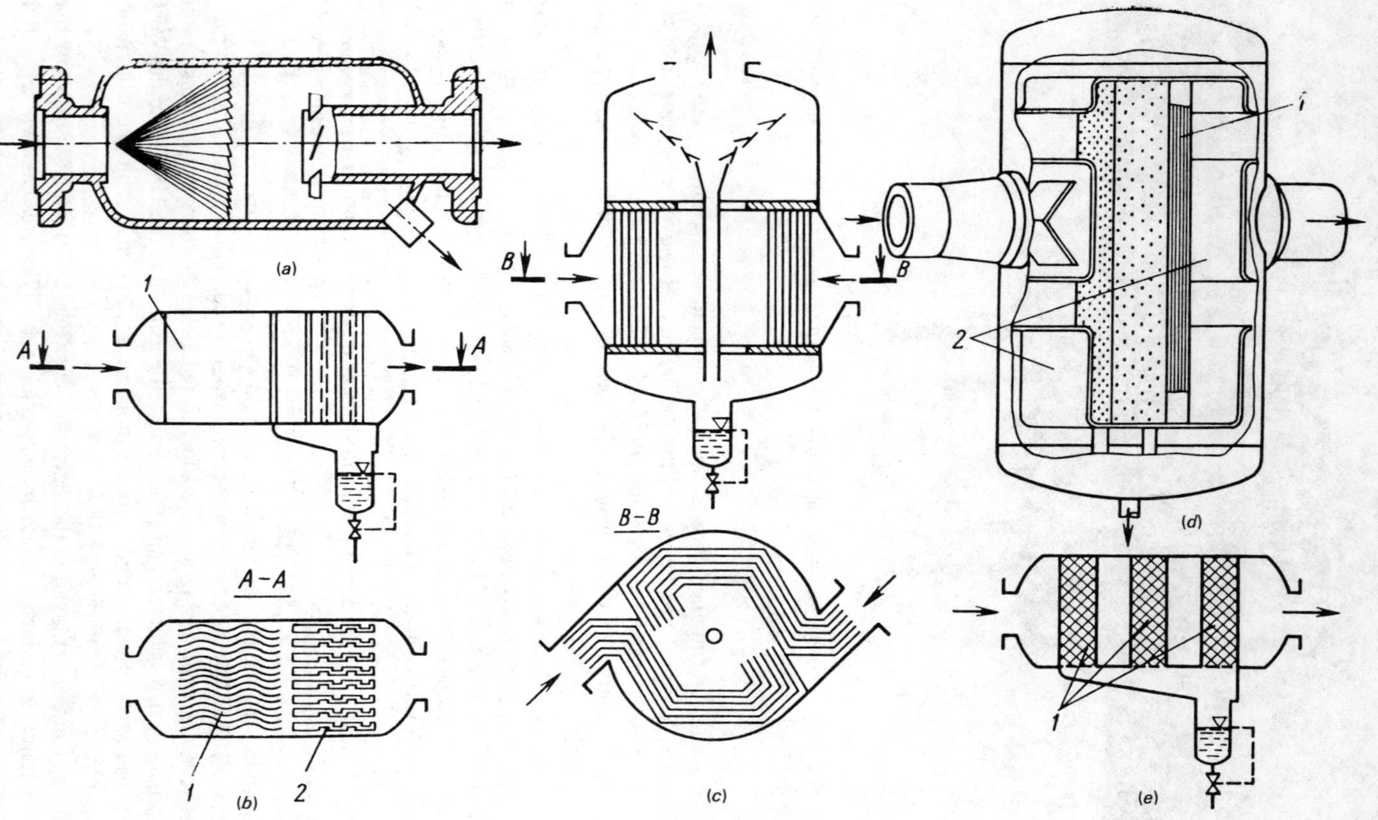

Figure 8.27 Types of external separators. (*a*) Vortex type. (*b*) Plate-type (1, plates; 2, draining grooves). (*c*) Herringbone (louvered) made by the SS company. (*d*) Louvered, Pearless type, made by the DE company (1, chevrons; 2, shields). (*e*) Mesh-type separators (1, meshes).

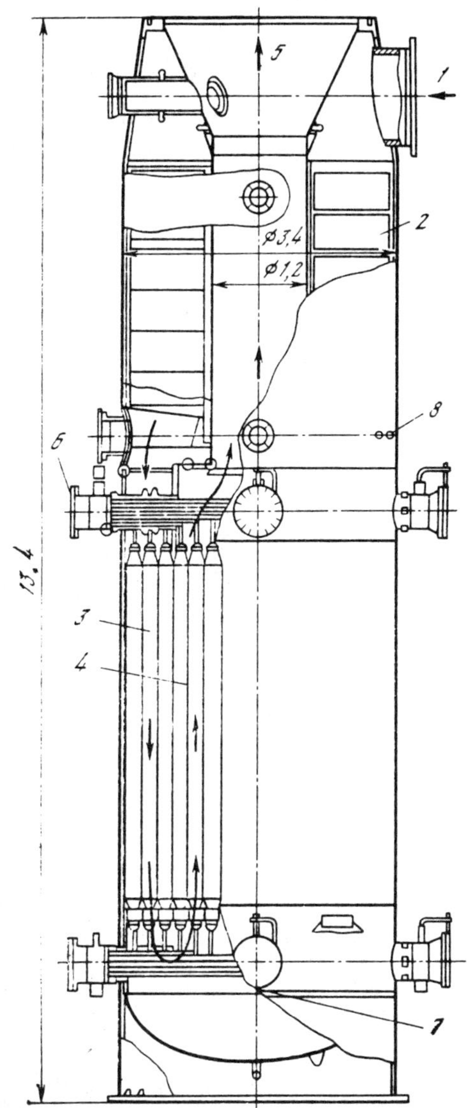

Figure 8.28 A separator reheater [8.12]. 1, Inlet of steam from the HP section; 2, louvered separation stacks; 3, first reheater stage; 4, second reheater stage; 5, discharge of superheated steam into the LP section; 6, heating-steam inlet; 7, heating-steam discharge; 8, moisture drain.

the direction of flow of the wet steam). Then, moisture is deposited on the surfaces of channels with sign-alternating curvature. Moisture flows as films by gravity and then into separation tank 8 (see Fig. 8.28). Each of the five stacks is an independent separation nit.

After leaving chevron stack 2 (Fig. 8.29) the dried steam passes a perforated plate 4, which ensures uniform distribution of steam over all the chevron channels. The velocity of steam in such a channel under nominal conditions of turbine operation does not exceed 3 m/sec.

The steam from the separation stacks enters profiled exhaust 5, whose variable

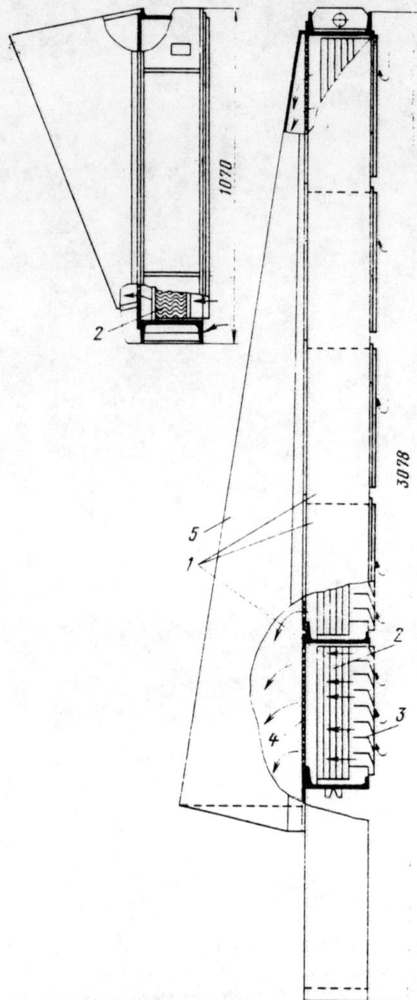

Figure 8.29 Separator unit of the SPP-220-1 [MSR] [8.12]. 1, Louvered separation stacks; 2, louvered elements; 3, guide plates; 4, perforated plate; 5, dry steam discharge.

cross sectional area ensures constant pressure of steam over the height past the perforated plate. This separator design can be used for a wide range of unit sizes. The unit capacity can be easily changed by using different numbers of stacks, which makes such a design rather universal. The steam leaving the separator (2 in Fig. 8.28) is directed to the first reheater stage 3, where it is heated by bled steam ($p = 0.19$ MPa, $t_0 \simeq 209\,°C$, $y_0 = 6.4\%$) to $190\,°C$ and then flows to the second reheater stage 4. Fresh heating steam ($p = 0.43$ MPa, $t_0 = 253.5\,°C$, $y_0 = 0.5\%$) passes through the pipes from the top down and then heats the wet steam under counterflow conditions. The reheated steam leaves the reheated at $241\,°C$ and flows to the LP section.

Lately, a modernized SPP-220M MSR using finned tubes is being produced by the Podol'sk Machinery Plant for K-200-44 turbines. The redesign involved improving the system for supplying the heating steam to the reheater and improving the manufacturing process and the inspection of the unit's components. The manifold employed for draining the condensate from the separation unit was radically redesigned. All these changes resulted in a highly reliable and efficient SPP-220M.

Two identical units are placed beneath the flow of the nuclear power plant turbine hall at zero elevation. The MSRs are interconnected by pipes and operate in parallel. The units employ external separated-moisture and condensate tanks, placed below the draining level from the MSR. The separated moisture and condensate are drained from the unit by gravity.

Data on the effectiveness of separators and reheaters are at present very scarce, since these units have not yet been tested under plant conditions. At the same time it is important to have information on the actual performance of MSRs under conditions of variable turbine load, and also on the effectiveness of parallel operation of several such units with a single turbine.

Two SPP-220-1 units were placed in operation with a K-220-44 turbine at the Kola Nuclear Power Plant, which made it possible to analyze the performance of the unit over a load range on the turbine $\bar{N} = N/N_{\mathrm{nom}}$ ranging between 0.25 and 1.1 [8.14].

The manner in which these MSRs were connected into the thermal system of the K-220-44 turbine is shown in Fig. 8.30. The observations were performed simultaneously on both units, in order to determine the features of their parallel operation. To be able to analyze the operation and to further refine the MSR one must know the actual percentage moisture of the steam entering the separator, the moisture extraction effectiveness (percentage moisture in the steam leaving the separator), the effectiveness of reheat in the first and second stage, and the feasibility of raising the load (velocity of the wet steam) on the separation units. However, due to absence of reliable methods for measuring the vapor quality there are at present virtually no data on the effectiveness of the separators within the MSR.

The investigations performed at the Kola Nuclear Power Plant employed a simple and (at the same time) rather reliable method (at $c_v^{\mathrm{wet}} < 100$ m/sec) for measuring the steam quality, namely, throttling of the wet steam from the region with $x < 1$ to the superheated steam region at constant steam enthalpy by means of a steam moisture content meter. This instrument was used for measuring the steam quality over the height of the pipe at different turbine loads (Fig. 8.31).

The fact that the upper limit of percentage moisture measurement is limited, made it possible to obtain curves of $y = f(\bar{d})$ only at $y < 8\%$ ($p_{\mathrm{fin}} = 0.0035$ MPa). Here one sees that a significant difference exists between the theoretical and actual values of moisture content in the steam in the inlet pipe of the MSR. This is, firstly, due to the nonuniform distribution of moisture over the height of the pipe, and secondly the fact that a part of the moisture is removed in the HP section by means of peripheral moisture separation and five intermediate steam bleeds.

The bulk of the wet steam flowing to the MSR has a quality between 0.92 and 0.93. It is likely that a significant part of the water is contained in the film which flows over the steam pipe wall. Measurements of the moisture content of steam directly at the inlet to the separation stacks by means of an IVP-2 (percentage moisture meter) also confirm the fact that virtually over the entire range of load on the

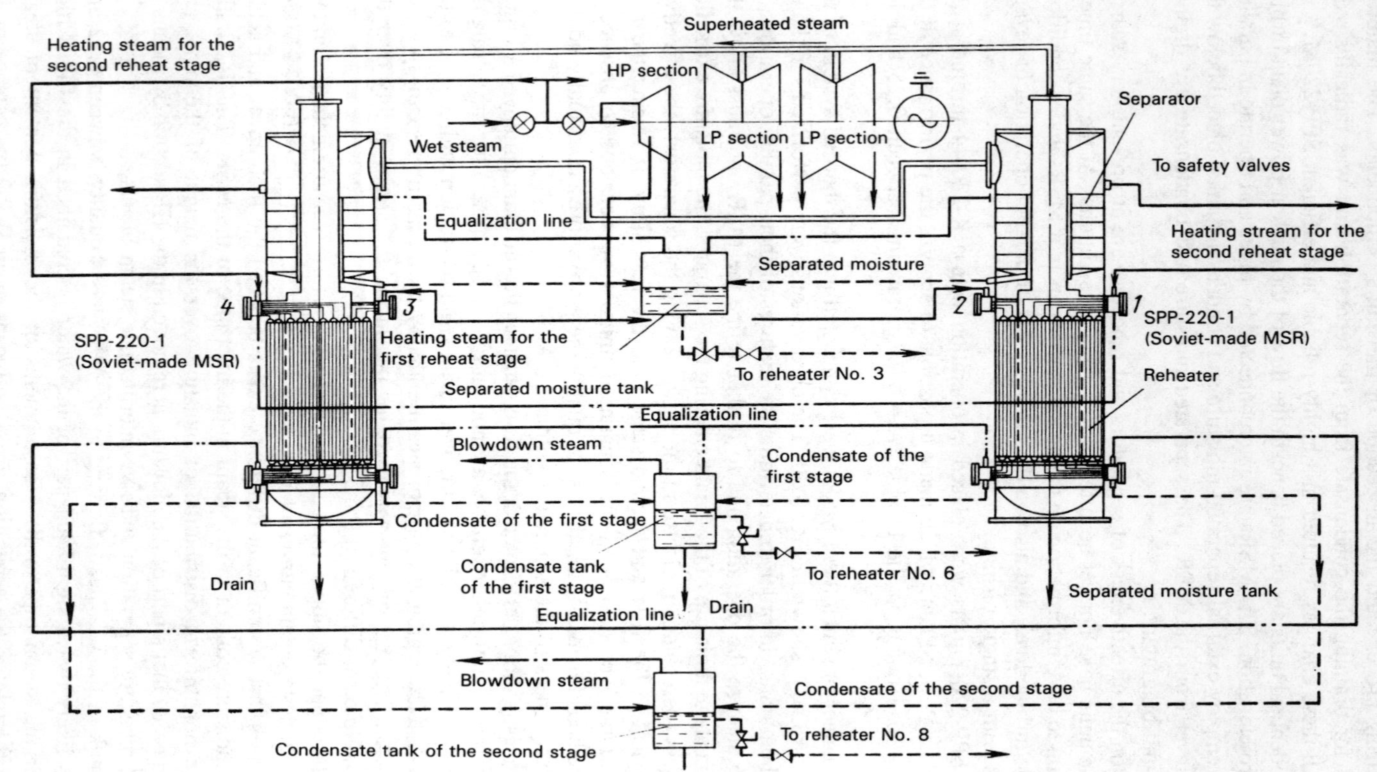

Figure 8.30 Operation of two moisture separator reheaters as a part of the thermal system of a turbine. 1–4, Designate the numbers of heating-steam receiving chambers.

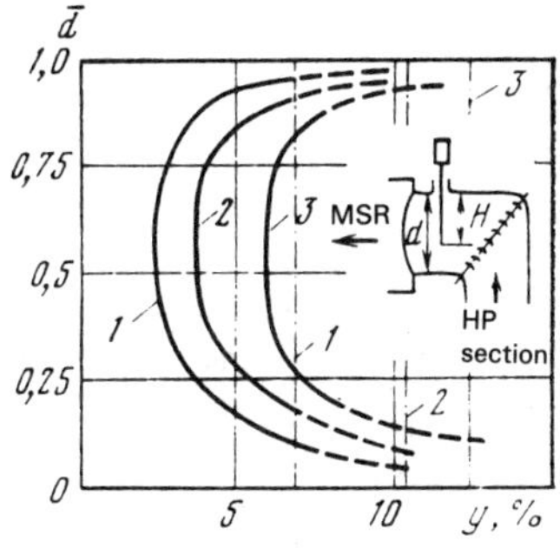

Figure 8.31 Distribution of percentage moisture y in the steam over the height of the inlet pipe to the SPP-200-1 [MSR]. 1, $\bar{N} + 0.35$; 2, 0.5; 3, 1.

turbine ($N < 150$ MW) the steam quality does not exceed 0.925. However, it was noted that the percentage moisture distribution over the stack is highly nonuniform and this is particularly so over the circumference at the separator inlet. Thus, for example, the moisture content in the region close to the steam inlet exceeded significantly the maximum range of measurements of the percentage moisture meter, i.e., $y > 0.93$.

The nonuniformity in the distribution of moisture content of the steam at the inlet to the separator stacks reduces the effectiveness of moisture extraction in individual parts of the separator, although the fact that the value of y at the MSR inlet is much lower than the theoretical should improve the unit's effectiveness.

Figure 8.32 shows the results of measurement of y in downstream of several stacks at variable turbine load. Over the entire range of $\bar{N}$ the experimental values of y virtually did not exceed 0.5%. This figure presents data on the mean values of moisture content of the steam in the packet, although the measurements point to some difference between the radial distribution of y from 0.30 to 1.15. Elevated values of y were noted at the inner and outer separator walls. Measurements of y over the circumference of the unit (in different separator stacks) showed that all the stacks perform quite effectively and provide a moisture removal level to $y = 0.4$–0.6%. If it is remembered that a significant nonuniformity in the velocity and moisture content of

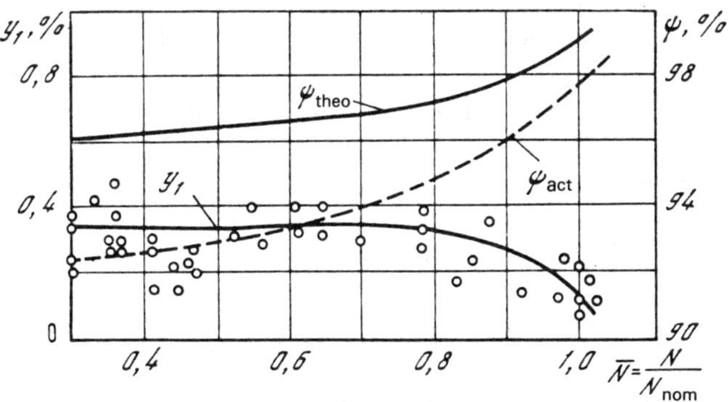

Figure 8.32 Mean value of percentage moisture y_1 in the steam leaving the separator and the dependence of the theoretical ψ_{theo} and actual ψ_{act} separation coefficients on the load on the turbine.

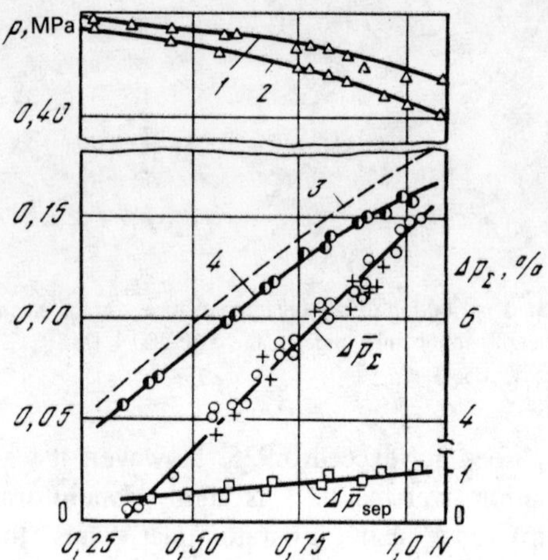

Figure 8.33 Relative drop in steam pressure $\overline{\Delta p}_\Sigma$ past the MSR and past the separator Δp_{sep} as a function of the load on the turbine and variation in the pressure of coolant streams in the MSR. 1, Fresh steam past the heating-steam intake; 2, heating steam at the inlet to the second reheat stage; 3, steam in the intake chamber past the second HP section; 4, heating steam at the inlet to the first reheat stage; the open circles denote data obtained at the Novovoronezhskaya Nuclear Power Plant, the remaining data were obtained at the Kola Nuclear Power Plant.

the steam exists over the circumference in the inlet to the MSR separator, then it is obvious that the actual velocities of steam flow in the louvered separator elements are significantly lower than the critical.

At nominal loads ($\overline{N}$ = 0.8–1.05) there is a tendency toward reduction in y past the separator, i.e., the MSR-220M provided a capability for improving the separation elements. Evidently, the designers used an excessive safety factor with respect to the velocity of the steam passing through the chevron elements. It is also possible that attaining the critical velocity in individual regions of limited area does not significantly increase the value of y downstream of the separator.

An important parameter of MSR efficiency is the energy loss of steam passing through the unit, i.e., the reduction in the pressure of steam entering the LP section. Measurements of the drop in the pressure of steam after passing and separator and reheater show that the resistance produced by the separator of the MSR-220M does not exceed 1%, and the measured losses of the entire unit (without allowance for the pressure drop in the inlet and discharge piping) amount to 8% of the pressure of the steam entering the MSR (Fig. 8.33). Naturally, these losses are a function of the steam velocity, for which reason the values of $\Delta \overline{p}$ drop with decreasing $\overline{N}$ and, for example, at $\overline{N}$ = 0.5 Δp ≅ 4%.

The effectiveness of using an MSR depends to a large extent on the reheater performance, since the latter is a part of the unit. Studies of the thermohydraulic performance of the MSR reheater were performed on one of the units operating in parallel, which was equipped by a full set of monitoring instrumentation. On the other hand, the measurements at the second unit were performed with a limited number of instruments.

Figure 8.34 shows data on the working fluids for the NSR-220 over a wide range of turbine loads.

The heating-steam pressures entering the MSR were decreased to about 10% of the fresh-steam pressure. It is seen from Fig. 8.33 that reducing the heating steam pressure entering the MSR reduces the heated-steam temperature below the design value. Obviously increasing the diameters of the heating-steam piping so as to attain the design values of pressure drop (2–2.5%) will raise the heated-steam temperature by 4–6 °C, which is equivalent to reducing the specific heat loss for the VVER-440 unit by 5 kcal/kWhr.

The temperature of heated steam leaving the MSR under rated conditions is 237 °C, whereas the difference between the heating-steam and heated-steam temperatures leaving the MSR does not exceed 13–15 °C, which points to the high effectiveness of the selected design of the heating surface. For comparison, in the MSR used in the West this difference amounts to 20 °C and above.

The heated-steam temperatures in the lower chamber of the MSR were observed to be highly nonuniformly distributed over the radius, from the wall to the axis. The heated-steam temperature was significantly lower at the wall of the casing and in the center of the unit. This is apparently due to shunt leakage of "cold" steam in the clearance between the stacks and the casing. Steam flowing through these clearances reaches the central part of the unit by moving over its bottom and apparently reaches the central part of the second reheater stage.

The operation of two MSRs with one nuclear power plant turbine requires that they perform in parallel under variable turbine load. The problem is even more complicated in the case of large turbines, which require four MSR for their operation.

Previous tests of MSRs of this type at the No. 4 unit of the Novovoronezhskaya Nuclear Power Plant showed that the heating-steam condensate drain line is subjected to significant pressure fluctuations, induced by nonuniformity in the flowrates of the

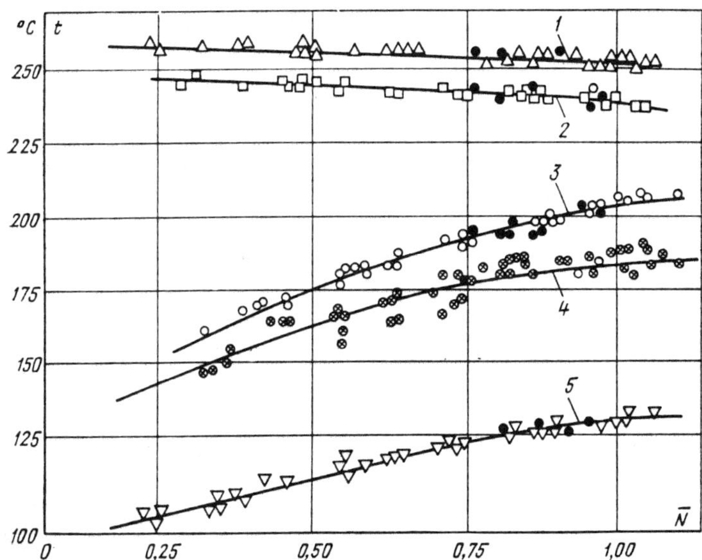

Figure 8.34 Temperatures of the heated and heating steam vs. the turbine load. 1, Fresh steam; 2, heated steam leaving the second stage; 3, heating steam of the first stage; 4, heated steam leaving the first stage; 5, heated steam entering the MSR; the open circles present data obtained at the Novovoronezhskaya, and the remaining data—at the Kola Nuclear Power Plant.

heating steam both by individual chambers of the unit as well as by individual stacks of cassettes.

The parallel operation of MSR-220M units at the Kola Nuclear Power Plant was checked when these were properly interconnected, i.e., employing equalization and breathing lines. The heating-steam condensate was drained in condensate tanks attached to the individual pipelines. The experiments were performed with the two units operating with a parallel system of heating-steam, heating-steam condensate drain and heated steam distributions. The measurements were performed over a turbine load range from 40 to 100%.

The heating-steam flowrate to the first stage of the reheater for the two units differs significantly at low turbine loads and becomes virtually equal at rated loads. At the same time, the nonuniformity of flowrates at the second reheater stage for both units prevails over the entire load range and amounts to 10–12%.

The main reason for the nonuniform distribution of steam over the chambers is the significantly asymmetric arrangement of the supply piping and of the condensate drain piping. The minimum heating-steam flowrate for the second stage devolves upon the cluster of cassettes closest to the steam inlet, whereas the maximum occurs at the last cluster of cassettes. Nonuniformity of flowrates is also confirmed by the measured pressure drops over the heating-steam supply and condensate drain chambers.

As previously noted, Western companies produce a large variety of MSRs, which are described in [8.9, 8.14, 8.15]. The most characteristic stages in the developed of horizontal-type MSRs can be followed for the case of the Westinghouse Company [8.15].

The Westinghouse-produced moisture separator reheaters went through three design generations. The first MSR was put on line by the company for operation with a 450 MW turbine at the Edison Nuclear Power Plant (in St. Enofree) and had the following dimensions: diameter 2.745 m, length 12.2 m, weight 40 tons (Fig. 8.5a). The steam leaving the HP section flows to the lower part of the cylindrical housing of the MSR, then is distributed uniformly between the separator units and is directed vertically upward into the space between tubes of the reheater, which are arranged crosswise to the flow, and then to the LP section of the turbine. Demisters (meshes) were used in first-generation MSR as separating elements.

The running-in of this type of MSR was accompanied by extensive testing of such units, both full scale and with a large-scale (1:2) model with air-water flow as the model fluid. Particular attention was paid here to the design of the inlet steam manifold, since the wire in the demistor (mesh) stacks underwent damage as early as after three months of operation of this MSR at the Edison plant. Analysis of the damage showed that it was due to impact of the steam flow. During the test the steam velocity in the inlet pipe was 45 m/sec and the energy losses amounted to 5 MPa (0.0525 kg/cm^2). It was noted in the experiments that, due to nonuniform supply of steam to the stacks of meshes the moisture extraction efficiency in certain zones past the separator was very low and the moisture content of the steam was as high as 10%. Following these tests the Westinghouse company designed a new steam distribution system in the MSR, which ensures uniform flowrate of steam through the meshes. It

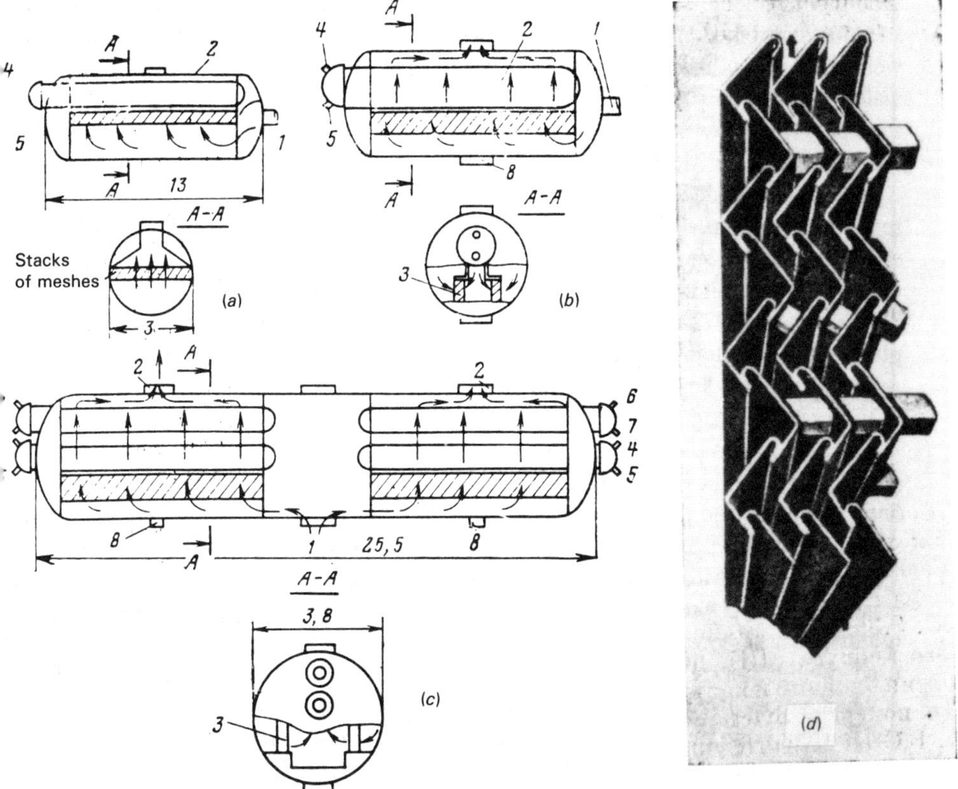

Figure 8.35 Design details of three types of Westinghouse MSRs. (*a*) First-generation MSR with demisters. (*b*) Second-generation MSR. (*c*) Latest type of MSR. (*d*) Louvered element employed in current MSR (1, inlet of wet steam from the HP section; 2, discharge of superheated steam to the LP section; 3, louvered separation stacks; 4, 5, inlet and discharge of heating steam to the first reheat state; 6, 7, same as above for the second reheat stage; 8, drain of separated moisture.

was found from these investigations that at velocity of steam through the demistors (meshes) of $c_v < 1.5$ m/sec virtually all the moisture is removed. All the subsequent first-generation MSR were produced by Westinghouse to the design shown in Fig. 8.35*a*.

However, the wire (mesh) demistors used in the MSR, while ensuring high moisture extraction efficiency, require a large amount of space, for which reason subsequent designs of MSRs at Westinghouse employed chevron (herringbone) elements in the separators (see Sec. 8.1). As early as 1970, second-generation MSRs were produced by Westinghouse with chevron separation elements shown in Fig. 8.35*b*. The use of these separation elements made it possible to increase the steam flowrate by 45% using the same stack dimensions as with demisters, i.e., it became possible to manufacture more compact units.

The use of chevron separating units was accompanied by modifications in the arrangement of the separator part of the MSR. In the lower part of the housing the steam is distributed symmetrically into two streams between the separating units. The wet steam passes through vertical chevron channels, and deposits moisture on their walls. The lower part of the MSR is equipped by a drainpipe through which the separated moisture is removed to regenerative preheaters. The reheater part of second-generation MSR may employ single or two-stage superheating.

Turbines produced for nuclear power plants by Westinghouse with first- and second-generation MSR usually are equipped with four or six such units (two units each per one LP section). This obviously takes up a large amount of space in the turbine hall and significantly complicates the turbine layout.

Starting in 1972 the Westinghouse Company started manufacturing nuclear power plant turbines with third-generation MSRs. These do not basically differ from second-generation units, since each of them is simply comprised of two such units (Fig. 8.35c). Such a design makes it possible to use one unit at each side of the turbine even in the case of two-pass LP sections.

Such a MSR has the following advantages: (1) the width of the turbine foundation level decreases and amounts to, depending on the layout, from 3.6 to 7.2 m; (2) the steam piping arrangement is simplified; (3) the control of cutoff seals is simplified.

In its latest MSR, Wesinghouse employs a more orderly system for transfer of steam from the HP to the LP sections. Thus, for example, the design of the deflecting (cutoff) seal has been made highly aerodynamic. Since guide vanes in pipeline turns used in the first MSR failed frequently, extensive studies were performed of the frequency response of these vanes, the reasons for their failure have been established and currently the most rigid and modern design of guide vanes is used.

Westinghouse has not made significant changes in the separator and reheater materials during this successive type development. The MSR casing, manifolds and internal supports are made of carbon steel and the tubesheets are made of forgings of the same steel. Reheater tubes may be constructed of 80% Cn plus 20% Ni or 90% Cn plus 10% Ni. The chevron separator elements (Fig. 8.35d) are fabricated of typical carbon steel for nuclear power plants with PWR reactors and from stainless steel for plants with boiling water reactors.

The Westinghouse MSRs have come into extensive use. Among others, they are used by the Brown-Boweri, Kraftwerk Union, and Mitsubisi companies [8.9], although these companies are presently manufacturing their own MSR. However, the latter are largely based on Westinghouse designs.

It should be noted in conclusion that separators and moisture separator reheaters are still in the design development stage. Obviously, after investigating details of the moisture separation processes in turbine flowpassages and of moisture arrangement in crossover piping, it will become possible to design more compact and reliable wet-steam turbines for nuclear power plants.

x, y	longitudinal and transverse coordinates	δ	thickness of film or boundary layer
w	velocity, u and c, relative and absolute in turbines (Chapters 7 and 8)	d, D	diameter
		R	radius
		L	length
j	ρw, velocity	t	time
p	pressure	$\tilde{D}$	precipitation flux
τ	shear stress	Re	Reynolds number
ρ	density	Re**	Reynolds number based on energy thickness
γ	specific weight		
h	enthalpy	Pr	Prandtl number
T	temperature	Pe	Peclet number
q	heat flux density	Nu	Nusselt number
q_v	volumetric heat release density	St	Stanton number
		Ψ, Ψ_s	relative friction and heat transfer laws
r	latent heat of vaporization		
c_p	specific heat at constant pressure	Fr	Foude number
		Q	volumetric flowrate
λ	thermal conductivity	c_f	friction factor
a	thermal diffusivity	C_D, ζ	drag factors
σ	surface tension	c	concentration
ν	kinematic viscosity	α	heat transfer coefficient
μ	dynamic viscosity	Π	perimeter
x	vapor quality	A	net cross sectional area
φ	void fraction	S	cross sectional area of side surface
β	vapor volumetric flow ratio		

Principal Subscripts

$0, \infty$	parameters at the flow axis		ev	evaporative
1	steam		lim	limiting
2	liquid film		act	actual
3	liquid droplets		lam	laminar
I	in, inlet		turb	turbulent
II	out, exit		r	rough
w	wall		exp	experimental
conv	convective		ef	effective
cr	critical		min	minimum
sm	smooth		max	maximum
b	boundary		D	diffusion
bend	bend		T	thermal
thr	threshold		int	internal
rel	relative		ext	external
i.b	incipiency of boiling		dyn	dynamic
d.an	dispersed annular		st	steady-state
mix	mixture		non.st	nonsteady
sub	subcooling		rew	rewetting
sup	superheating		t.ph	two-phase
econ	economizer		$d.o$	dryout
surf	surface		dr	droplet
ed	equilibrium			

REFERENCES

1.1 Proceedings of the 4th UN Conference on Peaceful Uses of Atomic Energy, Atomnaya energiya, *31*, No. 4, 1971.

1.2 Styrikovich, M. A. Vnutrikotlovyye protsessy (Boiler Processes). Gosenergoizdat Press, Moscow-Leningrad, 1954.

1.3 Styrikovich, M. A., Martynova, O. I., and Miropol'skiy, Z. L. Protsessy generatsii para na elektrostantsiyakh (Electric Power Plant Steam Generation Processes). Energiya Press, Moscow, 1969.

1.4 Styrikovich, M. A., Katkovskaya, K. Ya., and Dubrovskiy, I. Ya. Investigation of the Temperature Depression of Aqueous Solutions of Sodium Hydroxide. Trudy MEI, No. 238, 52–62, 1975.

1.5 Hodge, R. J., Le Surf, C. E., and Hilborn, J. W. Steam Generator Reliability. Canadian Practice. In: Proc. Intern. Symp. Application of Reliability Technology to Nuclear Power Plants, Vienna, 10–13 Oct., 1977.

1.6 Roofthoolt, R. VGB-Konferenz Chemie im Kraftwerk, 1977.

1.7 Collier, I. G., and Kennedy T. D. A. Harwell, AERE-R7203, 1972.

1.8 Cohen, P. Heat Transfer Res. and Design. AICHE Symp., 1971, vol. 70, p. 138.

1.9 Macbeth, R. V. Harwell. AERE-R705, AERE-R711, 1971.

1.10 Styrikovich, M. A., et al. Analysis of the Effect of Porous Scale on the Concentration of Impurities in Steam-Generating Channels. Teplofiz. vys. temp., *15*, No. 1, 109–114, 1977.

1.11 Styrikovich, M. A., et al. Experimental Study of Mass Transfer in Steam-Generating Channels with Porous Scale. Ibid., No. 2, 353–358.

1.12 Styrikovich, M. A., Leontiev, A. I., and Polonsky, V. S., et al. Influence of crude deposits on heat and mass transfer in steam generation channels. In: Proc. Intern Seminar "Two-phase momentum, heat and mass transfer in chemical process, and energy engineering systems." Hemisphere Publ. Co., N.Y., 1978, vol. 2, p. 549.

1.13 Styrikovich, M. A. Preprint of Intern. Conf. on Water Chemistry of Nuclear Reactor Systems, Bournemouth (England), 1977, 24–28 Oct.

1.15 Scroeder, H. I. Preprint Inter. Conf. on Water Chemistry of Nuclear Reactor Systems, Bournemouth (England), 1977, 24–28 Oct.

1.16 Martynova, O. I. Effect of Water-Chemistry of [Conventional] Thermal and Nuclear Power Plants on the Reliability of Steam-Turbine Operation. Energokhozyastvo za rubezhom, No. 1, 18–24, 1979.

1.17 Linday, W. T. Behavior of impurities in steam turbines. Power Eng., 1979, vol. 83, N 5, pp. 346–358.

2.1 Kosterin, S. I. Investigation of the Effect of Tube Diameter and Length on the Hydraulic Drag and Flow Structure in a Gas-Liquid Mixture. Izv. Akad. Nauk SSSR. Energetika i transport, 12, 1824–1835, 1949.

2.2 Styrikovich, M. A., and Surkov, A. V. Concerning the Flow Structure in a Dynamic Two-Phase Layer. Teploenergetika, 9, No. 4, 77–79, 1962.

2.3 Baker, O. Simultaneous Flow of Oil and Gas. Oil. Gas J., 1954, vol. 53, N 7.

2.4 Kozlov, B. K. Forms of Flow of Gas-Liquid Mixtures and the Limits of their Stability. Zhurn. tekhn. fiz., 24, 2285–2288, 1954.

2.5 Kutateladze, S. S., and Styrikovich, M. A. Gidrodinamika gazozhidkostnykh sistem (Fluid Mechanics of Gas-Liquid Systems). 2nd revised and supplemented edition. Energiya Press, Moscow, 1976.

2.6 Tobilevich, N. O., and Yeremenko, B. A. Investigation of the Aspects of Heat Transfer for Boiling in Tubes. In: Gidrodinamika i teploobmen pri kipenii v kotlakh vysokogo davleniya (Fluid Mechanics and Heat Transfer in Boiling in High-Pressure Boilers). USSR Academy of Sciences Press, Moscow, 1955.

2.7 Isachenko, V. P., Osipova, V. A., and Sukomel, A. S. Teploperedacha (Heat Transfer). 3rd revised and supplemented edition. Energiya Press, Moscow, 1975.

2.8 Bergles, A. E., and Suo, M. Boiling Water Flow Regime at High Pressure. In: Proc. 1966 Heat Trans. and Fluid Mech. Inst., 79–99. Stanford University Press, 1966.

2.9 Milashenko, V. I. Investigation of the Hydrodynamics of Dispersed Vapor-Liquid Flow. Author's abstract of Candidate thesis. Moscow Energetics Institute, Moscow, 1977.

2.10 Dzarasov, Yu. I., et al. Diagnostics of Phase Inclusions in Steam-Water Flows by Microphotography. In: Issledovaniya kriticheskikh teplovykh potokov v puchkakh sterzhney (Investigations of Critical Heat Fluxes in Rod Bundles). Atomizdat Press, Moscow, 411–418, 1974.

2.11 Dzarasov, Yu. I., Kol'chugin, B. A., and Liverant, E. I. An Experimental Study of Characteristics of a Two-Phase Flow with Heat Addition. In: Teplo- i massoobmen (Heat and Mass Transfer [Proceedings of the 4th All-Union Conference on Heat and Mass Transfer]), 2, Pt. 1, 97–101, 1972 [Engl. transl. HEAT TRANSFER, Soviet Research, 5, No. 4, 97–100, 1973].

2.12 Milashenko, V. I., et al. Experimental Study of Certain Characteristics of Two-Phase Flows. Trudy Khar'k. aviats. in-ta, No. 3, 31–36, 1976.

2.13 Tippets, F. E. Analysis of the Critical Heat-Flux Condition in High-Pressure Boiling Water Flows. J. Heat Transfer, 86, No. 1, 23–38, 1964.

2.14 Tippets, F. E. Critical Heat Fluxes and Flow Patterns in High-Pressure Boiling Water Flows. Ibid., 12–22.

2.15 Dzarasov, Yu. I., Kol'chugin, B. A., and Liverant, E. I. Investigation of the Relationship between Heat and Mass Transfer Characteristics in Evaporating Channels. In: Teploobmen, 1974. Sovetskiye issledovaniya (Heat Transfer, 1974. Soviet Investigations [Collection of Soviet Contributions at the 1974 (Fifth) International Heat Transfer Conference]). Nauka Press, Moscow, 1975, 259–268.

2.16 Bennet, A. W., Kearsey, H. A., and Keeys, R. K. F. Flow visualisation studies of boiling at high pressure. AERE-R 4874, Harwell, 1965.

2.17 Hosler, E. R. Visual study of boiling at high pressure. In: Chem. Eng. Progr. Symp. Ser., 1965, vol. 61, N 57, pp. 269–279.

2.18 Groeneveld, D. C. The thermal behavior of a heated surface at and beyond dryout. AECL-4309, 1972.

2.19 Fedorovich, Ye. D., et al. The Temperature Regime of Steam-Generating Tubes, Heated by a High-Temperature Liquid, at DNB. In: Teplo- i massoobmen, 2, Pt. 1, 346–351. Nauka i Tekhnika Press, Minsk, 1972.

2.20 Cumo, M. Elementi di termotecnica del reattore. Roma: CNEN, 1971.

2.21 Forslund, R. P., and Rohsenow, W. M. Dispersed Film Boiling. J. Heat Transfer, *90*, No. 4, 399, 1968.

2.22 In'kov, A. P., et al. Investigation of Nonequilibrium in Dispersed Film Boiling of Hydrogen in Tubes. In: Voprosy sovremennoy kriogeniki (Problems of Modern Cryogenics). Vneshtorgizdat Press, Moscow, 1975.

2.23 Cumo, M., Forello, G. E., and Ferrari, Q. Post-burnout heat transfer up to the critical pressure. Trans. Amer. Nucl. Soc., 1969, vol. 12, N 2, pp. 810–811.

2.24 Kasayev, N. O. Investigation of the Enhancement of Heat Transfer in Film Boiling of Cryogenic Fluids in Tubes. Author's abstract of Candidate thesis. Moscow Aviation Institute, Moscow, 1976.

2.25 Cumo, M., et al. On Two-Phase Highly Dispersed Flows. J. of Heat Transfer, *96*, No. 4, 496, 1974.

2.26 Grueneveld, D. C. Post-dryout heat transfer physical mechanisms and survey of predictions methods. Nucl. Eng. and Design, 1975, vol. 32, N 3, pp. 283–294.

2.27 Cumo, M., Farello, G. E., and Gerrari, Q., et al. On two-phase highly dispersed flows. In: Advance Heat Transfer, Roma, 1974, vol. 2, pp. 251–259.

2.28 Bergles, A. E., Fuller, W. D., and Hynek, S. J. Dispersed flow film boiling of nitrogen with swirl flow. Intern. J. Heat and Mass Transfer, 1971, vol. 14, N 9, pp. 1343–1354.

2.29 Mayinger, F. Scaling and modeling laws in two-phase flow and boiling heat transfer. In: Proc. of NATO Advanced Study Institute: Two-phase flow and heat transfer, Hemisphere Publ. Co., N.Y., 1976, vol. 1.

2.30 Landau, L. D., and Lifshits, E. M. Mekhanika sploshnykh sred (Mechanics of Continuous Media). 2nd revised and supplemented edition. Gostekhizdat Press, Moscow, 1954 [Engl. transl. Addison-Wesley, 1954].

2.31 Frankl', F. I. Experience with the Semiempirical Theory of the Motion of Suspended Sediments in Nonuniform Flow. Dokl. Akad. Nauk SSR, *102*, No. 6, 1093–1096, 1955.

2.32 Teletov, S. G. Aspects of Hydrodynamics of Two-Phase Mixtures. I. Hydronamic and Energy Equations. Vestn. MGU. Ser. matem., mekh., astron., fiz., khim., No. 2, 15, 1958.

2.33 Slezkin, N. A. Differential Equations of the Motion of Pulp. Dokl. Akad. Nauk SSR, *86*, No. 2, 235–237, 1952.

2.34 Rakhmatulin, Kh. A. Principles of Gasdynamics of Interpenetrating Motions of Compressible Fluids. Prikl. mat. i mekh., *20*, No. 2, 184–195, 1956.

2.35 Styrikovich, M. A., and Kutateladze, S. S. Gidravlika gazozhidkostnykh sistem (Hydraulics of Gas-Liquid Systems). Gosenergoizdat Press, Moscow-Leningrad, 1958 [Wright-Field transl. F-75-9814/V-58].

2.36 Krayko, A. N., and Sternin, L. Ye. On the Theory of Two-Velocity Flows of a Continuous Medium with Solid or Liquid Particles. Prikl. mat. i mekh., *29*, No. 3, 418, 1965.

2.37 Nigmatulin, R. I. Equations of Fluid Mechanics and Compression Waves in a Two-Velocity and Two-Temperature Continuous Medium in the Presence of Phase Transitions. Izv. Akad. Nauk SSSR. Mekh. zhidk. i gaza, No. 5, 33, 1967.

2.38 Vernier, P., and Delhaye, J. M. Equation générales des écoulements diphasiques appliquéé à la thermodynamique des reacteures. Energie Primaire, 1968, vol. IV, N1/2.

2.39 Boure, J., and Reocreux, M. General Equations of Two-Phase Flows. Application to Critical and Transient Regimes. In: IV Vsesoyuz. konf. po teploobmenu (Proceedings of the 4th All-Union Conference on Heat and Mass Transfer). Minsk, 1972.

2.40 Tsiklauri, G. V., Danilin, V. S., and Seleznev, L. I. Adiabatnyye dvukhfaznyye techeniya (Two-Phase Adiabatic Flows). Atomizdat Press, Moscow, 1973.

2.41 Proudman, J., and Pearson, J. R. A. Expansion at small Reynolds for the flow past a sphere and circular cylinder. J. Fluid Mech., 1957, vol. 2, p. 237.

2.42 Stonecypher, T. E. Report NP-60-17. Hintsvill (Alabama): Rohm and Haas Co., 1960.

2.43 Fuks, N. A. Mekhanika aerozoley (The Mechanics of Aerosols). USSR Academy of Sciences Press, Moscow-Leningrad, 1955 &Engl. transl. Pergamon Press, 1964.

2.44 Levich, V. G. Fiziko-khimicheskaya gidrodinamika (Physico-Chemical Hydrodynamics). USSR Academy of Sciences Press, Moscow-Leningrad, 1952 &Engl. transl. Prentice Hall, Englewood Cliffs, 1962é.

2.45 Toborin, L. B., and Canvin, W. H. Canad. J. Chem. Eng., 1959, vol. 87, N 129, p. 224; 1960, vol. 38, N 142, p 189.

2.46 Hoglund, R. F. Recent Advances in Gas-Particle Nozzle Flows. ARS Journal, 32, No. 5, 662–671, 1962.

2.47 Rudinger, G. Effective Drag Coefficient for Gas-Particle Flow in Shock Tubes. J. of Basic Engineering, 92, No. 1, 165, 1970.

2.48 Prrem, L. L., and Parkins, W. E. A New Method of MHD Power Conversion Employing a Fluid Metal. International Symposium on Magnetohydrodynamic Electrical Power Generation. Paris, France. Paper No. 63, July 6–10, 1964.

2.49 Saltanov, Ye. A. Dvukhfaznyye sverkhzvukovyye potoki (Two-Phase Supersonic Flows). Vysshaya Shkola Press, Minsk, 1972.

2.50 Ingebo, R. D. Drag coefficients for droplets and solid spheres in cloud. NACA TN 3762, 1956, USA.

2.51 Kaye, B. N. Fluid resistance in accelerated motion. In: Proc. Symp. on intern. between fluids and particles. Inst. of Chem. Eng., 1967, p. 17.

2.52 Ergun, D. Evaporation of fuel spray in airstream and effect of turbulence. Chem. Eng. Progr., 1952, vol. 8, p. 89.

2.53 Raushenbakh, V. V. Fizicheskiye osnovy rabochego protsessa v kamerakh sgoraniya vozdushno-reaktivnykh dvigateley (Physical Principles of the Working Process in Combustion Chambers of Air-Breathing Jet Engines). Mashinostroyeniye Press, Moscow, 1964.

2.54 Gol'dshtik, M. A., and Sorokin, V. N. Concerning the Motion of a Particle in a Vortex Chamber. Prikl. mat. i tekhn. fiz., No. 6, 149–152, 1968.

2.55 Beker, J. L. Flow regime transition in vertical two-phase flow. ANL R7093, USA, 1965.

2.56 Frenkel', Ya. I. Kineticheskaya teoriya zhidkostey (Kinetic Theory of Liquids). USSR Academy of Sciences Press, Moscow-Leningrad, 1945 [Engl. transl. Dover, 1955].

2.57 Ranz, W. E., and Marhsall, W. H. Evaporation of small droplets. Chem. Eng. Progr., 1958, vol. 48, p. 14.

2.58 Eckert, E. R. G., and Drake, R. M. Heat and Mass Transfer. McGraw-Hill, 1959 [Rus. transl. 1961].

2.59 Sampson, R., and Springer, C. Condensation and evaporation from droplet by a moment method. J. Fluid Mech., vol. 36, pp. 577–584.

2.60 Numann, M. Ein Beitrag zum Abscheidemechanismus im Venturiwächer: Diss. Techn. Univ. Hannover, 1978.

2.61 Miura, K., Miura, T., and Ohtavi, S. Heat and mass transfer to and from droplets. AIChE Symp. Ser. 73, 1977, vol. 163, pp. 95–102.

2.62 Schmidt-Traub, H. Impuls und Stoffanstausch an umströmten Kugeln unter Berücksichtigung der inneren Zirkulation: Diss TV. B., 1970.

2.63 Bankoff, S. G. J. Heat Transfer. Ser. C., 1960, vol. 82.

2.64 Armand, A. A. Drag in Two-Phase Systems in Horizontal Tubes. Izv. VTI [All-Union Institute of Heat Engineering], No. 1, 16, 1946.

2.65 Levy, S. Steam slip-theoretical prediction from momentum model. J. Heat Transfer. Ser. C, 1960, vol. 82, pp. 113–124.

2.66 Baroczy, C. A systematic correlation for two-phase pressure drop. Chem. Eng. Progr. Ser. 64, 1966, vol. 62, pp. 232–238.

2.67 Chien, S. F., and Ibele, W. Pressure Drop and Liquid Film Thickness of Two-Phase Annular and Annular Mist Flow. J. of Heat Transfer, 86, No. 1, 89–96, 1964.

2.68 Chisholm, D. Heat transfer and pressure drop for heated tubes. Intern. J. Heat and Mass Transfer, 1973, vol. 16, pp. 347–358.

2.69 Haberstroh, R. E., and Griffith, P. Slug annular two-phase flow regime transition. ASME, 1965, 65-HT-52.

2.70 Keeys, R. K. F., Ralph, J. C., Roberts, D. N. The effect of Heat flux on liquid entrainment in steam-water flow in a vertical tube at 1000 psia, 1970, UKAEA Rept. 6294.

2.71 Subbotin, V. I., Sorokin, D. N., and Nigmatulin, B. I. Integrated investigation into hydrodynamic characteristics. In: Proc. of VI International Heat Transfer Conference, Toronto, Canada, 1978, vol. 1, p. 327.

2.72 Nigamtullin, B. I., Milashenko, V. I., and Shugayev, Yu. Z. Investigation of the Distribution of

Liquid between the Core and Film and Dispersed-Annular Steam-Water Flows. Teploenergetika, No. 5, 77–79, 1976.

2.73 Gardner, G. C. Deposition of particles from a gas flowing parallel to a surface. Intern. J. Multi-Phase Flow, 1975, N 2, p. 213.

2.74 Hutchinson, P., Hewitt, G. F., and Dukler, A. E. Deposition of liquid or solid dispersion from turbulent gas stream: A stochastic model, AERE-R6657, Harwell, 1970.

2.75 Kirillov, P. L., and Smogalev, I. P. Effect of Droplet Size on Mass Transfer in Two-Phase Flows. Teplofiz. vys. temp., *11*, No. 6, 1312–1313, 1974.

2.76 Gardner, G. C. Deposition of particles from a gas flowing parallel to a surface. Intern. J. Multi-Phase Flow, 1975, N 2, p. 213.

2.77 Yanovskiy, L. P., and Zverets, N. I. Effect of Liquid Fluctuations on the Average Motion of a Suspended Particle. Inzh.-fiz. zhurn., *19*, No. 5, 837–943, 1970.

2.78 Cousin, L. B., and Hewitt, G. F. Liquid phase mass transfer in annular two-phase flow: Droplet deposition and liquid entrainment, AERE-R5657, Harwell, 1968.

2.79 Farmer, R., and Griffith, P., Rohsenow, W. H. Liquid droplets deposition in two-phase Flow, ASME paper, N 70-HT, 1970.

2.80 Namie, S., and Veda, T. Droplet transfer in two-phase annular mist flow. Bull. J.S.M.E., 1972, vol. 15, N 91, pp. 1472–1484.

2.81 Rachkov, V. I. Experimental Study of Moisture Transfer in Steam-Water Dispersed-Annular Flows. Author's abstract of Candidate thesis. All-Union Correspondence Polytechnic Institute, 1978.

2.82 McCoy, D., and Hanratty, T. Droplet Mixing and Deposition in a Turbulent Flow. In: Proc. of intern. seminar two-phase momentum, heat and mass transfer in chemical process and energy engineering systems. Hemisphere Publ. Co., N.Y., Sept. 4–9, 1978, vol. 1, p. 119.

2.83 Akagawa, K. Creation and deposition of entrained droplets in swirling annular-mist two-phase flow. Bull. of J.S.M.E., 1974, vol. 17, N 93, pp. 350–354.

2.84 Ivandayev, S. I. On the Determination of Laws Governing the Interaction between Components of Gas-Liquid Dispersed Flows. In: Nelineynyye volnovyye protsessy v dvukhfaznykh sredakh (Non-linear Wave Processes in Two-Phase Flows). Novosibirsk, 1977.

2.85 Alexander, I., and Coldien, C. Droplet transfer from suspending air to duct walls. J. Chem. Eng., 1951, vol. 42, pp. 1325–1336.

2.86 Jagota, A. K. Tracer measurement in two-phase annular flow to obtain interchange and entrainment. Canad. J. Chem. Eng., 1973, vol. 51, pp. 49–58.

2.87 Kapitsa, P. L. Wave Flow of Thin Layers of Viscous Fluids. Zhurn. eksp. i tekhn. fiz., *18*, No. 1, 3 and 19, 1948.

2.88 Nakoryakov, V. Ye., and Shreyber, I. R. Waves on the Surface of a Thin Layer of a Viscous Fluid. Prikl. mat. i tekhn. fiz., No. 2, 109, 1973.

2.89 Ishigai, S. Effect of transient flow on premature dryout in tube. Bull. of J.S.M.E., 1977, pp. 178–182.

2.90 Shkadov, V. Ya. Wave Modes of Flow of a Thin Layer of a Viscous Liquid due to Gravity. Izv. Akad. Nauk SSSR. Mekh. zhidk. i gaza, No. 1, 43, 1976.

2.91 Zhavoronkov, N. M., and Malyusov, V. A. Investigation of Hydrodynamics and Mass Transfer in Absorption and Rectification at High Flow Velocities. Teor. osn. khim. tekhn., *1*, No. 5, 562–577, 1961.

2.92 Vasil'chenko, Ye. G., et al. Investigation of Wave Characteristics of a Liquid Film in Cocurrent Vapor Flow. In: Nelineynyye volnovyye protsessy v dvukhfaznykh sredakh. Novosibirsk, 1977, 227–237.

2.93 Hanratty, T., and Engen, J. Interaction between a turbulent air stream and moving water surface. AIChE J., 1974, vol. 3, N 3, p. 299.

2.94 Maksimov, V. V., Kulov, N. N., and Malyusov, V. A. Investigation of the Wave Characteristics of Falling Liquid Films. Ibid, 153–158.

2.95 Chu, K. J., and Dukler, A. E. Statistical characteristics of thin wavy films. AIChE J., 1975, vol. 21, N 3, p. 583; 1974, vol. 20, N 4, p. 695.

2.96 Hewitt, G. F., and Wallis, G. B. Flooding and associated phenomena in falling film flow in a tube. In: Proc. of ASME Multi-Phase Symp. Philadelphia, 1963, p. 62.

2.97 Krylov, V. S., Vorotilin, V. P., and Levich, V. G. On the Theory of Wave Motion of Thin Liquid Films. Teor. osn. khim. tekh., *3*, no. 4, 499, 1969.

2.98 Ruston, E., and Davis, G. AIChE J., 1971, vol. 17, p. 671.

2.99 Miya, M., Woodmansce, D. E., and Hanratty, T. A model for roll waves in gas-liquid flow. Chem. Eng. Sci., 1971, vol. 26, p. 1915.

2.100 Styrikovich, M., Tsiklauri, G., Sohno, V., and Baryshev, J. Wave pattern and liquid film burnout with cocurrent vapor flow. In: Proc. of VIth Intern. Heat Transfer Conf., Toronto, 1978, vol. 1, p. 399.

2.101 Tsiklauri, G., Besfamilniny, P., and Barychev, Yu. Experimental Study of Hydrodynamic Processes for Wavy Water Film in a Cocurrent Air Flow. In: Intern. Proc. of Intern. Seminar (Two-Phase momentum, heat and mass transfer in chemical process and energy engineering systems), Hemisphere Publ. Co., N.Y., 1978, vol. 1, p. 367.

2.102 Cohen, L. S., and Hanratty, T. Effect of waves at a gas-liquid interface in a turbulent air flow. J. Fluid Mech., 1968, vol. 31, p. 467.

2.103 Moeck, E. O. Annular-dispersed two-phase flow and critical heat flux, 1970, AECh-3656, pp. 3–8.

2.104 Wallis, G. B. One-Dimensional Two-Phase Flow. McGraw-Hill, 1969 [Rus. transl. 1972].

2.105 Hewitt, G. F., and Hall-Taylor, N. S. Annular Two-Phase Flow. Pergamon Press, 1970 [Rus. transl. 1974].

2.106 Paleev, I. I., and Philipovich, B. S. Phenomena of liquid transfer in two-phase dispersed annular flow. Inter. J. Heat and Mass Transfer, 1966, vol. 9, N 10, pp. 1089–1093.

2.107 Quand, E. Measurement of some basic parameters in two-phase annular flow. AIChE J., 1965, vol. 11, pp. 311–319.

2.108 Mayinger, F. Heat and mass transfer at liquid—gas interphases. In: Proc. of Intern. Seminar (Two-Phase momentum, heat and mass transfer in chemical process and energy engineering systems), Hemisphere Publ. Co., N.Y., 1978, vol. 2, p. 1340.

2.109 Petrovichev, V. I., Deyev, V. I., and Andreyev, V. K. Effect of Heating Surface Orientation and Pressure on the Critical Heat Flux. Teplofiz. vys. temp., *14*, 436, 1976.

2.110 Bennet, A. W., and Hewitt, G. F., et al. Studies of burnout in boiling heat transfer to water in round tubes with nonuniform heating. UKAEA Rept. 1966, N 5076.

2.111 Doroshchuk, V. Ye. Krizis teploobmena pri kipenii v trubakh (Boiling Crisis in Pipes). Energiya Press, Moscow, 1970.

2.112 Kirillov, P. L., and Smogalev, I. P. Measurement of Certain Features of Wet-Steam Flows in a Circular Pipe at 68.6 bar. Trudy. Fiz.-energ. in-ta, p. 72, 1973.

2.113 Miropol'skiy, Z. L., Shitsman, M. Ye., and Shneyerova, R. I. Effect of Heat Flux and Velocity on the Hydraulic Drag of Pipe Flow of Wet Steam. Trudy TsKTI [Polzunov Boiler and Turbine Institute], No. 59, 31–41, 1965.

2.114 Tarasova, N. V. Hydraulic Drag in the Boiling of Water and Wet Steam in Heated Pipes and Annuli. Ibid, 47–58.

2.115 Tarasova, N. V., and Orlov, V. M. Investigation of the Hydraulic Drag in Surface Boiling in Water in a Pipe. Teploenergetika, No. 6, 48–52, 1962.

2.116 Semenov, P. I., and Tochilin, A. A. Void Fraction in Wet Steam Flows in Nonheated Vertical Tubes. Inzh.-fiz. zhurn., No. 7, 19–24, 1961.

2.117 Dolinin, I. V. Experimental Study of Local Parameters of Dispersed-Annular Wet-Steam Flows Under Adiabatic Conditions. Author's abstract of Candidate thesis. All-Union Correspondence Polytechnic Institute, Moscow, 1978.

2.118 Gaertner, R. F., and Westwater, J. M. Population of Active Sites in Nucleate Boiling Heat Transfer. Chem. Eng. Prog. Symp. Ser. 56, *30*, 39–48, 1960.

2.119 Kirillov, P. L., and Smogalev, I. P. A Technique of Hydraulic Calculation of a Vertical Steam-Generating Channel. Teploenergetika, No. 2, 71–74, 1980.

2.120 Tabulated Data for Calculating the Boiling Crisis in Water in Uniformly Heated Circular Pipes. Ibid., No. 9, 90–92, 1976.

2.121 Baldina, O. M., and Lokshin, V. A. Correlation of Experimental Data on the Void Fraction in Uptake and Downtake Pipes and the Resistance to Motion of Wet Steam in Pipes. Trudy TsKTI, No. 59, 72–78, 1965.

2.122 Krasyakova, L. Yu. Investigation of the Hydraulic and Temperature Regime in Components of a Coil with Uptake-Downtake Two-Phase Mixture Flow. Ibid., 12–16.

2.123 Przhiyalkovskiy, M. M., and Petrova, I. N. Hydraulic Drag of Tubes in Uptake High Velocity Wet-Steam Flows at High Pressures. Teploenergetika, No. 6, 25-28, 1961.

2.124 Tong, L. Boiling Heat Transfer and Two-Phase Flow. Wiley, 1965 [Rus. transl. 1969].

2.125 Borishanskiy, V. M., Andriyevskiy, A. A., and Fokin, B. S. Effect of Channel Wall Roughness on the Hydraulic Drag. In: Dostizheniya v oblasti issledovaniya teploobmena i gidravliki dvukhfaznykh potokov (Advances in the Study of Heat Transfer and Hydraulics of Two-Phase Flows), 119–122. Nauka Press, Leningrad, 1973.

3.1 Kutateladze, S. S., and Styrikovich, M. A. Gazodinamika gazozhidkostnykh sistem (Gasdynamics of Gas-Liquid Systems). Energiya Press, Moscow, 1976.

3.2 Kutateladze, S. S. Osnovy teorii teploobmena (Fundamentals of Heat Transfer). Mashgiz Press, Moscow-Leningrad, 1952 [Engl. transl. Academic Press, 1963].

3.3 Styrikovich, M. A., Martynova, O. I., and Miropol'skiy, Z. L. Protsessy generatsii para na elektrostantsiyakh (Electric Power Plant Steam Generation Processes). Energiya Press, Moscow, 1969.

3.4 Tarasova, N. V., and Orlov, V. M. Heat Transfer and Hydraulic Drag in Surface Boiling of Water in Annuli. In: Konvektivnaya teploperedacha v dvukhfaznom i odnofaznom potokakh (Convective Heat Transfer in Two-Phase and One-Phase Flows), 48–52. Energiya Press, Moscow-Leningrad, 1964 [Engl. transl. Israel Program for Scientific Translations, 1969].

3.5 Borishanskiy, V. M., et al. Correlation for Calculating Heat Transfer in Two-Phase Flows in Pipes and Ducts. In: Dostizheniya v oblasti issledovaniya teploobmena i gidravliki dvukhfaznykh potokov v elementakh energooborudovaniya (Advances in the Study of Heat Transfer and Hydraulics of Two-Phase Flow in Power-Generating Equipment), 194–201. Nauka Press, Leningrad, 1973.

3.6 Shvetsov, R. S. An Experimental Study of Boiling Heat Transfer of Water in Annuli. Author's abstract of Candidate thesis. Moscow Energetics Institute, Moscow, 1967.

3.7 Andriyevskiy, A. A., et al. Cooling of Heating Surfaces of Two-Phase Wet Steam Flows. Trudy TsKTI [Polzunov Boiler and Turbine Institute], No. 86, 3-25, 1968.

3.8 Tarasova, N. V., Armand, A. A., and Kon'kov, A. S. Investigation of Heat Transfer in a Pipe in the Boiling of Subcooled Water and Wet Steam. In: Teploobmen pri vysokikh teplovykh nagruzkakh i drugikh spetsial'nykh usloviyakh (Heat Transfer at High Heat Fluxes and Other Special Conditions), 6–22. Moscow GEI Press, 1959.

3.9 Dzarasov, Yu. I. Investigation of Conditions of Heat Transfer to Dispersed-Annular Wet Steam Flow in an Annulus. Author's abstract of Candidate thesis. Krzhizhanovskiy Power Engineering Institute of the USSR Academy of Sciences. Moscow, 1973.

3.10 Collier, J. G. A Review of Two-Phase Heat Transfer. AERE CE/R 2496, HMSO, 1958 [Rus. transl. 1962].

3.11 Metodika i zavisimosti dlya teoreticheskogo rascheta teploobmena i gidravlicheskogo soprotivleniya teploobmennogo oborudovaniya AEA: Rukovodyashchiy tekhnicheskiy material, RTM 24.031.05-72 (Methods and Relationships for Analytic Calculation of Heat Transfer and Pressure Drop in Heat-Exchange Equipment of Nuclear Power Plants. Instructional Manual RTM 24.031.05-72). Heavy Power and Transportation Machinery Building Ministry, 1972.

3.12 Przhiyalkovskiy, M. M., and Petrova, I. N. Hydraulic Drag of Tubes in Uptake Flows of Wet Steam at High Velocities and High and Superhigh Pressures. Teploenergetika, No. 6, 25–28, 1961.

3.13 Brantov, V. G. The Void Fraction and Boiling Crisis in Pipes. Author's abstract of Candidate thesis. Moscow Energetics Institute, Moscow, 1975.

3.14 Doroshchuk, V. Ye. Krizisy teploobmena pri kipenii vody v trubakh (Boiling Crisis with Water in Pipes). Energiya Press, Moscow, 1970.

3.15 Styrikovich, M. A., et al. Boiling Crisis in a Liquid Film on a Plate with Cocurrent Flow of Saturated Vapor. In: Teploobmen-78. Sovetskiye issledovaniya (Heat Transfer—78. Soviet Investigations [Collection of Soviet Contributions at the 1978 (Sixth) International Heat Transfer Conference]). Nauka Press, Moscow, 1980.

3.16 Labuntsov, D. A. Nonequilibrium Effects in Evaporation and Condensation. In: Parozhidkostnyye potoki (Vapor-Liquid Flows). Lykov Heat and Mass Transfer Institute, Minsk, 1977.

3.17 Bankoff, S. G. Entrainment of gas in the spreading of a liquid over a rough surface. AIChE J., 1958, vol. 4, pp. 24–26.

3.18 Hsu, Y. Y., and Graham, R. Transport processes in boiling and two-phase systems. Hemisphere Publ. Co., Washington, USA, 1976.

3.19 Hewitt, G. F., and Hall-Taylor, N. S. Annular Two-Phase Flow. Pergamon Press, 1970 [Rus. transl. 1974].

3.20 Bergles, A. E., and Rohsenow, W. M. The determination of forced convection surface-boiling heat transfer. J. Heat Transfer, 1964, vol. 86C, pp. 365–372.

3.21 Davis, E. J., and Anderson, G. H. The Incipience of Nucleate Boiling in Forced Convection Flow. AIChE J., *12*, No. 4, 774–780, 1966.

3.22 Dukler, A. E. Fluid mechanics and heat transfer in falling film system. In: Paper presented at the ASME-AIChE 3rd Nat. Heat Transfer Conf., Chicago, 1959, Aug.

3.23 Nishikawa, K., Kusuda, H., and Yamasaki, A. Growth and collapse of bubbles in nucleate boiling. Bull. J. SME, 1965, vol. 8, pp. 205–210.

3.24 Treshchev, G. G. Stability of Wet-Steam Flow in a Heated Pipe in a Single-Loop System. Trudy TsKTI, No. 101, 148–152, 1970.

3.25 Bankoff, S. G., and Mason, J. P. Heat Transfer from a surface of a steam bubble in a turbulent subcooled liquid stream. AIChE, 1962, vol. 8, pp. 30–33.

3.26 Willie, G. Evaporation and surface structure of liquids. Proc. Roy. Soc. London, 1972, vol. 197, p. 383.

3.27 Labuntsov, D. A., and Kryukov, A. P. High-Rate Evaporation Processes. Teploenergetika, No. 4, 8, 1977.

3.28 Griffith, P., and Wallis, J. D. The role of surface condition in nucleate boiling. Chem. Eng. Progr. Symp. Ser. 56, 1960, vol. 30, pp. 49–63.

3.29 Nakoryakov, V. Ye., Pokusayev, B. G., and Alekseyenko, S. V. Two-Dimensional Waves on a Vertical Liquid Film. In: Nelineynyye volny v dvukhfaznykh sredakh (Nonlinear Waves in Two-Phase Media). Novosibirsk, 1977.

3.30 Tsikla'ri, G. V., Danilin, V. S., and Seleznev, L. I. Adiabatnyye dvukhfaznyye techeniya (Two-Phase Adiabatic Flows). Atomizdat press, Moscow, 1973.

3.31 Subbotin, V. I., et al. Gidrodinamika i teploobmen v atomnykh energeticheskikh ustanovkakh (Fluid Mechanics and Heat Transfer in Nuclear Power Plants). Atomizdat press, Moscow, 1975.

3.32 Petukhov, B. S., Gening, L. G., and Kovalev, S. A. Teploobmen v yadernykh energeticheskikh ustanovkakh (Heat Transfer in Nuclear Power Plants). Atomizdat, Moscow, 1974.

3.33 Remizov, O. V. Boiling Crisis in Pipes. Trudy FEI [Fiz. energ. int], E-11, 06–09. Obninsk, 1975.

3.34 Tong, L. S. Boiling Crisis and Critical Heat Flux, AEC Critical Review Series, TID-25887, 1972 [Rus. transl. 1976].

3.35 Cumo, M. Elementi di termotecnica del reattore. Roma: Comit. Nat. Energ. Nucl., 1969, 437 p.

3.36 Miropol'skiy, Z. L., and Mostinskiy, I. L. Critical Heat Fluxes in Uniform and Nonuniform Heating of the Perimeter of Steam-Generating Tubes. Teploenergetika, No. 11, 64–68, 1958.

3.37 Doroshchuk, V. Ye. Concerning the Boiling Crisis in an Evaporator Tube. Teplofiz. vys. temp., *4*, 552–561, 1966.

3.38 Miropol'skiy, Z. L., and Shitsman, M. Ye. Permissible Heat Fluxes and Heat Transfer to Water Boiling in Tubes. In: Issledovaniya teplootdachi k paru i vode, kipyashchey v trubakh pri vysokikh davleniyakh (Studies of Heat Transfer to Steam and Water Boiling in Pipes at High Pressures, 24–53). Atomizdat Press, Moscow, 1958.

3.39 Smolin, V. I., Polyakov, V. K., and Yesikov, V. I. An Experimental Study of the Boiling Crisis in Steam-Generating Tubes. Atomnaya energiya, *16*, No. 5, 417–423, 1964.

3.40 Subbotin, V. I., et al. Critical Heat Fluxes in Forced Convection of Subcooled Water in Pipes at 140 to 220 atm. In: Issledovaniya teplootdachi k paru i vode, kipyashchey v trubakh pri vysokikh davleniyakh, 95–119. Atomizdat Press, Moscow, 1958.

3.41 Bennett, A. W., and Hewitt, G. F., et al. Heat transfer to steam-water mixture flowing in uniformly heated tubes in which the critical heat flux has been exceeded. J. Mech. Eng., pap. 27, Bristol, 1968, p. 27.

3.42 Jensen, A., and Manno, V. G. Measurements of burnout film flow and pressure drop in a concentric annulus 3500 26 17 with heated rod and tube. In: European two-phase flow group meet. Harwell, 1974.

3.43   Janssen, E. Sixteen rod heat flux in investigation in two-phase flow and heat transfer in rod bundles. ASME, 1969.

3.44   Kinneir, J. M., et al. Burnout power and pressure drop measurement on 7-rod clusters cooled by freon 12 at 125 ap. In: European two-phase flow group meeting. Karlsruhe, 1969.

3.45   Whalley, P. B., Hutchinson, Ph., and James, P. W. The calculation of critical heat flux complex situation using an annular flow model. In: Proc. of VIth Intern. Heat Transfer Conf., Toronto, 1978, vol. 5, p. 65.

3.46   Labuntsov, D. A. Critical Heat Fluxes in Forced Convection of Subcooled Water. Atomnaya ener-giya, *10*, No. 5, 523–525, 1961.

3.47   Miropol'skiy, Z. L., and Shitsman, M. Ye. Critical Heat Fluxes in the boiling of Water in Channels. Ibid., *11*, 6, 515–521, 1961.

3.48   Kirillov, P. L. Calculation of Critical Heat Fluxes in the Boiling of Subcooled Water in tubes (Uniformly Distributed Heat Flux). In: Krizis teploobmena pri kipenii v kanalakh (Boiling Crises in Channels), 100–173. Fiz. energ. int, Obninsk, 1974.

3.49   Tong, L. S., Cuvrin, H. B., and Thorn, A. G. New correlation predict DNB conditions. Nucleon-ics, 1963, vol. 21, N 5.

3.50   Thompson, B., and Macbeth, R. V. Boiling water heat transfer burnout in uniformly heated round tubes. A compilation of world date with accurate correlations. AEEW R-356, 1964.

3.51   Hewitt, G. F. A method of representing burnout data in two-phase heat transfer for uniformly heated round tubes. AERE-R4613, 1964.

3.52   Tong, L. S. Prediction of departure for nucleate boiling for an axially non-uniform heat flux distri-bution. J. Nucl. Energy, 1967, vol. 21, p. 241.

3.53   Tong, L. S., Currin, H. B., and Thorp, A. G. An evaluation of the departure from nucleate boiling in bundles of reactor tube rods. Nucl. Sci. Eng., 1968, vol. 33(1), pp. 7–15.

3.54   Lokshin, V. A., ed. Gidravlicheskiy raschet kotel'nykh agregatov Hormativnyye metod (Hydraulic Design of Boiler Units (Standard Method). Energiya Press, Moscow, 1978.

3.55   Rekomendatsii po raschetu krizisa teplootdachi pri kipenii vody v ravnomerno obogrevayemykh kruglykh trubakh. Nauch. soviet AN SSSR po kompleks. probl. "Teplofizika." Sektsiya teploob-mena (Recommendations on Calculating the Boiling Crisis in Water in Uniformly Heated Circular Pipe. Scientific Council of the USSR Academy of Sciences for Comprehensive Examination of Thermophysics Problems. Heat Transfer Section). Moscow, 1975.

3.56   Smolin, V. N. A Model of the Mechanism of the Boiling Crisis in Wet-Steam Flows and a Technique for Calculating Critical Conditions in Tubular Fuel Elements. Seminar TF-74. Issledovaniya kriti-cheskikh teplovykh potokov v puchkakh sterzhney (1974 Thermophysics Seminar. Studies of Criti-cal Heat Fluxes in Rod Bundles, 209–225). SEV Press, Moscow, 1974.

3.57   Peskov, O. L., et al. Critical Heat Fluxes in Forced Convection of Wet Steam in Tubes. In: Voprosy teplootdachi i gidravliki dvukhfaznykh sred (Problems of Heat Transfer and Hydraulics of Two-Phase Media). S. S. Kutateladze, ed., 44–56. Moscow, GEI Press, 1961 [Engl. transl. Pergamon, 1969].

3.58   Campolunghi F., Cumo, M., and Ferrari, G., et al. A burnout correlation for once-through L. M. F. B. R. steam generators. In: European two-phase flow group Meeting. Harwell, 1974.

3.59   Bertoletti, S., Gaspari, G. P., and Lombardi, C., et al. Heat transfer crisis with steam-water mixture. Energy Nucl., 1965, vol. 12(1), pp. 121–172.

3.60   Gaspari, G. P., et al. Ital. Rept. CJSE-R-276. Milan, 1968.

3.61   Gellerstedt, J. S., Lee, R. A., and Oberjohn, W. J., et al. Correlation of critical heat flux in a bundle cooled by pressurized water. In: Two-phase flow and heat transfer in rod bundles, 1969, p. 63.

3.62   Polonsky, V. S. Certain Aspects of Statics and Dynamics of the Boiling Crisis in Steam-Generating Channels. Author's abstract of Candidate thesis. Moscow Energetics Institute, Moscow, 1967.

3.63   Pikus, V. Yu. Investigation of Heat Transfer and Boiling Crisis in Curved Channels. Author's abstract of Candidate thesis. Moscow Energetics Institute. Moscow, 1967.

3.64   Campolunghi, F., Cumo, M., Ferrari, G., and Palazzi, G. Full scale jests and thermal design correlations for coiled once—through steam generators. In Adv. Heat Transfer, 1974, vol. 11, RT/ING (74) 23, pp. 149–158.

3.65   Gambill, W. K., and Greene, N. D. A study of burnout heat fluxes associated with forced convection subcooled and bulk nucleate boiling of water in source-vortex flow. CF-57-10-118, 1957, p. 15.

3.66   Gambill, W. K., Bundy, R. D., and Wansbrough, R. W. Heat transfer, burnout and pressure drop for water in swirl flow through tubes with internal twisted tapes. ORNL-2911, 1960, p. 105.

3.67   Hassid, A., Manzoni, G. C., and Ravetta, R. Heat Transfer crisis with steam-water mixtures; An experimental study on the increase of critical power with local swirl promoters in round tubes. Energia Nucl. 1966, vol. 13, N 11, p. 589.

3.68   Lamson, C. G., Kedl, R. G., and Donald, R. E. Enhanced heat transfer tubes for horizontal condensers with passible application in nuclear power plant design. Trans. ANS, 1966, vol. 9, N 2, pp. 585–586.

3.69   Barbarin, V. P., Sevast'yanov, R. I., and Alad'yev, I. G. A Special Hydrodynamic Effect on the Boiling Crisis in Tubes. In: Teploobmen, gidrodinamika i teplofizicheskiye svoystva veshchestv (Heat Transfer, Hydrodynamics and Thermophysical Properties of Substances), 45–53. Nauka Press, Moscow, 1968 [Engl. transl. Heat Transfer—Soviet Research, *1*, No. 4, 34–41, 1969].

3.70   Zarnett, G. D., and Charles, M. E. Cocurrent gas-liquid flow in horizontal tube with internal spiral ribs. Canad. J. Chem. Eng., 1969, vol. 47, N 3, pp. 238–241.

3.71   Durant, W. S., and Mirshak, S. Roughening of heat transfer surface as a method of increasing the heat flux at burnout. DP-380, 1959, p. 20.

3.72   Mirshak, S., and Towell, R. H. Heat transfer burnout a surface contacted by a spacer rib. DP-562, 1961, p. 22.

3.73   Kon'kov, A. S., and Sinitsyn, I. T. Heat Transfer and Pressure Drop in the Flow of Water and Wet Steam in Tubes with Enhancement Devices. In: Dokl. na III konf. NTO energetiki i elektrokhn, prom-sti pri VTI (Reports at the 3rd Conference of the Scientific and Engineering Society of Industrial Energetics and Electrical Engineering at the All-Union Heat Engineering Institute), 171–180. Moscow, 1970.

3.74   Brevi, R., Cumo, M., Palmieri, A., and Pitimada, D. Measurement of the effect of twisted tapes on the forced-convection heat transfer and burnout in heat exchangers. Nucl. Sci. and Eng., 1971, vol. 46, pp. 131–139.

3.75   Kirillov, P. L., Smogalev, I. P., and Masagutov, R. F. Calculation of the Boiling Crisis in a Vapor-Liquid Mixture Using a Model of Droplet Diffusion when Swirling the Flow in an Internally Finned Tube. Tr. FEI, *308*, 19, 1972.

3.76   Sevast'yanov, R. I., Zakharov, Yu. V., and Alad'yev, I. T. Concerning the Effect of Tube Length, Nonuniformity of Heating and "Auger" Type Turbulence Promoters on Critical Heat Fluxes in Pipes. Tr. TsKTI, No. 58, 201, 1965.

3.77   Rosuel, A., and Rouvillois, X. Nucl. Engl., 1968, vol. 13, N 140, pp. 43–45.

3.78   Bechtel von, G., and Ruth, H., et al. Kerntechnic, 1967, Bd. 9, N 7.

3.79   Moeck, E. O., Wiknammer, G. A., and MacDonald, J. P. L., et al. AECL-2109, 1964, p. 28.

3.80   Klochkova, L. F., and Sapankevich, A. P. Boiling Crisis in Rod Bundles. (An Analytic Survey). Tr. FEI, E-11-0B23. Obninsk, 1976.

3.81   Rosuel, A., and Rouvillois, X. Energ. Nucl., 167, vol. 9, N 8, p. 496.

3.82   Blagovestova, T. I., et al. Investigation of the Critical Power of Rod-Type Fuel Elements with Enhancement Devices for type RBM-K Reactors. In: Seminar TF-74. Issledovaniya kriticheskikh teplovykh potokov v puchkakh sterzhney, 291–306. Moscow, 1974.

3.83   Smolin, V. N., and Polyakov, V. K. Enhancement of Heat Transfer in Rod Bundles by Local Swirling Devices. Ibid., 307–312.

3.84   Subbotin, V. I., Kaznovskiy, S. P., and Sapankevich, A. P. An Experimental Study of Methods for Increasing the Critical Power of Steam Generators. Ibid., 313–322.

3.85   Barulin, Yu. D., et al. Investigation of Enhancement of Heat Transfer and Pressure Drop in a Model of a Boiling-Reactor Fuel-Rod Assembly. Ibid., 323–334.

3.86   Osmachkin, V. S., et al. Investigation of the Boiling Crisis in the Flow of Water in an Annulus. Ibid., 251–266.

3.87   Styrikovich, M. A., et al. Experimental Study of Heat Transfer under Post-crisis Conditions in Smooth and Rough Steam-Generating Channels. Teplofiz. vys. temp., *15*, No. 3, 566–572, 1977.

3.88   Homig, H. E. Phys. Grundlagen der Speisewasserchemie. Essen, 1963.

3.89   Richter, R., Schoch, W., and Schuster, H., et al. Magnetite formation and pressure-loss increase in a Benson boiler. In: VGB Kraftwerks technik Mitteilungen, 1972, s. 1–12.

3.90   Schoch, W. Allianz Berichte für Betriebstechnik und Schadenverhutung. In: Allianz Versicherungs-AG, München; Berlin, 1971, No. 16.

3.91   Schuster, H. Allianz-Berichte für Betriebstechnic und Schadenverhutung. In: Allianz Versicherungs-AG. München; Berlin, 1971, No. 16.

3.92   Macbeth, R. V. AEEW-R 711, 1971, 16 p.

3.93   Rassokhin, N. G., Kabanov, L. P., and Tevlin, S. A. On the Thermal Conductivity of Iron-Oxide Deposits. Teploenergetika, No. 9, 12–15, 1973.

3.94   Masterson, H. G., Gastle, G. E., and Mann, G. M. W. Waterside corrosion of power station boiler tubes. Chemistry and Industry, 1969, pp. 1261–1266.

3.95   Gastle, G. E., and Mann, G. M. W. The mechanism of formation of a porous oxide film on steel. Corr. Sci., 1966, N 6, pp. 253–262.

3.96   Pshemenskiy, A. A. Vodnyy rezhim teplofikatsionnogo oborudovaniya (The Water Regime of Cogeneration-Heating Equipment). In: Tr. III konf. NTO energetiki e elektrotekhn. prom-sti pri VTI, 55–59. ONTI VTI, 1970.

3.97   Kelen, T., and Arvesen, I. AE-459, Sweden; Studsvik: Actiebolaget Atomenergi, 1972, p. 30.

3.98   Cohen, P. The Water Regime of Pressurized-Water Cooled Reactors. In: Voprosy yadzymou tekhniki (Problems of Nuclear Engineering), No. 1, 88–102, 1958.

3.99   Kelen, T., and Gustafsson, R. The deposition kinetics of calcium hydroxy apatite on heat transfer surfaces at boiling. AE-498. Sweden; Studsvik: Actiebolaget Atomenergi, 1974, p. 30.

3.100 Klein, H. A., and Rice, J. K. Research Study on Internal Corrosion of High-Pressure Biolers. J. of Engineering for Power, *88*, No. 3, 232, 1966.

3.101 Goldstein, P., Dick, I. B., and Rice, J. K. Internal Corrosion of High-Pressure Boilers. Ibid., *89*, No. 3, 378, 1967.

3.102 Goldstein, P. A Research Study on Internal Corrosion of High-Pressure Boilers. Ibid., *90*, No. 1, 21, 1968.

3.103 Goldstein, P., and Burton, C. L. A Research Study on Internal Corrosion of High-Pressure Boilers. Final Report. Ibid., *9*, No. 2, 75, 1969.

3.104 Macbeth, R. V., Tranberth, R., and Wood, R. V. An investigation into effect of crude deposits on surface temperature, dryout and pressure drop with forced convection boiling of water at 69 bar in an annular test section. AEEW-R705, 1971, p. 31.

3.105 Rikhter, E. Experimental Study of the Temperature Regime and Boiling Crisis in an Annulus under Different Physico-Chemical Conditions. Author's abstract of Candidate thesis. Moscow Energetics Institute, Moscow, 1974.

3.106 Tomas, D., and Heitman, H. Untersuchungen über Magnetitablagerungen und deren Einflub auf den Warmeubergang in Beheizten Dampfenzeuger-rohren. DBR, Siemens, 1969.

3.107 Balabanov, Ye. D. Investigation of the Boiling Crisis in the Presence of Iron-Oxide Deposits. Author's abstract of Candidate thesis. Moscow Energetics Institute, Moscow, 1977.

3.108 Styrikovich, M., Leontiev, A., and Polonsky, V., et al. Influence of crude deposits on heat and mass transfer in steam generation channels. In: Intern. Seminar Momentum, Heat and Mass Transfer in Two-Phase Energy and Chemical Systems. Yugoslavia; Dubrovnik, 1978, pp. 549–560.

4.1   Styrikovich, M. A., Shitsman, M. Ye., and Miropol'skiy, Z. L. Certain Data on the Temperature Regime of a Vertical Boiling Tube at Postcritical Pressures. Teploenergetika, No. 12, 32–36, 1955.

4.2   Miropol'skiy, Z. L. Film Boiling Heat Transfer from Wet Steam in Steam-Generating Tubes. Ibid., No. 5, 49–52, 1963.

4.3   Kon'kov, A. S. An experimental Study of the Conditions for DNB in Wet Steam Flowing in Heated Tubes. Tr. TsKTI, No. 58, 170–178, 1965.

4.4   Smolin, V. N., Polyakov, V. K., and Yesikov, V. I. Concerning the Boiling Crisis in Forced Convection of the Coolant in Steam-Generating Tubes. Ibid., 128–138.

4.5   Schmidt, K. R. Warmetechniche untersuchungen an hoch belasteten kesselheizflachen. Mitt. Ver. Crosskesselbesitzer, 1959, N 63.

4.6   Swenson, H. S., Carver, J. K., and Szocka, G. The effects of nuclear boiling versus film boiling on heat transfer in power boiling tubes (61-WA-201). Trans. ASME. Ser. A., 1962, vol. 84, N 4, pp. 365–371.

4.7   Lokshin, V. A., Semenovker, I. Ye., and Vikhrev, Yu. V. Temperature Regime of Steam-Generating Tubes. Tr. TsKTI, No. 58, 112–122, 1965.

4.8   Tobilevich, N. O., and Yeremenko, B. A. Investigation of Aspects of Boiling Heat Transfer in Tubes. In: Gidrodinamika i teploobmen pri kipenii v kotlakh vysokogo davleniya (Fluid Mechanics and Heat Transfer in Boiling in High-Pressure Boilers), 186–205. USSR Academy of Sciences Press, Moscow, 1955.

4.9   Kon'kov, A. S., and Zuperman, D. A. Experimental Study of Heat Transfer to Wet Steam. Teploenergetika, No. 3, 54–56, 1967.

4.10   Cumo, M. Elementi di termotecnica del reattore. Roma: CNEN, 1969, 437 p.

4.11   Belyakov, N. I., et al. Investigation of Threshold Voidages in Steam-Generating Tubes at High Pressures. Teploenergetika, No. 10, 69–71, 1976.

4.12   Subbotin, V. I., Remizov, O. V., and Vorob'yev, V. A. The Temperature Regime and Heat Transfer at DNB. Teplofiz. vys. temp., 11, No. 6, 1220–1226, 1973.

4.13   Tong, L. S., and Young, J. D. A phenomenological transition and film boiling heat transfer correlation. In: Heat Transfer 1974; Proc. 5th Intern. Heat Transfer Conf. Tokyo, 1974, vol. 4, pp. 120–124.

4.14   Hewitt, G. F., and Hall-Taylor, N. S. Annular Two-Phase Flow. Pergamon Press, 1970 [Rus. transl. 1974].

4.15   Polonsky, V. S. Certain Aspects of Statics and Dynamics of the Boiling Crises in Steam-Generating Channels. Author's abstract of Candidate thesis. Moscow Energetics Institute, Moscow, 1967.

4.16   Vorob'yev, V. A. Investigation of Steady and Nonsteady Temperature Fields of a Steam-Generating Surface at DNB. Author's abstract of Candidate thesis. Moscow Energetics Institute, Moscow, 1972.

4.17   Remizov, O. V., and Vorob'yev, V. A. Heat Transfer in Transcritical Region. Analytic Survey. Preprint FEI E-11, OB-24. Obninsk, 1976.

4.18   Plummer, D. N. Post critical heat transfer to flowing liquid in a vertical tube. Ph.D. Thesis M.I.T., 1974.

4.19   Kistemaher, G. The spheroidal state of a waterdrop. Phisika, vol. 29, 1963, pp. 96–104.

4.20   Wachters, L. H. The heat transfer from a hot horizontal plate to sessile water drops in the spheroidal state. Chem. Eng. Sci., 1966, vol. 21, pp. 923–936.

4.21   Cumo, M., and Farello, P. Notes on droplet heat transfer. Chem. Eng. Progr. Symp. Ser., 1969, vol. 65, N 92, pp. 175–197.

4.22   Emmerson, G. S. The effect of pressure and surface material on the Leidenfrost point of discret drops of water. Intern. Heat Mass Transfer, 1975, vol. 18, pp. 381–386.

4.23   Kutateladze, S. S. Osnovy teorii teploobmena (Fundamentals of Heat Transfer). Mashgiz Press, Moscow-Leningrad, 1952 [Engl. transl. Academic Press, 1963].

4.24   Rao, P. S. V. K., and Sarma, P. K. Effect of subcooling on film-boiling heat transfer. J. Chem. Eng. Jap., vol. 7, N 5, pp. 341–346.

4.25   Moriyama, A. Evaporation rate of a single water droplet on hot. Trans. IRPN and Steel Inst. Jap., Nippon, 1971, vol. 14, pp. 1024–1026.

4.26   Bell, K. J. The Leidenfrost phenomenon a survey. Chem. Eng. Progr. Symp. Ser., 1967, vol. 63, N 79, pp. 73–82.

4.27   Boumeister, K. J. A generation correlation of vaporization times of drops in film boiling on a flat plate. In: Proc. of the Third Inter. Heat Transfer Conf., Chicago, A. J. Ch. Eng., 1966, Aug., vol. 4, pp. 66–75.

4.28   Rao, P. S. v. K., and Sarma, P. K. Effect of interfacial shear on evaporation rates of liquid patches. Canad. J. Chem. Eng., vol. 53, August 1975, pp. 456–459.

4.29   Nishio, S., and Hirata, M. Direct contact phenomenon between a liquid droplet and high temperature solid surface. In: Proc. VI intern. Heat Transfer Conf., Toronto, Hemisphere Publ. Co., Washington, 1978, vol. 1, p. 245.

4.30   Styrikovich, M. A., Baryshev, Yu. V., and Tsiklauri, G. V. The mechanisms of heat and mass transfer a water drop on a heated surface. In: VIth Intern. Heat Transfer Conf. Hemisphere Publ. Co., Washington, Toronto, 1978, PB-22.

4.31   Sahurai, A., and Shiotsu, M. Temperature-controlled pool-boiling heat transfer. In: Fifth Intern. Heat Transfer Conf., Hemisphere Publ. Co., Washington, 1974, Sept., vol. 4, pp. 81–85.

4.32   Wachters, L. H. J., and Westerling, N. A. The heat transfer from a hot water to impinging water drops in the spheroidal state. Chem. Eng. Sci., 1966, vol. 21, pp. 1047–1056.

4.33   Wachters, L. H. J., and Smulders, L. The heat transfer from a hot wall to impinging mist droplets in the spheroidal state. Chem. Eng. Sci., 1966, vol. 21.

4.34   Geist, J. M. Ind. Eng. Chem., 1951, vol. 43, p. 1372.

4.35   Corman, J. C. Water cooling of a thin, high temperature metal strip, Ph.D. Thesis. Carnegie Inst. Thecnol., 1966.

4.36   Pedersen, C. A. An experimental study of the dynamic behavior and heat transfer characteristics of water droplets impinging upon a heated surface. Intern. J. Heat Mass. Transfer, 1970, vol. 13, N 2, pp. 369–381.

4.37   Camia, ?. Impulse Theory of Thermal Conductivity. Moscow, 1972 [translated from French].

4.38   Kokorev, L. S., Ushakov, V. M., and Dubrovskiy, G. P. Effektivnyy krayevoy ugol pri kipenii zhidkostey (The Effective Wetting Angle in the boiling of Liquids). Atomizdat Press, Moscow, 1976.

4.39   Predvoditelev, A. S. Concerning Molecular Heat Transfer in Liquids. Dokl. Akad. Nauk SSSR, p. 623, 1950.

4.40   Missenard, A. Le conductivite thermique des solides, liquides, gaz et der melanges. Paris, Editions Eyrolles, 1965 [Rus. transl. 1968].

4.41   Skripov, V. P., Metastabil'naya zhidkost' (Metastable Liquid). Nauka Press, Moscow, 1972 [Engl. transl. Israel Program for Scientific Translations, for Wiley and sons, 1974].

4.42   Laverty, W. F., and Rohsenow, W. M. Film Boiling of Saturated Nitrogen Flowing in a Vertical Tube. J. of Heat Transfer, 89, No. 1, 90, 1967.

4.43   Forslund, R. P., and Rohsenow, W. M. Dispersed Flow Film Boiling. J. of Heat Transfer, 90, No. 4, 399, 1968.

4.44   Gukhman, A. A., et al. Investigation of Heat Transfer at the Stability Limit of the Metastable State of the Liquid. In: Teplo- i massoperenos (Heat and Mass Transfer [Proceedings of the 3rd All-Union Conference on Heat and Mass Transfer), 1, 792–800. Energiya Press, Moscow, 1968.

4.45   Rohsenow, W., Fedorovich, E. Post burnout heat transfer to mist flow; Heat and mass transfer in turbulent boundary layers. Beograd (Yugoslavia), 1970, pp. 683–699.

4.46   Cumo, M. Elementi di termotecnica del reattore. Roma: CNEN, 1969. 437 p.

4.47   Bergles, A. E., Fuller, W. D., and Hynek, S. J. Dispersed flow film boiling of nitrogen with swirl flow. Intern. J. Heat and Mass Transfer, 1971, vol. 14, N 9, pp. 1434–1454.

4.48   Marinov, M. I. Investigation of Postcritical Heat Transfer under Post-LOCA Cooldown Conditions in a Nuclear Reactor. Author's abstract of Candidate thesis. Moscow Energetics Institute, Moscow, 1977.

4.49   Kalinin, E. K., Korolev, A. L., and Yarkho, S. A. Investigation of Two-Phase Flow Parameters in Dispersed Film Boiling in Uptake Pipe Flow. Trudy VZMI (All-Union Correspondence Institute of Mechanical Engineering), No. 1, 115–126, 1972.

4.50   Subbotin, V. I., Remizov, O. V., and Vorob'yev, V. A. Calculation of the Wall Temperature Profile at DNB. Teplofiz. vys. temp., 12, No. 4, 785–789, 1974.

4.51   Dzarasov, V. I., Kol'chugin, B. A., and Liverant, E. I. Investigation of the Relationship between Heat and Mass Transfer Variables in Evaporation Channels. In: Teploobmen-1974: Sovetskiye issledovaniya (Heat Transfer 1974. Soviet Investigations [Soviet Contributions at the 1974 (Fifth) International Heat Transfer Conference), 259–268. Nauka Press, Moscow, 1975.

4.52   Pron'ko, V. G. A Two-Stage Model of Heat Transfer Applicable to Cooling of Cryogenic Equipment under Film Boiling Conditions Tr. NPO (Scientific and Industrial Society) Kriogenmash, No. 17, 28–43, 1975.

4.53   Iloeje, O. C., et al. An Investigation of the Collapse and Surface Rewet in Film Boiling in Forced Vertical Flow. J. of Heat Transfer, 97, No. 2, 166, No. 3, 462, 1975.

4.54   Khasayev, N. O. Investigation of Heat Transfer Enhancement in Film Boiling of Cryogenic Liquids in Tubes. Author's abstract of Candidate thesis. Moscow Aviation Institute, Moscow, 1976.

4.55   In'kov, A. P. Investigation of Heat Transfer and Fluid Mechanics in Two-Phase Nonequilibrium Flows of Cryogenic Fluids in Channels with Local Resistances. Author's abstract of Candidate thesis. Kriogenmash (Scientific and Industrial Society), 1977.

4.56 Ganic, E. N., and Rohsenow, W. M. Dispersed flow transfer. I. J. Heat and Mass Transfer, 1977, vol. 20, N 8, pp. 855–866.

4.57 Kalinin, E. K., et al.. Teploobmen pri plenochnom kipenii v elementakh energeticheskikh apparatov (Obshchiye i teoreticheskiye voprosy teploenergetiki; Gelioenergetika) (Film Boiling Heat Transfer in Power-Generating Machinery) (General and Theoretical Problems of Heat Engineering. Solar Energy Engineering). VINITY Press, Moscow, 1972.

4.58 Kalinin, E. K., Dreytser, G. A., and Yarkho, S. A. Intensifikatsiya teploobmena v kanalakh (Enhancement of Heat Transfer in Channels). Mashinostroyeniye Press, Moscow, 1972.

4.59 Yarkho, S. A., and Stat'yev, A. A. Technique of Engineering Calculation of Transient Cooling of Pipelines by Cryogenic Fluids in Dispersed Film Flow. Tr. VZMI, No. 10, 63–75, 1974.

4.60 Yarkho, S. A. Dispersed Film Boiling of Cryogenic Liquids in Tubes. In: Voprosy sovremennoy kriogeniki (Problems of Modern Cryogenics), 277–281. Vneshtorgizdat Press, Moscow, 1975.

4.61 Yarkoho, S. A., et al. Heat Transfer, Fluid Dynamics and Thermal Inequilibriumin Dispersed Flow Film Boiling of Hydrogen, Nitrogen and Argon in Vapor Generators. In: Teplo i massoobmen-V [sic] (Heat and Mass Transfer—V [Proceedings of the 5th All-Union Conference on Heat and Mass Transfer]), 3, Pt. 2, 80–89. Minsk, 1976 [Engl. transl. Heat Transfer—Soviet Research, 9, No. 2, 10–17, 1977].

4.62 Cumo, M., Ferrari, G., and Farello, G. E. A photographic study of two-phase highly dispersed flow. Termotecnica, 1971, vol. 25, N 9, pp. 450–458.

4.63 Styrikovich, M. A., et al. Investigation of Transcritical Heat Transfer in Smooth and Rough Steam-Generating Tubes. In: Teplomassoobmen-V, 3, 41–48. Minsk, 1976 [Engl. transl. Heat Transfer—Soviet Research, 9, No. 1, 123–131, 1977].

4.64 Polonsky, V. S., and Malashkin, I. I. Experimental Study of the Rate of Heat Transfer in the Postcritical Region of Steam-Generating Channels. Teplofiz. vys. temp., 16, No. 2, 360–364, 1978.

4.65 Kutateladze, S. S., and Leont'yev, A. I. Teplomassoobmen i treniye v turbulentnom pogranichnom sloye (Heat and Mass Transfer and Friction in Turbulent Boundary Layers). Energiya Press, Moscow, 1972.

4.66 Gukhman, A. A. Primeneniye teorii podobiya k issledovaniyu protssessov teplomassoobmena (Application of the Theory of Similitude to the Study of Heat and Mass Transfer). Vysshaya Shkola Press, Minsk, 1974.

4.67 Gomelauri, V. I., Kandelaki, R. D., and Kitmidze, M. Ye. Enhancement of Convective Heat Transfer by Artificial Roughness. In: Voprosy konvektivnogo teploobmena i chistoty vodyanogo para (Problems of Convective Heat Transfer and Purity of Steam, 98–131). Metsniereba Press, Tbilisi, 1970.

4.68 Doroshchuk, V. Ye. Krizisy teploobmena pri kipenii vody v trubakh (Boiling Crises in Water Flows in Tubes). Energiya Press, Moscow, 1970.

4.69 Styrikovich, M., and Polonsky, V. Experimental study of the effect of artificial roughness on post dryout heat transfer conference. In: 5th Intern. Heat Transfer Conf. Tokyo, 1974, vol. 4, pp. 145–149.

4.70 Miropol'skiy, Z. L., Bezrukov, Ye. K., and Grebennikov, V. N. Relationship between Heat and Mass Transfer and Pressure Drop in Steam-Generating Channels at DNB. Trudy MEI [Moscow Energetics Institute], No. 239, 1975.

4.71 Borishanskiy, V. M., et al. Heat Transfer in the Postcritical Zone of a Steam-Generating Channel. In: Kirzisy teploobmena i okolokriticheskaya oblast' (Boiling Crises and the Transcritical Region), 16–24. Nauka Press, Leningrad, 1977.

4.72 Coller, J. Y. Heat transfer and fluid dynamic research as applied to fog cooled power reactors. AECL-1631, 1962

4.73 Lee, D. H. Studies of heat transfer and pressure drop relevant to sul-critical once-through evaporators. FAEA-SM-130156. IAEA Symp. on Progress in Sodium-Cooling Fast Reactors Engineering. Monaco, 1970, pp. 681–701.

4.74 Mattson, R. J., Condie, K. V., Bengston, S. J., and Obenhain, C. F. Regression analysis of past-CH flow boiling data. Heat Transfer 1974: Proc. 5th Int. Heat Transfer Conf. Tokyo, 1974, vol. 4, pp. 115–119.

4.75 Subbotin, V. I., Remizov, O. V., and Vorob'yev, V. A. Temperature Regime and Heat Transfer at DNB. Teplofiz. vys. temp., *11*, No. 6, 1973.

4.76 Remizov, O. V., Vorob'yev, V. A., and Gal'chenko, E. F. Limits of the Onset of DNB Conditions and Heat Transfer in the Transcritical Region. Preprint FEI-653. Obninsk, 1975.

4.77 Cumo, M., and Urbani, G. Anomalies in post-dryout heat transfer at high pressure. Trans. Amer. Nucl. Soc., 1971, vol. 14, N 1, pp. 245–246.

4.78 Remizov, O. V. Investigation of the Thermal Conditions of Operation of a Steam-Generating Surface in the Course of the Boiling Crisis. Teploenergetika, No. 2, 16–21, 1978.

4.79 Kearsey, H. A. Steam water heat transfer post-burnout conditions. Chemical and Process Engineering, 1965, vol. 46, N 8, pp. 455–459.

4.80 Miropol'skiy, Z. L. Heat Transfer in Film Boiling of Wet Steam in Steam Generating Tubes. Teploenergetika, No. 5, 49–52, 1963.

4.81 Bishop, A. A. High temperature supercritical pressure water loop. P. V. Forced convection heat transfer to water after the critical heat flux at high supercritical pressure. WCAP-2056, 1964.

4.82 Polomik, E. E., Levy, S., and Sawochka, S. G. Film Boiling of Steam-Water Mixtures in Annular Flow at 800, 100, and 1400 psi. J. of Heat Transfer, *86*, No. 1, 81–88, 1964.

4.83 Bishop, A. A., Sandberg, R. P., and Tong, L. S. Forced convection heat transfer at high pressures after the critical heat flux. ASME, 1965, pap. 65-HT-31.

4.84 Quinn, E. P. Forced flow transfer to high pressure water belong the critical heat flux. ASME, 1966, pap. 66, WA/HT-36.

4.85 Keeys, R. K. F., Ralph, J. C., and Roberts, D. N. Post-burnout heat transfer in high pressure steam-water mixtures in a tube with cosine heat flux distribution. AERE-R6411, 1971.

4.86 Tong, L. S. Heat transfer in water cooling nuclear reactors. Nucl. Eng. and Design, 1967, vol. 6, pp. 301–324.

4.87 Brevi, R., Cumo, M., Palmeri, A., and Pitimada, D. Post-dryout heat transfer with stew mixture. Trans. Amer. Nucl. Soc., 1969, vol. 12, N 2, pp. 809–810.

4.88 Herkenrath, H., and Mork-Morkeinstein, P. Die Warmeubergangskrise von Wasser bei erzwangener Stromung under hohen Drüchen. T. 1. Der Warmeubergang im Bereich der Krise. Atomkernergie, 1969, Bd. 6, N 14, S. 403–407.

4.89 Greunevald, D. C. An investigation of heat transfer in the liquid deficient regime. AECL-3257, 1969.

4.90 Slanghterlek, D. S. Flow film boiling heat transfer correlations: A parameteric study data compressions. ASME, 1973, pap. 73-HT-50.

4.91 Greunvald, D. C. Post-dryout heat transfer at reactor operating conditions. AECL-4208, 1973.

4.92 Tong, L. S., and Young, J. D. A phenomenological transition and film boiling heat transfer correlation. In: Heat Transfer 1974: Proc. 5th Intern. Heat Transfer Conf. Tokyo, 1974, vol. 4, pp. 120–124.

4.93 Grehev, N. S., Ivashkevich, A. A., and Kirillov, P. L., et al. Proceedings of the USA/USSR seminar on the development of sodium-cooled fast breeder reactor, steam generators. USA; Los Angeles, 1974, vol. 1, pp. 302–331.

4.94 Greunevald, D. C., and Delorme, G. G. L. Prediction of thermal nonequilibrium in the post-dryout regime. Nucl. Eng. and Design, 1976, vol. 36, N 1, pp. 17–26.

4.95 Vorob'yev, V. A., Remizov, O. V., and Sergeyev, V. V. Heat Transfer to Wet Steam at DNB. Teploenergetika, No. 2, 27–28, 1978.

4.96 Remizov, O. V., and Vorob'yev, V. A. Heat Transfer in the Supercritical Region (Analytic Survey). Tr. FEI E-11, OB-24. Obninsk, 1976.

4.97 Brevi, R., and Cumo, M. Quality influence in psot-burnout. Intern. J. Heat and Mass Transfer, 1971, vol. 14, N 3, pp. 483–489.

4.98 Takagi, T., and Ogasawara, M. Some characteristics of heat and mass transfer in binary mist flow. In; Heat Transfer 1974: Proc. 5th Intern. Heat Transfer Conf. Tokyo, 1974, vol. 4, pp. 350–354.

4.99 Greunevald, D. S. Post-dryout heat transfer physical mechanisms and survey of predictions methods. Nucl. Eng. and Design, 1975, vol. 32, N 3, pp. 253–294.

4.100 Kalinin, E. K., and Yarkho, S. A. Effect of the Reynolds and Prandtl Numbers on the Effectiveness of Heat Transfer Augmentation in Pipes. Inzh.-fiz. zhurn., *11*, No. 4, 426–431, 1966.

4.101 Quinn, L. S. Single rod forced flow transition boiling heat transfer from smooth and liquid surfaces. GEAP-4786, 1965.

4.102 Tong, L. S., Kitzes, A. S., and Green, L., et al. Departure from nucleate boiling and finned surface heater rod. Nucl. Eng. and Design, 1967, vol. 5, N 4, pp. 386–390.

4.103 Greunevald, D. C. Effect of surface roughness on the post-dryout heat transfer coefficient. CRNL-647, 1971.

4.104 Styrikovich, M. A., et al. Experimental Study of the Influence of Calcium Sulfate Deposits on a Vapor-Generating Surface on the Post-dryout Temperature Regime. In: Teploobmen i gidrodinamika odno- i dvukhfaxnykh teplonositeley (Heat Transfer and Fluid Mechanics of Single- and Two-Phase Coolants). Trudy MEI, No. 81, 81–96. Moscow, 1971 [Engl. transl. Heat Transfer—Soviet Research, *4*, No. 4, 175–182, 1972].

4.105 Kutateladze, S. S., and Styrikovich, M. A. Gidrodinamika gazozhidnostnykh sistem (Fluid Mechanics of Gas-Liquid Systems). Energiya Press, Moscow, 1976.

4.106 Nichigava, K., Tujii, T., and Yoshida, S., et al. Flow boiling crisis in grooved boiling tubes. In: Proc. 5th Intern. Heat Transfer Conf. Tokyo, 1974, vol. 4, pp. 270–274.

4.107 Grachev, N. S., Kirillov, P. A., and Prokhorova, V. A. Experimental Study of Heat Transfer in an Internally-Finned Steam-Generating Tube. Teplofiz, vys. temp., *14*, No. 6, 1234–1240, 1976.

4.108 Rogers, T. F., and Mesler, R. B. Amer. Inst. Chem. Eng. J., 1964, vol. 10, N 5, p. 656.

4.109 Styrikovich, M. A., et al. Investigation of Transcritical Heat Transfer in Smooth and Rough Steam-Generating Tubes. In: Teplomassoobmen-V, *3*, 41–48. Minsk, 1976 [Engl. transl. Heat Transfer—Soviet Research, *9*, No. 1, 123–131, 1977].

4.110 Styrikovich, M. A., et al. Experimental Study of Heat Transfer in Smooth and Rough Steam-Generating Channels. Teplofiz. vys. temp., *15*, No. 3, 556–572, 1977.

4.111 Polonsky, V. S., and Malashkin, I. I. Experimental Study of Heat Transfer Augmentation in the Supercritical Region of Steam-Generating Channels. Ibid., *16*, No. 2, 360–364, 1976.

4.112 Dreytser, G. A., Kuz'menkov, V. A., and Neverov, A. S. Estimate of the Effectiveness of Heat Transfer Augmentation by Turbulizing the Flow in Tubular Heat Exchangers. In: Tr. vsesoyuz. zaochn. mashinostr. in-ta, Gidravlika (Hydraulics), *10*, No. 3, 102–104, 1974.

4.113 In'kov, A. P., et al. Investigation of the Parameters of Two-Phase Flow of Hydrogen in Dispersed Film Boiling in a Tube. Ibid., 49–62.

4.114 Petukhov, B. S., et al. Investigation of Boiling Heat Transfer on an Open Surface. In: Teplo- i massoobmen pri fazovykh prevrashcheniyakh (Heat and Mass Transfer in Phase Transitions), Pt. 1, 179–189. Minsk, 1974.

4.115 Koshkin, V. K., et al. Investigation of Film Boiling in Turbulent Pipe Flow of Subcooled Liquid Nitrogen. In: Teplo- i massoperenos (Heat and Mass Transfer [Proceedings of the 3rd All-Union Conference on Heat and Mass Transfer]), *2*, 153–161, 1968.

5.1 Styrikovich, M., Nevstrueva, E., and Romanovsky, J., et al. Interconnection between mass and heat transfer in boiling. In: 4th Intern. Heat Transfer Conf. Versailles, 1970, vol. 4, pp. 1–11.

5.2 Rohsenov, W., and Fedorovich, E. Post burnout heat transfer mist flow. International summer school on heat and mass transfer, 1968, Sept., Herceg-Novi. Yugoslavia. Beograd: 1970, vol. 11, pp. 683–699.

5.3 In'kov, A. P., et al. Technique of Experimental Study of Dispersed Film Boiling of Hydrogen and Nitrogen in Forced Motion in Tubes. Tr. Vsesoyuz. zaoch. mashinostr. in-ta, *10*, No. 3, 76–88, 1974.

5.4 Polonsky, V. S. Certain Aspects of Statics and Dynamics of the Boiling Crisis in Steam-Generating Channels. Author's abstract of Candidate thesis. Moscow Energetics Institute, Moscow, 1967.

5.5 Romanovsky, I. M. Experimental Study of Mass Transfer in Boiling in Steam-Generating Channel. Author's abstract of Candidate thesis. Moscow, Institute of High Temperatures of the USSR Academy of Sciences, 1968.

5.6 Bezrukov, Ye. K. Investigation of Heat and Mass Transfer in Steam-Generating Tubes. Author's abstract of Candidate thesis. Moscow Energetics Institute, Moscow, 1972.

5.7 Styrikovich, M. A., Polonsky, V. S., and Bezrukov, Ye. K. Experimental Study of the Limits of the Start of Deposition of Calcium Sulfate in Steam-Generating Tubes. Teplofiz. vys. temp., *9*, No. 1, 129–134, 1971.

5.8   Styrikovich, M. A., Polonsky, V. S., and Bezrukov, Ye. K. Investigation of Mass Transfer in Steam-Generating Channels by the "Salt Method." Teplofiz. vys. temp., *9*, No. 3, 1971.

5.9   Styrikovich, M. A., et al. Experimental Study of Mass Transfer Conditions in Steam-Generating Channels with Cosine Heat Release. Ibid., *16*, No. 3, 548–555, 1978.

5.10   Kutateladze, S. S., and Leont'yev, A. I. Turbulentnyy pogranichnyy sloy szhimayemogo gaza (The Turbulent Boundary Layer in Compressible Gases). Siberian Division of the USSR Academy of Sciences Press, Novosibirsk, 1962 [Engl. trnasl. Academic Press, 1964].

5.11   Styrikovich, M. A., Nevstruyeva, Ye. I., and Tyutyayev, V. V. Void Fraction and Structures of Nonequilibrium Two-Phase Flows in Nonheated Channels. Teplofizika, No. 9, 18–22, 1974.

5.12   Nevstruyeva, E. I., and Gonzales, H. Distribution of Vapor Content in Surface Boiling of Water by the beta Raying Method. Teploenergetika, No. 9, 34–39, 1960.

5.13   Hewitt, G. F., and Hall-Taylor, N. S. Annular Two-Phase Flow. Pergamon Press, 1970 [Rus. transl. 1974].

5.14   Tong, L. S. Boiling Crisis and Critical Heat Flux, AEC Critical Review Series, TID-25887, 1972 [Rus. transl. 1976].

5.15   Nigmatullin, B. I., Milashenko, V. I., and Shugayev, Yu. Z. Investigation of the Distribution of Liquid between the Core and Film in Dispersed-Annular Wet-Steam Flows. Teploenergetika, No. 5, 77–79, 1976.

5.16   Nigmatullin, B. I., and Nikolayev, V. B. Development of a Hydrodynamic Model of Dispsersed-Annular Flow for Investigating the Boiling Crisis in Rod Bundles. In: Voprosy gazotermodinamiki energoustanovok (Problems of Gas-Thermodynamics of Power Plants). Tr. Khar'k. aviats. in-ta, No. 4, 3–7. Khar'kov, 1977.

5.17   Dolinin, I. V., and Rachkov, V. I. Dropwise Mass Transfer in Adiabatic Dispersed-Annular Wet-Steam Flows. Ibid.

5.18   Jacobs, J. D., and Shade, A. H. Measurement of the Temperature Associated with Bubbles in Subcooled Pool. Boiling. J. of Heat Transfer, *91*, No. 2, 123, 1969.

5.19   Rogers, T. P., and Mesler, R. B. Amer. Inst. Chem. Eng. J., 1964, vol. 10, N 5, p. 656.

5.20   Bankoff, S. G., and Mason, I. P. Amer. Inst. Chem. Eng. J., 1964, vol. 8, N 1, p. 30.

5.21   Moore, F. D., and Mesler, R. B. Amer. Inst. Chem. Eng. J., 1964, vol. 7, N 4, p. 620.

5.22   Snyder, N. W., and Robin, T. T. Mass-Transfer Model in Subcooled Nucleate Boiling. J. of Heat Transfer, *91*, No. 3, 404–412, 1969.

5.23   Tarasova, N. I., Armand, A. A., and Kon'kov, A. S. Investigation of Heat Transfer in a Tube in the Boiling of Subcooled Water and Wet Steam. In: Teploobmen pri vysokikh teplovykh nagruzkakh i drugikh spetsial'nykh usloviyakh (Heat Transfer at High Heat Fluxes and Other Special Conditions). GEI Press, Moscow, 1959.

5.24   Styrikovich, M. A., Leontiev, A. I., and Polonsky, V. S., et al. Heat and mass transfer interrelation. In: Sixth Intern. Heat Transfer Conf., 1978, vol. 2, pp. 347–352.

5.25   Kutateladze, S. S., and Leont'yev, A. I. Teplomassoobmen i treniye v turbulentnom pogranichnom sloye (Heat and Mass Transfer and Friction in Turbulent Boundary Layers—). Energiya Press, Moscow, 1972.

5.26   Styrikovich, M. A., Martynova, S. I., and Miropol'skiy, Z. L. Protsessy generatsii para na elektrostantsiyakh (Electric Power Plant Steam Generation Processes). Energiya Press, Moscow, 1969.

5.27   Stermeh, L. S. On the Theory of Heat Transfer in the Boiling of Liquids. Zhurn. tekh. fiz., No. 2, 341–350, 1953.

5.28   Tarasova, N. V. Hydraulic Drag in Boiling of Water and Wet Steam in Heated Pipes and Annuli. Trudy TsKTI [Polzunov Boiler and Turbine Institute], No. 59, 1965.

5.29   Metodika i zavisimosti dlya teoreticheskogo rascheta teploobemena i gidravlicheskogo soprotivleniya teploobmennogo oborudovaniya AES: Rukovodyashchiy tekhnicheskiy material, RTM 24.031.05-72 (Methods and Relationships for Analytic Calculation of Heat Transfer and Pressure Drop in Heat-Exchange Equipment of Nuclear Power Plants. Instructions Manual RTM 24.031.05-72). Heavy Power and Transportation Machinery Building Ministry, 9172.

5.30   Miropol'skiy, Z. L. Voidage in Nonequilibrium Wet-Steam Flows with Heating and Cooling. Teplofiz. vys. temp., *9*, No. 3, 135–140, 1971.

5.31   Dzarasov, Yu. I. Investigation of Conditions for Heat Transfer to Dispersed-Annular Wet-Steam Flow in an Annulus. Author's abstract of Candidate thesis. Krzhizhanovskiy Power Engineering Institute of the USSR Academy of Sciences. Moscow, 1973.

5.32   Mann, G. M. W. Distribution of sodium chloride and sodium hydroxide between steam and water at dry-out in an experimental once-through boiler. Chem. Eng. Sci., 1975, Feb., vol. 30, N 2, pp. 249–260.

5.33   Bakker, N. A., and Hawtin, P. Solute concentration in highly rated high pressure steam generators, 11. A model for the concentration of salt in an evaporating film. AERE-R7224, Harwell, 1977, Sept. 10 p.

5.34   Cohen, P. Chemical thermohydraulics in steam generating surfaces. In: 17th Nat. Heat Transfer Conf. USA, 1977, 8 p.

5.35   Pritchard, A. M., Peakall, K. A., and Smart, E. F. A miniature high-pressure loop to study corrosion and deposition processes in high performance boilers. AERE-R8669. Harwell (Oxfordshire), 1977, Feb. 66 p.

5.36   Styrikovich, M. A. The role of two-phase flows in nuclear power plants. In: Intern. seminar Momentum, Heat and mass transfer in two-phase energy and chemical systems. Yugoslavia; Dubrovnik, 1978, 4–9th Sept. 20 p.

5.37   Deddens, I. C., and Montgomery, D. W. Review of 1977 operations and product improvement programs. In: Nuclear operating experience seminar. Hershey (Pennsylvania), 1978, Mar. 6–7, 9 p.

5.38   Sarver, L. W., and Rigdon, M. A. Tube damage once-through steam generators. In: Corrosion advisory committee meet. EPRI, 1978, Feb. 7. Alliance (Ohio). 12 p.

5.39   Significant steam generators inspections relative to OTSG tube leaks—1978: Report to Electric Power Research Institute, Corrosion Advisory Committee. Palo Alto (Calif.), 1978, 14 p.

6.1   Picone, L. F., Waite, D. D., and Taylor Q. R. Radiotracer Studies of Hide-out at High Temperature and Pressure, Westinghouse Electric Corporation. Atom. Power Div., WCAP-3731, 1963. 66 p.

6.2   Klein, H. A., and Rice, J. K. Research Study on Internal Corrosion of High-Pressure boilers. J. of Engineering for Power, 88, No. 3, 232, 1966 [sic].

6.3   Goldstein, P., Dick, I. B., and Rice, J. K. Internal Corrosion of High-Pressure Boilers. Ibid., 89, No. 3, 378, 1967.

6.4   Goldstein, P. A Research Study on Internal Corrosion of High-Pressure Boilers. Ibid., 90, No. 1, 21, 1968.

6.5   Goldstein, P., and Burton, C. L. A Research Study on Internal Corrosion of High-Pressure Boilers. Final. Report. Ibid., 91, No. 2, 75, 1969.

6.6   Castle, J. E., and Mann, G. M. W. The mechanism of formation of a porous oxide film on steel. Corr. Sci., 1966, No. 6, pp. 253–262.

6.7   Styrikovich, M. A., et al. Tr. MEI, No. 128, 25–32, 1972.

6.8   Styrikovich, M. A., et al. Analysis of the Effect of Porous Deposits on the Concentration of Impurities in Steam-Generating Channels. Teplofiz. vys. temp., 15, No. 1, 109–114, 1977.

6.9   Styrikovich, M. A., et al. Experimental Study of Mass Transfer in Steam-Generating Channels with Porous Deposits. Ibid., No. 2, 353–358.

6.10   Styrikovich, M. A., et al. Experimental Study of Conditions of Mass Transfer in Steam-Generating Channels with Cosine Heat Release. Ibid., 16, No. 3, 548–555, 1977.

6.11   Styrikovich, M. A., and Martynova, O. I. Certain Aspects of Steam Generation on Two-Loop Nuclear Power Plants. Teploenergetika, No. 6, 84–86, 1974.

6.12   Solomon, J. An overview water chemistry for pressurized water nuclear reactors. In: First Announcement Intern. Conf. Water Chem. of Nuclear Reactor Systems. England: Bournemouth, 1977, pp. 75–86.

6.13   Sarver, L. W., and Rigdon, M. A. Tube damage once-through steam generators. In: Corros. advisory committee meet. EPRI, 1978, Feb. 7, Alliance (Ohio). 12 p.

6.14   Significant steam generators inspections relative to OTSG tube leaks—1978: Report to Electric Power Research Institute, Corrosion Advisory Committee. Palo Alto (Calif.), 1978, 14 p.

6.15   Roofthooft, R. VGB-Speisewassertagung. In: VGB-Konferenz Chemie im Kraft-Werk 1977, 29–30 Sept. Essen., 1977. 27 s.

6.16   Macbeth, R. V., Tranberth, R., and Wood, R. V. An investigation into the effect of crude deposits on surface temperature, dry-out and pressure drop with forced convection boiling of water at 69 bar in annular test section. AEEW-R-705, 1971.

6.17 Rassokhin, N. G., et al. Concerning the Thermal Conductivity of Iron-Oxide Deposits. Teploenergetika, No. 9, 12–15, 1973.

6.18 Styrikovich, M. A., Polonsky, V. S., and Dvoretskiy, A. I. Experimental Studies of Mass Transfer in Boiling in Thick Capillary-Porous Structures. Dokl. Akad. Nauk SSSR, *245*, No. 1, 101–103, 1979.

6.19 Macbeth, R. V. Boiling on surface overlaid with a porous deposit: Heat transfer rates obtainable by capillary actions. AEEW-R-711. Winfrith, 1971.

6.20 Cohen, P. The chemistry of water and solution at high temperatures for application to corrosion in power systems. Ermounville, 1972.

6.21 Cohen, P. Heat and mass transfer in porous deposits with boiling. WARD-5836, 1972.

6.22 Cohen, P. Heat and mass transfer for boiling in porous deposits with chimneys. In: Heat Transfer-Research and Design, AIChE Symp. Ser., 1974, vol. 70, No. 138.

6.23 Collier, I. G., and Kennedy, T. D. A. Solute concentration in highly rated high pressure steam generators. AERE-R-7203. Harwell, 1972.

6.24 Styrikovich, M. A., Leont'yev, A. I., and Malyshenko, S. P. Concerning the Mechanism of Transfer of Nonvolatile Impurities in Boiling on Surfaces Coated by a Porous Structure. Teplofiz. vys. temp., *15*, No. 5, 998–1006, 1976.

6.25 Polonsky, V. S., et al. A Model of the Concentration Process in Boiling in Capillary-Porous Structures. Dokl. Akad. SSSR, *241*, No. 3, 579–582, 1978.

6.26 Andrianov, A. B., et al. Investigation of the Mechanism of Evaporation in the Immediate Proximity of the Heating Surface in Boiling. Preprint No. 2-001 of the Institute of High Temperatures of the USSR Academy of Sciences. Moscow, 1975.

6.27 Nesis, Ye. I. Kipeniye zhidkosti (Boiling of Liquids). Nauka Press, Moscow, 1973.

7.1 Filippov, G. A., Povarov, O. A., and Pryakhin, V. V. Issledovaniya i taschety turbin vlazhnogo para (Testing and Design of Wet-Steam Turbines). Energiya Press, Moscow, 1973.

7.2 Troyanovskiy, B. M. Turbiny dlya atomnykh elektrostantsiy (Turbines for Nuclear Power Plants). Energiya Press, Moscow, 1978.

7.3 Filippov, G. A., and Povarov, A. O. Separatsiya vlagi v turbinakh AES (Moisture Separation in Nuclear Power Plant Turbines). Energiya Press, Moscow, 1980.

7.4 Tsiklauri, G. V., Seleznev, L. I., and Danilin, V. S. Adiabatnyye dvukhfaznyye techeniya (Two-Phase Adiabatic Flows). Atomizdat Press, Moscow, 1973.

7.5 Filippov, G. A., et al. Study of Condensation in a Turbine Stage. Teploenergetika, *15*, No. 9, 1974.

7.6 Deych, M. Ye., Kazintsev, F. V., and Povarov, O. A. Concerning the Motion of a Moisture Particle in a Turbine Stage. Tr. MEI, 248–272, 1967.

7.7 Kiryukhin, V. I., et al. Investigation of the Structure of Wet Steam in a Multistage Turbine. Teploenergetika, No. 5, 1975.

7.8 Kosyak, Yu. F., Nakhman, yu. V., and Zil'ber, T. M. Determination of the Effectiveness of the Moisture-Removal System in Low-Pressure Flowpassages. NIIinformtyazhmash, ser. 3-67-7, 45–48.

7.9 Zhukovskiy, M. I. Calculation of Flow over Turbine Airfoils. Mashgiz Press, Moscow, 1960.

7.10 Tabakoff, W., Hamed, A., and Hussein, M. Experimental investigation of gas-particle flow trajectories and velocities in an axial flow turbine stage. In: ASME Winter Annu. Meeting, USA, 1973, pp. 42–47.

7.11 Stiastny, M., and Taš, L. Experimental Study of the Liquid Phase in the Last Stage of an 200 MW Steam Turbine. In: Proceedings of the 6th Conference on Steam Turbines, 40–44. Plzen. Czechoslovakia, 1976.

7.12 Povarov, O. A., et al. Measurement of the Local Flow Variables of Liquid Films by the Electric Method. Izv. Vuz. Energetika, No. 1, 141–145, 1976.

7.13 Filippov, G. A., and Povarov, A. O. On the Design of High Fan Ratio Stages of Wet-Steam Turbines. Teploenergetika, No. 5, 87–91, 1976.

7.14 Abramov, V. I., Filippov, G. A., and Frolov, V. V. Teplovoy raschet turbin (Thermal Design of Turbines). Mashinostroyeniye Press, Moscow, 1974.

7.15 Deych, M. Ye., and Filippov, G. A. Gazodinamika dvukhfaznykh sred (Gas-dynamics of Two-Phase Flows). Energiya Press, Moscow, 1968.

7.16 Styrikovich, M. A., Martynova, O. I., and Miropol'skiy, Z. A. Protsessy generatsii para na elektrostantsiyakh (Electric Power Plant Steam Generation Processes). Energiya Press, Moscow, 1969.

7.17 Garnsey, F. In: First Intern. Conf. on Water Chemistry of Nuclear Reactor Systems, Bournemouth, 1977, Oct., pp. 19–27.

7.18 Martynova, O. I. Effect of Water Chemistry of Power Generating Units of Thermal and Nuclear Power Plants on the Reliability of Steam Turbine Operation. Energokhozyastvo za rubezhom, No. 1, 1979.

7.19 Lindsay, W. T. Behavior of impurities in steam turbines. Power Eng., 1979, vol. 83, N 5.

7.20 Pocock, I. J. Understanding the turbine steam environment. In: IX Intern. Conf. on the Properties of Steam, München, 1979, Sept.

8.1 Povarov, A. O., Vasil'chenko, Ye. G., and Randin, V. N. The Moisture-Separating Capacity of a Duct Equipped with Louver-Type Baffles. Izv. Vuzov. Energetika, No. 10 [sic] 73–78, 1976 [Engl transl. Fluid Mechanics—Soviet Research, 6, No. 2, 104–110, 1977].

8.2 Ryzhkov, S. V., Yershov, V. V., and Albotov, A. K. Investigation of Heat and Mass Transfer in the Motion of Dispersed Gas-Liquid Mixture in a Curved Separator Channel. Teploenergetika, No. 9, 1974.

8.3 Demidova, L. I., and Sorokin, Yu. L. Permissible Velocities of Steam and Gas for Vertical and Inclined Louver-Type Separators. Energomashinostroyeniye, No. 3, 1972.

8.4 Kutateladze, S. S., and Styrikovich, M. A. Gazodinamika gazozhidkostnykh sistem (Gasdynamics of Gas-Liquid Systems). Energiya Press, Moscow, 1976.

8.5 Kigur, Yu. N. Investigation of Exit Separators of Nozzle Chambers of Air Conditioners. Vodosnabzheniye i sanitarnaya tekhnika, No. 4, 1969.

8.6 Glushchenko, N. N. Investigation of a Louver-Type Separator. Energomashinostroyeniye, No. 5, 1972.

8.7 Kiryukhin, V. I., et al. Investigation of a Louver-Type Annular Separator. Teploenergetika, No. 9, 1976.

8.8 Andreyev, P. A., Grinman, M. I., and Smolkin, Yu. V. Optimizatsiya teploenergeticheskogo oborudovaniya AES (Optimization of Thermal Power Equipment of Nuclear Power Plants). Atomizdat Press, 1975.

8.9 Andreyev, P. A. Moisture Separator-Reheater for Turbines of the Leningrad Nuclear Power Plant. energomashinostroyeniye, No. 10, 1974.

8.10 Andreyev, P. A. Separators and Superheaters for Nuclear Power Plants with Water-Cooled Reactors. Energomashinostroyeniye, No. 5, 1977.

8.11 Artemov, L. N. A Vertical Separator. USSR Patent 494,563. Otkrytiya, izobreteniya, promyshlennyye obraztsy, tovarnyye znaki, No. 45, 1975.

8.12 Volkov, A. P. Testing of Moisture Separator Reheaters of K-22044/3000 Turbines on the Kola Nuclear Power Plant. Teploenergetika, No. 7, 1977.

8.13 Coit, R. L., Ritlands, P. D., and Rabas, T. J. Moisture separator reheaters entering the second decade. WEC Techn. Pap., 1975, pp. 1–19.

8.14 Troyanovskiy, B. M., and Povarov, O. A. External Separators and Reheaters of Nuclear Power Plant Turbines. Energokhozyaystvo za rubezhom, No. 6, 1976.

8.15 Bald, A. Turbosätze in konventionelen und nuklearen Kraftwerken. Techn. Mitt. 1975, N 1/2, S. 22–36.